HANDBOOK of GRAPH GRAMMARS and COMPUTING by GRAPH TRANSFORMATION

HANDBOOK OF GRAPH GRAMMARS AND COMPUTING BY GRAPH TRANSFORMATION

Vol. 1: Foundations

Forthcoming:

Vol. 2: Specifications and Programming

Vol. 3: Concurrency

Volume 1

FOUNDATIONS

HANDBOOK of GRAPH GRAMMARS and COMPUTING by GRAPH TRANSFORMATION

Edited by

Grzegorz Rozenberg

Leiden University, The Netherlands

World Scientific

Singapore • New Jersey • London • Hong Kong

Published by

World Scientific Publishing Co. Pte. Ltd.
5 Toh Tuck Link, Singapore 596224
USA office: 27 Warren Street, Suite 401-402, Hackensack, NJ 07601
UK office: 57 Shelton Street, Covent Garden, London WC2H 9HE

Library of Congress Cataloging-in-Publication Data
Handbook of graph grammars and computing by graph transformation /
edited by Grzegorz Rozenberg.
p. cm.
Includes bibliographical references and index.
Contents: v. 1. Foundations.
ISBN-13 978-981-02-2884-2 -- ISBN-10 981-02-2884-8
1. Graph grammars. 2. Graph theory -- Data processing.
I. Rozenberg, Grzegorz.
QA267.3.H364 1997
511'.5--dc21 96-37597
CIP

British Library Cataloguing-in-Publication Data
A catalogue record for this book is available from the British Library.

Preface

Graph grammars originated in the late 60's, motivated by considerations about pattern recognition, compiler construction, and data type specification. Since then the list of areas which have interacted with the development of graph grammars has grown impressively. Besides the aforementioned areas it includes software specification and development, VLSI layout schemes, database design, modelling of concurrent systems, massively parallel computer architectures, logic programming, computer animation, developmental biology, music composition, visual languages, and many others. Graph grammars are interesting from the theoretical point of view because they are a natural generalization of formal language theory based on strings and the theory of term rewriting based on trees.

The wide applicability of graph grammars is due to the fact that graphs are a natural way of describing complex situations on an intuitive level. Moreover, the graph transformations associated with a graph grammar bring "dynamic behaviour" to such descriptions, because they model the evolution of graphical structures. Therefore graph grammars and graph transformations become attractive as a "programming paradigm" for software and graphical interfaces.

Over the last 25-odd years graph grammars have developed at a steady pace into a theoretically sound and well-motivated research field. In particular, they are now based on solid foundations, presented in this volume. It includes a state-of-the-art presentation of the foundations of all basic approaches to graph grammars and computing by graph transformations.

The two most basic choices for rewriting a graph are *node replacement* and edge replacement, or, in the more general setting of hypergraphs, *hyperedge replacement*. In a node replacement graph grammar a node of a given graph is replaced by a new subgraph which is connected to the remainder of the graph by new edges depending on how the node was connected to it. In a hyperedge replacement graph grammar a hyperedge of a given hypergraph is replaced by a new subhypergraph which is glued to the remainder of the hypergraph by fusing (identifying) some nodes of the subhypergraph with some nodes of the remainder of the hypergraph depending on how the hyperedge was glued to it. Chapter 1 surveys the theory of node replacement graph grammars concentrating mainly on "context-free" (or "confluent") node replacement graph grammars, while Chapter 2 surveys the theory of hyperedge replacement graph grammars. Both types of graph grammars naturally generalize the context-free string grammars.

The gluing of graphs plays also a central role in the *algebraic approach* to graph transformations, where a subgraph, rather than a node or an edge only, can be replaced by a new subgraph (generalizing in this way arbitrary type-0 Chomsky string grammars). Originally, the rewriting of graphs based on gluing has been formulated by the so-called *double pushout* in the category of graphs and total graph morphisms. More recently a *single pushout* in the category of graphs and partial graph morphisms has been used for this purpose. Chapter 3 gives an overview of the double pushout approach, and Chapter 4 gives an overview of the single pushout approach; it also presents a detailed comparison of the two approaches.

Graphs may be considered as *logical structures* and so one can express formally their properties by logical formulas. Consequently classes of graphs may be described by formulas of appropriate logical languages. Such logical formulas are used as finite devices, comparable to grammars and automata, to specify classes of graphs and to deduce properties of such classes from their logical descriptions. Chapter 5 surveys the relationships between monadic second-order logic and the context-free graph grammars of Chapters 1 and 2.

The research on graph transformations leads to a careful re-thinking of the formal framework that should be used for the specification of graphs. The theory of *2-structures*, a specific relational framework, has turned out to be fruitful for the investigation of graphs especially in their relationship to graph transformations. The "static part" of the theory of 2-structures allows to obtain rather strong decomposition results for graphs, while the "dynamic part" of the theory employs group theory to consider transformations of graphs as encountered in networks. Chapter 6 presents the basic theory of 2-structures.

In order to specify classes of graphs and graph transformations as they occur in various applications (e.g. databases and database manipulations) one often needs quite powerful extensions of the basic mechanisms of graph replacement systems. One such extension is to control the order of application of the graph replacement rules. Chapter 7 presents a basic framework for *programmed graph replacement* systems based on such a control.

We believe that this volume together with the two forthcoming volumes - on specification and programming, and on concurrency - provide the reader with a rather complete insight into the mathematically challenging area of graph grammars which is well motivated by its many applications to computer science and beyond.

My thanks go first of all to the graph grammar community - the enthusiastic group of researchers, often with very different backgrounds, spread all over the world - for providing a very stimulating environment for the work on graph grammars. Perhaps the main driving force in this community during the last 7 years was the ESPRIT Basic Research Working Group COMPUGRAPH (Computing by Graph Transformation) in the period March 1989 - February 1992, followed by the ESPRIT Basic Research Working Group COMPUGRAPH II in the period October 1992 - March 1996. The initial planning of the handbook took place within the COMPUGRAPH project, and most of Volume I of the handbook has been written within the COMPUGRAPH period. The European Community is also founding now the TMR network GETGRATS (General Theory of Graph Transformation Systems) for the period from September 1996 to August 1999. We hope to complete volumes II and III of the handbook within the GETGRATS project. The gratitude of the graph grammar community goes to the European Community for the generous support of our research. We are also indebted to H. Ehrig and his Berlin group for managing the COMPUGRAPH and COMPUGRAPH II Working Groups, and to A. Corradini and U. Montanari for their work in preparing the GETGRATS project.

I am very grateful to all the authors of the chapters of this volume for their cooperation, and to the Advisory Board: B. Courcelle, H. Ehrig, G. Engels, D. Janssens, H.-J. Kreowski and U. Montanari, for their valuable advice.

Finally, very special thanks go to Perdita Löhr for her help in transforming all the separate chapters into a homogeneous and readable volume as it is now. Neither she nor myself could foresee how much work it would be - but due to her efforts we could bring the project to a happy end.

G. Rozenberg
Managing Editor
Leiden, 1996

Contents

Chapter 1

NODE REPLACEMENT GRAPH GRAMMARS

J. ENGELFRIET
Department of Computer Science, Leiden University
P.O.Box 9512, 2300 RA Leiden, The Netherlands

G. ROZENBERG
Department of Computer Science, Leiden University
P.O.Box 9512, 2300 RA Leiden, The Netherlands
and
Department of Computer Science, University of Colorado at Boulder
Boulder, Co 80309, U.S.A.

In a node-replacement graph grammar, a node of a given graph is replaced by a new subgraph, which is connected to the remainder of the graph by new edges, depending on how the node was connected to it. These node replacements are controlled by the productions (or replacement rules) of the grammar. In this chapter mainly the "context-free" (or "confluent") node-replacement graph grammars are considered, in which the result of the replacements does not depend on the order in which they are applied. Although many types of such grammars can be found in the literature, the emphasis will be on one of them: the C-edNCE grammar. Basic notions (such as derivations, associativity, derivation trees, normal forms, etc.) are discussed that facilitate the construction and analysis of these grammars. Properties of the class of generated graph languages, such as closure properties, structural properties, and decidability properties, are considered. A number of quite different characterizations of the class are presented, thus showing its robustness. This robustness of the class of C-edNCE graph languages, together with the fact that it is one of the largest classes of "context-free" graph languages, motivates the choice of the C-edNCE grammar to be central in this chapter.

Contents

1.1 Introduction

Graph grammars provide a mechanism in which local transformations on graphs can be modelled in a mathematically precise way. The main component of a graph grammar is a finite set of productions; a production is, in general, a triple (M, D, E) where M and D are graphs (the "mother" and "daughter" graph, respectively) and E is some embedding mechanism. Such a production can be applied to a ("host") graph H whenever there is an occurrence of M in H. It is applied by removing this occurrence of M from H, replacing it by (an isomorphic copy of) D, and finally using the embedding mechanism E to attach D to the remainder H^- of H.

Two main types of embedding can be distinguished: gluing and connecting. In the *gluing* case, certain parts (i.e., nodes and edges) of D are identified with certain parts of H^-. To be more precise, the latter are the parts of H^- that were previously identified with certain parts of M. The gluing parts of D are usually defined in terms of the gluing parts of M. In the *connecting* case, certain new edges are used as bridges that connect D to H^-, i.e., edges of which one node belongs to D and the other to H^-. When M is removed from H, all bridges between nodes of M and nodes of H that do not belong to M are removed too. The new bridges are usually defined in terms of the old ones.

Based on these two types of embedding there are two main approaches to graph grammars: the gluing approach and the connecting approach. Unfortunately, they are usually called the algebraic approach and the algorithmic (or set theoretic) approach, which names refer to the mathematical techniques that are used. In this chapter we describe the connecting approach to graph rewriting, and, apart from Section 1.2.3 we discuss context-free graph grammars only. For the general case we refer to Chapter 7. The gluing approach is described in Chapters 4, 5, 2.

In the connecting approach, a *context-free* graph grammar has productions of the form (M, D, E) where M consists of one node only. Thus, nodes are replaced by graphs. Moreover, we require that the result of a sequence of node replacements does not depend on the order in which they are applied. This makes the context-free graph grammar well suited to describe recursively defined sets of graphs, i.e., recursive graph properties. And so, the context-free graph grammar is the appropriate generalization of the usual context-free grammar for strings. However, as opposed to the case of strings, there is not just one canonical way of defining context-free graph grammars. In the connecting approach, there are several natural ways of defining the embedding mechanism E and there are several natural restrictions on the form of the right-hand side D.

In the gluing approach, context-free graph grammars rewrite edges (or even hyperedges) instead of nodes, see Chapter 2. In recent years it has turned out that for each embedding approach there is one main class of context-free graph grammars that is of most interest: the HR grammars in the gluing approach (where HR stands for 'hyperedge replacement'), and the so-called C-edNCE grammars in the connecting approach. The C-edNCE grammar is the most powerful context-free graph grammar of the connecting approach; it has a quite general embedding mechanism and no restrictions on right-hand sides. We also note that the C-edNCE grammar has more generating power than the HR grammar. Due to the importance of the C-edNCE grammars they are also called VR (vertex replacement) grammars. For a comparison between HR and VR sets of graphs see also Chapter 3.

In Sections 1.2.1 and 1.2.2 we describe the connecting approach to context-free graph grammars in an informal way (as in [53]). We start with the easiest types of embedding and end with the C-edNCE grammar. In Section 1.3 we define the C-edNCE grammar formally, show a number of its elementary properties, and discuss some natural subclasses. In Section 1.4 we present four different characterizations of the class of C-edNCE graph languages: by regular path languages in regular tree languages, by monadic second-order logic translation of trees, by handle-replacement graph grammars, and by regular sets of graph expressions. Section 1.4.2 also contains a discussion of closure properties and decidability properties of that class. Finally, in Section 1.5, we take a brief look at the complexity of recognizing C-edNCE graph languages.

The selection of the results that are presented is of course our personal choice. We assume the reader to be familiar with basic notions from formal language theory (in particular context-free grammars, to be used as a point of reference) and graph theory.

1.2 From NLC to edNCE

In this section we give an informal introduction to node replacement graph grammars. The reader who prefers formal definitions can skip this section and proceed to Section 1.3.

1.2.1 Node replacement and the NLC methodology

The usual way of rewriting a graph H into a graph H' is to *replace* a subgraph M of H by a graph D and to *embed* D into the remainder of H, i.e., into the graph H^- that remains after removing M from H; H' is the resulting graph. We say that H is the *host graph*, M is the *mother graph*, and D is the *daughter graph*. In the connecting approach (see Section 1.1), M is an induced subgraph of H and the removal of M from H consists of the removal of all nodes of M and of all edges of H that are incident with nodes of M. The embedding process connects D to the remainder H^- by building bridges, i.e., by establishing edges between certain nodes of D and certain nodes of H^-.

In the restricted case of *node replacement* the mother graph consists of one node only, a "local unit" of the host graph. Thus, if also the embedding is done in a local fashion (by connecting the daughter graph to host nodes that are "close" to the mother node), then a node rewriting step is a local graph transformation. The iteration of such node rewriting steps leads to a *global* transformation of a graph into a graph that is based on *local* transformations. This is the underlying idea of graph grammars based on node rewriting. In such a grammar the replacement of nodes is specified by a finite number of *productions* and the embedding mechanism is specified by a finite number of *connection instructions*. In many cases, each production has its own, private, set of connection instructions. The productions are the main constituent of the graph grammar, which is a finite specification of a graph rewriting system.

A typical, very simple, example of a node-replacement mechanism is the *Node Label Controlled* mechanism, or NLC mechanism. In the NLC framework one rewrites undirected node-labeled graphs. The productions are node-replacing productions and the embedding connection instructions connect the daughter graph to the neighbourhood of the mother node — hence the rewriting process is completely local. In the NLC approach "everything" is based on node labels. An *NLC production* is of the form $X \to D$, where X is a (nonterminal) node label, and D is an undirected graph with (terminal or nonterminal) node labels. Such a production can be applied to *any* node m in the host graph that has label X (i.e., there are no application conditions); its application results in the replacement of mother node m by daughter graph D. All productions share the same connection instructions. A *connection instruction* is of the form (μ, δ), where both μ and δ are (terminal or nonterminal) node labels. The meaning of such an instruction is that the embedding process should establish an edge between *each* node labeled δ in the daughter graph D and *each* node labeled μ that is a neighbour of the mother node m. Note that the presence of edges cannot be tested; if m has no μ-labeled neighbour, then the connection instruction (μ, δ) remains unused. Since a connection instruction is an ordered pair, a finite number of connection instructions is a relation, called a *connection relation*. Warning: in the literature a connection instruction is often given as (δ, μ) rather than (μ, δ).

Example 1.1
Consider the production $X \to D$ in Fig. 1.1, and consider the connection instructions $(c, a), (b, b)$, and (b, X), where we assume that X is a nonterminal node label and a, b, c are terminal node labels. Consider also the host graph H in Fig. 1.2(a), and let m be the node of H with label X. Note that graphs are drawn as usual, except that, for clearness sake, nonterminal nodes are represented by boxes. The application of $X \to D$ to mother node m is shown in two steps in Figs. 1.2(b) and 1.3. Figure 1.2(b) shows the result of replacing m by D; m is removed, together with its three incident edges, and D is added (disjointly) to the remainder H^- of H. Figure 1.3 shows the resulting graph H', i.e., the result of embedding D in the remainder H^-; the five edges that are added by the connecting process are drawn fatter than the other edges. Thus, H is transformed into H' by the application of $X \to D$ to m. □

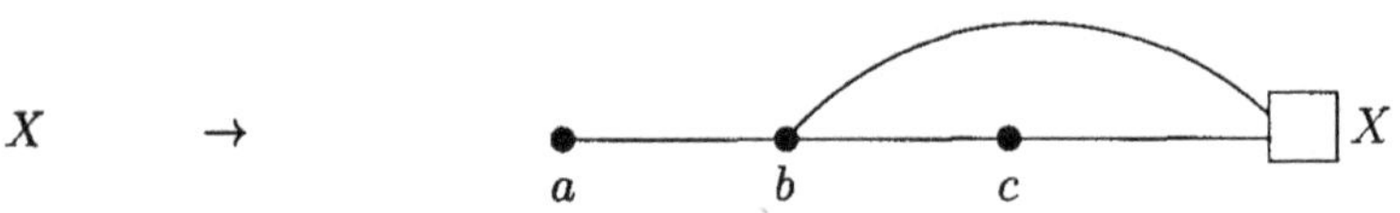

Figure 1.1: A production of an NLC grammar.

An *NLC graph grammar* is a system $G = (\Sigma, \Delta, P, C, S)$ where $\Sigma - \Delta$ and Δ (with $\Delta \subseteq \Sigma$) are the alphabets of nonterminal and terminal node labels, respectively, P is a finite set of NLC productions, C is a connection relation, i.e., a binary relation over Σ, and S is the initial graph (usually with a single node). As usual, the graph language generated by G is $L(G) = \{H \in GR_\Delta \mid S \Rightarrow^* H\}$, where GR_Δ is the set of undirected graphs with node labels in Δ, $\Rightarrow$ represents one rewriting step, and $\Rightarrow^*$ represents a derivation, i.e., a sequence of rewriting steps.

To simplify the description of examples, we will use strings to denote certain "string-like" graphs. More precisely, a string $\alpha_1\alpha_2 \cdots \alpha_n$, with $\alpha_i \in \Delta$ for some alphabet Δ and $n \geq 1$, denotes the undirected node-labeled graph with nodes $x_1, x_2, \ldots, x_n$ and edges $\{x_i, x_{i+1}\}$ for every $1 \leq i \leq n-1$, where x_i is labeled by α_i for every $1 \leq i \leq n$. In particular, $\alpha \in \Delta$ denotes the graph with one α-labeled node and no edges.

Example 1.2
Let us consider an NLC graph grammar $G = (\Sigma, \Delta, P, C, S)$ such that $L(G)$ is the set of all strings in $(abc)^+$ with additional edges between all nodes with label b. G is defined as follows: $\Sigma = \{X, a, b, c\}$, $\Delta = \{a, b, c\}$, $S = X$, P consists of the production in Fig. 1.1 and the production $X \to abc$, and $C =$

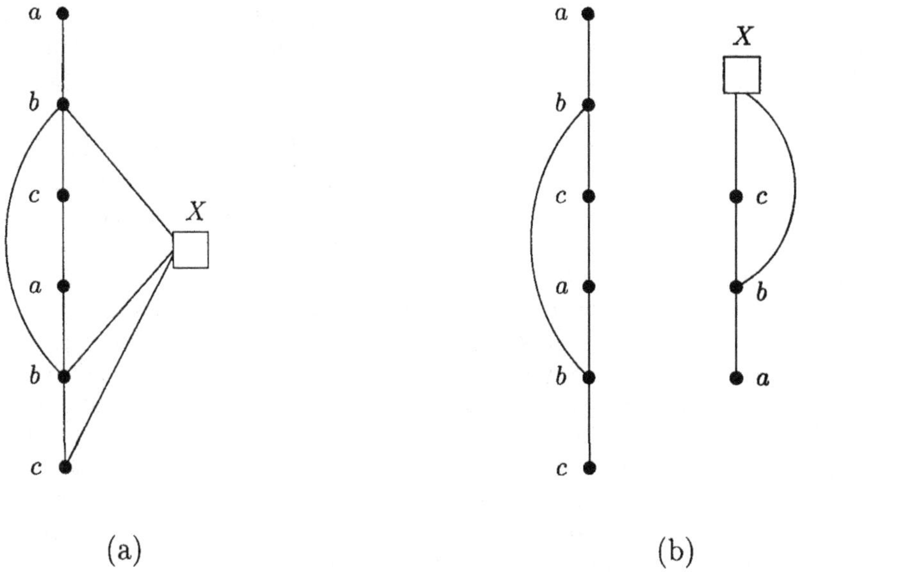

Figure 1.2: (a) A host graph. (b) Replacement of the mother node by the daughter graph.

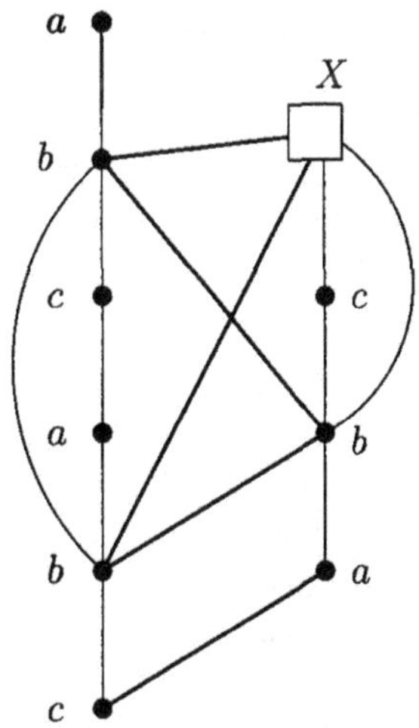

Figure 1.3: Embedding of the daughter graph.

$\{(c,a),(b,b),(b,X)\}$. The intermediate graphs, or *sentential forms*, generated by this grammar are the strings in $(abc)^*X$ with additional edges between all nodes with label b or label X. One rewriting step of G was considered in Example 1.1. □

NLC graph grammars are one attempt to define a class of "context-free" graph grammars, i.e., graph grammars that are similar to the usual context-free grammars for strings. Such an attempt is useful if one wishes to carry over the nice properties of context-free grammars to the case of graphs. In particular, context-free graph grammars are meant to be a framework for the description of recursively defined properties of graphs. NLC graph grammars are context-free in the sense that they are completely local and have no application conditions. This allows their derivations to be modelled by derivation trees (which embody the recursive nature of context-free grammars). However, in general, NLC grammars do not have the desirable context-free property that the yield of a derivation tree is independent of the order in which the productions are applied. A graph grammar that *has* this property is said to be *confluent* or to have the finite Church-Rosser property (not to be confused with the related but different notions in term rewriting systems).

Example 1.3
Consider an NLC grammar G with initial graph S and productions $S \to AB$, $A \to a$ and $B \to b$ (where S, A, B are nonterminal node labels and a, b are terminal node labels). Viewed as a context-free string grammar, $L(G) = \{ab\}$ and G has exactly one derivation tree. Suppose now that G has connection relation $C = \{(B,a),(a,b)\}$. Then $L(G) = \{ab, a+b\}$ where $a+b$ is the graph with two nodes, labeled a and b, and no edges. Note that, intuitively, G still has one derivation tree, but the two ways of traversing this tree produce different graphs: ab is produced when $A \to a$ is applied before $B \to b$, whereas $a+b$ is produced when they are applied in the reverse order. Thus, G is not confluent. □

Since, in this chapter, we are mainly interested in context-free grammars, we restrict attention to confluent NLC (or C-NLC) graph grammars. There is a natural structural restriction on the productions of the NLC grammar that is equivalent with confluence (see Section 1.3, Definition 1.3.5). There are other natural structural restrictions on the NLC grammar that guarantee confluence (but are not equivalent to it). An attractive example of such a restriction is the "boundary" restriction: in a *boundary* NLC (or B-NLC) graph grammar no two nodes with a nonterminal label are connected by an edge (in the right-hand sides of productions and in the initial graph). In fact, an important feature of NLC rewriting is the following: if, in a derivation, one obtains a sentential form with two nodes x and y that are *not* connected by an edge, then, whatever will happen later to these nodes in the derivation, no node descending from x will ever be connected to a node descending from y: connections can be broken, but cannot be re-established. In a boundary NLC grammar this feature ensures that in every sentential form the nodes with a nonterminal label are not connected to each other, and this in turn ensures that the grammar is confluent. It should be clear that the grammar of Example 1.2 is a B-NLC grammar; in fact, it is even *linear*: there is at most one node with a nonterminal label in every derived graph.

1.2.2 *Extensions and variations: the edNCE grammar*

The NLC approach described in the previous section can be extended and modified in various ways. Here we discuss some of these generalizations which are also called "NLC-like" or "NCE-like" graph grammars. The last generalization to be considered is the edNCE grammar, which is much more powerful than the NLC grammar.

It is often convenient to be able to refer in the connection instructions *directly* to nodes in the daughter graph rather than through their labels. Hence each connection instruction will now be of the form (μ, x), where x is a node in the daughter graph (i.e., in one of the right-hand sides of the productions of the grammar), and, as before, μ is a node label. The meaning of this instruction is that node x should be connected to each node labeled μ that is a neighbour of the mother node. This gives us a convenient way of distinguishing between *individual nodes in the daughter graph*. NLC-like grammars with this type of connection instructions are called *NCE graph grammars*, i.e., graph grammars with *Neighbourhood Controlled Embedding*. This acronym stresses the locality of the embedding process, which NCE grammars inherit from NLC grammars. Of course, NCE grammars are still "node label controlled" as far as replacement is concerned; the embedding process is node label controlled with respect to the neighbourhood of the mother node only, because the nodes of the daughter graph are accessed directly (independently of their labels).

In the formal definition of an NCE graph grammar one could stick to one connection relation and require the right-hand sides of productions to be mutually disjoint. However, it is more natural to define an NCE grammar as a system (Σ, Δ, P, S), such that Σ, Δ, and S are as before, and P is a finite set of productions, where each production is now of the form $X \to (D, C)$ such that $X \to D$ is an NLC production and C is a connection relation "for D", i.e., $C \subseteq \Sigma \times V_D$ (where V_D is the set of nodes of D).

Example 1.4
Consider the graph language L that is obtained from the one of Example 1.2 by the assumption that $b = a$. Thus, the nodes of the graphs in L have labels a and c. An NCE grammar (similar to the one of Example 1.2) that generates L, has productions $X \to (D_1, C_1)$ and $X \to (D_2, C_2)$, where $X \to D_1$ is shown in Fig. 1.1 (with $b = a$), $C_1 = \{(c, x_1), (a, x_2), (a, x_4)\}$, $D_2 = aac$, and $C_2 = \{(c, x_1), (a, x_2)\}$, with the nodes of D_1 and D_2 numbered from left to right. Figs. 1.2 and 1.3 (with $b = a$) show a derivation step of this grammar. □

One might think of extending the connection instructions to be of the form (y, x), where x is a node of the daughter graph and y is a neighbour of the mother node. However, this does not work in general, because the number of neighbours of the mother node may be unbounded, which makes it impossible to access individual neighbours. A way to distinguish between neighbours in a better way than just through their labels will be discussed later.

It is natural to extend the domain of graph rewriting to graphs more general than undirected node-labeled graphs. In particular we will discuss directed graphs and graphs with edge labels.

To extend the NLC approach to *directed* node-labeled *graphs* is quite easy. The connection relation C now consists of triples (μ, δ, d), where $d \in \{\text{in}, \text{out}\}$, to deal with the incoming edges and the outgoing edges of the mother node, respectively. These connection instructions are used in an obvious way. Thus, a connection instruction (μ, δ, in) means that the embedding process should establish an edge to each node labeled δ in the daughter graph D from each node labeled μ that is an "in-neighbour" of the mother node m (where the in-neighbours of m are all nodes n for which there is an edge from n to m in the host graph). Similarly, a connection instruction $(\mu, \delta, \text{out})$ means that an edge should be established from each node labeled δ in the daughter graph to each node labeled μ that is an out-neighbour of m. Note that the grammar is direction preserving, in the sense that edges leading to m (from m, respectively) are replaced by edges leading to D (from D, respectively). There is also a variation that allows directions to be changed: the connection instructions are then of the form (μ, δ, d, d') with $d, d' \in \{\text{in}, \text{out}\}$, where d is the old direction and d' the new one.

As observed before, neighbours of the mother node can be distinguished by the embedding process of an NLC grammar when they have distinct labels only. The extension to directed graphs already gives some more discerning power: the grammar can also distinguish between out-neighbours and in-neighbours of the mother node. Extending the NLC approach to (undirected) graphs that, in addition to node labels, also have *edge labels* gives even more discerning power in the neighbourhood of the mother node.

Example 1.5
For the host graph in Fig. 1.4, when replacing the mother node m, the embedding process can distinguish between neighbours x and y (both labeled a), because x is a p-neighbour of m (i.e., is connected to m by an edge with label p) and y is a q-neighbour of m. Note that it is still impossible to distinguish between neighbours x and u. □

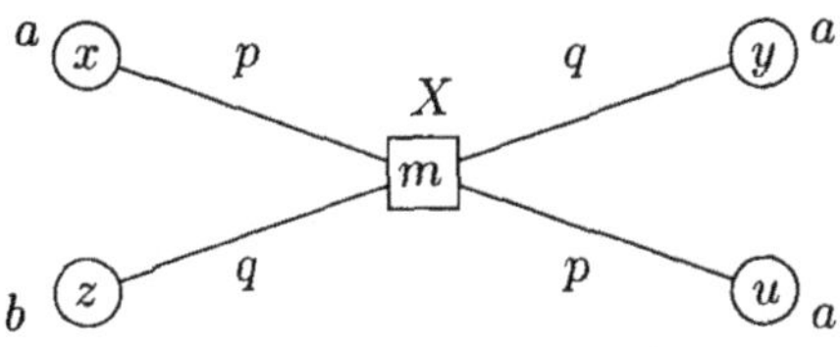

Figure 1.4: A neighbourhood of the mother node.

Thus, intuitively, the neighbours of the mother node m are divided into several distinct types, depending on the label of the edge that connects the neighbour to m. A natural, and powerful, idea is to allow the embedding process to change the type of these neighbours: when the embedding process connects a daughter node x to a neighbour n of m, the type of n with respect to x may differ from its type with respect to m. This leads to connection instructions of the form $(\mu, p/q, \delta)$, where p and q are edge labels, and μ and δ are node labels as before. The meaning of this connection instruction is that the embedding process should establish an edge with label q between each μ-labeled p-neighbour of the mother node and each δ-labeled node in the daughter graph. Thus, edge label p is changed into edge label q. This feature is also called *dynamic edge relabeling*.

Even if one is not interested in generating graphs with edge labels, dynamic edge relabeling can be used as a natural additional control of the rewriting and embedding process during the derivation of graphs. In many cases this facilitates the construction and the understanding of the grammar.

We use a small letter d to indicate that an NLC-like graph grammar generates directed graphs, and a small letter e to indicate that the generated graphs have edge labels in addition to node labels. Thus, one has, e.g., dNLC grammars, eNCE grammars, and edNLC grammars.

The following example shows the use of dynamic edge relabeling.

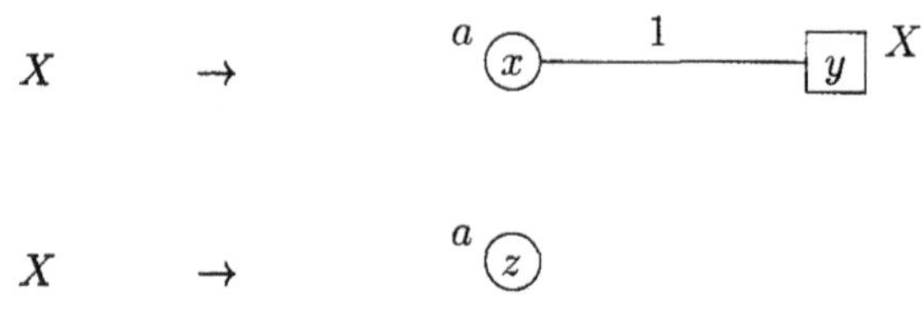

Figure 1.5: Productions of an eNCE grammar (without connection relations).

Example 1.6

(a) Consider an eNCE grammar G that generates the edge complements of all strings in a^+, i.e., all graphs with a-labeled nodes $x_1, \ldots, x_n$ ($n \geq 1$) and edges $\{x_i, x_j\}$ with $i \leq j-2$; all edges have label q. Note that the connection instructions of an eNCE grammar have the form $(\mu, p/q, x)$, where μ is a node label, p and q are edge labels, and x is a daughter node. $G = (\Sigma, \Delta, \Gamma, \Omega, P, S)$ is defined as follows: $\Sigma = \{X, a\}$, $\Delta = \{a\}$, the edge label alphabet Γ is $\{1, 2, q\}$ with "final" edge label alphabet $\Omega = \{q\} \subseteq \Gamma$, $S = X$, and P consists of the two productions $X \to (D_1, C_1)$ and $X \to (D_2, C_2)$, where $X \to D_1$ and $X \to D_2$ are shown in Fig. 1.5, $C_1 = \{(a, 2/q, x), (a, 2/2, y), (a, 1/2, y)\}$ and $C_2 = \{(a, 2/q, z)\}$. G generates sentential forms of the form shown in Fig. 1.6.

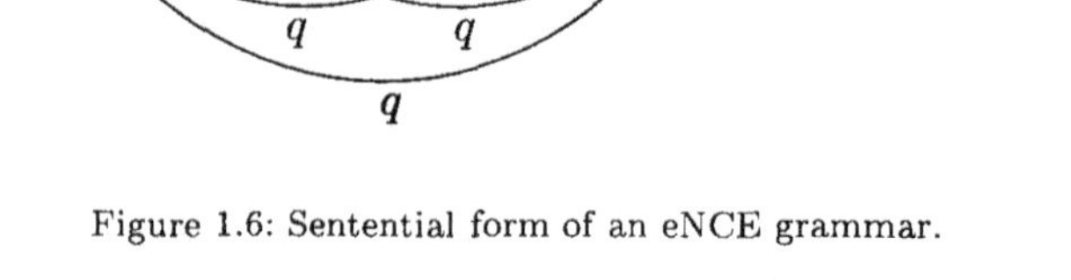

Figure 1.6: Sentential form of an eNCE grammar.

The edge labels 1 and 2 divide the neighbours of the nonterminal node into two types: the first node to its "left", and all the other nodes. Application of the production $X \to (D_1, C_1)$ leads to a change of type of the 1-neighbour of X: after application it becomes of type 2; this dynamic change of type is caused by the connection instruction $(a, 1/2, y)$. Note that G is linear and hence confluent, i.e., a C-eNCE grammar.

(b) Consider the graph language L consisting of all strings in $(aaa)^+$ with additional edges (labeled q) between any two nodes x_i such that $i \equiv 2(\text{mod } 3)$; in fact, L is obtained from the graph language of Examples 1.2 and 1.4 by the assumption that $c = b = a$. It can easily be seen that L can also be generated by a linear eNCE grammar G. We just note that G has a production as in Fig. 1.1, such that the edges incident with X have distinct ("nonfinal") labels. In every sentential form, X has two types of neighbours: the first type is the a-labeled node that is generated last, and the second type consists of all generated nodes x_i with $i \equiv 2(\text{mod } 3)$. □

Combining all features discussed above naturally leads to the class of edNCE graph grammars. As argued in Section 1.2.1, of particular interest are the C-edNCE grammars: the confluent edNCE grammars. Also of interest is the subclass of B-edNCE grammars: the boundary edNCE grammars.

Each production of an edNCE grammar is of the form $X \to (D, C)$, and each connection instruction in C is of the form $(\mu, p/q, x, d)$, where μ is a node label, p and q are edge labels, x is a node of D, and $d \in \{\text{in}, \text{out}\}$. If, say, $d = \text{in}$, then this instruction is interpreted as follows: the embedding process should establish an edge with label q to node x of D from each μ-labeled p-neighbour of m that is an in-neighbour of m. Note that the grammar is direction preserving; it can be shown that allowing the directions of edges to change (with connection instructions of the form $(\mu, p/q, x, d/d'))$ does not increase the power of the edNCE grammar. Also, allowing the connection instructions to make use of multiple edges does not increase the power of the edNCE grammar (in that case a connection instruction is of the form $(\mu, B/q, x, d)$ where B is a set of edge labels; it is applicable only to a neighbour of m that is a p-neighbour of m for every $p \in B$). On the other hand, it can be shown that the information concerning the labels of neighbours can be coded into the labels of the connecting edges. Thus, one might in fact assume the connection instructions to be of the simpler form $(p/q, x, d)$ and disregard the label μ of the neighbour of the mother node.

In going from the class of NLC grammars to the class of edNCE grammars we have clearly made the embedding mechanism much more involved. Thus, since in analyzing derivations of edNCE grammars one has to keep track additionally of edge labels, directions, and individual nodes, it would seem that it is more involved to analyze edNCE grammars than NLC grammars. However, there is a definite trade-off here: it turns out that the additional features of the edNCE grammar can often be used in a straightforward, easily understandable fashion to show that they have certain desirable properties that are not possessed by NLC grammars. As a simple example, every B-edNCE grammar has an equivalent B-edNCE grammar in Chomsky Normal Form (appropriately defined, see Theorem 1.3.31); the analogous result does not hold for B-NLC grammars. As another example, it turns out that the class of C-edNCE graph languages can be characterized in several different, natural ways (see Section 1.4); no such characterizations are known for the class of C-NLC languages.

1.2.3 Graph replacement grammars

For an arbitrary graph grammar that uses the connecting approach to embedding, the productions of the grammar are of the form (M, D, C) where M and D are graphs (the mother and the daughter graph, respectively) and C is a set of connection instructions. Such an instruction is applied to a graph H by removing from H an induced subgraph (isomorphic to) M, replacing it by (a copy of) D, and embedding D in the remainder H^- of H by the connection instructions from C.

Each of the node-replacement grammar types discussed in the previous sections has its counterpart in the graph-replacement case. As an example, for NCE grammars the connection instructions are of the form (m, μ, x), where m is a node of M, and, as in the node-replacement case, μ is a node label and x is a node of D. The meaning of the instruction is that daughter node x should be connected to each μ-labeled node of H^- that is a neighbour of mother node m. Similarly, for edNCE grammars a connection instruction is of the form $(m, \mu, p/q, x, d)$ with obvious meaning: a q-labeled edge should be established between x and every μ-labeled node of H^- that is a p-neighbour of m (preserving direction d).

Up to now we have only considered the neighbourhood controlled embedding mechanism (NCE) that is completely *local*, in the sense that the nodes of the daughter graph are always connected to former neighbours of the nodes of the mother graph (which seems to be necessary in the context-free case). In general one can of course consider embedding mechanisms of arbitrary complexity that connect a daughter node to any node of H^-. Thus, in the case of edNCE grammars one might have connection instructions of the form $(m, \Pi/q, x, d)$ where Π specifies a binary relation between nodes of M and nodes of H^-; the meaning of such an instruction is: a q-labeled edge should be established between x and every node y of H^- that is in the relation Π with m (preserving direction d). The graph grammars of [84,85,86] are of this type (see [3] for a special case). In the grammars of Nagl, the admissible binary relations Π are, roughly speaking, defined recursively as follows: (1) for every edge label p and every node label μ, the relations edge_p, consisting of all (x, y) with a p-labeled edge from x to y, and lab_μ, consisting of all (x, x) such that x has label μ, are admissible, and (2) if Π, Π_1, and Π_2 are admissible, then so are $\Pi_1 \cup \Pi_2$, Π^c (the complement of Π), Π^{-1}, $\Pi_2 \circ \Pi_1$, and Π^*. Admissible relations are, e.g., $\text{edge}_q^{-1} \circ \text{lab}_\mu \circ \text{edge}_p$ which holds between y and m if there is a p-labeled edge from y to a q-out-neighbour of m with label μ, and $(\text{edge}_p \cup \text{edge}_q)^c$ which holds between y and m if y is neither a p-in-neighbour nor a q-in-neighbour of m. Since the admissible relations can be written as expressions, this is said to be an expression-oriented approach to embedding.

For more information on arbitrary graph grammars that use the connecting approach to embedding we refer to Chapter 7.

1.2.4 Bibliographical comments

Graph grammars were introduced in [88] and [95], both based on the connecting approach (see the Introduction). In the seventies, several variations of these grammars were considered in, e.g., [2,9,80,84,89], as described in [85,86]. The graph grammars of Nagl subsume almost all these variations.

The (node replacement) edNCE graph grammars were introduced (as a special case) in [84,86], where they are called depth-1 context-free graph grammars, and investigated in, e.g., [13,39,40,41,43,44,52,70,71,72]. Confluence was introduced explicitly for edNCE grammars in [70], and the class of C-edNCE languages was studied in [14,15,16,18,20,27,28,46,47,49,51,96,99,100,101]. NLC graph grammars are a very special case of edNCE graph grammars. They were introduced in [60] as a more fundamental variant of the web grammars of [88,89], and investigated in, e.g., [12,36,38,45,58,61,62,65,69,73,75,102]. NCE graph grammars were introduced in [64] and used, e.g., in [42]; they are even closer to the web grammars of [89] than NLC grammars.

Boundary NLC grammars were introduced in [92], and studied, e.g., in [93,94,104,105,106,108,109]. Boundary grammars are confluent. Confluence was introduced explicitly for NLC grammars in [36]. The notion of confluence for context-free graph grammars in general was stressed in [23], where also the C-NLC grammars were investigated. Another class of confluent grammars was considered in [66,23]: the neighbourhood-uniform NCE (or NLC) grammars. They were studied before, under other names, in [2,61,64,86].

The extension of the NLC grammar to directed graphs, i.e., the dNLC graph grammar, was introduced in [63] and investigated, e.g., in [1,4,5]. The extension to graphs with edge labels, i.e., the eNLC and edNLC graph grammar, was first considered in [68]. Some other extensions of NLC grammars are considered in [54,81,82]. The (confluent) edNCE grammars are extended to hypergraphs in [74].

The parallel rewriting of NLC grammars and edNCE grammars (similar to the L-systems for strings) is studied, e.g., in [36,68,86,90].

Graph replacement edNCE grammars (that disregard the labels of the neighbours of the mother graph) are used in [67] to formalize actor systems, a model of concurrent computation.

A proposal for a categorical framework for NCE rewriting (and hyperedge rewriting) is presented in [6].

1.3 Node replacement grammars with Neighbourhood Controlled Embedding

1.3.1 Formal definition of edNCE graph grammars

In this subsection we give formal definitions for the edNCE graph grammars, and in particular for the confluent edNCE (C-edNCE) graph grammars. These grammars generate directed graphs with labeled nodes and labeled edges. Most proofs of results in the literature on eNCE grammars (which generate undirected graphs) can be adapted to edNCE grammars in a straightforward way (but not always the other way around!); thus, in the sequel, we will quote results for edNCE grammars from the literature even if, in the literature, they are stated for the eNCE case only.

Let Σ be an alphabet of node labels and Γ an alphabet of edge labels. A *graph* over Σ and Γ is a tuple $H = (V, E, \lambda)$, where V is the finite set of nodes, $E \subseteq \{(v, \gamma, w) \mid v, w \in V, v \neq w, \gamma \in \Gamma\}$ is the set of edges, and $\lambda : V \to \Sigma$ is the node labeling function. The components of H are also denoted as V_H, E_H, and λ_H, respectively. Thus, we consider directed graphs without loops; multiple edges between the same pair of nodes are allowed, but they must have different labels. These restrictions are convenient, but not essential, for the considerations of this chapter. A graph is undirected if for every $(v, \gamma, w) \in E$, also $(w, \gamma, v) \in E$. Graphs with unlabeled nodes and/or edges can be modeled by taking $\Sigma = \{\#\}$ and/or $\Gamma = \{*\}$, respectively.

As usual, two graphs H and K are *isomorphic* if there is a bijection $f : V_H \to V_K$ such that $E_K = \{(f(v), \gamma, f(w)) \mid (v, \gamma, w) \in E_H\}$ and, for all $v \in V_H$, $\lambda_K(f(v)) = \lambda_H(v)$. For a graph H, the set of all graphs isomorphic to H is denoted $[H]$. In general, for any notion of isomorphism between objects, we use $[x]$ to denote the set of objects isomorphic to x. Sometimes, x is called a 'concrete' object and $[x]$ an 'abstract' object. As usual, we will not always distinguish carefully between concrete and abstract graphs.

The set of all (concrete) graphs over Σ and Γ is denoted $GR_{\Sigma,\Gamma}$, and the set of all abstract graphs is denoted $[GR_{\Sigma,\Gamma}]$. A subset of $[GR_{\Sigma,\Gamma}]$ is called a *graph language*.

A production of an edNCE grammar will be of the form $X \to (D, C)$ where X is a nonterminal node label, D is a graph, and C is a set of connection instructions. A rewriting step according to such a production consists of removing a node labeled X (the "mother node") from the given "host" graph H, substituting D (the "daughter graph") in its place, and connecting D to the remainder of H in a way specified by the connection instructions in C. The pair (D, C) can be viewed as a new type of object, and the rewriting step can be viewed as the substitution of this object (D, C) for the mother node in the host graph. Intuitively, these objects are quite natural: they are graphs ready to be embedded in an environment.

Formally, a *graph with* (neighbourhood controlled) *embedding* over Σ and Γ is a pair (H, C) with $H \in GR_{\Sigma,\Gamma}$ and $C \subseteq \Sigma \times \Gamma \times \Gamma \times V_H \times \{\text{in}, \text{out}\}$. C is the *connection relation* of (H, C), and each element $(\sigma, \beta, \gamma, x, d)$ of C (with $\sigma \in \Sigma$, $\beta, \gamma \in \Gamma$, $x \in V_H$, and $d \in \{\text{in}, \text{out}\}$) is a *connection instruction* of (H, C). To improve readability, a connection instruction $(\sigma, \beta, \gamma, x, d)$ will always be written as $(\sigma, \beta/\gamma, x, d)$. Warning: in the literature the elements of a connection instruction are often listed in another order. Two graphs with embedding (H, C_H) and (K, C_K) are isomorphic if there is an isomorphism f from H to K such that $C_K = \{(\sigma, \beta/\gamma, f(x), d) \mid (\sigma, \beta/\gamma, x, d) \in C_H\}$.

The set of all graphs with embedding over Σ and Γ is denoted $GRE_{\Sigma,\Gamma}$. Every ordinary graph is also viewed as a graph with empty embedding (i.e., $C = \emptyset$); thus $GR_{\Sigma,\Gamma} \subseteq GRE_{\Sigma,\Gamma}$. In what follows we will also say 'graph' instead of 'graph with embedding'; this should not lead to confusion.

Intuitively, for a graph with embedding (D, C), a connection instruction $(\sigma, \beta/\gamma, x, \text{out})$ of C means that if there was a β-labeled edge from the mother node v for which (D, C) is substituted to a node w with label σ, then the embedding process will establish a γ-labeled edge from x to w. And similarly for 'in' instead of 'out', where 'in' refers to incoming edges of v and 'out' to outgoing edges of v. Note in particular that the edge label is changed from β into γ (which explains the notation β/γ).

An example is now given of a graph with embedding. At the same time we discuss an elegant graphical specification of graphs with embedding (and hence of productions of edNCE grammars) that was introduced in [70], and implemented in the graph editor Graph$^{\rm Ed}$ of [56].

Example 1.7
Let $\Sigma = \{X, Y, a, b, \sigma, \sigma'\}$ and $\Gamma = \{\alpha, \alpha', \beta, \beta', \gamma, \gamma', \gamma_1, \gamma_2, \delta, \delta'\}$. Consider the graph with embedding $(D, C_D) \in GRE_{\Sigma,\Gamma}$ with $V_D = \{x, y\}$, $\lambda_D(x) = X$, $\lambda_D(y) = b$, $E_D = \{(y, \alpha, x)\}$, and C_D consists of the tuples $(\sigma, \gamma/\delta, y, \text{out})$, $(\sigma', \gamma'/\delta', x, \text{in})$, $(Y, \beta/\gamma_1, y, \text{in})$, $(Y, \beta/\gamma_2, x, \text{in})$, and $(a, \alpha/\alpha', x, \text{out})$. In Fig. 1.7(b) a graphical specification of (D, C_D) is given. The graph D is drawn within a large box. Graphs are drawn in the usual way, except that nodes that are labeled by a capital (which is usually a nonterminal symbol), are represented by a small box. The area outside the large box around D symbolizes the environment of D, and the large box itself may be viewed as a (nonterminal) node, to be rewritten by (D, C_D). The connection instructions are represented by directed lines that connect nodes (of D) inside the box with node labels outside the box. Such a directed line has two arrows (in the same direction) and two labels: the one outside the box is the "old" label and the one inside the box is the "new" label. Lines outside the box may be shared, in an obvious way.

As another example, Fig. 1.7(a) represents the graph with embedding (H, C_H) in $GRE_{\Sigma,\Gamma}$ where $V_H = \{u, v, w\}$, $\lambda_H(u) = Y$, $\lambda_H(v) = X$, $\lambda_H(w) = a$, $E_H = \{(u, \beta, v), (v, \alpha, w), (w, \beta, u)\}$, and $C_H = \{(\sigma, \beta/\gamma, u, \text{in}), (\sigma, \beta/\gamma, v, \text{out}), (\sigma', \beta'/\gamma', v, \text{in}), (\sigma', \beta'/\alpha, w, \text{in})\}$. □

We now define formally how to substitute a graph with embedding for a node of a graph (see [20]). It is convenient (and natural) to allow the host graph to be a graph with embedding too; in the frequent case that the host graph is an ordinary graph, the result of the substitution will also be an ordinary graph. Substitution is defined for disjoint graphs with embedding only, i.e., for graphs with embedding that have disjoint sets of nodes.

Definition 1.3.1
Let (H, C_H) and (D, C_D) be two graphs with embedding, in $GRE_{\Sigma,\Gamma}$, such that H and D are disjoint, and let v be a node of H. The *substitution* of (D, C_D)

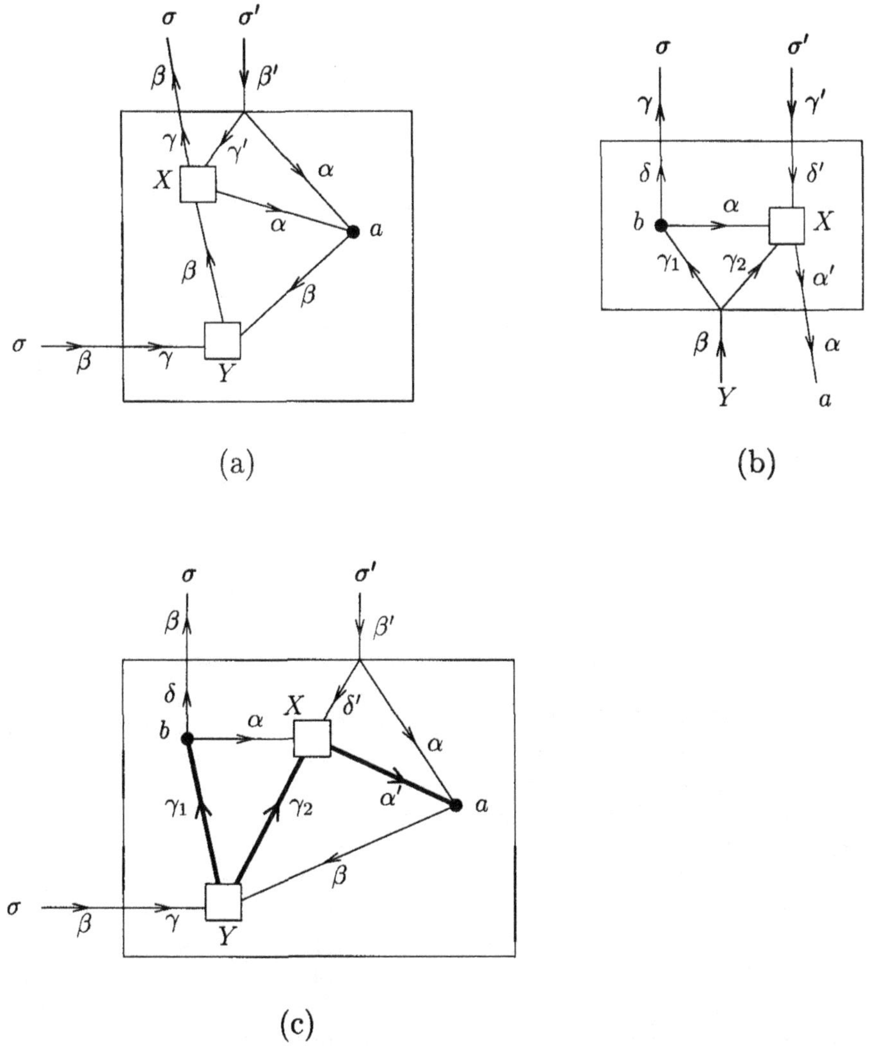

Figure 1.7: Two graphs with embedding, and the result of their substitution: (a) is (H, C_H), (b) is (D, C_D), and (c) is $(H, C_H)[v/(D, C_D)]$, where v is the node of H with label X.

for v in (H, C_H), denoted $(H, C_H)[v/(D, C_D)]$, is the graph with embedding (V, E, λ, C) in $GRE_{\Sigma,\Gamma}$ such that

$$\begin{aligned} V &= (V_H - \{v\}) \cup V_D, \\ E &= \{(x,\gamma,y) \in E_H \mid x \neq v, y \neq v\} \cup E_D \\ &\quad \cup \{(w,\gamma,x) \mid \exists \beta \in \Gamma : (w,\beta,v) \in E_H, (\lambda_H(w), \beta/\gamma, x, \text{in}) \in C_D\} \\ &\quad \cup \{(x,\gamma,w) \mid \exists \beta \in \Gamma : (v,\beta,w) \in E_H, (\lambda_H(w), \beta/\gamma, x, \text{out}) \in C_D\}, \\ \lambda(x) &= \lambda_H(x) \text{ if } x \in V_H - \{v\}, \text{ and } \lambda(x) = \lambda_D(x) \text{ if } x \in V_D, \\ C &= \{(\sigma, \beta/\gamma, x, d) \in C_H \mid x \neq v\} \\ &\quad \cup \{(\sigma, \beta/\delta, x, d) \mid \exists \gamma \in \Gamma : (\sigma, \beta/\gamma, v, d) \in C_H, (\sigma, \gamma/\delta, x, d) \in C_D\}. \end{aligned}$$

Example 1.8
Let (H, C_H) and (D, C_D) be the graphs with embedding of Figs. 1.7(a) and (b) respectively (as considered in Example 1.7). Let v be the node of H with label X. The graph with embedding $(H, C_H)[v/(D, C_D)]$ is drawn in Fig. 1.7(c). The three edges that are the result of the embedding process are drawn fatter than the other edges. Formally, $(H, C_H)[v/(D, C_D)]$ is the graph (V, E, λ, C) with $V = \{u, w, x, y\}$, $\lambda(u) = Y$, $\lambda(w) = a$, $\lambda(x) = X$, $\lambda(y) = b$, $E = \{(w, \beta, u), (y, \alpha, x), (x, \alpha', w), (u, \gamma_1, y), (u, \gamma_2, x)\}$, and $C = \{(\sigma', \beta'/\alpha, w, \text{in}), (\sigma, \beta/\gamma, u, \text{in}), (\sigma, \beta/\delta, y, \text{out}), (\sigma', \beta'/\delta', x, \text{in})\}$. Note that, pictorially, the parts of (H, C_H) and $(H, C_H)[v/(D, C_D)]$ outside their boxes are the same; a line inside the box of (H, C_H) that represents part of a connection instruction is treated "in the same way" as an edge of H. □

The edges of $(H, C_H)[v/(D, C_D)]$ that are established by the embedding process, i.e., that are not in E_H or E_D, are sometimes called *bridges* (between D and the remainder). It is easy to see that the number of bridges is at most $2 \cdot \#\Sigma \cdot \#\Gamma \cdot \delta^2$, where δ is the maximal degree of the nodes of $(H, C_H)[v/(D, C_D)]$. In fact, for each $\sigma \in \Sigma$ and $\beta \in \Gamma$, if $(\sigma, \beta/\gamma, x, \text{out}) \in C_D$, then x is connected to *every* w with $(v, \beta, w) \in E_H$ and $\lambda_H(w) = \sigma$, and so the number of such w's is at most the degree of such an x (and similarly for 'in'); thus, the number of neighbours of v that have bridges to D is at most $2 \cdot \#\Sigma \cdot \#\Gamma \cdot \delta$. This basic property of substitution was first used for boundary NLC grammars in the proof of Theorem 6.3 of [92], and then observed and exploited for confluent edNCE grammars in [96] (see also [14,15]); for linear edNCE grammars it was used in the proof of Theorem 15 of [39].

The connection relation C of the graph $(H, C_H)[v/(D, C_D)]$ that results from the substitution is defined in such a way that *the substitution operation is associative*. This is expressed in the following lemma that can easily be checked.

Lemma 1.3.2
Let K,H,D be three mutually disjoint graphs with embedding. Let w be a node of K and v a node of H. Then $K[w/H][v/D] = K[w/H[v/D]]$. □

Associativity is one of the two natural properties that is required for context-free rewriting in the general framework of [23]. The other property is confluence, to be discussed later in this section.

After these preliminaries, we turn to the definition of the edNCE graph grammar. As mentioned already in Section 1.2.2, NCE stands for *neighbourhood controlled embedding*, the d stands for "directed graphs", and the e means that not only the nodes but also the edges of the graphs are labeled; in particular, the e stresses the fact that the edNCE grammar allows for *dynamic edge relabeling*. Thus, edNCE grammars are *graph grammars with neighbourhood controlled embedding and dynamic edge relabeling*. They were introduced in [84,85,86], as depth-1 context-free graph grammars (which also allow the embedding to change the direction of edges and to react to multiple edges, cf. the discussion at the end of Section 1.2.2).

Definition 1.3.3
An *edNCE grammar* is a tuple $G = (\Sigma, \Delta, \Gamma, \Omega, P, S)$ where Σ is the alphabet of node labels, $\Delta \subseteq \Sigma$ is the alphabet of terminal node labels, Γ is the alphabet of edge labels, $\Omega \subseteq \Gamma$ is the alphabet of final edge labels, P is the finite set of productions, and $S \in \Sigma - \Delta$ is the initial nonterminal. A production is of the form $X \to (D, C)$ with $X \in \Sigma - \Delta$ and $(D, C) \in GRE_{\Sigma,\Gamma}$. □

Elements of $\Sigma - \Delta$ are called nonterminal node labels, and elements of $\Gamma - \Omega$ nonfinal edge labels. A node with a terminal or nonterminal label is said to be a terminal or nonterminal node, respectively, and similarly for final and nonfinal edges. A graph is *terminal* if all its nodes are terminal (but its edges need not be final), i.e., it belongs to $GRE_{\Delta,\Gamma}$. For a production $p : X \to (D, C)$, X is the left-hand side of p, (D, C) is the right-hand side of p, and C is its connection relation. We write $\text{lhs}(p) = X$ and $\text{rhs}(p) = (D, C)$. Two productions $X_1 \to (D_1, C_1)$ and $X_2 \to (D_2, C_2)$ are called isomorphic if $X_1 = X_2$ and (D_1, C_1) and (D_2, C_2) are isomorphic graphs with embedding. We will assume that P does not contain distinct isomorphic productions. By copy(P) we denote the (infinite) set of all productions that are isomorphic to a production in P; an element of copy(P) will be called a *production copy* of G.

The process of rewriting in an edNCE grammar is defined through the notion of substitution, in a standard language theoretic way, as follows.

Let $G = (\Sigma, \Delta, \Gamma, \Omega, P, S)$ be an edNCE grammar. Let H and H' be graphs in $GRE_{\Sigma,\Gamma}$, let $v \in V_H$, and let $p : X \to (D, C)$ be a production copy of G such that D and H are disjoint. Then we write $H \Rightarrow_{v,p} H'$, or just $H \Rightarrow H'$, if $\lambda_H(v) = X$ and $H' = H[v/(D,C)]$. $H \Rightarrow_{v,p} H'$ is called a *derivation step*, and a sequence of such derivation steps is called a *derivation*. A derivation

$$H_0 \Rightarrow_{v_1,p_1} H_1 \Rightarrow_{v_2,p_2} \cdots \Rightarrow_{v_n,p_n} H_n,$$

$n \geq 0$, is *creative* if the graphs H_0 and $\text{rhs}(p_i)$, $1 \leq i \leq n$, are mutually disjoint. *We will restrict ourselves to creative derivations.* Thus, we write $H \Rightarrow^* H'$ if there is a *creative* derivation as above, with $H_0 = H$ and $H_n = H'$. Let $sn(S, z)$ denote the (ordinary) graph with a single S-labeled node z, no edges, and empty connection relation. A *sentential form* of G is a graph H such that $sn(S, z) \Rightarrow^* H$ for some z; note that H is an ordinary graph, i.e., $H \in GR_{\Sigma,\Gamma}$. The *graph language generated by* G is

$$L(G) = \{[H] \mid H \in GR_{\Delta,\Omega} \text{ and } sn(S,z) \Rightarrow^* H \text{ for some } z\}.$$

Thus, sentential forms of G are concrete graphs, but the language generated by G consists of abstract graphs. It is not difficult to show that if H and H' are isomorphic and $sn(S, z) \Rightarrow^* H$, then $sn(S, z') \Rightarrow^* H'$ for some z'. Two edNCE grammars G and G' are *equivalent* if $L(G) - \{\Lambda\} = L(G') - \{\Lambda\}$, where Λ is the empty graph.

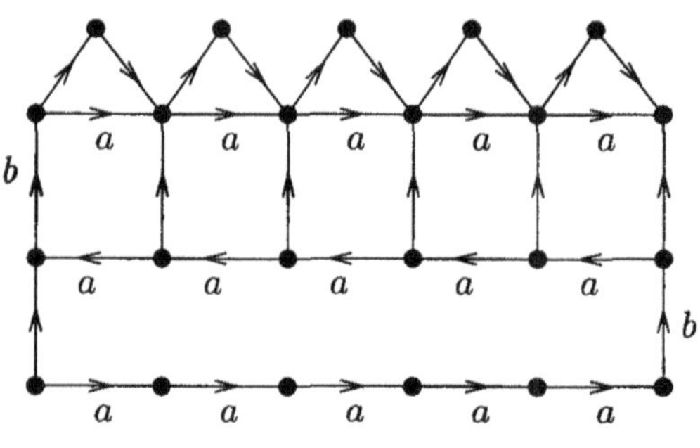

Figure 1.8: A street.

Assumption. *From now on we will assume, for every edNCE grammar* $G = (\Sigma, \Delta, \Gamma, \Omega, P, S)$, *that if* $S \to (D, C)$ *is in* P, *then* $C = \emptyset$. To change an arbitrary edNCE grammar into an equivalent one satisfying this assumption, take a new initial nonterminal S' and add all productions $S' \to (D, \emptyset)$ with $S \to (D, C)$ in P for some C.

As observed before, we could have generalized the connection instructions of an edNCE grammar to be of the form $(\sigma, \beta/\gamma, x, d/d')$ with $d, d' \in \{\text{in}, \text{out}\}$, in which case edges with direction d are turned into edges with direction d' by the embedding process (see, e.g., [15,16,52]). It is easy to show that for each such grammar there is an equivalent edNCE grammar (for every edge (v, γ, w) for which v and w are not both terminal, add an edge $(w, \overline{\gamma}, v)$, see [17] for details).

An example of an eNCE grammar (i.e., an edNCE grammar in which all edges are undirected), showing the use of dynamic edge relabeling, was given in Example 1.6(a). In that example, a connection instruction $(\mu, p/q, x)$ abbreviates the two connection instructions $(\mu, p/q, x, \text{in})$ and $(\mu, p/q, x, \text{out})$. We now give some more examples of edNCE grammars; they are all confluent, i.e., C-edNCE grammars (to be defined in Definition 1.3.5). Each of these grammars is or will be used to illustrate certain aspects of C-edNCE grammars.

Example 1.9
(1) Let us consider an edNCE grammar G_1 such that $L(G_1)$ is the set of all "streets", of the form shown in Fig. 1.8, where unlabeled edges (nodes) have label $*$ ($\#$ respectively). $G_1 = (\Sigma, \Delta, \Gamma, \Omega, P, S)$ with $\Sigma = \{S, X, \#\}$, $\Delta = \{\#\}$, $\Gamma = \{h, r, a, b, *\}$ (where h stands for 'house' and r for 'road'), $\Omega = \{a, b, *\}$, and P contains the three productions shown in Fig. 1.9.
In Fig. 1.9, the right-hand side of a production (which is a graph with embedding) is drawn as explained in Example 1.7, and the left-hand side is given as a label of the large box (usually at the upper left corner); this suggests that the large box is a nonterminal node that can be rewritten by the represented graph with embedding. Thus, the first production has empty connection relation, and the other two productions both have the connection instructions $(\#, h/*, x_1, \text{in})$, $(\#, h/a, x_2, \text{in})$, $(\#, h/a, x_3, \text{out})$, and $(\#, r/a, x_4, \text{in})$ where x_1, x_2, x_3, and x_4 are the terminal nodes of the right-hand side, numbered from above. The grammar G_1 generates a street with n houses, $n \geq 1$, by starting with the first production, adding one house with each application of the second production, $n-1$ times, and adding the nth house with the last production. An example of a derivation for $n = 3$ is given in Fig. 1.10, where, at each derivation step, the edges that are established by the embedding are drawn fatter.

If the grammar G_1 is changed into grammar G_1' by removing all unlabeled edges and both terminal nodes x_1, then $L(G_1')$ is the set of all chains (or paths) of which the edges are labeled by the strings $a^n b a^n b a^n$, $n \geq 1$. This shows that, if one codes strings as chains in the obvious way, (confluent) context-free graph

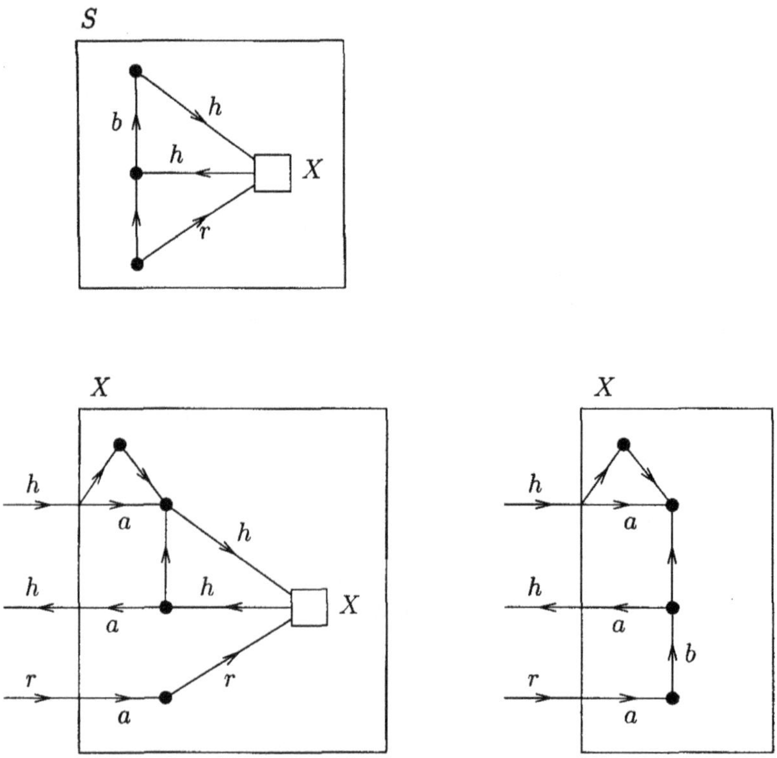

Figure 1.9: Productions of G_1.

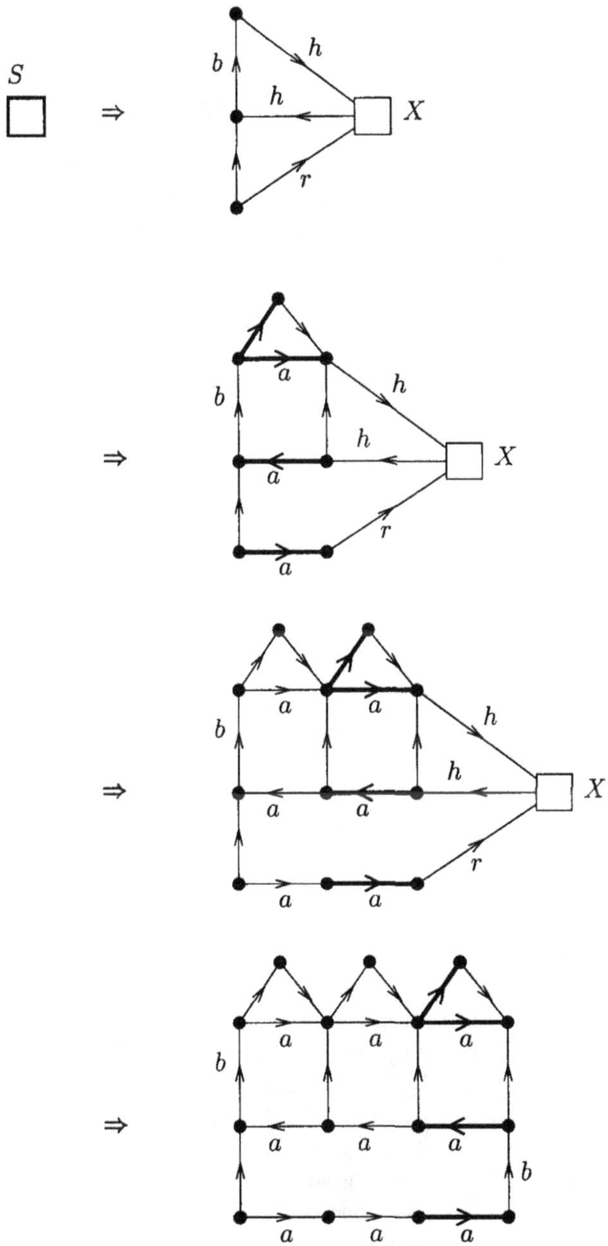

Figure 1.10: A derivation of G_1.

grammars can generate non-context-free string languages (see Section 2.5.2 of Chapter 2).

(2) The next example is an edNCE grammar G_2 that generates all rooted binary trees (with edges from a parent to its children) such that there is an additional edge from each leaf to the root. $G_2 = (\Sigma, \Delta, \Gamma, \Omega, P, S)$ with $\Sigma = \{S, X, \#\}$, $\Delta = \{\#\}$, $\Gamma = \{r, *\}$, and $\Omega = \{*\}$. Furthermore $P = \{p_a, p_b, p_c\}$, where p_a, p_b, and p_c are the three productions that are shown in Fig. 1.11 without their connection relations (and again, unlabeled nodes/edges have la-

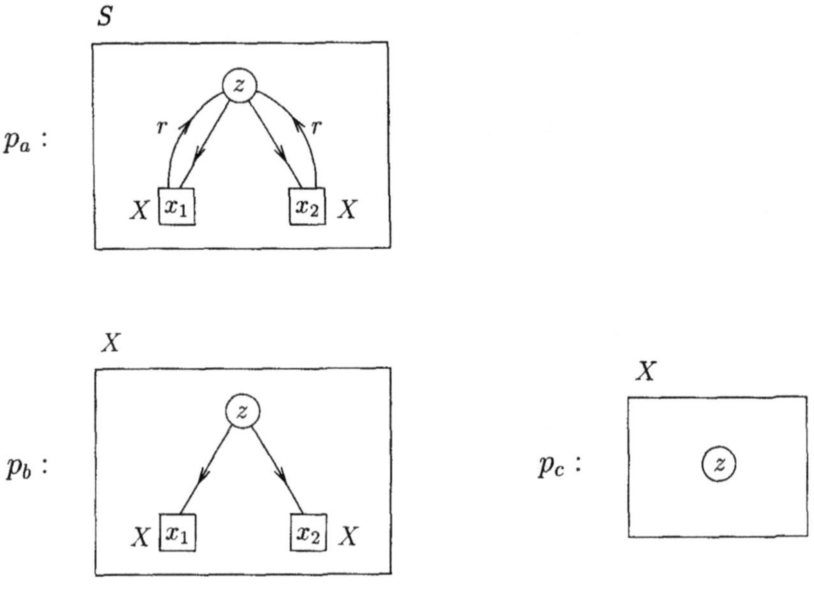

Figure 1.11: Productions of G_2, G_3, G_4 (without connection relations).

bel $\#/*$, respectively). The connection relation C_a of p_a is empty, the connection relation C_b of p_b is $\{(\#, */*, z, \text{in}), (\#, r/r, x_1, \text{out}), (\#, r/r, x_2, \text{out})\}$, and the connection relation C_c of p_c is $\{(\#, */*, z, \text{in}), (\#, r/*, z, \text{out})\}$. The r-labeled edges to the root are created by p_a, they are "passed" from nonterminal to nonterminal by p_b, and they are finally attached to the leaves (and changed into ordinary unlabeled edges) by p_c. Edge label r is not really needed, but is used for clearness sake. An example of a derivation of a graph of $L(G_2)$ is given in Fig. 1.12, where (again) embedding edges are drawn fatter; the applied productions are p_a, p_c, p_b, p_b, p_c, p_c, p_c, respectively. Note that the addition of

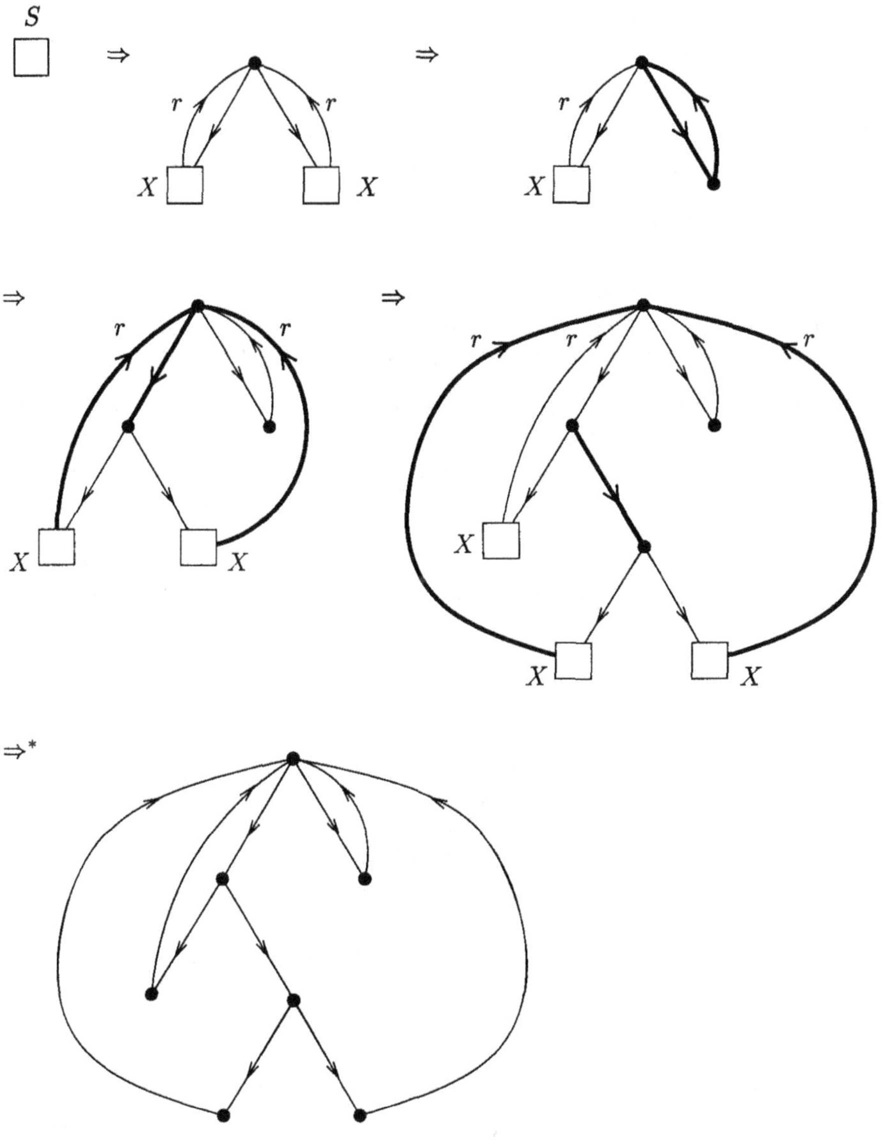

Figure 1.12: A derivation of G_2.

connection instruction $(\#, r/*, z, \text{out})$ to p_b would result in an additional edge to the root from each node of the tree. Note also that if we delete the nodes x_2 (and their incident edges) from p_a and p_b, we obtain a grammar G_2' that generates all cycles.

(3,4) Let G_3 be the edNCE grammar that is obtained from G_2 above by dropping the connection instructions that contain r. Thus, $C_b = C_c = \{(\#, */*, z, \text{in})\}$. Then $L(G_3)$ is the set of all rooted binary trees. Let G_4 be the same grammar with $C_a = \emptyset$, $C_b = \{(\#, */*, z, \text{in}), (\#, */*, x_1, \text{in}), (\#, */*, x_2, \text{in})\}$ and $C_c = \{(\#, */*, z, \text{in})\}$. Then $L(G_4)$ is the set of all transitive binary trees, i.e., all binary trees with an additional edge from each node to each of its descendants. Note that, in these two grammars, the r-labeled edges of p_a can be dropped.

(5) Let G_5 be the edNCE grammar that is obtained from one of the grammars G_2, G_3, or G_4, by adding edge labels l to $\Gamma - \Omega$ and f to Ω, adding edge (x_1, l, x_2) to the right-hand sides of p_a and p_b, adding connection instructions $(\sigma, l/l, x_1, \text{in})$ and $(\sigma, l/l, x_2, \text{out})$ to C_b, with $\sigma \in \{X, \#\}$, and adding connection instructions $(X, l/l, z, d)$ and $(\#, l/f, z, d)$ to C_c, with $d \in \{\text{in}, \text{out}\}$. Then the leaves of the trees generated by G_5 are chained by f-labeled edges "from left to right" (where the order is obtained by taking node x_1 to the left of node x_2).

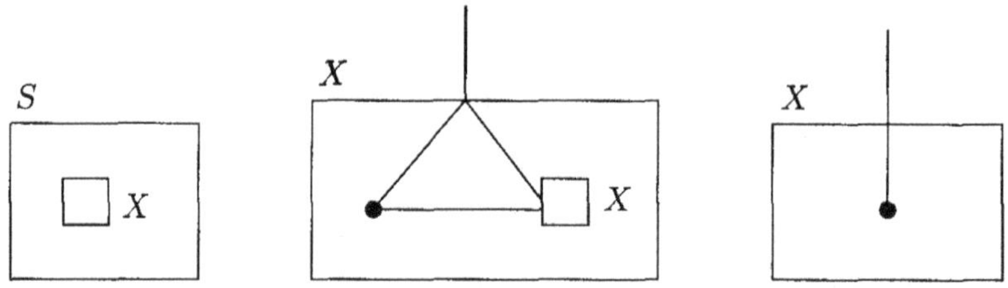

Figure 1.13: Productions of G_6.

(6) The edNCE grammar G_6 with the three productions shown in Fig. 1.13 generates all (undirected) complete graphs K_n, with $n \geq 1$. As usual, an undirected edge stands for two directed edges. Similarly, two connection instructions $(\sigma, \beta/\gamma, x, \text{in})$ and $(\sigma, \beta/\gamma, x, \text{out})$ are represented by an undirected line. Moreover, as usual, we drop the labels $\#$ and $*$ (also from connection instructions). Thus, the second production is $X \to (D, C)$ with

$V_D = \{y, x\}$, $\lambda_D(y) = \#$, $\lambda_D(x) = X$, $E_D = \{(y, *, x), (x, *, y)\}$, and $C = \{(\#, */*, y, \text{in}), (\#, */*, y, \text{out}), (\#, */*, x, \text{in}), (\#, */*, x, \text{out})\}$.

(7) Let $K_{m,n}$ be the (undirected) complete bipartite graph on m and n nodes, i.e., the graph (V, E, λ) such that $V = \{u_1, \ldots, u_m, v_1, \ldots, v_n\}$, $E = \{(u_i, *, v_j), (v_j, *, u_i) \mid 1 \leq i \leq m, 1 \leq j \leq n\}$, and $\lambda(x) = \#$ for every $x \in V$. The edNCE grammar G_7 with the five productions shown in Fig. 1.14

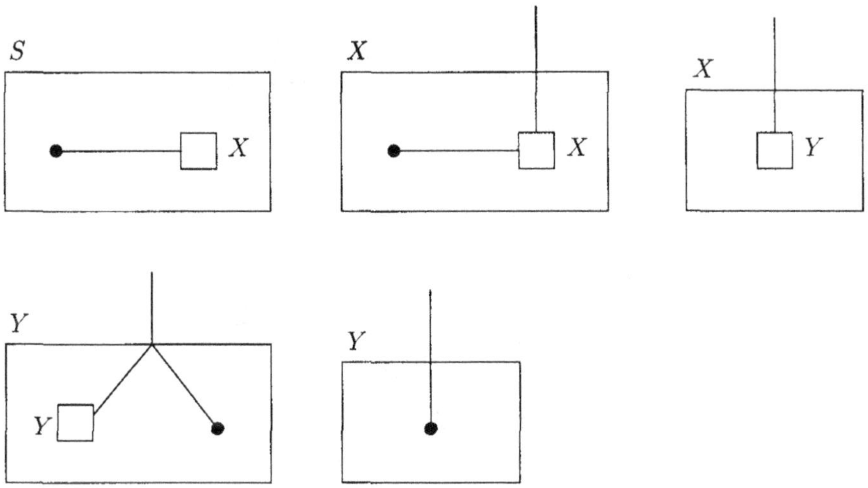

Figure 1.14: Productions of G_7.

generates all graphs $K_{m,n}$ with $m, n \geq 1$. A derivation of $K_{2,3}$ is shown in Fig. 1.15 (with fat embedding edges).

(8,9) A *star* is a graph $K_{1,n}$ (see the previous example). The edNCE grammar G_8 with the three productions shown in Fig. 1.16 generates all stars $K_{1,n}$ with $n \geq 1$. The *wheel* W_n (with n spokes) is the undirected unlabeled graph with nodes $u, v_1, \ldots, v_n$ and edges $\{u, v_i\}$ for $1 \leq i \leq n$, $\{v_i, v_{i+1}\}$ for $1 \leq i \leq n-1$, and $\{v_n, v_1\}$. The edNCE grammar G_9 with the three productions shown in Fig. 1.17 generates all wheels W_n with $n \geq 3$.

(10) The edNCE grammar G_{10} with the four productions shown in Fig. 1.18 generates all *cographs.* The grammar reflects the recursive definition of cographs: a cograph is either a single node, or the disjoint union of two

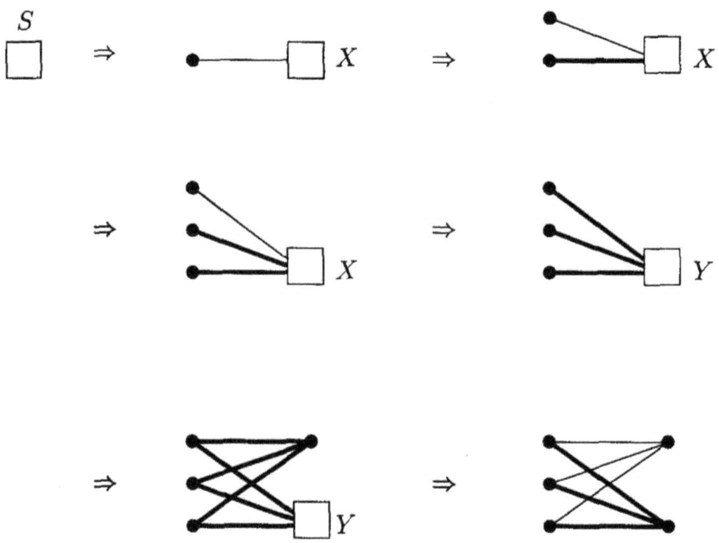

Figure 1.15: A derivation of $K_{2,3}$ in G_7.

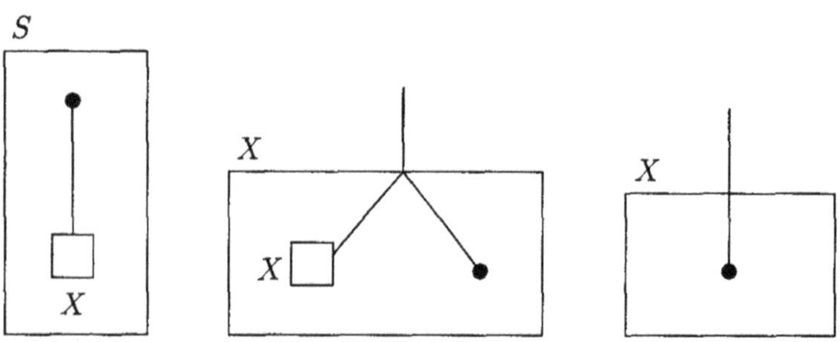

Figure 1.16: Productions of G_8.

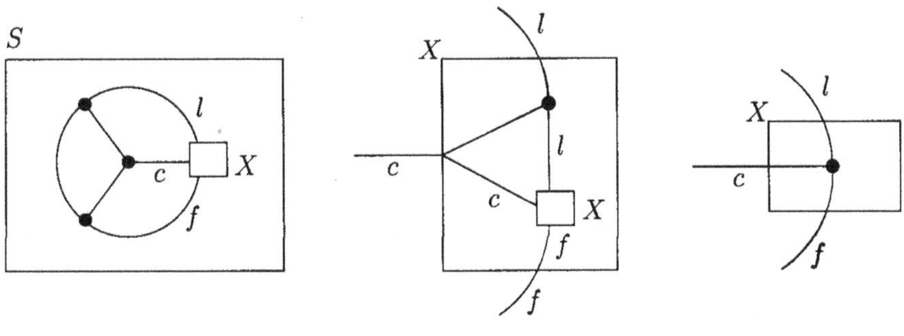

Figure 1.17: Productions of G_9.

cographs, or the join of two cographs (see, e.g., [21]).

(11) For this example the reader is assumed to be familiar with 2-structures (see Chapter 6). Consider the edNCE grammar $G_{11} = (\Sigma, \Delta, \Gamma, \Omega, P, S)$ with $\Sigma = \{S, X, \#\}$, $\Delta = \{\#\}$, $\Gamma = \Omega$ (where Ω is an arbitrary alphabet), and $P = \{p_1, p_2\} \cup \{p_{a,b} \mid a, b \in \Omega\}$. The productions are shown in Fig. 1.19 without their connection relations. For each production $X \to (D, C)$, the connection relation is "hereditary", i.e., $C = \{(\sigma, \gamma/\gamma, x, d) \mid \sigma \in \Sigma, \gamma \in \Gamma, x \in V_D, d \in \{\text{in}, \text{out}\}\}$. Grammar G_{11} generates all uniformly non-primitive (labeled) 2-structures over Ω, see Section 4.4.2 of Chapter 6 (and see Section 3.3 of [34] for the correctness of the construction). For a finite set F of primitive labeled 2-structures, let Prim(F) be the set of all labeled 2-structures H such that every quotient of the shape of H is complete, linear, or in F (see Sections 4.2 and 4.3 of [34]). Add to grammar G_{11} all productions $X \to (D, C)$ such that $D \in F$ (with, additionally, every node labeled by X) and C is hereditary (as above). The resulting edNCE grammar generates Prim(F). □

As mentioned in Section 1.2.1, the edNCE grammar has certain undesirable non-context-free properties. For instance, there is an edNCE grammar G_b that generates the set of all stars $K_{1,m}$ with $m + 1 = 2^n + 1$ nodes, $n \geq 0$, see Example 13 of [44]. This grammar has sentential forms H with some nonfinal edges between terminal nodes. Such edges are called *blocking edges* (see p.38 of [86], or p.129 of [84] where they are called "forbidden" edges), because they cannot be removed by rewriting H and, thus, no graph in $L(G_b)$ can be generated from H (note that the edges of a graph in $L(G_b)$ have to be final). This means that blocking edges can be used to filter out of the language some of the terminal graphs. It is shown in Theorem I.3.17 of [86], for a related class

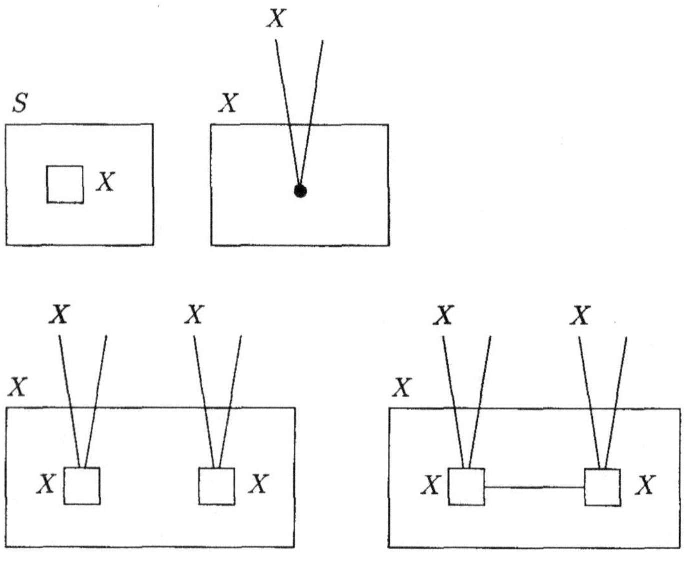

Figure 1.18: Productions of G_{10}.

of node-replacement graph grammars (see Section 1.2.3), that blocking edges allow the simulation of arbitrary context-sensitive graph grammars.

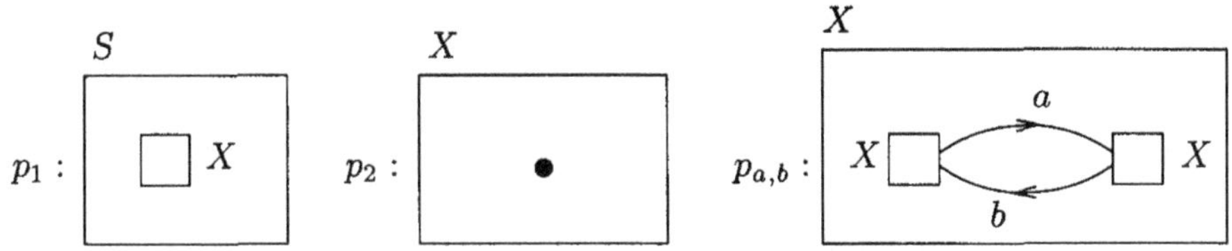

Figure 1.19: Productions of G_{11} (without connection relations).

Definition 1.3.4
An edNCE grammar G is *nonblocking* if every terminal sentential form H of G has final edges only (i.e., $[H]$ is in $L(G)$).

For a nonblocking grammar G, the string language $\{a^n \mid n = \#V_H \text{ for some } [H] \in L(G)\}$ is context-free (proof: turn each production $X \to (D, C)$, of which the nodes of D are labeled $\sigma_1, \ldots, \sigma_k$, into the string production $X \to \sigma'_1 \cdots \sigma'_k$ where $\sigma'_i = a$ if $\sigma_i \in \Delta$ and $\sigma'_i = \sigma_i$ otherwise). This shows that the above-mentioned graph language $L(G_b)$ cannot be generated by a nonblocking edNCE grammar. However, to forbid blocking edges is not a solution to the problem. In fact, let $G = (\Sigma, \Delta, \Gamma, \Omega, P, S)$ be an edNCE grammar, and consider now the grammar $G' = (\Sigma, \Delta, \Gamma, \Gamma, P, S)$ that has no nonfinal edge labels and hence no blocking edges. Clearly, $L(G) = L(G') \cap [GR_{\Delta,\Omega}]$; hence, every edNCE language can be obtained from a nonblocking edNCE language by intersecting with some language $[GR_{\Delta,\Omega}]$. This implies that the class of nonblocking edNCE languages is not closed under intersection with languages of the form $[GR_{\Delta,\Omega}]$, which is also undesirable.

The real cause of the non-context-free properties of *edNCE grammars* is that they *need not be confluent*, i.e., that the result of a derivation may depend on the order in which the productions are applied. This problem turns up in sentential forms that have edges between two nonterminal nodes, as in the following example (which is the edNCE version of Example 1.3).

Example 1.10
Consider an edNCE grammar $G = (\Sigma, \Delta, \Gamma, \Omega, P, S)$ with $\Sigma = \{S, A, B, a, b\}$, $\Delta = \{a, b\}$, $\Gamma = \{\alpha, \beta, \delta, \delta'\}$, $\Omega = \{\delta, \delta'\}$, and the following three productions

$X \to (D, C)$:

p_S : $X = S, V_D = \{u, v\}, E_D = \{(u, \alpha, v)\}, \lambda_D(u) = A, \lambda_D(v) = B$, and $C = \emptyset$,

p_A : $X = A, V_D = \{x\}, E_D = \emptyset, \lambda_D(x) = a$, and $C = \{(B, \alpha/\beta, x, \text{out}), (b, \beta/\delta', x, \text{out})\}$,

p_B : $X = B, V_D = \{y\}, E_D = \emptyset, \lambda_D(y) = b$, and $C = \{(A, \alpha/\beta, y, \text{in}), (a, \beta/\delta, y, \text{in})\}$.

The application of productions p_S, p_A, p_B (in that order), gives a derivation $sn(S, z) \Rightarrow_{z,p_S} H \Rightarrow_{u,p_A} H_1 \Rightarrow_{v,p_B} H_{12}$, where H is rhs(p_S), and H_{12} is the graph with two nodes x and y, labeled a and b, respectively, and one edge (x, δ, y). However, interchanging the application of p_A and p_B to nodes u and v, respectively, gives a derivation $sn(S, z) \Rightarrow_{z,p_S} H \Rightarrow_{v,p_B} H_2 \Rightarrow_{u,p_A} H_{21}$, where H_{21} is the same as H_{12} except that its edge is (x, δ', y). Since the order of application of productions results in two different graphs H_{12} and H_{21}, the grammar is not confluent. □

In the literature it is customary to give a dynamic definition of confluence. Here we propose a static one that can easily be checked on the embedding relations of the productions of the grammar (cf. Definition 5.1 of [36] and Lemma 3.11 of [23]).

Definition 1.3.5
An edNCE grammar $G = (\Sigma, \Delta, \Gamma, \Omega, P, S)$ is *confluent*, or a *C-edNCE* grammar, if for all productions $X_1 \to (D_1, C_1)$ and $X_2 \to (D_2, C_2)$ in P, all nodes $x_1 \in V_{D_1}$ and $x_2 \in V_{D_2}$, and all edge labels $\alpha, \delta \in \Gamma$, the following equivalence holds:

$$\exists \beta \in \Gamma : (X_2, \alpha/\beta, x_1, \text{out}) \in C_1 \text{ and } (\lambda_{D_1}(x_1), \beta/\delta, x_2, \text{in}) \in C_2$$
$$\iff$$
$$\exists \gamma \in \Gamma : (X_1, \alpha/\gamma, x_2, \text{in}) \in C_2 \text{ and } (\lambda_{D_2}(x_2), \gamma/\delta, x_1, \text{out}) \in C_1.$$

By C-edNCE we denote the class of graph languages generated by C-edNCE grammars. To stress the importance of this class within the area of node replacement graph grammars, it is also called VR, and C-edNCE grammars are also called VR grammars (where VR stands for 'vertex replacement'). This is analogous to the HR grammars which are the most important class of (hyper)edge replacement grammars.

As observed before, all grammars discussed in Example 1.9 are C-edNCE grammars. Many sets of graphs with "tree-like" graph theoretic properties can be defined by C-edNCE grammars. For example trees, cographs, series-parallel graphs, transitive VSP graphs, uniformly non-primitive 2-structures, complete bipartite graphs, acyclic tournaments, (maximal) outerplanar graphs, edge complements of trees, and for fixed k, k-trees, graphs of treewidth $\leq k$, pathwidth $\leq k$, cutwidth $\leq k$, bandwidth $\leq k$, cyclic bandwidth $\leq k$, and topological bandwidth $\leq k$ (see, e.g., [92,39,46]).

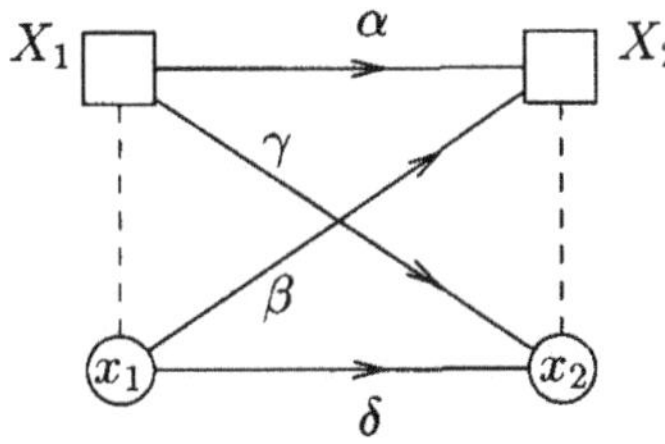

Figure 1.20: Confluence.

A symbolic picture of Definition 1.3.5 is given in Fig. 1.20 (where x_1 and x_2 may also be boxes). Intuitively the definition means that if the two productions are applied to a graph with a single edge (u_1, α, u_2), where u_i is labeled X_i, then the same edges (x_1, δ, x_2) are established between nodes of their right-hand sides, independent of the order in which the productions are applied. From this intuition the following characterization of confluence easily follows: the result of a derivation does not depend on the order in which the productions are applied.

Proposition 1.3.6
An edNCE grammar $G = (\Sigma, \Delta, \Gamma, \Omega, P, S)$ is confluent if and only if the following holds for every graph $H \in GR_{\Sigma,\Gamma}$ (or $H \in GRE_{\Sigma,\Gamma}$): if $H \Rightarrow_{u_1,p_1} H_1 \Rightarrow_{u_2,p_2} H_{12}$ and $H \Rightarrow_{u_2,p_2} H_2 \Rightarrow_{u_1,p_1} H_{21}$ are (creative) derivations of G with $u_1, u_2 \in V_H$ and $u_1 \neq u_2$, then $H_{12} = H_{21}$.

Proof
(Only if) Let p_i be $X_i \to (D_i, C_i)$. Note that H, D_1, and D_2 are mutually disjoint. It has to be shown that

$$H[u_1/(D_1, C_1)][u_2/(D_2, C_2)] = H[u_2/(D_2, C_2)][u_1/(D_1, C_1)].$$

Using Definition 1.3.1 it can easily be verified that these two graphs have the same nodes with the same labels, the same connection relation, and the same edges (v,γ,w) with $(v,w),(w,v) \notin V_{D_1} \times V_{D_2}$. It follows from Definition 1.3.5 that they also have the same edges (v,γ,w) with $v \in V_{D_1}, w \in V_{D_2}$ or $v \in V_{D_2}, w \in V_{D_1}$.

(If) To prove the property in Definition 1.3.5, consider the graph H with nodes u_1 and u_2, where u_i is labeled X_i, a single edge (u_1,α,u_2), and empty connection relation, and let p_i be $X_i \to (D_i, C_i)$ (or, more precisely, copies of these productions with disjoint right-hand sides, disjoint with H). The equivalence in Definition 1.3.5 then follows from the equality of H_{12} and H_{21}. □

It should be noted that all HR grammars are confluent in the sense of Proposition 1.3.6 (see Section 2.2.2 of Chapter 2).

We now give the definition of confluence from the literature (where it is also called the finite Church-Rosser, or fCR, property), see, e.g., [70,14,23,46,74]. It is exactly the same as the previous proposition, except that H should be a sentential form of G; we will call this "dynamic confluence".

Definition 1.3.7
An edNCE grammar $G = (\Sigma,\Delta,\Gamma,\Omega,P,S)$ is *dynamically confluent* if the following holds for every sentential form H of G:
if $H \Rightarrow_{u_1,p_1} H_1 \Rightarrow_{u_2,p_2} H_{12}$ and $H \Rightarrow_{u_2,p_2} H_2 \Rightarrow_{u_1,p_1} H_{21}$ are (creative) derivations of G with $u_1,u_2 \in V_H$ and $u_1 \neq u_2$, then $H_{12} = H_{21}$.

Thus, every confluent edNCE grammar is dynamically confluent, but, clearly, not the other way around. However, we will show later (Theorem 1.3.11) that for every dynamically confluent edNCE grammar there is an equivalent confluent grammar. Hence, the dynamically confluent edNCE grammars generate the same languages as the C-edNCE grammars. We note that dynamic confluence is decidable, because the set of all 2-node subgraphs of sentential forms is computable (see [70]), but we will not need this fact. In [59] an even weaker (and decidable) notion of confluence is investigated: an edNCE grammar is called *order independent* if, in Definition 1.3.7, the conclusion is not that $H_{12} = H_{12}$, but that for every $H' \in GR_{\Delta,\Omega}$, $H_{12} \Rightarrow_{v_1,q_1} \cdots \Rightarrow_{v_n,q_n} H'$ if and only if $H_{21} \Rightarrow_{v_1,q_1} \cdots \Rightarrow_{v_n,q_n} H'$. In [23] another, weaker notion of confluence is used implicitly (see the Correction to [23]): an edNCE grammar is *abstractly confluent* if, in Definition 1.3.7, the conclusion is that H_{12} and H_{21} are isomorphic (rather than equal); it is open whether this is decidable. It will follow from Theorem 1.3.11 that also order-independent or abstractly confluent edNCE grammars have the same power as C-edNCE grammars.

It is shown in [99] that blocking edges are not a problem in confluent grammars any more. Thus, it is not necessary (although quite feasible) to explicitly require C-edNCE grammars to be nonblocking (as, e.g., in [46,47]). We will give the proof in Theorem 1.3.21.

The class C-edNCE of graph languages generated by C-edNCE grammars has a large number of nice closure properties, see Theorem 1.4.7. Here we prove a very simple special case: closure under edge relabeling. Let ρ be an edge relabeling, i.e., a mapping $\rho : \Omega \to \Omega'$, where Ω and Ω' are edge label alphabets. For a graph $H \in GR_{\Delta,\Omega}$ we define $\rho(H) \in GR_{\Delta,\Omega'}$ to be the graph (V_H, E, λ_H) with $E = \{(v, \rho(\gamma), w) \mid (v, \gamma, w) \in E_H\}$. For a graph language L, $\rho(L) = \{[\rho(H)] \mid [H] \in L\}$.

Proposition 1.3.8
C-edNCE is closed under edge relabelings, i.e., if ρ is an edge relabeling and $L \in$ C-edNCE, then $\rho(L) \in$ C-edNCE.

Proof
Let $G = (\Sigma, \Delta, \Gamma, \Omega, P, S)$ be a C-edNCE grammar, and let $\rho : \Omega \to \Omega'$ be an edge relabeling. It can be assumed that $\Gamma \cap \Omega' = \emptyset$. We construct the edNCE grammar $G' = (\Sigma, \Delta, \Gamma \cup \Omega', \Omega', P', S)$ with $P' = \{X \to \rho(D,C) \mid X \to (D,C)$ is in $P\}$, where, for a graph $(D,C) \in GRE_{\Sigma,\Gamma}$, $\rho(D,C) \in GRE_{\Sigma,\Gamma\cup\Omega'}$ is defined to be the graph (V_D, E, λ_D, C') with embedding, such that

$$\begin{aligned} E &= \{(v,\rho(\gamma),w) \mid && (v,\gamma,w) \in E_D, \lambda_D(v) \in \Delta, \text{ and } \gamma \in \Omega, \\ & && \text{and } \lambda_D(w) \in \Delta\} \\ &\quad \cup \{(v,\gamma,w) \in E_D \mid && \lambda_D(v) \in \Sigma - \Delta, \text{ or } \gamma \in \Gamma - \Omega, \\ & && \text{or } \lambda_D(w) \in \Sigma - \Delta\}, \\ C' &= \{(\sigma,\beta/\rho(\gamma),x,d) \mid && (\sigma,\beta/\gamma,x,d) \in C, \lambda_D(x) \in \Delta, \text{ and } \gamma \in \Omega, \\ & && \text{and } \sigma \in \Delta\} \\ &\quad \cup \{(\sigma,\beta/\gamma,x,d) \in C \mid && \lambda_D(x) \in \Sigma - \Delta, \text{ or } \gamma \in \Gamma - \Omega, \\ & && \text{or } \sigma \in \Sigma - \Delta\}. \end{aligned}$$

Thus, only the final edges between terminal nodes are relabeled.

Using the confluence of G, it is easy to verify that G' is still confluent (in Definition 1.3.5, first consider the case that x_1 and x_2 are terminal and δ is final, and then the remaining case). It is also easy to see that the derivations of G' starting with $sn(S,z)$ are of the form $sn(S,z) \Rightarrow \rho(H_1) \Rightarrow \cdots \Rightarrow \rho(H_n)$ where $sn(S,z) \Rightarrow H_1 \Rightarrow \cdots \Rightarrow H_n$ is a derivation of G. Formally this can be proved by induction on n. It shows that $L(G') = \rho(L(G))$. □

C-edNCE is also closed under node relabelings (as opposed to NLC); see, e.g., Theorem 9 of [44] (and Theorem 4.1 of [93], respectively).

1.3.2 Leftmost derivations and derivation trees

An important property of a context-free string grammar, based on its confluence, is that every derivation is equivalent with a leftmost derivation. Also, there is a one-to-one correspondence between leftmost derivations and derivation trees. For edNCE grammars it turns out that this correspondence even holds in the non-confluent case, and that the language generated by leftmost derivations is in C-edNCE. Thus, restricting the derivations in edNCE grammars to be leftmost is an alternative solution to the problem of the non-context-free properties of edNCE grammars.

Leftmost derivations

To define leftmost derivations of an edNCE grammar G, we have to put a linear order on the nodes of the right-hand sides of the productions of G (since there is no natural linear order between them as in the case of strings). This order induces a linear order on the nodes of the sentential forms of G in a natural way. Note that a linear order on the nonterminal nodes would be sufficient; also, it would suffice to order the right-hand sides of the production copies, rather than the productions. However, the present choice seems to be more natural (compared to context-free string grammars) and technically convenient.

An *ordered graph* (possibly with embedding) is a graph together with a linear order of its nodes. If H and D are disjoint graphs (with embedding) that are ordered, and v is a node of H, then the order of the substitution $H[v/D]$ is obtained, informally, by substituting the order of D for v in the order of H, just as for strings. Formally, if the nodes of H are ordered as $(v_1, \ldots, v_h)$ with $v = v_i$, and those of D are ordered as $(w_1, \ldots, w_d)$, then the order of $H[v/D]$ is $(v_1, \ldots, v_{i-1}, w_1, \ldots, w_d, v_{i+1}, \ldots, v_h)$.

Let G be an edNCE grammar of which the right-hand sides of its productions are made into ordered graphs (by choosing a linear order on their nodes). The right-hand sides of the production copies inherit this order in the obvious way (i.e., they are ordered in such a way that the isomorphism is order preserving). Since the derivation steps of G are defined through the notion of substitution (Definition 1.3.1), the sentential forms of G also become ordered graphs, as explained above. It should be clear that this ordering of the sentential forms has no influence on the language $L(G)$ generated by G (where we disregard the order of a generated terminal graph). For a derivation $H_0 \Rightarrow_{v_1,p_1} H_1 \Rightarrow_{v_2,p_2} \cdots \Rightarrow_{v_n,p_n} H_n$ where H_0 is an ordered graph, we will also assume that the orders of H_1 to H_n are determined by the above.

Assumption. *From now on we will assume that the right-hand sides of the productions of an edNCE grammar are ordered graphs, and that its sentential forms are ordered graphs, as explained above.*

Let G be an edNCE grammar. For an ordered graph H, a derivation step $H \Rightarrow_{v,p} H'$ of G is a *leftmost derivation step* if v is the first nonterminal node in the order of H. A derivation is leftmost if all its steps are. We write $H \Rightarrow^*_{\text{lm}} H'$ if there is a leftmost derivation from H to H'. The *graph language leftmost generated by* G is

$$L_{\text{lm}}(G) = \{[H] \mid H \in GR_{\Delta,\Omega} \text{ and } sn(S,z) \Rightarrow^*_{\text{lm}} H \text{ for some } z\},$$

where the abstract graph $[H]$ does not involve an order any more (i.e., we disregard the order of the graphs in $L(G)$; the order is just used as a means to define leftmost derivations). Note that $sn(S,z)$ can be viewed as an ordered graph in one way only.

By L-edNCE we denote the class of graph languages leftmost generated by edNCE grammars.

As in the case of context-free string grammars (which are all confluent), every derivation of a confluent edNCE grammar can be transformed into a leftmost derivation, by repeated application of the dynamic confluence property (see Proposition 1.3.6 and Definition 1.3.7).

Proposition 1.3.9
For every C-edNCE grammar G, $L_{lm}(G) = L(G)$.

Proof
The proof is similar to the case of context-free string grammars. Consider a non-leftmost derivation $sn(S,z) \Rightarrow^* H$ where H is a terminal graph. This derivation can be written as

$$sn(S,z) \Rightarrow \cdots \Rightarrow H_{i-1} \Rightarrow_{v_i,p_i} H_i \Rightarrow \cdots \Rightarrow H_{j-1} \Rightarrow_{v_j,p_j} H_j \Rightarrow \cdots \Rightarrow H$$

where $H_{i-1} \Rightarrow H_i$ is the first non-leftmost derivation step, and v_j is the leftmost nonterminal node of H_{i-1} (note that since H is terminal, the leftmost nonterminal node of H_{i-1} has to be rewritten in the derivation). Clearly, by the definition of a (creative) derivation, v_j is a node of $H_i, \ldots, H_{j-1}$, different from $v_i, \ldots, v_{j-1}$. Hence, by applying the dynamic confluence property of Proposition 1.3.6 $j-i$ times, starting with $H_{j-2} \Rightarrow_{v_{j-1},p_{j-1}} H_{j-1} \Rightarrow_{v_j,p_j} H_j$, a derivation of H of the same length is obtained, in which the first leftmost derivation step occurs after the ith step (because the ith step is now of the form $H_{i-1} \Rightarrow_{v_j,p_j} H_i'$ for some graph H_i'). Repetition of this procedure leads to a leftmost derivation of H. □

This shows that for a C-edNCE grammar G the linear order on the right-hand sides of the productions of G can be chosen arbitrarily.

Proposition 1.3.9 implies that C-edNCE $\subseteq$ L-edNCE. It is shown in [20,87,46] that also L-edNCE $\subseteq$ C-edNCE, i.e., for every edNCE grammar G there exists a C-edNCE grammar G' such that $L(G') = L_{\mathrm{lm}}(G)$. This means that the restriction to leftmost derivations is equivalent with the restriction to confluent grammars. The proofs in [20,87,46] use the alternative characterizations of the class C-edNCE discussed in Sections 1.4.3 and 1.4.1, respectively. Here we give a direct proof. To simplify the proof, we first give a natural sufficient condition for an edNCE grammar to be confluent.

Lemma 1.3.10
Let $G = (\Sigma, \Delta, \Gamma, \Omega, P, S)$ be an edNCE grammar such that $\Gamma = \Gamma_1 \times \Gamma_2$ for certain sets Γ_1 and Γ_2, and such that for every production $X \to (D, C)$,

*(1) if $(\sigma, (\beta_1, \beta_2)/(\gamma_1, \gamma_2), x, \mathrm{in}) \in C$,
then $\beta_1 = \gamma_1$ and
$(\sigma', (\rho_1, \beta_2)/(\rho_1, \gamma_2), x, \mathrm{in}) \in C$ for every $\sigma' \in \Sigma$ and $\rho_1 \in \Gamma_1$.*

*(2) if $(\sigma, (\beta_1, \beta_2)/(\gamma_1, \gamma_2), x, \mathrm{out}) \in C$,
then $\beta_2 = \gamma_2$ and
$(\sigma', (\beta_1, \rho_2)/(\gamma_1, \rho_2), x, \mathrm{out}) \in C$ for every $\sigma' \in \Sigma$ and $\rho_2 \in \Gamma_2$.*

Then G is confluent.

Proof
Intuitively, the conditions on G mean that every edge in a sentential form H of G is of the form $(u, (\gamma_1, \gamma_2), v)$. If u is rewritten, then the edge label changes into (γ_1', γ_2) for some γ_1', and the embedding process does not inspect γ_2 or $\lambda_H(v)$. Similarly, if v is rewritten, it changes into (γ_1, γ_2') for some γ_2', and the embedding process does not look at γ_1 or $\lambda_H(u)$. As a consequence, if both u and v are rewritten, the edge label will be (γ_1', γ_2'), independent of the rewriting order. Thus, each node incident with an edge has a "private part" of the edge label; if the edge is outgoing, it is the first part of the label, and if the edge is incoming it is the second part.

Formally, to prove the only-if direction of the equivalence in Definition 1.3.5, observe that $\alpha_2 = \beta_2$ and $\beta_1 = \delta_1$, and take $\gamma = (\alpha_1, \delta_2)$ (where the subscripts 1 and 2 of course denote the first and second component of the edge label, respectively). Similarly, in the if direction, note that $\alpha_1 = \gamma_1$ and $\gamma_2 = \delta_2$, and take $\beta = (\delta_1, \alpha_2)$. □

Using this lemma, we now show that the classes L-edNCE and C-edNCE are equal (as opposed to the NLC case, see [36]).

Theorem 1.3.11
L-edNCE = C-edNCE.

Proof
Let $G = (\Sigma, \Delta, \Gamma, \Omega, P, S)$ be an edNCE grammar. We will construct a C-edNCE grammar $G' = (\Sigma, \Delta, \Gamma', \Omega', P', S)$ and an edge relabeling ρ such that $L_{\text{lm}}(G) = \rho(L(G'))$. By Proposition 1.3.8 this implies that $L_{\text{lm}}(G) \in$ C-edNCE. The grammar G' will have the properties stated in the previous lemma. In what follows we will say that a node u of an ordered graph H is "to the left" of a node v of H, if u comes before v in the linear order of the nodes of H.

The intuitive idea is to replace every edge (u, γ, v) between nonterminal nodes u and v such that u is to the left of v, by all edges $(u, ((\gamma, a, \beta, \sigma), (a, \beta)), v)$ with $a \in \Delta$, $\beta \in \Gamma$, and σ is the label of v. In a leftmost derivation, first the node u is completely rewritten into terminal nodes x, and then v is rewritten. The pair (a, β) represents an edge (x, β, v) where a is the label of x. In the new grammar G', v only sees the (a, β) part of the edge label and acts as if u has already been rewritten. Node u only sees the $(\gamma, a, \beta, \sigma)$ part of the edge label and acts (based on the γ and σ) as if v has not yet been rewritten; moreover, at the moment that x is produced, the correct edge is selected on the basis of the a and β, and the first part of the edge label is replaced by ε (where ε indicates that the first part is "empty"). A label $(\varepsilon, (a, \beta))$ of an edge in a terminal graph (with $\beta \in \Omega$) is turned into β by the edge relabeling. In the case of an edge (v, γ, u), with u to the left of v, everything is "reversed", i.e., the edge is replaced by all edges $(v, ((a, \beta), (\gamma, a, \beta, \sigma)), u)$.

We now start the formal definitions. Let $\Gamma' = \Gamma_1 \times \Gamma_2$ with

$$\Gamma_1 = \Gamma_2 = (\Delta \times \Gamma) \cup (\Gamma \times \Delta \times \Gamma \times \Sigma) \cup \{\varepsilon\}, \text{ and}$$

$$\Omega' = \{(\varepsilon, (a, \beta)) \mid a \in \Delta, \beta \in \Omega\} \cup \{((a, \beta), \varepsilon) \mid a \in \Delta, \beta \in \Omega\}.$$

The edge relabeling is defined by $\rho(\varepsilon, (a, \beta)) = \rho((a, \beta), \varepsilon) = \beta$ for all $a \in \Delta$ and $\beta \in \Omega$.

For an ordered graph H without embedding, in $GR_{\Sigma,\Gamma}$, we define the ordered graph $\text{sim}(H) \in GR_{\Sigma,\Gamma'}$ to be $\text{sim}(H) = (V_H, E, \lambda_H)$ with the same order on V_H and with E obtained as follows. If (u, γ, v) is an edge of H with u to the left of v, and u is a nonterminal node of H, then E contains all edges

$$(u, ((\gamma, a, \beta, \sigma), (a, \beta)), v)$$

with $a \in \Delta$, $\beta \in \Gamma$, and $\sigma = \lambda_H(v)$; if u is a terminal node of H, then E contains the edge

$$(u, (\varepsilon, (a, \beta)), v)$$

with $a = \lambda_H(u)$ and $\beta = \gamma$. Analogously, if (v, γ, u) is an edge of H with u to the left of v, and u is a nonterminal node of H, then E contains all edges

$$(v, ((a, \beta), (\gamma, a, \beta, \sigma)), u)$$

with $a \in \Delta$, $\beta \in \Gamma$, and $\sigma = \lambda_H(v)$; and if u is a terminal node of H, then E contains the edge

$$(v, ((a, \beta), \varepsilon), u)$$

with $a = \lambda_H(u)$ and $\beta = \gamma$.

The set P' of productions of G' is now defined as follows: if $X \to (D, C)$ is a production of P, then P' contains the production $X \to (\text{sim}(D), C')$ where C' is obtained as follows.

If $(\sigma, \gamma/\gamma', x, \text{out})$ is in C, then C' contains the following connection instructions:

(1) if x is a nonterminal node, then C' contains all

$$(\sigma', ((\gamma, a, \beta, \sigma), \pi)/((\gamma', a, \beta, \sigma), \pi), x, \text{out})$$

with $\sigma' \in \Sigma$, $a \in \Delta$, $\beta \in \Gamma$, and $\pi \in \Gamma_2$ (intuitively, for an edge (u, γ, v), u is to the left of v, and u is rewritten into x; in the first part of the edge label, γ is changed into γ'),

(2) if x is a terminal node, then C' contains all

$$(\sigma', ((\gamma, \lambda_D(x), \gamma', \sigma), \pi)/(\varepsilon, \pi), x, \text{out})$$

with $\sigma' \in \Sigma$ and $\pi \in \Gamma_2$ (again, for an edge (u, γ, v), u is to the left of v, and u is rewritten into x; but now the correct edge is selected, with $a = \lambda_D(x)$ and $\beta = \gamma'$, and the first part of the edge label is changed into ε),

(3) if σ is in Δ, then C' contains all

$$(\sigma', ((\sigma, \gamma), \pi)/((\sigma, \gamma'), \pi), x, \text{out})$$

with $\sigma' \in \Sigma$ and $\pi \in \Gamma_2$ (now, for an edge (v, γ, u), u is to the left of v, and v is rewritten into x; this case would be more intuitive if σ would be denoted by a, γ by β, and γ' by β': then (a, β) is changed into (a, β')).

Analogously, if $(\sigma, \gamma/\gamma', x, \text{in})$ is in C, then C' contains all connection instructions as specified above, in which the first and second part of all edge labels are interchanged (note that $\pi \in \Gamma_1 = \Gamma_2$).

This ends the formal definition of G'. Since it is obvious that G' satisfies the requirements of Lemma 1.3.10, G' is a C-edNCE grammar. To prove that $L_{\text{lm}}(G) = \rho(L(G'))$, it suffices to prove that $L_{\text{lm}}(G) = \rho(L_{\text{lm}}(G'))$ by Proposition 1.3.9. Thus, it suffices to compare the leftmost derivations of the two grammars. In fact, it is straightforward to verify that the mapping 'sim' defined above turns the leftmost derivations of G into those of G'. More precisely, the leftmost derivations of G' starting with $sn(S, z)$ are of the form $sn(S, z) \Rightarrow \text{sim}(H_1) \Rightarrow \cdots \Rightarrow \text{sim}(H_n)$ where $sn(S, z) \Rightarrow H_1 \Rightarrow \cdots \Rightarrow H_n$ is a leftmost derivation of G (which can be proved by induction on n). This shows that the terminal graphs leftmost generated by G' are all $\text{sim}(H)$ where H is a terminal graph leftmost generated by G. Hence $L_{\text{lm}}(G) = \rho(L_{\text{lm}}(G'))$, because for a terminal graph H, $\text{sim}(H)$ is obtained from H by changing every edge label γ into some label $(\varepsilon, (a, \gamma))$ or $((a, \gamma), \varepsilon)$. Such a label is final if and only if γ is final, and if it is, then ρ replaces it by γ. Thus, if H has final edges only, then $\rho(\text{sim}(H)) = H$. □

We now come back to the discussion after the definition of dynamic confluence (Definition 1.3.7). It should be clear that Proposition 1.3.9 also holds for a dynamically confluent, and even for an order independent or abstractly confluent, grammar G (with the same proof). Thus, for such a grammar, $L(G) = L_{\text{lm}}(G)$, and hence, by the previous theorem, a C-edNCE grammar can be constructed that is equivalent with G. This shows that the dynamically confluent (and the order independent and the abstractly confluent) edNCE grammars generate the same class of languages as the (statically) confluent edNCE grammars, viz. C-edNCE.

Theorem 1.3.11 can be viewed as a first characterization of the class C-edNCE. Other characterizations will be discussed in Section 1.4.

Derivation trees

Due to our assumption that the right-hand sides of the productions of an edNCE grammar are ordered, the notion of derivation tree can be defined in the same way as for context-free string grammars. Thus, they are rooted, ordered trees, which we view here (as usual) as a special type of graph: each vertex of the tree has directed edges to each of its k children, $k \geq 0$, and the order of the children is indicated by using the numbers $1, \ldots, k$ to label these edges. A slight difference with the string case is that the vertices of the tree will be labeled by productions (or production copies) rather than symbols. Furthermore, for the sake of precision, we will distinguish here between two types of derivation trees: "c-labeled" derivation trees (in which the vertices are labeled by, mutually disjoint, production copies) and "p-labeled" derivation trees (in which they are labeled by productions). Note that the nodes of derivation trees are called 'vertices', in order not to confuse them with the nodes of the graphs involved. For c-labeled derivation trees, the nonterminal graph nodes will also be used as vertices of the tree, in a natural way. We now first consider c-labeled derivation trees.

Definition 1.3.12
Let $G = (\Sigma, \Delta, \Gamma, \Omega, P, S)$ be an edNCE grammar. A *c-labeled derivation tree* of G is a rooted, ordered tree t of which the vertices are labeled by production copies in copy(P), such that

(1) the right-hand sides of all the production copies that label the vertices of t are mutually disjoint, and do not contain the root of t as a node, and

(2) if vertex v of t has label $X \to (D, C)$, then the children of v are the nonterminal nodes of D, and their order in t is the same as their order in D; moreover, for each child w, the left-hand side of the label of w in t equals its label in D.

If the root of a derivation tree t is labeled with production $X \to (D, C)$, then we also say that t is a derivation tree *for* X (note that not necessarily $X = S$).

Example 1.11
Figure 1.21 shows a c-labeled derivation tree for S of edNCE grammar G_2 of Example 1.9. The large boxes are the vertices of the tree, labeled by production copies (of which the connection relations are not shown). The order of the right-hand sides of the productions is such that x_1 is before x_2 (see Fig. 1.11). The dotted lines between the large boxes are the edges of the tree. To show that the nonterminal nodes of the right-hand sides of the production copies are also used as vertices of the tree, the tree edges are drawn from such a nonterminal node (represented by a small box) to the corresponding large box. Note that since the figure does not show connection relations, it can also be viewed as a derivation tree of grammars G_3 and G_4 of Example 1.9. □

We note here that c-labeled derivation trees are closely related to shapes of 2-structures (compare Fig. 1.21 with, e.g., Fig. 6.3.6 of Chapter 6). In particular, the c-labeled derivation trees of the 2-structures generated by grammar G_{11} are closely related to their shapes.

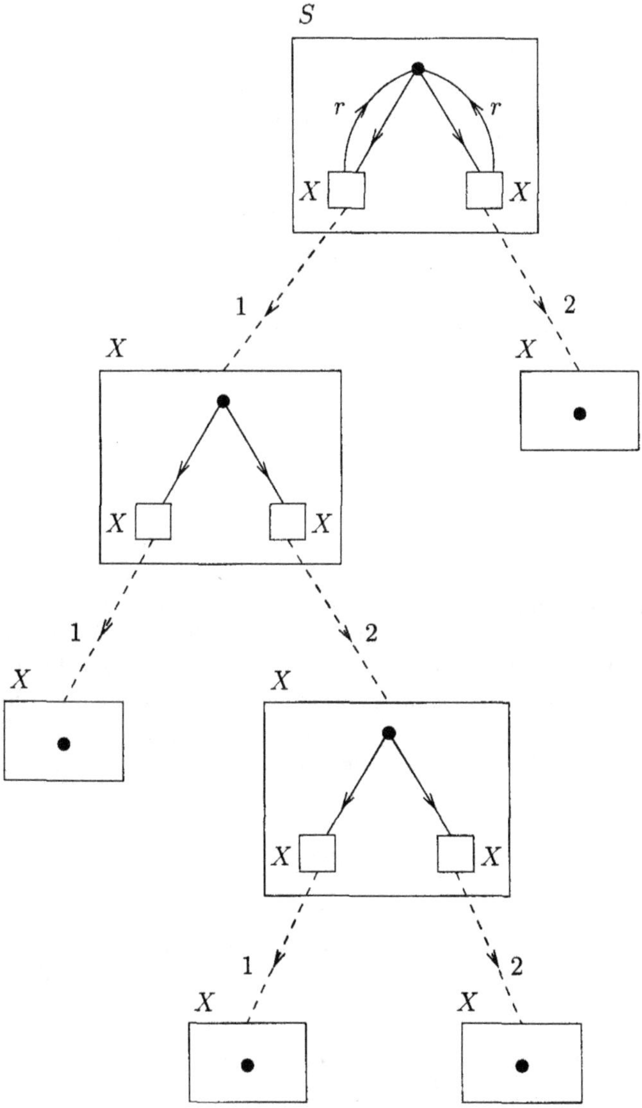

Figure 1.21: A c-labeled derivation tree.

As in the case of context-free string grammars, a derivation tree t for X will correspond to a derivation that starts from X, i.e., from a graph with a single node labeled X. Since, intuitively, such a derivation will be part of a larger one, this single-node graph should generate a graph with embedding and, hence, should have an embedding itself. Clearly, this must be the "hereditary embedding", which just "takes over" the edges of the environment.

By $sin(X, x)$ we denote the graph with embedding (V, E, λ, C) with $V = \{x\}$, $E = \emptyset$, $\lambda(x) = X$, and $C = \{(\sigma, \gamma/\gamma, x, d) \mid \sigma \in \Sigma, \gamma \in \Gamma, d \in \{\text{in}, \text{out}\}\}$. Note that if $sin(X, x) \Rightarrow_{v,p} (H, C_H)$, then $v = x$ and p equals $X \to (H, C_H)$. Thus, due to our Assumption in Section 1.3.1, we can use $sin(S, z)$ instead of $sn(S, z)$ in the definition of $L(G)$ and $L_{\text{lm}}(G)$, i.e., $L(G) = \{[H] \mid H \in GR_{\Delta,\Omega}, sin(S, z) \Rightarrow^* H \text{ for some } z\}$, and similarly for $L_{\text{lm}}(G)$. Note also that if K has a node v with label X, then $K[v/sin(X, x)]$ is isomorphic to K.

It is straightforward to *associate derivation trees with derivations.* With a derivation $sin(X, v_1) \Rightarrow_{v_1,p_1} H_1 \Rightarrow_{v_2,p_2} \cdots \Rightarrow_{v_n,p_n} H_n$ such that H_n is a terminal graph with embedding, we associate the c-labeled derivation tree t with vertices $v_1, \ldots, v_n$, labeled $p_1, \ldots, p_n$, respectively, such that the children of v_i are the nonterminal nodes of $\text{rhs}(p_i)$, in the same order. Since H_n is terminal, the nonterminal nodes of $\text{rhs}(p_i)$ do appear among $v_{i+1}, \ldots, v_n$. It is easy to verify the requirements of Definition 1.3.12: (1) holds because the derivation is creative (note that v_1 is the root of t), and (2) holds by definition and because the left-hand side of p_i is the label of v_i. Note that the construction of t depends on $(v_1, p_1), \ldots, (v_n, p_n)$ only; thus, derivations with the same set $\{(v_1, p_1), \ldots, (v_n, p_n)\}$ have the same derivation tree. In the case that the derivation is leftmost, it should be clear that the sequence $v_1, \ldots, v_n$ is the pre-order of the vertices of t. As an example, the derivation tree of Fig. 1.21 is associated with the (non-leftmost) derivation of Fig. 1.12.

Note that the association of derivation trees with derivations is even more straightforward than in the case of context-free string grammars, where one does not have a convenient notation for occurrences of (nonterminal) symbols, which are similar to the nodes $v_1, \ldots, v_n$.

The other way around, it is also easy to *associate leftmost derivations with derivation trees.* For a c-labeled derivation tree t, let $v_1, \ldots, v_n$ be the sequence of its vertices in pre-order, let p_i be the label of v_i, and let $X = \text{lhs}(p_1)$. Then it can be shown in exactly the same way as for context-free string grammars, that there is a leftmost derivation $sin(X, v_1) \Rightarrow_{v_1,p_1} H_1 \Rightarrow_{v_2,p_2} \cdots \Rightarrow_{v_n,p_n} H_n$ for certain $H_1, \ldots, H_n$ with H_n terminal. Note that this leftmost derivation is uniquely determined by v_1 and the sequence $p_1, \ldots, p_n$. For Example 1.11 this is the sequence $p_a^1, p_b^1, p_c^1, p_b^2, p_c^2, p_c^3, p_c^4$, where the superscripts indicate the fact that these are production *copies*.

As in the case of context-free string grammars it is straightforward to prove that both compositions of the above two mappings between leftmost derivations and c-labeled derivation trees are the identity, and hence that they are bijections. This shows the one-to-one correspondence between leftmost derivations and c-labeled derivation trees for an arbitrary edNCE grammar.

We will now define the yield of a derivation tree and show that the yield of the derivation tree corresponding to a leftmost derivation is the result of that derivation. It is natural to define the yield in a bottom-up fashion, as for context-free string grammars.

Definition 1.3.13
Let t be a c-labeled derivation tree of an edNCE grammar $G = (\Sigma, \Delta, \Gamma, \Omega, P, S)$. Let $p \in \text{copy}(P)$ be the label of the root of t, and let $x_1, \ldots, x_k$ be the nonterminal nodes of rhs(p), ordered as in rhs(p). Let $t_1, \ldots, t_k$ be the direct subtrees of t (i.e., the subtrees rooted in $x_1, \ldots, x_k$, respectively). Then the *yield* of t is the graph with embedding, in $GRE_{\Delta,\Gamma}$, defined recursively as $\text{yield}(t) = \text{rhs}(p)[x_1/\text{yield}(t_1)] \cdots [x_k/\text{yield}(t_k)]$.

By $Y(G)$ we denote the set $\{[\text{yield}(t)] \in [GR_{\Delta,\Omega}] \mid t \text{ is a c-labeled derivation tree of } G \text{ for } S\}$.

Note that the yield of a derivation tree need not be in $GRE_{\Delta,\Omega}$, due to the possible presence of blocking edges (see Definition 1.3.4).

Thus, the yields of the subtrees are substituted for the nonterminal nodes of the right-hand side of the root production, from left to right. It is easy to prove by induction that the set of nodes of yield(t) equals the set of terminal nodes of the right-hand sides of the labels of t; this ensures, by condition (1) of Definition 1.3.12, that the substitution is well defined. It also shows that each node u of yield(t) "belongs" to exactly one vertex $\text{vertex}_t(u)$ of t, in the sense that it belongs to the right-hand side of the label of $\text{vertex}_t(u)$. The yield of the derivation tree t of Fig. 1.21 (Example 1.11) is shown in Fig. 1.22(b) (see also Fig. 1.12); in this case the mapping vertex_t from the nodes of yield(t) to the vertices of t is a bijection (because each production contains exactly one terminal node), but in general it need not be injective or surjective.

For a given derivation tree, the corresponding leftmost derivation executes substitutions in a top-down fashion, whereas the yield function executes them bottom-up. The fact that these two methods produce the same graph must therefore be based on the associativity of substitution (Lemma 1.3.2). In the following lemma we formulate an easy consequence of associativity (cf. Lemma 2.12 of [23]).

Lemma 1.3.14
Let $sin(X,x) \Rightarrow_{v_1,p_1} H_1 \Rightarrow_{v_2,p_2} \cdots \Rightarrow_{v_n,p_n} H_n$ be a leftmost derivation of an edNCE grammar G (with $v_1 = x$ and $rhs(p_1) = H_1$), and let H be an ordered graph such that x is the first nonterminal node of H and $\lambda_H(x) = X$. Then $H \Rightarrow_{v_1,p_1} H[x/H_1] \Rightarrow_{v_2,p_2} H[x/H_2] \Rightarrow \cdots \Rightarrow_{v_n,p_n} H[x/H_n]$ is a leftmost derivation of G.

Proof
Clearly $H \Rightarrow_{v_1,p_1} H[x/H_1]$, because $\text{rhs}(p_1) = H_1$. Consider the leftmost derivation step $H_{j-1} \Rightarrow_{v_j,p_j} H_j$, for $2 \le j \le n$. Thus, $H_j = H_{j-1}[v_j/\text{rhs}(p_j)]$. Then $H[x/H_{j-1}] \Rightarrow_{v_j,p_j} H[x/H_{j-1}][v_j/\text{rhs}(p_j)] = H[x/H_{j-1}[v_j/\text{rhs}(p_j)]]$ by associativity, which equals $H[x/H_j]$. Clearly, this is also a leftmost derivation step. □

We now show that the yield of a derivation tree equals the result of the corresponding leftmost derivation.

Theorem 1.3.15
Let $sin(X,v_1) \Rightarrow_{v_1,p_1} H_1 \Rightarrow_{v_2,p_2} \cdots \Rightarrow_{v_n,p_n} H_n$, with H_n terminal, be a leftmost derivation of an edNCE grammar, and let t be the corresponding c-labeled derivation tree. Then $yield(t) = H_n$.

Proof
In this proof we will say that the sequence $(v_1,p_1)(v_2,p_2)\cdots(v_n,p_n)$ is the *trace* of a derivation $K_0 \Rightarrow_{v_1,p_1} K_1 \Rightarrow_{v_2,p_2} \cdots \Rightarrow_{v_n,p_n} K_n$.

Since there is only one leftmost derivation corresponding to t, it suffices to show that there is a derivation $sin(X,v_1) \Rightarrow^*_{\text{lm}} \text{yield}(t)$ with trace $s = (v_1,p_1)(v_2,p_2)\cdots(v_n,p_n)$. We prove this by induction on the structure of t. The root v_1 of t has label p_1, and $H_1 = \text{rhs}(p_1)$. Let H_1 have nonterminal nodes $x_1,\ldots,x_k$, in that order, with labels $X_1,\ldots,X_k$ respectively. Then t has direct subtrees $t_1,\ldots,t_k$. Since trace s is the pre-order sequence of the vertices of t, with their labels, it can be written as $s = (v_1,p_1)s_1\cdots s_k$, where s_i is the pre-order sequence of t_i. By induction there is a derivation $sin(X_i,x_i) \Rightarrow^*_{\text{lm}} \text{yield}(t_i)$ with trace s_i, for every $1 \le i \le k$. By Lemma 1.3.14, there is a derivation $H_1 \Rightarrow^*_{\text{lm}} H_1[x_1/\text{yield}(t_1)]$ with trace s_1. Applying that lemma again, we obtain a derivation $H_1[x_1/\text{yield}(t_1)] \Rightarrow^*_{\text{lm}} H_1[x_1/\text{yield}(t_1)][x_2/\text{yield}(t_2)]$ with trace s_2. Repeating this argument k times, and putting the derivations together, we obtain a derivation $H_1 \Rightarrow^*_{\mathrm{lm}} H_1[x_1/\mathrm{yield}(t_1)][x_2/\mathrm{yield}(t_2)]\cdots[x_k/\mathrm{yield}(t_k)] = \mathrm{yield}(t)$ with trace $s_1 \cdots s_k$, and hence a derivation $sin(X, v_1) \Rightarrow_{v_1,p_1} H_1 \Rightarrow^*_{\mathrm{lm}} \mathrm{yield}(t)$ with trace $(v_1, p_1)s_1 \cdots s_k = s$. □

As a corollary we obtain the fact that the language leftmost generated by an edNCE grammar equals the set of yields of its c-labeled derivation trees.

Theorem 1.3.16
For every edNCE grammar G, $L_{lm}(G) = Y(G)$. □

Thus, in particular, by Proposition 1.3.9, $L(G) = Y(G)$ for every C-edNCE grammar G. In fact, for C-edNCE grammars Theorem 1.3.15 holds for arbitrary derivations (because derivations starting with some $sin(X, v)$ can be turned into leftmost derivations, with the same derivation tree, as in the proof of Proposition 1.3.9). For a C-edNCE grammar G one might wonder whether, in the definition of yield(t), it is relevant that the graphs yield(t_1), ..., yield(t_k) are substituted from left to right. The following lemma shows that it is not.

Lemma 1.3.17
Let $G = (\Sigma, \Delta, \Gamma, \Omega, P, S)$ be a C-edNCE grammar. Let u_1 and u_2 be two distinct nonterminal nodes of $H \in GRE_{\Sigma,\Gamma}$ with labels X_1 and X_2, respectively. Let $K_1, K_2 \in GRE_{\Sigma,\Gamma}$ be such that $sin(X_i, u_i) \Rightarrow^* K_i$.

Then $H[u_1/K_1][u_2/K_2] = H[u_2/K_2][u_1/K_1]$.

Proof
The proof is by induction on the sum of the lengths of the two derivations. If that sum equals 2, then we are ready by Proposition 1.3.6, because K_i is the right-hand side of a production. Otherwise, consider an additional derivation step, say, $K_1 \Rightarrow_{v,p} K_1'$ with rhs(p) $= M$ and so $K_1' = K_1[v/M]$. Then $H[u_1/K_1'][u_2/K_2] = H[u_1/K_1[v/M]][u_2/K_2] = H[u_1/K_1][v/M][u_2/K_2] = H[u_1/K_1][u_2/K_2][v/M] = H[u_2/K_2][u_1/K_1][v/M] = H[u_2/K_2][u_1/K_1[v/M]] = H[u_2/K_2][u_1/K_1']$ by associativity (Lemma 1.3.2), induction (twice), and associativity again. □

This leads to the following basic lemma on C-edNCE grammars (Lemma 2.14 of [23]). It is sometimes called the *context-freeness lemma* because it expresses the existence of derivation trees without explicitly mentioning them (see Section 2.3 of Chapter 2).

Lemma 1.3.18
Let $G = (\Sigma, \Delta, \Gamma, \Omega, P, S)$ be a C-edNCE grammar. Let the graph $H \in GRE_{\Sigma,\Gamma}$ have nonterminal nodes $v_1, \ldots, v_k$ with labels $X_1, \ldots, X_k$ respectively, and let $F \in GRE_{\Delta,\Gamma}$ be a terminal graph. Then $H \Rightarrow^* F$ if and only if there exist $F_1, \ldots, F_k \in GRE_{\Delta,\Gamma}$ such that $F = H[v_1/F_1] \cdots [v_k/F_k]$ (in any order) and $sin(X_i, v_i) \Rightarrow^* F_i$ for all $1 \le i \le k$.
Moreover, the length of the derivation $H \Rightarrow^* F$ is the sum of the lengths of the derivations $sin(X_i, v_i) \Rightarrow^* F_i$, $1 \le i \le k$.

Proof
By the previous lemma (and the confluence of the given grammar), we may as well prove the statement for leftmost derivations of an arbitrary edNCE grammar G, and for an ordered graph H such that $v_1, \ldots, v_k$ are in the correct order. Consider the edNCE grammar G' that is obtained from G by adding the new production $X \to H$, where X is a new nonterminal. The result now follows from Theorem 1.3.15 by considering c-labeled derivation trees for X (of which the root is necessarily labeled with the new production). Clearly, F is the yield of such a tree, and the F_i are the yields of its direct subtrees. □

If, in particular, H is a sentential form of G with one nonterminal node v_1 (i.e., $k = 1$), then $F = H[v_1/F_1]$ and the number of bridges between F_1 and $F - F_1$ is at most $2 \cdot \#\Delta \cdot \#\Gamma \cdot \delta_F^2$, where δ_F is the maximal degree of the nodes of F (see the remark following Definition 1.3.1). Thus, in a derivation $sin(S, v_1) \Rightarrow_{v_1,p_1} H_1 \Rightarrow_{v_2,p_2} \cdots \Rightarrow_{v_n,p_n} F$ the part of F that is generated from v_i (which is the yield of the subtree with root v_i of the corresponding c-labeled derivation tree) has at most $2 \cdot \#\Delta \cdot \#\Gamma \cdot \delta_F^2$ bridges to the remainder of F. This fact is used in [96,14] to prove a *separator theorem* for C-edNCE languages. Roughly speaking, for every C-edNCE language L there is a constant c (with $c = 2 \cdot \#\Delta \cdot \#\Gamma$) such that every graph $F \in L$ can be separated into two parts of almost the same size by removing at most $c \cdot \delta_F^2$ edges (intuitively, this can be done by cutting the derivation tree t of F into two parts of almost the same size by removing one edge of t, i.e., by cutting a subtree from t). A similar result for linear edNCE grammars is proved in Theorem 15 of [39].

We now turn to p-labeled derivation trees.

Definition 1.3.19
Let $G = (\Sigma, \Delta, \Gamma, \Omega, P, S)$ be an edNCE grammar. A *p-labeled derivation tree* of G is a rooted, ordered tree t of which the vertices are labeled by productions in P, such that

if vertex v of t has label p, and rhs(p) has nonterminal nodes $x_1, \ldots, x_k$, in that order, then v has children $v_1, \ldots, v_k$, in that order, and the left-hand side

of the label of v_i in t equals the label of x_i in rhs(p).
By $D(G)$ we denote the set of all abstract p-labeled derivation trees of G for S.

Abstract p-labeled derivation trees can be written as ordinary expressions, viewing a production p as k-ary operator, where k is the number of nonterminal nodes of rhs(p). Thus, they are trees in the sense of tree language theory [55]; in fact, it is easy to see that $D(G)$ is a regular tree language (see Definition 1.4.1 and the discussion following it).

From a given p-labeled derivation tree t, a c-labeled derivation tree ct can be obtained by choosing mutually disjoint copies of the productions that label the vertices of t (satisfying the conditions of Definition 1.3.12). Note that all c-labeled derivation trees can be obtained in this way. By definition, the *yield* of the p-labeled derivation tree t is the abstract graph yield(t) = [yield(ct)]. This definition does not depend on the choice of ct by the following argument: if H and H' are isomorphic ordered graphs in $GRE_{\Sigma,\Gamma}$ with nonterminal nodes $x_1, \ldots, x_k$ and $x'_1, \ldots, x'_k$ in that order, respectively (where x_i corresponds to x'_i by the isomorphism), and, for $1 \leq i \leq k$, H_i and H'_i are isomorphic graphs in $GRE_{\Delta,\Gamma}$, then $H[x_1/H_1]\cdots[x_k/H_k]$ and $H'[x'_1/H'_1]\cdots[x'_k/H'_k]$ are isomorphic. In fact, for *a production* $p \in P$, the operator p *can be interpreted as an operation on abstract graphs* (viz. the substitution of its argument graphs into the nonterminal nodes of its right-hand side, from left to right), and then the yield of an abstract p-labeled derivation tree t is the value of the expression t. This gives the following result.

Proposition 1.3.20
For every edNCE grammar G, $Y(G)$ = yield($D(G)$) $\cap$ $[GR_{\Delta,\Omega}]$, and hence $L_{lm}(G)$ = yield($D(G)$) $\cap$ $[GR_{\Delta,\Omega}]$. □

Example 1.12
Figure 1.22(a) shows a p-labeled derivation tree t of edNCE grammar G_2 of Example 1.9. From t the c-labeled derivation tree ct of Fig. 1.21 can be obtained as indicated above. These trees have the same yield, shown in Fig. 1.22(b) (cf. Fig. 1.12). Viewing the productions of G_2 as operators, t corresponds to the expression $p_a(p_b(p_c, p_b(p_c, p_c)), p_c)$ or, in parenthesis-free notation, $p_ap_bp_cp_bp_cp_cp_c$. The value of this expression is the graph yield(t). □

From Theorem 1.3.15 (or Lemma 1.3.18) and our considerations concerning p-labeled derivation trees, a *least fixed point result* for edNCE grammars easily

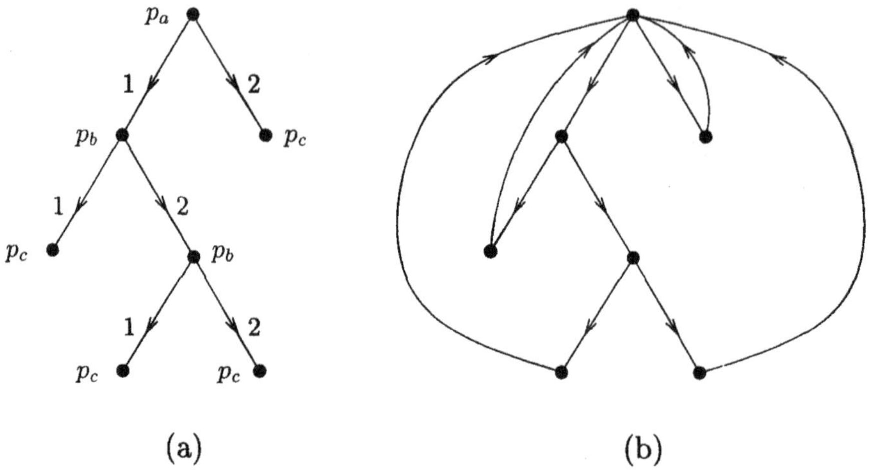

Figure 1.22: (a) A p-labeled derivation tree, and (b) its yield.

follows (see Section 2.4.1 of Chapter 2 for the analogous case of HR grammars). To be more precise, the graph languages $L_{\rm lm}(G, X)$ with $X \in \Sigma - \Delta$, defined by $L_{\rm lm}(G, X) = \{[H] \in [GRE_{\Delta,\Omega}] \mid sin(X, x) \Rightarrow^*_{\rm lm} H$ for some $x\}$, are the least solution of the set of all inclusions $p(X_1, \ldots, X_k) \subseteq X$, where p is a production of G (viewed as operation on graphs as explained above, and extended to graph languages in the obvious way), with lhs$(p) = X$ and $X_1, \ldots, X_k$ are the labels of the nonterminal nodes of rhs(p), in that order. For C-edNCE grammars the subscripts 'lm' can, of course, be dropped in the above. For confluent grammars this result was proved, in a general framework, in Theorem 2.16 of [23]. The above least fixed point result for leftmost derivations of non-confluent grammars can be carried over to that general framework.

Derivation trees (and Theorem 1.3.15) can in particular be used in formal proofs on edNCE grammars that need a bottom-up argument, as opposed to top-down proofs that follow the steps of a derivation (such as the proofs of Proposition 1.3.8 and Theorem 1.3.11). This is a general fact, which is formally expressed, e.g., in Proposition 1.10 of [24] (see also Theorem 5.7.3 of Chapter 3). Instead of derivation trees, one may also use Lemma 1.3.18, the "context-freeness lemma", see also Sections 2.4 and 2.6 of Chapter 2. To illustrate this idea here, we will now prove that blocking edges can be removed from C-edNCE grammars, as shown in [99].

Theorem 1.3.21
For every C-edNCE grammar an equivalent nonblocking C-edNCE grammar can be constructed.

Proof
Let $G = (\Sigma, \Delta, \Gamma, \Omega, P, S)$ be a C-edNCE grammar. We will construct a new grammar $G' = (\Sigma', \Delta, \Gamma, \Omega, P', S')$ with nonterminals of the form $(X, (b, c))$, where X is a nonterminal of G and b is a boolean that contains the information whether or not $sin(X, x)$ generates a terminal graph with blocking edges. This boolean can be computed bottom-up (on the derivation tree corresponding to the derivation), but to update it we also need some knowledge c about the connection relation of the generated terminal graph.

To be precise, we define, for a terminal graph $H \in GRE_{\Delta,\Gamma}$, $\text{info}(H) = (b, c)$ such that b is a boolean which is *true* if and only if $H \in GRE_{\Delta,\Omega}$ and $c = \{(\sigma, \beta/\gamma, \lambda_H(x), d) \mid (\sigma, \beta/\gamma, x, d) \in C_H \text{ for some } x \in V_H\}$ where C_H is the connection relation of H.

Grammar G' will be constructed in such a way that if $sin(X, x)$ generates a terminal graph H, and $\text{info}(H) = (b, c)$, then $sin((X, (b, c)), x)$ generates H', where H' is the same as H except that, due to the additional information in the nonterminals, each connection instruction $(X', \beta/\gamma, x, d)$ of H (where X' is a nonterminal) has to be replaced by all connection instructions $((X', (b', c')), \beta/\gamma, x, d)$. This implies that $sin((S, (\mathit{true}, \emptyset)), x)$ generates all terminal sentential forms of G without blocking edges, whereas $sin((S, (\mathit{false}, \emptyset)), x)$ generates those with blocking edges. Taking $S' = (S, (\mathit{true}, \emptyset))$, G' generates the same language as G, but does not generate any blocking edges any more.

Let $Q = \{(b, c) \mid b \in \{\mathit{true}, \mathit{false}\}, c \subseteq (\Sigma \times \Gamma \times \Gamma \times \Delta \times \{\text{in}, \text{out}\})\}$, let $\Sigma' = ((\Sigma - \Delta) \times Q) \cup \Delta$, and let $S' = (S, (\mathit{true}, \emptyset))$.

Let $p : X \to (D, C)$ be a production of G, and let $x_1, \ldots, x_k$ be the nonterminal nodes of D, in that order, with labels $X_1, \ldots, X_k$, respectively. It is straightforward (but tedious) to construct a function $\delta_p : Q^k \to Q$ such that for all terminal graphs $H_1, \ldots, H_k$:

$$\text{info}((D, C)[x_1/H_1] \cdots [x_k/H_k]) = \delta_p(\text{info}(H_1), \ldots, \text{info}(H_k)).$$

Thus, according to the definition of 'yield', the information can be computed bottom-up on the derivation trees using the δ_p (in fact, the δ_p define a bottom-up tree automaton on the p-labeled derivation trees of G, with set of states Q, see [55]).

The productions P' of G' are defined as follows. For every production p of G as above, and all $q_1, \ldots, q_k \in Q$, P' contains the production prod$(p, q_1, \ldots, q_k)$ which has left-hand side $(X, \delta_p(q_1, \ldots, q_k))$ and of which the right-hand side is obtained from the one of p by replacing the label X_i of x_i with (X_i, q_i) and replacing each connection instruction $(X', \beta/\gamma, x, d)$ with all connection instructions $((X', q'), \beta/\gamma, x, d)$, $q' \in Q$.

Since the productions of G' are a simple variation of those of G (and, in particular, their connection relations correspond in a straightforward way), it is easy to verify that G' is confluent.

It can now be proved formally that $sin((X, q), x) \Rightarrow^* H'$ in G' if and only if there exists H such that $sin(X, x) \Rightarrow^* H$ in G, info$(H) = q$, and H' is the same as H except that it has connection relation $C_{H'} = \{(\sigma', \beta/\gamma, x, d) \mid (\phi(\sigma'), \beta/\gamma, x, d) \in C_H\}$ where $\phi(X', q') = X'$ and $\phi(a) = a$ for $a \in \Delta$.

The proof is by structural induction on the c-labeled derivation trees that correspond to the derivations, using Theorem 1.3.15. Or equivalently on the lengths of the derivations, using Lemma 1.3.18. This is, of course, a bottom-up induction argument. The details are left to the reader. Taking $X = S$, this implies that $L(G') = L(G)$. □

The argument in the above proof is quite general. In fact, for every C-edNCE grammar G, and every bottom-up finite tree automaton A (see [55]), another C-edNCE grammar G' can be constructed, as in the above proof, such that G' generates only those terminal graphs that are the yield of a p-labeled derivation tree of G that is accepted by A. If, in particular, A accepts t iff yield(t) is in a certain set Π (i.e., has a certain desirable property), then G' generates only those terminal graphs that are generated by G and have property Π, and so $L(G') = L(G) \cap \Pi$ (note that for the nonblocking property the generated language remained the same). This generalizes the well-known result on string languages that the class of context-free string languages is closed under intersection with regular languages. Many properties of graphs can be recognized by finite tree automata that work on p-labeled derivation trees, in particular all properties that are compatible with substitution (see Section 2.6.1 of Chapter 2, and [24], for the HR case), including all properties that are expressible in monadic second order logic (see Proposition 3.19 and Corollary 4.6 of [23] for C-NLC grammars, and see Theorem 1.4.7 in Section 1.4).

Theorem 1.3.21 is of importance, because it allows one to decide emptiness and finiteness of C-edNCE languages in the same way as for context-free string grammars. In fact, as observed after Definition 1.3.4, for a nonblocking grammar G, the string language $S(G) = \{a^n \mid n = \#V_H \text{ for some } [H] \in L(G)\}$ is context-free. Clearly, $S(G)$ is empty (finite) if and only if $L(G)$ is empty (finite, respectively). This shows the result. For nonblocking C-edNCE grammars the algorithms take polynomial time as usual, but for arbitrary C-edNCE grammars the emptiness problem is exponential time complete, as shown in [99] (which means that the removal of blocking edges needs exponential time, in general).

Proposition 1.3.22
It is decidable, for an arbitrary C-edNCE grammar G, whether or not $L(G)$ is empty, and also whether or not $L(G)$ is finite. Moreover, if $L(G)$ is finite, it can be constructed effectively. □

Together with the above remarks, this shows that it is decidable for an arbitrary C-edNCE grammar G whether $L(G) \cap \Pi \neq \emptyset$, where Π is a graph property that can be recognized by a finite tree automaton on derivation trees. In general this problem has a high complexity (as shown in [105]); for a methodology to find efficient algorithms, see [106]. In the special case that $L(G)$ contains exactly one graph H, the grammar G can be viewed as a succinct hierarchical representation of H, and the problem is to decide whether or not H has the property Π; for this case the decision algorithm is often more efficient (see [105,106]).

Other decidability properties of C-edNCE grammars are discussed at the end of Section 1.4.2.

1.3.3 Subclasses

Several natural subclasses of the C-edNCE grammars have been investigated in the literature. We mention a number of the most important ones. Whenever we define an X-edNCE grammar for some X, X-edNCE will denote the class of graph languages generated by X-edNCE grammars. X·Y-edNCE grammars are edNCE grammars that are both X and Y.

Definition 1.3.23
An edNCE grammar $G = (\Sigma, \Delta, \Gamma, \Omega, P, S)$ is *boundary*, or a B-edNCE grammar, if, for every production $X \to (D, C)$,

(B1) D does not contain edges between nonterminal nodes, and

(B2) C does not contain connection instructions $(\sigma, \beta/\gamma, x, d)$ where σ is nonterminal.

Grammars G_5, G_{10}, and G_{11} in Example 1.9 are not boundary, but all the others are.

In the definition of B-edNCE grammar either one of conditions (B1) and (B2) can be dropped, in the sense that for every edNCE grammar satisfying one of the conditions an equivalent grammar satisfying both conditions can easily be constructed: just remove either the "wrong" edges or the "wrong" connection instructions (see Theorem 2.2 of [92]). For this reason, we will assume from now on that every edNCE grammar that satisfies one of the conditions (B1) and (B2), also satisfies the other (i.e., we will disregard edNCE grammars that satisfy only one of the two conditions).

Boundary grammars were introduced (in [92]) and studied because they are a natural example of confluent grammars, and because they are easy to "program". The idea of a boundary grammar is that nonterminal nodes are not connected, i.e., every nonterminal node has a "boundary" of terminal neighbours around itself. Note that it follows from (B1) that no sentential form of G contains edges between nonterminal nodes; hence, it does not need the connection instructions mentioned in (B2). From (B2) it immediately follows that every B-edNCE grammar is a C-edNCE grammar. Since the construction in the proof of Theorem 1.3.21 preserves (B1), blocking edges can be removed from any B-edNCE grammar (and note that the construction is even easier, because the connection relations remain the same; thus, only the labels of the nonterminal nodes are changed); the same holds for all other subclasses to be discussed.

In terms of derivation trees, the boundary condition means the following: for a c-labeled derivation tree t of a B-edNCE grammar, if (u, γ, v) is an edge of yield(t), then either vertex$_t(v)$ is a descendant of vertex$_t(u)$ in t, or the other way around (see the discussion after Definition 1.3.13); thus, edges are established "along paths" of the derivation tree only.

Definition 1.3.24
A B-edNCE grammar $G = (\Sigma, \Delta, \Gamma, \Omega, P, S)$ is *nonterminal neighbour deterministic*, or a B_{nd}-edNCE grammar, if, for every production $X \to (D, C)$, every nonterminal node $x \in V_D$, and all $\gamma \in \Gamma$ and $\sigma \in \Delta$,

(1) $\{y \in V_D \mid (y, \gamma, x) \in E_D \text{ and } \lambda_D(y) = \sigma\} \cup \{\beta \in \Gamma \mid (\sigma, \beta/\gamma, x, \text{in}) \in C\}$ is a singleton or empty, and, analogously for 'out',

(2) $\{y \in V_D \mid (x, \gamma, y) \in E_D \text{ and } \lambda_D(y) = \sigma\} \cup \{\beta \in \Gamma \mid (\sigma, \beta/\gamma, x, \text{out}) \in C\}$ is a singleton or empty (where we assume that $V_D \cap \Gamma = \emptyset$).

Of the grammars of Example 1.9, $G_1, G_1', G_2, G_2', G_3, G_8$, and G_9 are $\mathrm{B_{nd}}$-edNCE grammars.

The idea of a $\mathrm{B_{nd}}$-edNCE grammar G is that the neighbours y of a nonterminal node v of a sentential form of G are uniquely determined by the label and direction of the edge between v and y, and by the label of y. This means that, when rewriting v, the embedding process can distinguish each neighbour of v separately. For this reason, $\mathrm{B_{nd}}$-edNCE grammars are close to the HR grammars of Chapter 2: a nonterminal node of a $\mathrm{B_{nd}}$-edNCE grammar, with its different edges, is similar to a nonterminal hyperedge of an HR grammar, with its different tentacles. In fact, it is shown in [52] (see [104] for closely related results) that $\mathrm{B_{nd}}$-edNCE grammars and HR grammars have the same generating power. To express this formally it has to be taken into account that HR grammars generate another type of graphs: they allow arbitrary multiple edges, and even hyperedges, but they do not allow node labels. Thus, it is necessary to encode each of the two types of graphs into the other one.

Here, for reasons of comparison, and in order to stress explicitly the relationship between the node replacement and the hyperedge replacement grammars, we will restrict attention to HR grammars that generate our type of graphs, encoded in an obvious way (simulating node labels by hyperedges of rank 1). Moreover, we will denote the corresponding class of graph languages by HR. With this convention, we have the result that $\mathrm{B_{nd}}$-edNCE = HR (see Lemma 8, Theorem 2 of [52], where the encoding is called 'hyp').

It is shown in Theorem 1 of [52] that also the other way around, encoding hypergraphs as bipartite graphs in the obvious way, $\mathrm{B_{nd}}$-edNCE grammars and HR grammars have the same hypergraph generating power; in fact, even C-edNCE grammars do not have more hypergraph generating power, as shown in Theorem 3.13 of [18].

A C-edNCE grammar is *of bounded nonterminal degree*, or a $\mathrm{C_{bntd}}$-edNCE grammar, if there is a bound on the degree of the nonterminal nodes of its sentential forms (see [52,104]). The *nonterminal degree* of such a grammar is the maximal degree of these nodes. It is decidable whether a C-edNCE grammar is of bounded nonterminal degree, and if so, its nonterminal degree can be computed (see the end of Section 1.4.2). Clearly, every $\mathrm{B_{nd}}$-edNCE grammar is of bounded nonterminal degree; its nonterminal degree is the maximal number of tentacles of the nonterminal hyperedges of the corresponding HR grammar (also called its 'order'). It is not difficult to show that, the other way around, every B-edNCE grammar of bounded nonterminal degree can be turned into an equivalent $\mathrm{B_{nd}}$-edNCE grammar, with the same nonterminal degree (Lemma 8 of [52]). Even, for every C_{bntd}-edNCE grammar there is an equivalent B_{nd}-edNCE grammar (see the discussion after Theorem 1.3.30). Hence C_{bntd}-edNCE = HR.

Definition 1.3.25
An edNCE grammar $G = (\Sigma, \Delta, \Gamma, \Omega, P, S)$ is *apex*, or an A-edNCE grammar, if for every production $X \to (D, C)$ and every connection instruction $(\sigma, \beta/\gamma, x, d) \in C$, x and σ are terminal.

In Example 1.9, grammar G_1, G'_1, and G_3 are apex, and the others are not.

The idea of an apex grammar G (introduced in Definition I.4.9 of [86] and studied in, e.g., [41,42]) is that nonterminal nodes cannot "pass neighbours" to each other. When a nonterminal node v is replaced by a daughter graph D, only terminal nodes of D can be connected to the neighbours of v (which are terminal nodes too). This easily implies that the set of sentential forms of G, and in particular the language generated by G, is of bounded degree (which means that there is a bound on the degree of all nodes of all sentential forms): the edges of a node x are generated when x itself is generated in a daughter graph D, and when the nonterminal neighbours of x in D are rewritten. This way of building graphs is riminiscent of playing the domino game, and is particularly easy to visualize. Since every A-edNCE grammar is a B-edNCE grammar of bounded nonterminal degree, A-edNCE is a subset of B_{nd}-edNCE, and hence of HR. There is a natural restriction on HR grammars that corresponds to the apex property, see [52,33].

In terms of c-labeled derivation trees, the apex condition means that if $(\sigma, \beta/\gamma, x, d)$ is a connection instruction of yield(t), then $\text{vertex}_t(x)$ is the root of t. This implies for an edge (u, γ, v) of yield(t), that $\text{vertex}_t(u) = \text{vertex}_t(v)$, or $\text{vertex}_t(v)$ is a child of $\text{vertex}_t(u)$, or the other way around.

Definition 1.3.26
An edNCE grammar $G = (\Sigma, \Delta, \Gamma, \Omega, P, S)$ is *linear*, or a LIN-edNCE grammar, if for every production $X \to (D, C)$, D has at most one nonterminal node.

Grammars $G_1, G'_1, G'_2, G_6, G_7, G_8$, and G_9 of Example 1.9 are linear. The grammar of Example 1.6(a) is also linear.

Linear graph grammars are studied for the same reasons as linear string grammars (see, e.g., [86]). Their derivations are particularly easy to understand.

Clearly, every LIN-edNCE grammar is a B-edNCE grammar. Since the results of [52] hold for linear grammars too, LIN·B_{nd}-edNCE = LIN-HR, and LIN·A-edNCE = LIN·A-HR, where linearity for HR grammars is defined in the obvious way.

Linear edNCE grammars are maybe more powerful than one would expect: a fixed number of nonterminal nodes can be simulated by one (as opposed to the case of context-free string grammars). An edNCE grammar is *nonterminal bounded* if there is a bound on the number of nonterminal nodes in every sentential form of the grammar. It is shown in [39] that for every nonterminal bounded edNCE grammar there is an equivalent LIN-edNCE grammar (and the same holds for B_{nd} and for apex grammars). This result does not hold for NLC grammars (see [38]).

An inclusion diagram of the subclasses defined above, is given in Fig. 1.23 (where, in order to improve the readability of the diagram, we have consistently omitted the suffix edNCE). The set of all binary trees is in A-edNCE (see G_3 of Example 1.9) but not in LIN-edNCE (Theorem 16 of [39], see also Lemma I.4.18 of [86]). The set of all edge complements of all binary trees is in C-edNCE (see the end of Section 1.4.1, and Theorem 1.4.7) but not in B-edNCE (Theorem 35 of [44]); in fact, for any graph language L that is in A-edNCE but not in LIN-edNCE, the set of edge complements of the graphs in L is in C-edNCE but not in B-edNCE. The set of all complete graphs is in LIN-edNCE (see grammar G_6 of Example 1.9) but not in HR (see Theorem 1.3.27 below). The set of all stars is in LIN-HR (see G_8 of Example 1.9) but not in A-edNCE (because apex languages are of bounded degree).

The classes HR and A-edNCE can be characterized within the class C-edNCE: there are simple, and decidable, conditions on the graphs of a language $L \in$ C-edNCE that express membership of L in HR and in A-edNCE. To present these characterizations we need some more terminology.

Recall that $K_{m,n}$ is the complete bipartite graph on m and n nodes (see G_7 of Example 1.9). For a graph H we denote by und(H) the underlying undirected, unlabeled graph (i.e., $V = V_H$, $E = \{(u, *, v) \mid (u, \gamma, v) \in E_H \text{ or } (v, \gamma, u) \in E_H \text{ for some } \gamma\}$, and $\lambda(x) = \#$ for every $x \in V$). We will say that a graph language L is *of bounded bi-degree* if there exists a natural number n such that $K_{n,n}$ is not a subgraph of any graph und(H), $H \in L$ (and hence the same holds for $K_{m,m}$ with $m > n$). This means that it is impossible to find a graph in L with $2n$ nodes $u_1, \ldots, u_n, v_1, \ldots, v_n$ such that there is an edge (in any direction) between each u_i and each v_j, $1 \leq i, j \leq n$. As usual, a graph language L is *of bounded degree* if there exists n such that all nodes in all graphs of L have degree at most n. This is equivalent to saying that there exists n such that the star $K_{1,n}$ is not a subgraph of any graph und(H),

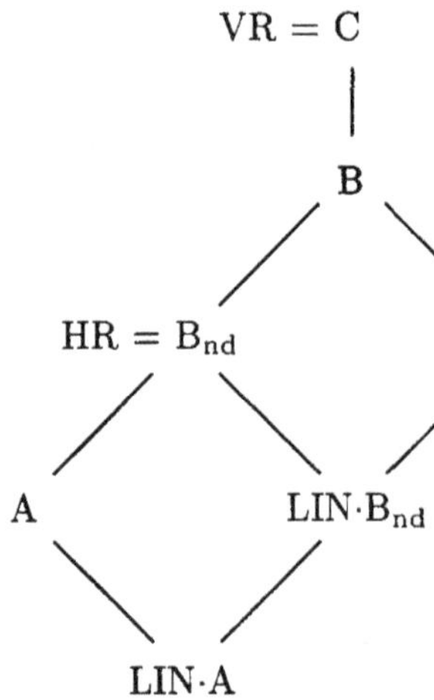

Figure 1.23: Inclusion diagram of C-edNCE and its subclasses.

$H \in L$. A graph language L is *of bounded average degree* (or *sparse*) if there exists n such that $\#E_H \leq n \cdot \#V_H$ for every $H \in L$, i.e., if the number of edges of a graph of L is linearly bounded by its number of nodes. Of course, every language of bounded degree is also of bounded bi-degree and of bounded average degree, but, in general, the latter two properties are independent (see, e.g., Theorem VI.2.8 of [10]: there is a graph language $\{H_k \mid k \geq k_0\}$, for some natural number k_0, such that $\#V_{H_k} = k$, $\#E_{H_k} \geq 1/2(k^{3/2} - k^{4/3})$, and H_k has no subgraph $K_{2,2}$).

The following result is shown in [28]; see [108] for a related earlier result.

Theorem 1.3.27
For every graph language $L \in$ C-edNCE the following three properties are equivalent:
$L \in HR$, L is of bounded bi-degree, and L is of bounded average degree.
It is decidable, for a C-edNCE grammar G, whether or not $L(G) \in HR$. □

Thus, the set of all complete graphs is not in HR; and, e.g., the set of all graphs $K_{m,n}$ (which is in LIN-edNCE, see G_7 of Example 1.9) is not in HR.

It follows from Theorem 1.3.27 that every C-edNCE language of bounded degree is in HR. This was proved first for B-NLC languages in [76] and for

B-edNCE languages in Theorem 7 of [52], and then for C-edNCE languages in [15,31]. The following grammatical characterization of the HR languages of bounded degree is shown in [33] (see also [48]).

Theorem 1.3.28
For every graph language $L \in$ C-edNCE,
$L \in$ A-edNCE if and only if L is of bounded degree.
It is decidable, for a C-edNCE grammar G, whether or not $L(G) \in$ A-edNCE.
□

It is open whether similar results hold for B-edNCE, LIN-edNCE, LIN·B_{nd}-edNCE, and LIN·A-edNCE. It is proved in Theorems 12 and 18 of [39] that LIN·A-edNCE is properly included in the class of all linear edNCE languages of bounded degree.

1.3.4 Normal forms

In this subsection we discuss a number of normal forms of edNCE grammars. Some of the previous results can already be seen as normal form results. For instance, the C-edNCE grammar can be viewed as a normal form of the edNCE grammar, when only leftmost derivations are considered (Theorem 1.3.11), and Theorem 1.3.21 shows that blocking edges can be removed from C-edNCE grammars. Theorems 1.3.27 and 1.3.28 can also be viewed as normal form results, for languages of bounded (bi-)degree only.

There are some normal forms that are similar to those for context-free string grammars. First of all, it is easy to see that every C-edNCE grammar can be *reduced*, in the sense that nonterminals that do not occur in derivations $sin(S, z) \Rightarrow^* H$ with H terminal, can be removed; these nonterminals can be determined in exactly the same way as for context-free string grammars. Also, Λ-productions and chain productions can be removed from C-edNCE grammars. A production p of a C-edNCE grammar is a *Λ-production* if rhs(p) is the empty graph, and a *chain production* if rhs(p) has a single node, with a nonterminal label (for an example of a chain production, see the third production of Fig. 1.14).

Lemma 1.3.29
For every C-edNCE grammar an equivalent C-edNCE grammar can be constructed without Λ-productions or chain productions. The same holds for the subclasses of Section 1.3.3.

Proof

The proof for Λ-productions is fully analogous to the one for context-free string grammars: substitute, in the right-hand sides of the productions, the empty graph Λ for some of the nonterminal nodes x that can generate Λ (i.e., for which $sin(X,x) \Rightarrow^* \Lambda$, where $X = \lambda(x)$). Now let $G = (\Sigma, \Delta, \Gamma, \Omega, P, S)$ be a C-edNCE grammar without Λ-productions. The proof for chain productions is also the same as for context-free string grammars, except that we have to compute the correct connection relations. We construct $G = (\Sigma, \Delta, \Gamma, \Omega, P', S)$, where P' is the smallest set of productions such that

(1) P' contains all non-chain productions of P, and

(2) if p' is in P', and $p \in P$ is a chain production (copy, disjoint with p') such that the single node x of $\text{rhs}(p)$ has label $\text{lhs}(p')$, then P' contains the production $\text{lhs}(p) \to \text{rhs}(p)[x/\text{rhs}(p')]$.

The correctness of this construction can be understood from the following fact. Let p and p' be as above, and let H be a graph with a node v that is labeled with $\text{lhs}(p)$. Then $H[v/\text{rhs}(p)[x/\text{rhs}(p')]] = H[v/\text{rhs}(p)][x/\text{rhs}(p')]$, by associativity of substitution. This implies that every derivation of G' can be turned into one of G with the same result, and, the other way around, that every leftmost derivation of G can be turned into one of G' (note that in a leftmost derivation an application of chain production p has to be followed by a derivation step that rewrites x); hence $L_{\text{lm}}(G') = L(G') = L(G)$. It also implies that G' is confluent, using the characterization of Proposition 1.3.6 (or, alternatively, note the obvious fact that the construction in the proof of Theorem 1.3.11 does not introduce Λ-productions or chain productions).

It can easily be verified that the above construction preserves all subclasses of Section 1.3.3. □

From this lemma we obtain the following enumeration property of C-edNCE languages, as shown in a general framework in Corollary 10 of [107].

Theorem 1.3.30

For every C-edNCE language L there is a constant c such that for every natural number n, $\#\{[H] \in L \mid \#V_H = n\} \leq 2^{c \cdot n}$.

Proof

Let G be a C-edNCE grammar generating L, without Λ-productions or chain productions. Then every derivation of a graph H with $\#V_H = n$ is of length $\leq 2n$, and hence every (c- or p-labeled) derivation tree of such a graph has at most $2n$ vertices. Clearly, there are at most d^{2n} abstract p-labeled derivation trees with $2n$ vertices, where d is the number of productions of G. ◻

As observed in [107] this implies that the set $[GR_{\Delta,\Omega}]$ of all graphs (over any Δ and Ω) is not in C-edNCE. This is an unexpected weakness of context-free graph grammars that one has to live with; intuitively, the graphs in a context-free graph language are always "tree-like". There does exist a non-blocking edNCE grammar G with $L(G) = [GR_{\Delta,\Omega}]$ (see, e.g., [46]). The fact that $[GR_{\Delta,\Omega}]$ is not in C-edNCE also follows from the separator theorem of [96,14] mentioned after Lemma 1.3.18. Other graph languages that are not in C-edNCE as a consequence of Theorem 1.3.30 are, e.g., the set of all connected graphs and the set of all tournaments (where a tournament is a graph with exactly one edge between every two nodes); note that the set of acyclic tournaments (or linear orders) is in LIN-edNCE, by an obvious modification of grammar G_6 of Example 1.9.

It can now be shown that $\mathrm{C_{bntd}}$-edNCE = HR (cf. the discussion before Definition 1.3.25). Let G be a C-edNCE grammar of bounded nonterminal degree. By the proof of Lemma 1.3.29, we may assume that G has no Λ- or chain productions. This implies that $L(G)$ is of bounded average degree (see Theorem 1.3.27). In fact, if $H \in L(G)$ has n nodes, then it has a derivation of length $\leq 2n$ (as in the proof of Theorem 1.3.30). At each step of the derivation, at most $2gm^2$ daughter edges are added and at most $2gmk$ new edges are established by the embedding, where g is the number of edge labels, m is the maximal number of daughter nodes, and k is the bound on the nonterminal degree. Hence, H has at most $4g(m^2 + m)k \cdot n$ edges. It follows from Theorem 1.3.27 that $L(G)$ is in HR.

We now turn to an analogue of Chomsky Normal Form: a C-edNCE grammar G is *in Chomsky Normal Form* if for every production p of G, either rhs(p) has two nodes, not both with a terminal label, or rhs(p) has one node, with a terminal label. The following result is stated in Lemma 4 of [14]; see also the results on Chomsky Normal Form in [84,86].

Theorem 1.3.31
For every C-edNCE grammar an equivalent C-edNCE grammar in Chomsky Normal Form can be constructed. The same holds for the subclasses B, B_{nd}, LIN, and LIN·B_{nd}.

Proof
Again, the proof is similar to the one for context-free string grammars. We use a variation of the proof in [87]. Let G be a C-edNCE grammar without Λ-productions or chain productions, and let $p : X \to (D, C)$ be a production of

G that does not have the required form. We will show that this production can be simulated by two productions $p_1 : X \to (D_1, C_1)$ and $p_2 : Y \to (D_2, C_2)$, where p_1 is as required, and D_2 is the result of removing one node from D. Repetition of this construction produces a grammar in Chomsky Normal Form.

Let x_0 be a fixed (but arbitrary) node of D, with label σ_0. Take a new nonterminal Y, and new edge labels (x, γ) and (γ, x) for every (old) edge label γ and every node x of D, $x \neq x_0$. The edge label (x, γ) will be used on an outgoing edge of a Y-labeled node y, to indicate that when y produces D_2, the edge should be connected to x (and similarly for (γ, x) and incoming edges).

Production p_1 is defined as follows: D_1 is a graph with two nodes y and x_0 (where y is new), labeled Y and σ_0, respectively, and all edges $(y, (x, \gamma), x_0)$ with $(x, \gamma, x_0) \in E_D$, and $(x_0, (\gamma, x), y)$ with $(x_0, \gamma, x) \in E_D$; C_1 contains all connection instructions $(\sigma, \beta/\gamma, x_0, d)$ that are in C, and all connection instructions $(\sigma, \gamma/\gamma, y, d)$.

Production p_2 is defined as follows: D_2 is the subgraph of D induced by $V_D - \{x_0\}$; C_2 contains, for $x \in V_D - \{x_0\}$, all connection instructions $(\sigma, \beta/\gamma, x, d)$ that are in C, and all connection instructions $(\sigma, (x, \gamma)/\gamma, x, \mathrm{out})$ with $(x, \gamma, x_0) \in E_D$, and $(\sigma, (\gamma, x)/\gamma, x, \mathrm{in})$ with $(x_0, \gamma, x) \in E_D$.

Let G' be the grammar obtained from G by replacing p with p_1 and p_2. Since $(D_1, C_1)[y/(D_2, C_2)] = (D, C)$, it can be shown as in the proof of Lemma 1.3.29 that $L_{\mathrm{lm}}(G') = L_{\mathrm{lm}}(G)$ (where we take x_0 to be the last node in both productions p and p_1; this forces an application of p_2 directly after one of p_1 in a leftmost derivation of G').

In general, G' is not confluent (to see this, consider the application of productions to the two nodes of D_1, assuming that x_0 is nonterminal). Since the above construction works for arbitrary edNCE grammars (with leftmost derivations), it would suffice to observe that the construction in the proof of Theorem 1.3.11 does not change the nodes (and their labels) of the right-hand side of a production. However, it is not difficult to turn G' into a confluent grammar: just add to each production $X' \to (D', C')$ of G' all connection instructions $(\sigma, (\beta, z)/(\gamma, z), x, \mathrm{out})$ with $(\sigma, \beta/\gamma, x, \mathrm{out}) \in C'$, and all $(\sigma, (z, \beta)/(z, \gamma), x, \mathrm{in})$ with $(\sigma, \beta/\gamma, x, \mathrm{in}) \in C'$.

Also, if G is a B-edNCE grammar, then G' need not be one. The problem is that if x_0 is a nonterminal node, then p_1 violates the boundary condition. However, the construction works if one chooses x_0 to be a terminal node whenever D has one. Note that if D has nonterminal nodes only, then it has no edges, and hence D_1 has no edges. It is easy to verify that this construction also preserves the B_{nd} and the LIN properties. It does not preserve the apex property, due to the connection instructions $(\sigma, \gamma/\gamma, y, d)$ in C_1. □

It is more usual to require for Chomsky Normal Form that right-hand sides with two nodes have nonterminal nodes only. It is not difficult to show that, for a C-edNCE grammar, this requirement can be met by a similar trick as for context-free string grammars. For B-edNCE grammars, it would mean that only discrete graphs can be generated.

Note that LIN-edNCE grammars in Chomsky Normal Form are very close to the right-linear context-free string grammars: at every derivation step, the unique nonterminal node generates one terminal node (see the grammar in Example 1.6(a)).

For apex grammars there is no Chomsky Normal Form. Even worse, there does not exist a normal form for apex grammars with a fixed bound on the number of nonterminal nodes in the right-hand side of a production. This is shown in [48] for HR grammars (and nonterminal hyperedges, of course), but the argument can easily be carried over to A-edNCE grammars (it is based on the property of derivation trees mentioned just before Definition 1.3.26). The set of all k-ary trees can be generated by an A-edNCE grammar with at most k nonterminal nodes in each right-hand side, but not with less than k.

A C-edNCE grammar G is *in Operator Normal Form* (ONF) if the right-hand side of each production of G contains exactly one terminal node (see grammars $G_2, \ldots, G_5$ of Example 1.9). It seems to be likely that there exists an equivalent grammar in ONF for every C-edNCE grammar, but no proof can be found in the literature. For B-edNCE grammars this is proved in Theorem 24 of [44] (where it is called Greibach Normal Form). The construction in that proof preserves the B_{nd} property, and a similar construction can be used to show the existence of ONF for A-edNCE grammars (note that A-edNCE productions that have nonterminal nodes only, are of a very simple type: the set of edges and the connection relation are both empty). For linear grammars we observe that every LIN-edNCE in Chomsky Normal Form (Theorem 1.3.31) is also in Operator Normal Form.

There are also normal forms that concern the embedding process, and in particular the edges that are incident with nonterminal nodes. An edge that is incident with a nonterminal node v of a sentential form, is usually needed in order to connect terminal nodes generated by v to neighbours of v, using the connection instructions. In the next normal form (introduced in Definition I.4.9 of [86]), it is required that all such edges are indeed used for that purpose, i.e., nonterminal nodes do not have superfluous edges. Intuitively it means that, when a production is applied, every edge that is incident with the mother node should be "passed" to at least one node of the daughter graph, i.e., edges are not "dropped" by the embedding process. This implies that every neighbour of the mother node is the neighbour of at least one daughter node (which, in general, is not equivalent with the above requirement, due to the presence of multiple edges).

Definition 1.3.32
An edNCE grammar G is *neighbourhood preserving* if, for every sentential form H of G, every nonterminal node v of H, and every production p with $\text{lhs}(p) = \lambda_H(v)$, the following holds:
if (w, β, v) is in E_H ((v, β, w) is in E_H), then $\text{rhs}(p)$ has a connection instruction of the form $(\lambda_H(w), \beta/\gamma, x, d)$ with $d = \text{in}$ ($d = \text{out}$, respectively).

All grammars in Example 1.9 are neigbourhood preserving. If one adds to G_2 a production p'_c which is the same as p_c except that it has connection relation $C'_c = \{(n, e/e, z, \text{in})\}$ instead of $C_c = \{(n, e/e, z, \text{in}), (n, r/e, z, \text{out})\}$, then a grammar is obtained that is *not* neighbourhood preserving. It generates the set of all binary trees such that the root is connected to *some* of the leaves (rather than all the leaves).

The neighbourhood preserving normal form was proved for B-NLC in [92], for B-edNCE in [52], and for C-edNCE in [101].

Theorem 1.3.33
For every C-edNCE grammar an equivalent neighbourhood preserving C-edNCE grammar can be constructed. The same holds for the subclasses of Section 1.3.3.

Proof
Let $G = (\Sigma, \Delta, \Gamma, \Omega, P, S)$ be a C-edNCE grammar. Similar to the proof of Theorem 1.3.21, we define, for a terminal graph $H \in GRE_{\Delta,\Gamma}$, $\text{info}(H) = \{(\sigma, \beta/\gamma, \lambda_H(x), d) \mid (\sigma, \beta/\gamma, x, d) \in C_H \text{ for some } x \in V_H\}$. First, we construct the equivalent C-edNCE grammar $G' = (\Sigma', \Delta, \Gamma, \Omega, P', S')$ as in the proof of Theorem 1.3.21 (but without the booleans). Note that $\Sigma' = ((\Sigma - \Delta) \times Q) \cup \Delta$, where Q is the powerset of $\Sigma \times \Gamma \times \Gamma \times \Delta \times \{\text{in}, \text{out}\}$, and $S' = (S, \emptyset)$. Thus, every nonterminal of G' is of the form (X, q), where X is a nonterminal of G and q is the info of the terminal graph generated by $sin(X, x)$ in G. From q we can see which incident edges the nonterminal is going to use; the others can be deleted. So, in G', whether or not an edge (v, γ, w) is useful, in this sense,

depends on $(\lambda(v), \gamma, \lambda(w))$ only. We define the triples in $\Sigma' \times \Gamma \times \Sigma'$ that are *useful*, as follows.

Any triple (a, β, b) with $a, b \in \Delta$ is useful. A triple $(a, \beta, (X, q))$ is useful if q contains some $(a, \beta/\gamma, b, \text{in})$ with $\gamma \in \Gamma$ and $b \in \Delta$, and, analogously, $((X, q), \beta, b)$ is useful if q contains some $(b, \beta/\gamma, a, \text{out})$ with $\gamma \in \Gamma$ and $a \in \Delta$. A triple $((X_1, q_1), \beta, (X_2, q_2))$ is useful if there exist tuples $(X_2, \beta/\gamma, a, \text{out}) \in q_1$ and $(a, \gamma/\delta, b, \text{in}) \in q_2$, for some $a, b \in \Delta$ and $\gamma, \delta \in \Gamma$. Note that the last definition can equivalently be stated with 1 and 2, and 'out' and 'in', interchanged (cf. Definition 1.3.5); this follows from Lemma 1.3.17.

From G', the desired grammar G'' is obtained by throwing away all useless edges and all useless connection instructions from the right-hand sides of the productions of G', where an edge (v, γ, w) is useless if $(\lambda(v), \gamma, \lambda(w))$ is not useful, a connection instruction $(\sigma', \beta/\gamma, x, \text{in})$ is useless if $(\sigma', \gamma, \lambda(x))$ is not useful, and a connection instruction $(\sigma', \beta/\gamma, x, \text{out})$ is useless if $(\lambda(x), \gamma, \sigma')$ is not useful.

It can be verified that G'' is a C-edNCE grammar, equivalent with G'. It is easy to see that the sentential forms of G'' contain useful edges only. From this it can be deduced that G'' is neighbourhood preserving. □

As an alternative (static) definition of the neighbourhood preserving property we could require that for every $\sigma \in \Sigma$, $\beta \in \Gamma$, $d \in \{\text{in}, \text{out}\}$, and every production p, rhs(p) has a connection instruction of the form $(\sigma, \beta/\gamma, x, d)$. In fact, missing connection instructions can be added (because they will never be used) in such a way that the resulting grammar is still confluent. The details are left to the reader.

For a node v of a graph H we define the *context* of v in H by $\text{cont}_H(v) = \{(\gamma, \lambda_H(w)) \mid (v, \gamma, w) \in E_H\} \cup \{(\lambda_H(w), \gamma) \mid (w, \gamma, v) \in E_H\}$. Thus, the context of a nonterminal node v indicates precisely which kinds of edges are incident with v, to be used by the embedding process. For boundary grammars it is possible to store the context of a nonterminal node in its label, as shown in Section 3 of [92]. An edNCE grammar G is *context consistent* if there is a mapping η from $\Sigma - \Delta$ to the set of all possible contexts, such that for every nonterminal node v of every sentential form H of G, $\text{cont}_H(v) = \eta(\lambda_H(v))$. For every B-edNCE grammar there is an equivalent context consistent B-edNCE grammar (Lemma 4 of [52]), and the same is true for the subclasses of Section 1.3.3 below B-edNCE. It should be clear that there cannot be a context consistent normal form for arbitrary C-edNCE grammars: the context of a node can change when one of its neighbours is rewritten. For B-edNCE grammars one could introduce productions of the form: $X \to (D, C)$ *provided* c, which can only be applied to a node v with label X if v has context c; thus, contexts are used as application conditions (see Theorem 3.3 of [92]). For C-edNCE grammars such productions would allow the generation of non-context-free graph languages such as the set of all stars $K_{1,m}$ with m a power of 2.

1.4 Characterizations

The class C-edNCE of graph languages generated by confluent edNCE grammars can be characterized in a number of different ways, showing the robustness and naturalness of the class. The characterizations we have in mind are the following: by leftmost derivations in edNCE grammars, i.e., C-edNCE = L-edNCE (Theorem 1.3.11), by regular tree and string languages, by Monadic Second Order logic translation of trees, by handle replacement graph grammars, and by regular sets of graph expressions. In this section these characterizations (except the first) will be discussed one by one.

1.4.1 Regular path characterization

As observed before, all graphs generated by a C-edNCE grammar are "tree-like". An alternative way of describing a set of "tree-like" graphs H is by taking a tree t from some regular tree language, defining the nodes of H as a subset of the vertices of t, and defining an edge between nodes u and v of H if the string of vertex labels on the shortest (undirected) path between u and v in t belongs to some regular string language. Such a description of a graph language will be called a *regular path description.* This idea was introduced in [109], for linear graph grammars, and investigated for LIN-edNCE and B-edNCE grammars in [44], and for C-edNCE grammars in [87,46,51].

Let us first define the regular tree languages. The following definition of a regular tree grammar (in normal form) as a special type of A-edNCE grammar, is equivalent with the usual one (see [55]). As usual, a *ranked alphabet* is an alphabet Δ together with a mapping rank : $\Delta \to \{0, 1, 2, \dots\}$. By m_Δ we denote the maximal number rank(σ), $\sigma \in \Delta$.

Definition 1.4.1
An edNCE grammar $G = (\Sigma, \Delta, \Gamma, \Omega, P, S)$ is a *regular tree grammar*, or a REGT grammar, if Δ is a ranked alphabet, $\Gamma = \Omega = \{1, \dots, m_\Delta\}$, and for every production $X \to (D, C)$ in P:
$V_D = \{x_0, x_1, \dots, x_k\}$ for some $k \geq 0$,
$\lambda_D(x_0) \in \Delta$ with rank($\lambda_D(x_0)$) $= k$, and $\lambda_D(x_i) \in \Sigma - \Delta$ for $1 \leq i \leq k$,
$E_D = \{(x_0, i, x_i) \mid 1 \leq i \leq k\}$, and
$C = \{(\sigma, i, i, x_0, \text{in}) \mid i \in \Gamma, \sigma \in \Delta\}$.

Thus, for a production p of a regular tree grammar, rhs(p) consists of a (terminal) parent x_0, and k (nonterminal) children $x_1, \ldots, x_k$. Edges lead from the parent to each child, and the children are ordered by numbering their edges from 1 to k. Such a production $X \to (D, C)$ can (modulo isomorphism) be denoted uniquely as $X \to \sigma X_1 \cdots X_k$ with $\sigma = \lambda_D(x_0)$ and $X_i = \lambda_D(x_i)$ for $1 \leq i \leq k$.

A graph language generated by a REGT grammar is a *regular tree language.* For a ranked alphabet Δ, we denote by T_Δ the set of all trees over Δ, i.e., the regular tree language generated by the REGT grammar with one nonterminal X and all productions $X \to \sigma X^k$ for every $\sigma \in \Delta$, where $k = \text{rank}(\sigma)$.

As observed in Section 1.3.2, the set $D(G)$ of (p-labeled) derivation trees of an edNCE grammar G is a regular tree language. In fact, it is generated by the regular tree grammar G_D which has the set P of productions of G as ranked alphabet (where, for $p \in P$, rank(p) is the number of nonterminal nodes in rhs(p)) and has the same nonterminals (and initial nonterminal) as G. For each production p of G, G_D has the production $X \to pX_1 \cdots X_k$, where $X = \text{lhs}(p)$ and $X_1, \ldots, X_k$ are the labels of the nonterminal nodes of rhs(p), in that order.

We now turn to the regular path description of graph languages. For an alphabet Σ, $\overline{\Sigma} = \{\overline{\sigma} \mid \sigma \in \Sigma\}$.

Definition 1.4.2
A *regular path description* is a tuple $R = (\Sigma, \Delta, \Omega, T, h, W)$, where Σ is a ranked alphabet, Δ and Ω are alphabets (of node and edge labels, respectively), T is a regular tree language in T_Σ, h is a partial function from Σ to Δ, and W is a mapping from Ω to the class of regular string languages, such that, for every $\gamma \in \Omega$, $W(\gamma) \subseteq \Sigma^*\overline{\Sigma}\Sigma^*$. The *graph language defined by* R is $L(R) = \{[gr_R(t)] \mid t \in T\}$, where $gr_R(t)$ is the graph $H \in GR_{\Delta,\Omega}$ such that
V_H is the set of vertices v of t for which $\lambda_t(v)$ is in the domain of h,
$\lambda_H(v) = h(\lambda_t(v))$ for $v \in V_H$, and
E_H is the set of all edges (u, γ, v) such that $\text{path}_t(u, v) \in W(\gamma)$, where $\text{path}_t(u, v)$ is the sequence of labels of the vertices on the shortest (undirected) path in t from u to v, in which the label σ of the least common ancestor of u and v is barred, i.e., replaced by $\overline{\sigma}$.

Note that $L(R) \subseteq [GR_{\Delta,\Omega}]$. Note that h is used both to determine which vertices of the tree t are nodes of the graph $gr_R(t)$, and to define their labels in that graph (on the basis of their labels in the tree). Note that for each edge label γ, $W(\gamma)$ is the regular string language that defines the graph edges with label γ. And note that the shortest path from vertex u to v first ascends from u to the least common ancestor z of u and v, and then descends from z to v.

Let RPD denote the class of graph languages that are defined by regular path descriptions. The main result of [51] is that RPD = C-edNCE.

To give an idea of the intuition behind this result, consider the inclusion C-edNCE $\subseteq$ RPD. One starts from a (nonblocking) C-edNCE grammar G such that each right-hand side of a production contains at most one terminal node, for instance by taking G to be in Chomsky Normal Form, see Section 1.3.4. A regular path description R of $L(G)$ then has the set $D(G)$ of p-labeled derivation trees of G as regular tree language T. Consider some p-labelled derivation tree t of G, and let t also denote a corresponding c-labeled derivation tree. R is defined in such a way that $gr_R(t) = \text{yield}(t)$. By the above assumption, the function vertex_t that maps each node u of yield(t) to a vertex $\text{vertex}_t(u)$ of t is injective. This means that the vertices of t that are labeled with a production p that has one terminal node in its right-hand side, can be used to represent the nodes of yield(t). Thus, the function h is defined for those productions p only, and $h(p)$ is the label of the terminal node of rhs(p). The remaining problem is how to express the graph edges (u, γ, v) in the regular language $W(\gamma)$, i.e., through the labels of the path from $\text{vertex}_t(u)$ to $\text{vertex}_t(v)$. This is done by simulating the embedding behaviour of G along that path by a finite automaton. For details see [51,46].

Example 1.13
Consider the grammar G_2 of Example 1.9. A regular path description $R = (\Sigma, \Delta, \Omega, T, h, W)$ of the graph language $L(G_2)$ (of all binary trees with edges from the leaves to the root) is defined as follows. $\Sigma = \{a, b, c\}$ with rank(a) = rank(b) = 2 and rank(c) = 0 (where a, b, and c are used instead of p_a, p_b, and p_c, for readability). $\Delta = \{\#\}$ and $\Omega = \{*\}$. $T = D(G_2)$, the regular tree language of all p-labeled derivation trees of G_2, which is generated by the regular tree grammar with productions $S \to aXX$, $X \to bXX$, and $X \to c$ (see Fig. 1.22(a) for an example of such a derivation tree). The node relabeling h has domain Σ, and $h(a) = h(b) = h(c) = \#$. Finally, $W(*) = W_1(*) \cup W_2(*)$ where $W_1(*) = \{\bar{a}b, \bar{a}c, \bar{b}b, \bar{b}c\}$ and $W_2(*) = cb^*\bar{a}$. $W_1(*)$ is used to keep the edges of the tree t in the graph $gr_R(t)$, and $W_2(*)$ is used to add the edges from the leaves to the root.

A regular path description of $L(G_3)$ (all binary trees) is the same as R above except that $W(*) = W_1(*)$. For $L(G_4)$ (all transitive binary trees) we take $W(*) = \bar{a}b^+ \cup \bar{a}b^*c \cup \bar{b}b^+ \cup \bar{b}b^*c$.

To obtain a regular path description for the graph language $L(G_5)$ (where the leaves are chained by f-labeled edges) is a bit more work, because we have hidden a technical detail from the above intuitive sketch of the proof. In fact, we need to be able to see from the label of a node whether it is the first or second child of its parent. Thus, we take the regular tree language generated by the productions $S \to aLR$, $L \to b_l LR$, $R \to b_r LR$, $L \to c_l$, and $R \to c_r$. And then we take $W(f) = (c_r b_r^* b_l \cup c_l)(\overline{a} \cup \overline{b_l} \cup \overline{b_r})(b_r b_l^* c_l \cup c_r)$. □

Let B-RPD be the subclass of RPD obtained by restricting every $W(\gamma)$ to be a subset of $\Sigma^*\overline{\Sigma} \cup \overline{\Sigma}\Sigma^*$ (see the examples above, except the last one). This means that graph edges are only established between tree vertices of which one is a descendant of the other (which is in accordance with the property of derivation trees of B-edNCE grammars mentioned just before Definition 1.3.24). It is shown in [44] that B-RPD = B-edNCE. Let LIN-RPD be the subclass of RPD obtained by restricting the symbols of the ranked alphabet Σ to have rank 1 or 0. This means that the trees in the regular tree language are in fact strings (and note that derivation trees of LIN-edNCE are strings). Thus, a linear RPD uses regular string languages only. As an example, a regular path description of the set of all cycles (see $L(G_2')$ of Example 1.9) is obtained from the one of $L(G_2)$ in the above example by taking $T = ab^*c$. It is shown in [44] that LIN-RPD = LIN-edNCE. Let A-RPD be the subclass of RPD obtained by restricting every $W(\gamma)$ to be finite. Thus, graph edges can only be established between tree vertices that are at a bounded distance from each other. It is shown in [51], using Theorem 1.3.28, that A-RPD = A-edNCE.

Altogether we have the following result. It is not clear whether a natural RPD characterization exists for the B_{nd}-edNCE graph languages

Theorem 1.4.3
C-edNCE = RPD, B-edNCE = B-RPD, A-edNCE = A-RPD, and LIN-edNCE = LIN-RPD. □

For a graph H, its *edge complement* is the graph $ec(H) = (V_H, E, \lambda_H)$ where E is the set of all (v, γ, w), $v \neq w$, that are not in E_H. Since the regular string languages are closed under complement, it follows from Theorem 1.4.3 that C-edNCE and LIN-edNCE are closed under edge complement (i.e., if $L \in$ C-edNCE then $\{ec(H) \mid H \in L\} \in$ C-edNCE). Thus, the set of edge complements of binary trees is in C-edNCE, and the set of edge complements of chains is in LIN-edNCE (see Example 1.6(a)). The class of B-edNCE languages is *not* closed under edge complement (see [44]).

1.4.2 Logical characterization

The most useful, and entirely grammar independent, characterization of C-edNCE is by Monadic Second Order (MSO) logic. In particular, it leads to a lot of closure and decidability properties of C-edNCE. The main idea is that of an MSO definable function on graphs, introduced in [47,26]. The characterization says that C-edNCE is the class of all graph languages $f(T_\Delta)$ where T_Δ is the set of trees over a ranked alphabet Δ and f is an MSO definable function from trees to graphs. Since f is defined on trees t, the graphs $f(t)$ are "tree-like". The usefulness of the MSO characterization is that MSO logic is a convenient language to talk about graphs, and hence can be used as a specification language for sets of graphs and functions on graphs. Many results on C-edNCE languages have a proof through the MSO characterization that is much easier than the one based on C-edNCE grammars. For more information concerning the relationship between VR (= C-edNCE) and MSO logic we refer to Chapter 3.

For alphabets Σ and Γ, we define a monadic second-order logical language $\mathrm{MSOL}(\Sigma, \Gamma)$, of which each closed formula expresses a property of the graphs in $GR_{\Sigma,\Gamma}$. The language has *node variables*, denoted $u, v, \ldots$, and *node-set variables*, denoted $U, V, \ldots$. For a given graph H, each node variable ranges over the elements of V_H and each node-set variable over the subsets of V_H. There are four types of atomic formulas in $\mathrm{MSOL}(\Sigma, \Gamma)$: $\mathrm{lab}_\sigma(u)$, for every $\sigma \in \Sigma$, $\mathrm{edge}_\gamma(u, v)$, for every $\gamma \in \Gamma$, $u = v$, and $u \in U$. Their meaning should be clear: node u has label σ, there is an edge with label γ from node u to node v, nodes u and v are the same, and node u is an element of node-set U, respectively. The formulas of the language are constructed from the atomic formulas through the propositional connectives $\wedge$, $\vee$, $\neg$, $\rightarrow$, $\leftrightarrow$, and the quantifiers $\forall$ and $\exists$, in the usual way. Note that not only node variables but also node-set variables may be quantified (which makes the logic monadic second-order rather than first-order). Note also that there are no edge or edge-set variables. As usual, a formula is closed if it has no free variables. For a closed formula ϕ of $\mathrm{MSOL}(\Sigma, \Gamma)$ and a graph H of $GR_{\Sigma,\Gamma}$ we write $H \models \phi$ if ϕ is true for H. If formula ϕ has free variables, say u, v, and U (and no others), then we also write the formula as $\phi(u, v, U)$; if graph H has nodes $x, y \in V_H$ and a set of nodes $X \subseteq V_H$, then we write $H \models \phi(x, y, X)$ to mean that ϕ is true for H when u, v, and U are given the values x, y, and X, respectively.

It is well known that there exists an MSOL formula $\mathrm{path}(u, v)$ that expresses the existence of a directed path from u to v (see Section 5.2.7 of Chapter 3). Thus, as a very simple example, $H \models \forall u \forall v : \mathrm{path}(u, v)$ means that H is strongly connected. As another example, bipartiteness of a graph is expressed by the formula $\exists U \exists V : (\text{part}(U,V) \wedge \forall u \forall v : \text{edge}(u,v) \rightarrow (u \in U \wedge v \in V) \vee (u \in V \wedge v \in U)$, where part$(U,V)$ expresses that U and V form a partition of the set of all nodes, and edge(u,v) is an abbreviation of the disjunction of all edge$_\gamma(u,v)$, $\gamma \in \Gamma$.

Definition 1.4.4
A graph language $L \subseteq [GR_{\Sigma,\Gamma}]$ is *MSO definable* if there is a closed formula ϕ in MSOL(Σ,Γ) such that $L = \{[H] \in [GR_{\Sigma,\Gamma}] \mid H \models \phi\}$.

Many sets of graphs that are *not* "tree-like", are MSO definable. Thus, as shown above, the set of strongly connected graphs and the set of bipartite graphs are MSO definable. Other examples of MSO definable sets of graphs are: all planar graphs, grids, chordal graphs, trees, and, for fixed k, k-colourable graphs, k-connected graphs, etc. (see Chapter 3).

For tree languages $L \subseteq T_\Delta$ (as defined in Section 1.4.1) the following classical result is well known: a tree language is MSO definable if and only if it is a regular tree language; the analogous classical result holds for string languages (cf. Section 5.7.4 of Chapter 3). For graph languages there is no generally accepted notion of regularity, but the MSO definable graph languages enjoy several properties that are similar to those of the regular tree and string languages (see, e.g., Theorem 1.4.7).

We now turn to MSO definable functions f. The idea is that, for a given input graph H, the nodes, edges, and labels of the output graph $H' = f(H)$ are described in terms of MSOL formulas on H. For each node label σ of H' there is a formula $\phi_\sigma(u)$ expressing that u will be a node of H' with label σ. Thus, the nodes of H' are a subset of the nodes of H. For each edge label γ of H' there is a formula $\phi_\gamma(u,v)$ expressing that there will be a γ-labeled edge from u to v in H'. Finally, to allow for partial functions, there is a closed formula ϕ_{dom} that specifies the domain dom(f) of f (which means that dom(f) is an MSO definable set of graphs).

Definition 1.4.5
Let Σ_i and Γ_i be alphabets, for $i \in \{1,2\}$. An *MSO definable function* $f : GR_{\Sigma_1,\Gamma_1} \rightarrow GR_{\Sigma_2,\Gamma_2}$ is specified by formulas in MSOL(Σ_1,Γ_1), as follows:

- a closed formula ϕ_{dom}, the *domain formula*,
- a formula $\phi_\sigma(u)$, for every $\sigma \in \Sigma_2$, the *node formulas*,
- a formula $\phi_\gamma(u,v)$, for every $\gamma \in \Gamma_2$, the *edge formulas*.

The domain of f is $\{H \in GR_{\Sigma_1,\Gamma_1} \mid H \models \phi_{\text{dom}}\}$, and for every $H \in \text{dom}(f)$, $f(H)$ is the graph $(V, E, \lambda) \in GR_{\Sigma_2,\Gamma_2}$ such that

- $V = \{x \in V_H \mid$ there is exactly one $\sigma \in \Sigma_2$ such that $H \models \phi_\sigma(x)\}$
- $E = \{(x, \gamma, y) \mid x, y \in V, x \neq y, \gamma \in \Gamma_2,$ and $H \models \phi_\gamma(x, y)\}$, and
- for $x \in V$, $\lambda(x) = \sigma$ where $H \models \phi_\sigma(x)$.

Note that an MSO definable function f also works on abstract graphs. Thus, we will also view f as a function $[GR_{\Sigma_1,\Gamma_1}] \to [GR_{\Sigma_2,\Gamma_2}]$ with $\text{dom}(f) \subseteq [GR_{\Sigma_1,\Gamma_1}]$.

In Chapter 3, the MSO definable functions defined above are the non-copying (1,1)-definable transductions without parameters.

Many functions on graphs are MSO definable, for instance, node relabelings, edge relabelings (cf. Proposition 1.3.8), edge complement, transitive closure, Hasse diagram, and the function that maps a cotree into the cograph it represents (where a cotree of a cograph is the tree representing an expression, in terms of the operations of disjoint union and join, which denotes the graph, cf. Example 1.9(10) and [21]).

Example 1.14
(1) Consider the function f that is defined for every acyclic graph in $GR_{\Sigma,\Gamma}$ and computes its transitive closure in $GR_{\Sigma,\Gamma\cup\{\tau\}}$, where τ is used to label the new edges. To show that f is MSO definable, we take ϕ_{dom} to be $\neg(\exists u \exists v : \neg(u = v) \wedge \text{path}(u, v) \wedge \text{path}(v, u))$, expressing that the input graph H is acyclic. For every $\sigma \in \Sigma$, $\phi_\sigma(u)$ is $lab_\sigma(u)$, i.e., every node of H is a node of $f(H)$ and has the same label. Finally, $\phi_\gamma(u, v)$ is $\text{edge}_\gamma(u, v)$, for every $\gamma \in \Gamma$, and $\phi_\tau(u, v)$ is $\text{path}(u, v)$. Thus, the edges of H remain in $f(H)$, but $f(H)$ also contains all edges that correspond to paths in H.

(2) In this example we show that the function yield, defined on the p-labeled derivation trees of the edNCE grammar G_2 of Example 1.9(2), is MSO definable. Note that yield : $[GR_{\{a,b,c\},\{1,2\}}] \to [GR_{\{\#\},\{*\}}]$ and $\text{dom}(\text{yield}) = D(G_2)$. For an example of an input and an output graph, see Fig. 1.22(a) and (b), respectively. For ϕ_{dom} we take $\phi_{\text{bt}} \wedge \forall u : (\text{lab}_a(u) \leftrightarrow \text{root}(u)) \wedge (\text{lab}_c(u) \leftrightarrow \text{leaf}(u))$, where ϕ_{bt} is a formula expressing that the input graph is a rooted, ordered binary tree, $\text{root}(u)$ is $\neg\exists v : \text{edge}(v, u)$, $\text{leaf}(u)$ is $\neg\exists v : \text{edge}(u, v)$, and $\text{edge}(u, v)$ is $\text{edge}_1(u, v) \vee \text{edge}_2(u, v)$. For $\phi_\#(u)$ we take $u = u$ (which is always true), and for $\phi_*(u, v)$ we take $\text{edge}(u, v) \vee (\text{leaf}(u) \wedge \text{root}(v))$. This shows that the yield function of G_2 is MSO definable. It is easy to see that the same holds for G_3 and G_4. For G_5 we

take $\phi_f(u,v)$ to be $\text{left}(u,v) \wedge \neg\exists w : \text{left}(u,w) \wedge \text{left}(w,v)$, where $\text{left}(u,v)$ is $\text{leaf}(u) \wedge \text{leaf}(v) \wedge \exists u', v', z : \text{edge}_1(z,u') \wedge \text{edge}_2(z,v') \wedge \text{path}(u',u) \wedge \text{path}(v',v)$, i.e., $\text{left}(u,v)$ expresses that u and v are leaves such that u is to the left of v in the given (ordered) derivation tree. □

The logical characterization of C-edNCE is the following (presented in [47], proved in [87], and generalized in [27]): C-edNCE = MSOF(TREES), where MSOF(TREES) denotes the class of all graph languages $f(T_\Delta)$ where T_Δ is the set of all trees over some ranked alphabet Δ and f is an MSO definable function from $GR_{\Delta,\{1,\ldots,m_\Delta\}}$ to some $GR_{\Sigma,\Gamma}$ (for the definition of T_Δ and m_Δ, see Section 1.4.1). Let MSOF(STRINGS) denote the class of all $f(T_\Delta)$ as above, with the restriction that the symbols of Δ all have rank 1 or 0. Then LIN-edNCE = MSOF(STRINGS). To obtain a characterization of B-edNCE, we define B-MSOF(TREES) to be the class of all $f(T_\Delta)$ as above, with the restriction on f that for every edge formula $\phi_\gamma(u,v)$ and every input tree $t \in \text{dom}(f)$, $t \models \forall u,v : \phi_\gamma(u,v) \to (\text{path}(u,v) \vee \text{path}(v,u))$; as in the case of B-RPD, this means that graph edges are only established between tree vertices of which one is a descendant of the other. These results are summarized in the following theorem.

Theorem 1.4.6
C-edNCE = MSOF(TREES), B-edNCE = B-MSOF(TREES), and LIN-edNCE = MSOF(STRINGS). □

In [18] a similar characterization is shown for B_{nd} (i.e., HR), of another nature: in this case, both the nodes *and* the edges are defined to be (disjoint) subsets of the set of vertices of the input tree, and the incidence relation between nodes and edges is defined by an MSOL formula. For details see Section 5.6.3 of Chapter 3.

In one direction, the proof of Theorem 1.4.6 in [87] is based on the fact that the yield function of an edNCE grammar is MSO definable (cf. Example 1.14(2)). In fact, this is quite straightforward using the regular path characterization of Theorem 1.4.3: roughly speaking, the regular tree language can be turned into the domain formula of the MSO definable function (because regular tree languages are MSO definable), and the regular string language $W(\gamma)$ can be turned into the edge formula $\phi_\gamma(u,v)$ (because regular string languages are MSO definable, and because the shortest path from u to v can be expressed in MSOL). In the other direction the proof is also based on Theorem 1.4.3.

From Theorem 1.4.6 a lot of closure properties and decidability results can be deduced, which we will now discuss (see [47]). We start with *closure properties*. It is not hard to see that the class of MSO definable functions is closed under composition. Note also that for every MSO definable graph language L, the identity on L is an MSO definable function. This gives the following result.

Theorem 1.4.7
C-edNCE and LIN-edNCE are closed under MSO definable functions. In particular, they are closed under intersection with MSO definable sets of graphs. □

Since taking edge complement is MSO definable, B-edNCE is not closed under MSO definable functions (see the end of Section 1.4.1). However, B-edNCE *is* closed under intersection with MSO definable sets (because the composition of an MSO definable function f with an MSO definable identity only restricts the domain formula of f). By Theorem 1.3.28, A-edNCE is also closed under intersection with MSO definable sets.

The closure under intersection with MSO definable sets was first proved by Courcelle in Corollary 4.8 of [24] (for the case of HR), which was the basis for all further developments in this area. For C-edNCE it was first shown in Theorem 6.9 of [20]. We now show how this closure result leads to examples of graph languages that are *not* in C-edNCE. Let $G_{m,n}$ be the $m \times n$ *grid*, i.e., the graph with #-labeled nodes $u_{i,j}$, $1 \leq i \leq m$, $1 \leq j \leq n$, and all edges $(u_{i,j}, h, u_{i,j+1})$ and $(u_{i,j}, v, u_{i+1,j})$ (where h and v indicate 'horizontal' and 'vertical' edges, respectively). A *square grid* is a grid $G_{n,n}$. It is not difficult to show that the set of all grids is MSO definable: a grid is an acyclic, connected graph such that (1) each node has at most one incoming/outgoing horizontal/vertical edge, (2) if (x, h, y_1), (y_1, v, z_1), (x, v, y_2), and (y_2, h, z_2) are edges, then $z_1 = z_2$, and (3) if there is a horizontal edge from x to y or from y to x, and x has an outgoing (incoming) vertical edge, then y has an outgoing (incoming, respectively) edge; and similarly with 'horizontal' and 'vertical' interchanged. The set of all square grids is also MSO definable: a grid is square if there is a path π from the node with no incoming edges to the node with no outgoing edges such that the sequence of edge labels of π is in $(hv)^*$, i.e., π is a "diagonal". The following result was proved in Corollary 4.3 of [94] for B-NLC, and extended to C-edNCE in Theorem 1.5.16 of [96] (see also [14]); see Section 4 of [24] for HR.

Proposition 1.4.8
No C-edNCE graph language contains infinitely many square grids.

Proof
Let L be a C-edNCE language and let $Lg = L \cap \{[G_{n,n}] \mid n \geq 2\}$. By Theorem 1.4.7 and the MSO definability of the set of square grids, Lg is in C-edNCE. Hence, the string language $S(Lg) = \{a^m \mid m = \#V_H \text{ for some } [H] \in Lg\}$ is context-free (see the discussion before Proposition 1.3.22). Note that the number of nodes of $G_{n,n}$ is n^2. Hence, if Lg would be infinite, then $S(Lg)$ would be an infinite subset of the string language $\{a^{n^2} \mid n \geq 2\}$, which is impossible by the pumping lemma for context-free string languages. □

This implies immediately that the graph language $[GR_{\Delta,\Omega}]$ (with $\#\Omega \geq 2$) is not in C-edNCE (as we already know from Theorem 1.3.30). Similarly the following sets of graphs in $[GR_{\Delta,\Omega}]$ are not C-edNCE graph languages: all connected graphs, all planar graphs, all grids, all graphs of degree ≤ 4, and all chordal graphs (for this last set, add all edges from $u_{i,j}$ to $u_{i+1,j+1}$ to the grids). With some more effort the same results can be shown in the case that all the edges have the same label and also for the undirected case (cf. the Section 5.2.9 of Chapter 3).

In [96,94] the proof of Proposition 1.4.8 is by the separator theorem mentioned after Lemma 1.3.18. Since grids are of bounded degree, the separator theorem would imply that there is a constant c such that the grids in L can be separated into two parts of almost the same size by removing at most c edges. However, it is well known that $\Omega(\sqrt{n})$ edges need to be removed for dividing $G_{n,n}$ into two (see, e.g., [91]). In Theorems 15 and 16 of [39] it is proved in a similar way that no LIN-edNCE language contains infinitely many binary trees.

The MSO definable functions of [26] are more general in two ways (see Chapter 3). First, each node of the input graph is used to represent (at most) k nodes of the output graph, where k is fixed. As an example, the function f that maps each graph H into the disjoint union of H with itself is then MSO definable (with $k = 2$); clearly this function is not MSO definable in our sense, because $f(H)$ has more nodes than H. Second, by admitting free variables in the defining formulas, MSO definable relations are obtained. Theorems 1.4.6 and 1.4.7 are also valid for these generalized MSO definable relations. A special case of this is the following. For a graph H and $X \subseteq V_H$, we denote by $H\langle X\rangle$ the subgraph of H induced by X. Let $\phi(U)$ be an MSOL formula. We say that $H\langle X\rangle$ is an ***induced ϕ-subgraph*** of H if $H \models \phi(X)$.

Theorem 1.4.9
Let $\phi(U)$ be an MSOL formula. If L is in C-edNCE, then the set of all induced ϕ-subgraphs of L is in C-edNCE. The same holds for B-edNCE and LIN-edNCE. □

Note that C-edNCE is not closed under taking subgraphs: the set of all complete graphs is in C-edNCE, but the set of all graphs (which is its set of subgraphs) is not in C-edNCE (see Theorem 1.3.30). On the other hand, HR *is* closed under taking subgraphs; moreover, if L is a C-edNCE language that is not in HR, then the set of subgraphs of L is not in C-edNCE (see Theorem 3.12 of [18]). This is a characterization of HR within C-edNCE similar to Theorem 1.3.27 (see also [108]).

We now turn to *decidability properties*. First, Proposition 1.3.22 and Theorem 1.4.7 can be combined in an obvious way: it is decidable for an MSO definable function f and a C-edNCE language L whether or not $f(L)$ is empty (and, whether or not it is finite). In particular, it is decidable whether or not $L \cap R = \emptyset$, and whether or not $L \subseteq R$, where R is an MSO definable set (see [105,106] for complexity aspects). This last fact means that the MSO theory of L is decidable; since no set of graphs of which the MSO theory is decidable contains infinitely many square grids (see Proposition 5.2.2 of Chapter 3), this gives still another proof of Proposition 1.4.8. Second, a similar combination of Proposition 1.3.22 and Theorem 1.4.9 gives the following result. Let $\phi(U)$ be an MSOL formula and H a graph. By $\text{size}_\phi(H)$ we denote the maximal number of nodes of an induced ϕ-subgraph of H, and by $\text{num}_\phi(H)$ we denote the number of induced ϕ-subgraphs of H (where isomorphic ones are *not* identified).

Theorem 1.4.10
Let $\phi(U)$ be an MSOL formula. It is decidable for a C-edNCE grammar G
(1) whether or not there exists a natural number b such that $size_\phi(H) \leq b$ for all $[H] \in L(G)$, and
(2) whether or not there exists a natural number b such that $num_\phi(H) \leq b$ for all $[H] \in L(G)$. ∎

The decidability of boundedness of numerical functions on graphs was first investigated (for HR) in [57] (see Section 2.6.2 in Chapter 2). Their result (which can be extended to C-edNCE, see [49]) is formulated in terms of compatible functions rather than MSO formulas. It is shown (for HR) in [22] that for every MSOL formula $\phi(U)$, size_ϕ and num_ϕ are compatible functions.

As an example, since the connected components of a graph are induced ϕ-subgraphs (for an appropriate $\phi(U)$), it is decidable for a C-edNCE graph language whether or not there is a bound on the size of the connected components of its graphs, and also, whether or not there is a bound on the number of connected components of its graphs. As another example, the cliques of a graph are induced ϕ-subgraphs. Hence, it is decidable for a C-edNCE language whether there is a bound on the size of the cliques in its graphs. This was shown for B-NLC in [94]. By Theorem 1.3.27, HR languages always have bounded clique size. As a last example, since the subgraph induced by a node and all its neighbours is a ϕ-subgraph (for appropriate ϕ), it is decidable whether a C-edNCE graph language is of bounded degree (see Theorem 1.3.28). For (arbitrary) NLC grammars this was first proved in [69]. It was recently shown in [100] that the problem is in NLOG for reduced nonblocking (not necessarily confluent) edNCE grammars. Since it is easy to see that the set of sentential forms of a C-edNCE grammar is a C-edNCE language L, it is also decidable whether a C-edNCE grammar is of bounded nonterminal degree (see Section 1.3.3); if so, its nonterminal degree can be computed, because, for fixed k, the set R of all graphs of nonterminal degree $\leq k$ is MSO definable (and decide $L \subseteq R$). Decidability of the bounded bidegree property (see Theorem 1.3.27) follows from a strengthened version of Theorem 1.4.10(1), involving MSOL formulas ϕ with an arbitrary number of free variables, shown in [28] (see also Section 5.7.3 of Chapter 3).

1.4.3 Handle replacement

The C-edNCE languages can be generated by a different type of graph grammar that is based on handle rewriting: the *separated handle hypergraph grammar*, or S-HH grammar, introduced in [20]. Handle rewriting generalizes both node rewriting and edge rewriting: a *handle* is an edge together with its nodes. When a handle is rewritten, the edge, its nodes, and the edges incident with those nodes are removed, and, as usual, replaced by the right-hand side of the production. The S-HH grammar can be viewed as a generalization of the HR grammar: the graphs that it rewrites also have (nonterminal) hyperedges, but the embedding is simpler than the one of edNCE grammars.

Let us describe the S-HH grammar in some more detail. A (directed) *hypergraph* is a graph with hyperedges. A *hyperedge* (of rank n) is a tuple $(v_1, \dots, v_n, \gamma)$, where $v_1, \dots, v_n$ are nodes and γ is a label; for $n = 2$, it is an ordinary edge. Note that we do not need multiple hyperedges, with the same nodes and the same label (as in the hypergraphs of Chapter 2). The sentential forms of an S-HH grammar are hypergraphs of which the edges are labeled by terminal and nonterminal symbols (node labels are simulated by terminal hyperedges (v, γ) of rank 1); they should be separated, which means that nonterminal hyperedges do not have nodes in common; and, moreover, since we wish the generated language to consist of ordinary graphs, we require that all hyperedges of rank ≥ 3 are nonterminal. The productions of an S-HH grammar are of the form $X \rightarrow D$, where X is a (ranked) nonterminal, say of rank n, and D is a hypergraph of the type just described, together with a mapping 'port' from $\{1, \ldots, n\}$ to the subsets of V_D. Application of this production to a sentential form H results in a sentential form H' that is obtained as follows. First, a hyperedge $(v_1, \ldots, v_n, X)$ is removed, together with its nodes $v_1, \ldots, v_n$ and the edges incident with these nodes. Second, (a disjoint copy of) D is added to the remainder H^- of H. And, third, D is embedded in H^-: each ordinary edge (v, v_i, γ) of H (with v a node of H^-) is replaced by all edges (v, w, γ) of H', where w is a node of D such that $w \in \text{port}(i)$; and similarly, (v_i, v, γ) is replaced by all (w, v, γ) (note that there is *no* dynamic edge relabeling). Intuitively, the nodes w of D that are labeled with number i (i.e., for which $w \in \text{port}(i)$) are the new connection points for all edges that are incident with the ith node v_i of the replaced nonterminal hyperedge. Thus, the embedding is based on the connecting approach (see the Introduction). It should be clear (to the reader familiar with HR grammars) that an S-HH grammar with the property that for each production $X \to D$, $\text{port}(i)$ is a singleton for every i, is an HR grammar. In that case the embedding can also be realized by the gluing approach: remove the nonterminal hyperedge (only), add D, and identify v_i with the (unique) node in $\text{port}(i)$, for every i. However, only separated HR grammars are obtained in this way (which are less powerful than arbitrary HR grammars, see Lemma 7.6 of [20]).

It is shown in [20] that S-HH grammars generate exactly the C-edNCE languages. An S-HH grammar is *boundary* if, in the right-hand sides of the productions, there is no edge (v, w, γ) such that v belongs to one nonterminal hyperedge and w to another (i.e., there is no connection between nonterminal handles). The boundary S-HH grammars generate the B-edNCE languages. The *linear* S-HH grammars generate the LIN-edNCE languages (where, as usual, an S-HH grammar is linear if there is at most one nonterminal hyperedge in each right-hand side of a production). It is open whether there is a natural characterization of the B_{nd}-edNCE (= HR!) graph languages. It should be clear that S-HH grammars can also be used to generate hypergraph languages. Unfortunately the classes of hypergraph languages generated by S-HH and HR grammars are incomparable, even for hypergraphs without multiple edges (see Theorem 7.5 of [20]). It would be nice to find a natural variation of the S-HH grammar that is able to generate all HR hypergraph languages. Another possibility is to extend edNCE rewriting to hypergraphs (as in [74]); this gives a more complicated embedding mechanism than the one of the S-HH grammar.

1.4.4 Graph expressions

As observed in Sections 1.3.2 and 1.4.1, the set $D(G)$ of abstract p-labeled derivation trees of an edNCE grammar G is a regular tree language, where each tree is a *graph expression*, i.e., an expression that denotes a graph when viewing each vertex label (i.e., each production of G) as an operation on graphs. This is a special case of a general approach (introduced in [83]) in which regular tree languages are used to define subsets of arbitrary algebras (see Section 1.4.1 or [55] for the definition of regular tree languages). To be precise, if Δ is a ranked alphabet, and A is an algebra which has an operation $\sigma_A : A^k \to A$ for every $\sigma \in \Delta$ of rank k, then every tree $t \in T_\Delta$ can be viewed as an expression that denotes an element $\text{val}(t)$ of A. A subset L of A is said to be *equational* if $L = \text{val}(R)$ for some regular tree language $R \subseteq T_\Delta$. It would also be appropriate to call them the "context-free" subsets of the algebra; the term "equational" comes from viewing the regular tree grammar that generates R as a system of equations of which $\text{val}(R)$ is the least fixed point (as shown in [83]). For graphs this approach was introduced in [8]. See Section 5.6 of Chapter 3, and [49], for more information.

If one now considers the algebra $AGRE$ of all (abstract) graphs with embedding, with an operation p_{AGRE} for every possible production p, then the considerations in Sections 1.3.2 and 1.4.1 show that C-edNCE is precisely the class of equational sets of graphs in this algebra. Since the operations p_{AGRE} are rather complicated substitution operations on graphs with embedding, it would be nice to have simpler operations on a simpler type of graphs. The existence of such simple operations was shown in [8] for HR grammars and in [20] for C-edNCE grammars. We now define the operations for C-edNCE.

Let $AGRP$ be the set of (abstract) *graphs with ports*, where a graph with ports is a graph H together with a finite set port_H of pairs (i, v), where i is a natural number and $v \in V_H$, cf. the right-hand sides of productions of S-HH grammars. The set $AGRP$ is made into an algebra with the following operations. First of all, it has the empty graph and all graphs with a single node as constants. Second, it has the (usual) *disjoint union* $H_1 + H_2$, where one just takes the union of the port sets. Third, it has operations of *edge creation*: for every edge label γ and natural numbers i and j, $\eta_{i,\gamma,j}(H)$ is the graph that is obtained from H by the addition of all edges (v, γ, w) such that $(i, v), (j, w) \in \text{port}_H$. Finally, to be able to redefine the port numbers of nodes, it has operations of *port selection*: for every finite set z of pairs (i, j) of natural numbers, $\pi_z(H)$ is obtained from H by changing its port set into $\{(i, v) \mid (i, j) \in z \text{ and } (j, v) \in \text{port}_H \text{ for some } j\}$. The operation of edge creation is typical for the connecting approach to embedding (see the Introduction): graphs can be built by putting them together (using disjoint union) and then connecting them by edges using edge creation.

It is shown in [20] that C-edNCE is the class of all equational sets of graphs in the algebra *AGRP* defined above. It is open whether similar characterizations exist for B-edNCE and LIN-edNCE. For HR a different set of operations (mainly based on gluing) is given in [8]. For a uniform approach to these two sets of operations see [49] (where also a characterization of A-edNCE can be found).

1.5 Recognition

In this final section we discuss the complexity of the recognition problem (or membership problem) for edNCE graph languages L. This is the problem, for a given (abstract) graph H, whether or not H belongs to L. As usual, we identify L with its recognition problem.

We first consider arbitrary edNCE languages. Every edNCE language is generated by an edNCE grammar without Λ-productions (because the removal of these productions, as discussed in Lemma 1.3.29, also works for arbitrary edNCE grammars). In a derivation of a grammar without Λ-productions the number of nodes of the sentential forms cannot decrease. Guessing such a derivation nondeterministically needs at most space n^2, where n is the number of nodes of the given graph. This means that every edNCE language can be recognized in NSPACE(n^2) and hence belongs to PSPACE ([11]). Since, as we have observed in Section 1.3.1, blocking edges can be used to simulate context-sensitive graph grammars (see [86]) and in particular context-sensitive string grammars (see Theorem 9 of [61]), there exist PSPACE-complete edNCE languages (see [13,14]). This stresses again the non-context-free nature of arbitrary edNCE languages (because one would expect recognition in polynomial time for context-free grammars).

We now turn to C-edNCE languages. By Lemma 1.3.29 every C-edNCE language can be generated by a grammar without Λ- or chain productions. Hence (as observed in the proof of Theorem 1.3.30) every derivation of a graph with n nodes has length at most $2n$. This implies that every C-edNCE language is in NPTIME. Surprisingly, in comparison with context-free string languages, there exist quite simple NP-complete C-edNCE (and even C-NLC) languages, as shown independently in [92,79] and [5] (and first shown in [13,102] for non-confluent NLC grammars without Λ- or chain productions). With 'simple' we mean here that they can be generated by linear apex edNCE (and hence HR) grammars.

Proposition 1.5.1
There is an NP-complete LIN·A-edNCE graph language.

Proof
In [92,102] the NP-complete problem whether a graph H has circular bandwidth ≤ 2 is used (which means that the nodes of H can be linearly ordered as $v_1, \ldots, v_n$ such that $|j - i| \leq 2 \bmod n$ for every $(v_i, v_j) \in E_H$). It is not difficult to write a linear apex edNCE grammar for the set of all graphs of circular bandwidth ≤ 2 (cf. the linear B-NLC grammar in Example 4 of [92] and the linear apex HR grammar in Example 2.2.4 of [77]). In [5] a graph language is given to which the strongly NP-complete 3-partition problem can be reduced (see also the proof of Theorem 3.5 in [1]). This language consists of all graphs with a-labeled nodes $x_1, \ldots, x_n$ and b-labeled nodes $y_1, \ldots, y_n$ (for some $n \geq 1$), and a set E of edges such that $E \subseteq \{(x_i, *, x_{i+1}) \mid 1 \leq i \leq n-1\} \cup \{(y_i, *, y_{i+1}) \mid 1 \leq i \leq n-1\}$ and if $(x_i, *, x_{i+1}) \in E$, then $(y_i, *, y_{i+1}) \in E$. It is generated by the linear apex edNCE (NLC) grammar G with productions $S \to (D, \emptyset)$, $X \to (D, C_1)$, $X \to (D, C_2)$, $X \to (D, \emptyset)$, and $X \to (\Lambda, \emptyset)$, where $D = (V, E, \lambda)$ with $V = \{x, y, z\}$, $E = \{(x, *, z), (y, *, z)\}$, $\lambda(x) = a$, $\lambda(y) = b$, $\lambda(z) = X$, and where $C_1 = \{(a, */*, x, \text{in}), (b, */*, y, \text{in})\}$, $C_2 = \{(b, */*, y, \text{in})\}$, and Λ is the empty graph. Note that the nonterminal degree of G is 2. $\square$

The graphs in the NP-complete graph languages mentioned in the proof above are of bounded degree, but they are not connected. By starting with one additional node, and connecting all other nodes to it, it is easy to obtain NP-complete LIN·B_{nd}-edNCE languages of which the graphs are connected, but not any more of bounded degree.

Proposition 1.5.1 has led to a search for large classes of context-free graph languages that can be recognized in deterministic polynomial time. A natural requirement that guarantees PTIME recognition is that all graphs of the language are connected and of bounded degree. Such a result was first shown in [98]. For B-NLC languages it was proved in [92]. It was then extended to B-edNCE in [44] and to C-edNCE in [11,96], and for HR languages it was shown in [77]. Since we now know that all C-edNCE languages of bounded degree are in HR (see Theorems 1.3.27 and 1.3.28), these latter results are in fact all equivalent. However, it should be noted that, in these papers, the graphs are not only recognized, but also parsed, i.e., a derivation tree is produced as output. From the parsing point of view the algorithms are, of course, not equivalent. In [44,77] it was shown (using the closure of NLOG under complement) that the graph languages considered are even in LOG(CFL), i.e., can be reduced to a context-free string language in logarithmic space. By a result of Ruzzo, LOG(CFL) is the class of problems that can be recognized by an alternating Turing machine that uses logarithmic space and has a computation tree of polynomial size. The graph language can be recognized by such a machine in such a way that the computation tree corresponds to the derivation tree of the input graph. Thus, if, in particular, the graph grammar is linear, then the machine is a nondeterministic log-space Turing machine. Hence the linear versions of the above graph languages are in NLOG (as first shown in [1,39]). In [77] these results were extended to certain graph languages that are not both connected and of bounded degree. We first state the results, and then the extensions.

Proposition 1.5.2
Every C-edNCE graph language L of which the graphs are connected and of bounded degree is in LOG(CFL), and hence also in PTIME. If, moreover, L is a LIN-edNCE graph language, then L is in NLOG. □

Since (representatives of) the connected components of a graph can be enumerated in nondeterministic logarithmic space, this proposition also holds for C-edNCE languages L of bounded degree such that a graph H is in L iff all connected components of H are in L (such as, for fixed k, the set of all graphs of bandwidth $\leq k$, of binary tree band-width $\leq k$, or of cutwidth $\leq k$).

We now turn to the first extension of [77] (see Section 2.7.2 of Chapter 2 for more details). For $r \geq 1$, a graph language L is *of bounded r-separability* if there exists $k \geq 1$ such that for every $H \in L$ and every subset V of V_H with $\#V = r$, the subgraph of H induced by $V_H - V$ has at most k connected components. Intuitively this means that if one takes away r nodes from a graph of the language, it falls apart into at most k pieces. The set of all stars is not of bounded r-separability for any r (taking away the center node gives arbitrarily many pieces). The set of all wheels (see grammar G_9 of Example 1.9) is of bounded r-separability for every r: taking away r nodes gives at most $r - 1$ pieces. Note that every graph language of which the graphs are connected and of bounded degree is of bounded r-separability for every r; in fact, if the degree is at most m, then the number of pieces is at most $1 + r(m - 1)$.

Recall from Section 1.3.3 that the nonterminal degree of a C-edNCE grammar is the maximal degree of the nonterminal nodes in its sentential forms. For a B_{nd}-edNCE (= HR) grammar it is the maximal number of tentacles of the nonterminals of the corresponding HR grammar. The following result was shown in Theorem 4.2.1 of [77].

Theorem 1.5.3
If G is a B-edNCE grammar of nonterminal degree $\leq r$ and $L(G)$ is of bounded r-separability, then $L(G)$ is in LOG(CFL), and hence also in PTIME. If, moreover, G is a LIN-edNCE grammar, then $L(G)$ is in NLOG.

Proof
(Sketch) The algorithm is similar to the usual Cocke-Younger-Kasami algorithm for context-free string grammars. For a given input graph H it computes all triples (X, K, F) where X is a nonterminal, K is a graph with exactly one nonterminal node v, with label X, and F is an induced subgraph of H such that, for some connection relation C_F, $sin(X, x) \Rightarrow^* (F, C_F)$ and $K[v/(F, C_F)] = H$. The main problem is how to represent such a triple in logarithmic space. Clearly, if F is given, then K is determined by v, the neighbours of v, and the edges between them; since G is of nonterminal degree $\leq r$, v has at most r neighbours, and so K can be stored in logarithmic space. Since, by the bounded r-separability of $L(G)$, removing the neighbours of v from H leads to at most k connected components, for some fixed k, F consists of a number of these components and can be represented by one node from each component, also in logarithmic space. □

Bounded r-separability of $L(G)$ is decidable: by Theorem 1.4.9 one obtains a grammar G' for the set of all induced subgraphs that are obtained from graphs of $L(G)$ by removing r nodes, and by Theorem 1.4.10(2) it can be decided whether there is a bound on the number of connected components of the graphs in $L(G')$.

Both Proposition 1.5.2 and Theorem 1.5.3 are optimal, in the following sense. Since, coding strings as chains, all context-free string languages can be generated by HR grammars (see Section 2.5.2 of Chapter 2), there are LOG(CFL)-complete C-edNCE graph languages of which all graphs are connected and of bounded degree. It is shown in Theorem 4.9 of [1] that there exist NLOG-complete LIN$\cdot$B$_{\mathrm{nd}}$-edNCE graph languages of which all graphs are connected and of bounded degree (see also Theorem 4.3 of [73]).

It is not clear whether Theorem 1.5.3 also holds for C-edNCE grammars (with the same restrictions). It certainly does not hold without the restriction of bounded nonterminal degree. In fact, let L be the language that is obtained from any NP-complete C-edNCE language (see Proposition 1.5.1) by adding a γ-labeled edge (where γ is a new label) between all nodes of all graphs. Then L is clearly still NP-complete, it is a C-edNCE language by Theorem 1.4.7 (because the function that adds the edges is MSO definable), and it is obviously of bounded r-separability for all r.

The PTIME result of Theorem 1.5.3 also holds if $L(G)$ is of "logarithmically bounded" r-separability, which means that the number of "pieces" is at most $k \cdot \log n$, where n is the number of nodes of the given graph H.

A second general LOG(CFL) recognition result is proved in [77] that covers HR languages such as the set of all discrete graphs and the set of all graphs of tree-width $\leq k$, for fixed k. A *connected component* of a graph with embedding (H, C_H) is, as one would expect, a graph with embedding (K, C_K) where K is a connected component of H and C_K is the restriction of C_H to K, i.e., $C_K = \{(\sigma, \beta/\gamma, x, d) \in C_H \mid x \in V_K\}$. An edNCE grammar $G = (\Sigma, \Delta, \Gamma, \Omega, P, S)$ has *componentwise derivations* if for every nonterminal $X \in \Sigma - \Delta$ and every terminal graph $H \in GRE_{\Delta,\Gamma}$ (and every x), $sin(X, x) \Rightarrow^* H$ if and only if $sin(X, x) \Rightarrow^* K$ for every connected component K of H. In other words, for each of the graph languages $L(G, X)$, defined after Proposition 1.3.20, H is in $L(G, X)$ iff all connected components of H are in $L(G, X)$. The following result was shown in Theorem 4.2.2 of [77].

Theorem 1.5.4
If G is a B_{nd}-edNCE grammar that has component-wise derivations, then $L(G)$ is in LOG(CFL), and hence also in PTIME. If, moreover, G is a LIN-edNCE grammar, then $L(G)$ is in NLOG. □

There is a B_{nd}-edNCE grammar with component-wise derivations that generates the set of all graphs of tree-width $\leq k$, for fixed k. Hence, by Theorem 1.5.4, the set of graphs of tree-width $\leq k$ is in LOG(CFL), see [77]. For similar reasons the set of graphs of path-width $\leq k$ is in NLOG. It is open whether it is decidable whether or not a B_{nd}-edNCE grammar has component-wise derivations.

The proof of Theorem 1.5.3 only depends on the fact that there is a fixed k such that, for every nonterminal X, each graph in $L(G, X)$ has at most k connected components. Thus, with the techniques from [77], it is not difficult to see that Theorems 1.5.3 and 1.5.4 can be combined as follows. Let G be a B_{nd}-edNCE grammar such that for every nonterminal X, either the graphs of $L(G, X)$ have a bounded number of connected components, or H is in $L(G, X)$ iff all connected components of H are in $L(G, X)$. Then $L(G)$ is in LOG(CFL) (and in NLOG if G is also linear). This covers for instance the LIN·B_{nd}-edNCE grammar G_8 of Example 1.9 that generates the set of all stars. In fact, for that grammar, $L(G_8, S) = L(G_8)$ is the set of all stars, which satisfies the first requirement, and $L(G_8, X)$ is the set of all discrete graphs H with connection relation $\{(\#, */*, x, d) \mid x \in V_H, d \in \{\text{in}, \text{out}\}\}$, which satisfies the second requirement. Hence the set of all stars is in NLOG (which is of course obvious).

A completely different way to obtain polynomial time recognition is to assume that the generation order of the nodes of the input graph (i.e., the linear order described in Section 1.3.2) is given a priori. This means that one views a C-edNCE grammar as generating a set of (abstract) ordered graphs, rather than graphs. Let us call this the *ordered recognition problem*. With each C-edNCE grammar G one can easily associate a new C-edNCE grammar G' such that G' generates the graphs in $L(G)$ to which the generation order of the nodes is explicitly added, as a chain of edges with a new edge label. Clearly, this reduces the ordered recognition problem of G to the (ordinary) recognition problem of G'. Since the resulting language $L(G')$ is of bounded r-separability for every r (because chains are), and since the construction preserves $\mathrm{B_{nd}}$-edNCE grammars, it follows from Theorem 1.5.3 that the ordered recognition problem for $\mathrm{B_{nd}}$-edNCE (= HR) grammars is in LOG(CFL), and hence in PTIME. It seems to be unknown, in general, whether the ordered recognition problem for C-edNCE grammars is in PTIME. The problem has been considered for some special cases, obtained by restricting the embedding. In Theorem 4 of [12] it was shown to be in PTIME for neighborhood uniform dNCE grammars (in which edge labels are allowed, but should not be changed by the embedding), and in Theorem 5.3 of [1] it was shown to be in DLOG for linear dNLC grammars in Chomsky Normal Form such that the nonterminal nodes have incoming edges only (called "regular" dNLC grammars). In other cases, the order of the input graph is not given explicitly, but can be (partially) obtained from the structure of the graph itself. Results of this type are shown in Section 3 of [96], Section 5 of [1], and Theorem 4.5 of [73]. In [103] it is shown for an edge replacement grammar generating 2-connected graphs, that the derivation tree can be obtained from the structure of the input graph, leading to recognition in $O(n^3)$ time (see Section 2.7.2 of Chapter 2 for details).

All the above-mentioned PTIME algorithms recognize the input graph by a bottom-up dynamic programming technique that is similar to the usual Cocke-Younger-Kasami algorithm for context-free string grammars. Since the recognition is syntax-directed, the algorithms in fact parse the input graph, i.e., produce its possible derivation trees. In the above, the concern is with PTIME parsing methods that are as general as possible. Efficient parsing methods of context-free graph grammars (generalizing the usual methods for context-free string grammars) are outside the scope of this chapter; we refer the reader to the literature, in particular to the precedence parsing of C-edNCE grammars in $O(n^2)$ time of [70,71].

In [16] edNCE grammars are enriched by a layout component that specifies constraints on the drawing of the generated graphs, i.e., their embedding in a grid. For such a *layout graph grammar* G, a polynomial time algorithm is given that produces, for a given graph $H \in L(G)$, a nice drawing that satisfies the given constraints. The algorithm is based on a PTIME parsing algorithm (for a subclass of C-edNCE grammars) as discussed above, and uses an attribute grammar that computes, in polynomial time, for each vertex v of the derivation tree t, the coordinates of the nodes u of H that "belong" to v, i.e., such that $\text{vertex}_t(u) = v$ (see the discussion after Definition 1.3.13). The resulting drawing of the input graph is "nice" because it is syntax-directed (and hence well structured, assuming that G is well structured) and because the algorithm produces a syntax-directed layout that is optimal with respect to certain measures such as area or maximal edge length. Such an algorithm is implemented in the HiGraD tool (see [7]) which is part of the Graph^{Ed} editor that draws and manipulates graphs and edNCE graph grammars (see [56]).

Acknowledgment

The authors are grateful to Bruno Courcelle, Dirk Janssens, and Hans-Jörg Kreowski for their comments on a previous version of this chapter.

References

[1] IJ.J.Aalbersberg, J.Engelfriet, G.Rozenberg; The complexity of regular DNLC graph languages, J. of Comp. Syst. Sci. 40 (1990), 376-404

[2] N.Abe, M.Mizumoto, J.-I.Toyoda, K.Tanaka; Web grammars and several graphs, J. of Comp. Syst. Sci. 7 (1973), 37-65

[3] K.Aizawa, A.Nakamura; Graph grammars with path-controlled embedding, Theor. Comput. Sci. 88 (1991), 151-170

[4] IJ.J.Aalbersberg, G.Rozenberg; Traces, dependency graphs and DNLC grammars, Discrete Appl. Math. 11 (1985), 299-306

[5] IJ.J.Aalbersberg, G.Rozenberg, A.Ehrenfeucht; On the membership problem for regular DNLC grammars, Discrete Appl. Math. 13 (1986), 79-85

[6] M.Bauderon; A uniform approach to graph rewriting: the pullback approach, in *Graph-Theoretic Concepts in Computer Science*, WG'95 (M.Nagl, ed.), Lecture Notes in Computer Science 1017, Springer-Verlag, Berlin, 1995, pp.101-115

[7] W.Bachl, F.-J.Brandenburg, T.Hickl; Hierarchical graph design using HiGraD, Technical Report MIP-9405, Universität Passau, 1994

[8] M.Bauderon, B.Courcelle; Graph expressions and graph rewritings, Math. Systems Theory 20 (1987), 83-127

[9] J.M.Brayer, K.S.Fu; Some properties of web grammars, Techn. Report TR-EE 74-19, Purdue University, 1974

[10] B.Bollobás; *Extremal Graph Theory*, Academic Press, London, 1978

[11] F.-J.Brandenburg; The computational complexity of certain graph grammars, Proc. 6th GI Conference on Theoretical Computer Science, Lecture Notes in Computer Science 145, Springer-Verlag, Berlin, 1983, pp.91-99

[12] F.-J.Brandenburg; On the complexity of the membership problem of graph grammars, in *Graph-Theoretic Concepts in Computer Science*, WG'83 (M.Nagl, J.Perl, eds.), Tauner-Verlag, Linz, 1983, pp.40-49

[13] F.-J.Brandenburg; On partially ordered graph grammars, in [50], 99-111

[14] F.-J.Brandenburg; On polynomial time graph grammars, Proc. STACS 88, Lecture Notes in Computer Science 294, Springer-Verlag, Berlin, 1988, pp.227-236

[15] F.-J.Brandenburg; The equivalence of boundary and confluent graph grammars on graph languages of bounded degree, in *Rewriting Techniques and Applications* (R.V.Book, ed.), Lecture Notes in Computer Science 488, Springer-Verlag, Berlin, 1991, pp.312-322

[16] F.-J.Brandenburg; Layout graph grammars: the placement approach, in [37], 144-156

[17] F.-J.Brandenburg; *Graph Grammatiken*, Lecture Notes (in German), Passau University, 1994

[18] B.Courcelle, J.Engelfriet; A logical characterization of the sets of hypergraphs defined by hyperedge replacement systems, Math. Systems Theory 28 (1995), 515-552

[20] B.Courcelle, J.Engelfriet, G.Rozenberg; Handle-rewriting hypergraph grammars, J. of Comp. Syst. Sci. 46 (1993), 218-270

[21] D.G.Corneil, H.Lerchs, L.Stewart Burlingham; Complement reducible graphs, Discrete Appl. Math. 3 (1981), 163-174

[22] B.Courcelle, M.Mosbah; Monadic second-order evaluations on tree decomposable graphs, Theor. Comput. Sci. 109 (1993), 49-82

[23] B.Courcelle; An axiomatic definition of context-free rewriting and its application to NLC graph grammars, Theor. Comput. Sci. 55 (1987), 141-181; Correction:
http://www.labri.u-bordeaux.fr/LaBRI/People/courcell.html

[24] B.Courcelle; The monadic second-order logic of graphs I: Recognizable sets of finite graphs, Inform. and Comput. 85 (1990), 12-75

[25] B.Courcelle; The monadic second-order logic of graphs III: Tree-decompositions, minors and complexity issues, RAIRO Theoretical Informatics and Applications 26 (1992), 257-286

[26] B.Courcelle; The monadic second-order logic of graphs V: On closing the gap between definability and recognizability, Theor. Comput. Sci. 80 (1991), 153-202

[27] B.Courcelle; The monadic second-order logic of graphs VII: Graphs as relational structures, Theor. Comput. Sci. 101 (1992), 3-33

[28] B.Courcelle; Structural properties of context-free sets of graphs generated by vertex replacement, Inform. and Comput. 116 (1995), 275-293

[31] J.Engelfriet, L.M.Heyker; Hypergraph languages of bounded degree, J. of Comp. Syst. Sci. 48 (1994), 58-89

[33] J.Engelfriet, L.M.Heyker, G.Leih; Context-free graph languages of bounded degree are generated by apex graph grammars, Acta Informatica 31 (1994), 341-378

[34] J.Engelfriet, T.Harju, A.Proskurowski, G.Rozenberg; Characterization and complexity of uniformly non-primitive labeled 2-structures, Theor. Comput. Sci. 154 (1996), 247-282

[36] H.Ehrig, D.Janssens, H.-J.Kreowski, G.Rozenberg; Concurrency of node-label controlled graph transformations, Report 82-38, University of Antwerp, U.I.A., 1982

[37] H.Ehrig, H.-J.Kreowski, G.Rozenberg (eds.); *Graph-Grammars and their Application to Computer Science*, Lecture Notes in Computer Science 532, Springer-Verlag, Berlin, 1991

[38] J.Engelfriet, G.Leih; Nonterminal bounded NLC graph grammars, Theor. Comput. Sci. 59 (1988), 309-315

[39] J.Engelfriet, G.Leih; Linear graph grammars: power and complexity, Inform. and Comput. 81 (1989), 88-121

[40] J.Engelfriet, G.Leih; Complexity of boundary graph languages, RAIRO Theoretical Informatics and Applications 24 (1990), 267-274

[41] J.Engelfriet, G.Leih, G.Rozenberg; Apex graph grammars, in [50], 167-185

[42] J.Engelfriet, G.Leih, G.Rozenberg; Apex graph grammars and attribute grammars, Acta Informatica 25 (1988), 537-571

[43] J.Engelfriet, G.Leih, G.Rozenberg; Nonterminal separation in graph grammars, Theor. Comput. Sci. 82 (1991), 95-111

[44] J.Engelfriet, G.Leih, E.Welzl; Boundary graph grammars with dynamic edge relabeling, J. of Comp. Syst. Sci. 40 (1990), 307-345

[45] A.Ehrenfeucht, M.G.Main, G.Rozenberg; Restrictions on NLC graph grammars, Theor. Comput. Sci. 31 (1984), 211-223

[46] J.Engelfriet; Context-free NCE graph grammars, Proc. FCT '89, Lecture Notes in Computer Science 380, Springer-Verlag, Berlin, 1989, pp.148-161

[47] J.Engelfriet; A characterization of context-free NCE graph languages by monadic second-order logic on trees, in [37], 311-327

[48] J.Engelfriet; A Greibach normal form for context-free graph grammars, Proc. ICALP '92 (W.Kuich, ed.), Lecture Notes in Computer Science 623, Springer-Verlag, Berlin, 1992, pp.138-149

[49] J.Engelfriet; Graph grammars and tree transducers, Proc. CAAP'94 (S.Tison, ed.), Lecture Notes in Computer Science 787, Springer-Verlag, Berlin, 1994, pp.15-36

[50] H.Ehrig, M.Nagl, G.Rozenberg, A.Rosenfeld (eds.); *Graph-Grammars and their Application to Computer Science*, Lecture Notes in Computer Science 291, Springer-Verlag, Berlin, 1987

[51] J.Engelfriet, V.van Oostrom; Regular description of context-free graph languages, Report 95-34, Leiden University, October 1995, to appear in J. of Comp. Syst. Sci.

[52] J.Engelfriet, G.Rozenberg; A comparison of boundary graph grammars and context-free hypergraph grammars, Inform. and Comput. 84 (1990), 163-206

[53] J.Engelfriet, G.Rozenberg; Graph grammars based on node rewriting: an introduction to NLC graph grammars, in [37], pp.12-23

[54] H.J.Genrich, D.Janssens, G.Rozenberg, P.S.Thiagarajan; Generalized handle grammars and their relation to Petri nets, EIK 4 (1984), 179-206

[55] F.Gécseg, M.Steinby; *Tree automata*, Akadémiai Kiadó, Budapest, 1984

[56] M.Himsolt; Graph^{Ed}: an interactive tool for developing graph grammars, in [37], 61-65

[57] A.Habel, H.-J.Kreowski, W.Vogler; Decidable boundedness problems for sets of graphs generated by hyperedge replacement, Theor. Comput. Sci. 89 (1991), 33-62

[58] J.Hoffmann, M.G.Main; Results on NLC grammars with one-letter terminal alphabets, Theor. Comput. Sci. 73 (1990), 279-294

[59] J.Jeffs; Order independent NCE grammars recognized in polynomial time, Inf. Proc. Letters 39 (1991), 161-164

[60] D.Janssens, G.Rozenberg; On the structure of node-label-controlled graph languages, Information Sciences 20 (1980), 191-216

[61] D.Janssens, G.Rozenberg; Restrictions, extensions, and variations of NLC grammars, Information Sciences 20 (1980), 217-244

[62] D.Janssens, G.Rozenberg; Decision problems for node label controlled graph grammars, J. of Comp. Syst. Sci. 22 (1981), 144-177

[63] D.Janssens, G.Rozenberg; A characterization of context-free string languages by directed node-label controlled graph grammars, Acta Informatica 16 (1981), 63-85

[64] D.Janssens, G.Rozenberg; Graph grammars with neighbourhood-controlled embedding, Theor. Comput. Sci. 21 (1982), 55-74

[65] D.Janssens, G.Rozenberg; A survey of NLC grammars, Proc. CAAP'83, Lecture Notes in Computer Science 159, Springer-Verlag, Berlin, 1983, pp.114-128

[66] D.Janssens, G.Rozenberg; Neighborhood-uniform NLC grammars, Computer Vision, Graphics, and Image Processing 35 (1986), 131-151

[67] D.Janssens, G.Rozenberg; Actor grammars, Math. Systems Theory 22 (1989), 75-107

[68] D.Janssens, G.Rozenberg, R.Verraedt; On sequential and parallel node-rewriting graph grammars, Computer Graphics and Image Processing 18 (1982), 279-304

[69] D.Janssens, G.Rozenberg, E.Welzl; The bounded degree problem for NLC grammars is decidable, J. of Comp. Syst. Sci. 33 (1986), 415-422

[70] M.Kaul; *Syntaxanalyse von Graphen bei Präzedenz-Graph-Grammatiken*, Dissertation, Universität Osnabrück, 1985

[71] M.Kaul; Practical applications of precedence graph grammars, in [50], 326-342

[72] C.Kim, D.H.Lee; Separating k-separated eNCE graph languages, Theor. Comput. Sci. 120 (1993), 247-259

[73] C.Kim, D.H.Lee; Node replacement graph languages squeezed with chains, trees, and forests, University of Oklahoma, 1993, to appear in Inform. and Comput.

[74] R.Klempien-Hinrichs; Node replacement in hypergraphs: simulation of hyperedge replacement, and decidability of confluence, in Graph-Grammars and their Application to Computer Science, (J.Cuny, H.Ehrig, G.Engels, G.Rozenberg, eds.), Lecture Notes in Computer Science 1073, Springer-Verlag, Berlin, 1996, pp.397-411

[75] H.-J.Kreowski, G.Rozenberg; Note on node-rewriting graph grammars, Inf. Proc. Letters 18 (1984), 21-24

[76] C.Lautemann; Efficient algorithms on context-free graph languages, Proc. 15th ICALP (T.Lepistö, A.Salomaa, eds.), Lecture Notes in Computer Science 317, Springer-Verlag, Berlin, 1988, pp.362-378

[77] C.Lautemann; The complexity of graph languages generated by hyperedge replacement, Acta Informatica 27 (1990), 399-421

[78] U.Lichtblau; Recognizing rooted context-free flowgraph languages in polynomial time, in [37], 538-548

[79] K.-J.Lange, E.Welzl; String grammars with disconnecting or a basic root of the difficulty in graph grammar parsing, Discrete Appl. Math. 16 (1987), 17-30

[80] U.G.Montanari; Separable graphs, planar graphs and web grammars, Inform. and Control 16 (1970), 243-267

[81] M.G.Main, G.Rozenberg; Handle NLC grammars and r.e. languages, J. of Comp. Syst. Sci. 35 (1987), 192-205

[82] M.G.Main, G.Rozenberg; Edge-label controlled graph grammars, J. of Comp. Syst. Sci. 40 (1990), 188-228

[83] J.Mezei, J.B.Wright; Algebraic automata and context-free sets, Inform. and Control 11 (1967), 3-29

[84] M.Nagl; Formal languages of labelled graphs, Computing 16 (1976), 113-137

[85] M.Nagl; A tutorial and bibliographical survey of graph grammars, in *Graph Grammars* (V.Claus, H.Ehrig, G.Rozenberg, eds.); Lecture Notes in Computer Science 73, Springer-Verlag, Berlin, 1980, pp.70-126

[86] M.Nagl; *Graph-Grammatiken*, Vieweg, Braunschweig, 1979

[87] V.van Oostrom; Graph grammars and 2nd order logic (in Dutch), M. Sc. Thesis, Leiden University, January 1989

[88] J.L.Pfaltz, A.Rosenfeld; Web grammars, Proc. Int. Joint Conf. Art. Intelligence, Washington, 1969, pp.609-619

[89] A.Rosenfeld, D.Milgram; Web automata and web grammars, Mach. Intelligence 7 (1972), 307-324

[90] H.-J.Röder; Parallel BNLC graph grammars, in *Developments in Language Theory* (G.Rozenberg, A.Salomaa, eds.), World Scientific, Singapore, 1994, pp.438-449

[91] A.L.Rosenberg, L.Snyder; Bounds on the cost of data encodings, Math. Systems Theory 12 (1978), 9-39

[92] G.Rozenberg, E.Welzl; Boundary NLC graph grammars - basic definitions, normal forms, and complexity, Inform. and Control 69 (1986), 136-167

[93] G.Rozenberg, E.Welzl; Graph theoretic closure properties of the family of boundary NLC graph languages, Acta Informatica 23 (1986), 289-309

[94] G.Rozenberg, E.Welzl; Combinatorial properties of boundary NLC graph languages, Discrete Appl. Math. 16 (1987), 59-73

[95] H.J.Schneider; Chomsky-Systeme für partielle Ordnungen, Arbeitsber. d. Inst. f. Math. Masch. u. Datenver. 3, Erlangen, 1970

[96] R.Schuster; Graphgrammatiken und Grapheinbettungen: Algorithmen und Komplexität, Technical Report MIP-8711, Universität Passau, 1987

[98] A.O.Slisenko; Context-free grammars as a tool for describing polynomial-time subclasses of hard problems, Inf. Proc. Letters 14 (1982), 52-56

[99] K.Skodinis, E.Wanke; Emptiness problems of eNCE graph languages, J. of Comp. Syst. Sci. 51 (1995), 472-485

[100] K.Skodinis, E.Wanke; The bounded degree problem for non-obstructing eNCE graph grammars, in Graph-Grammars and their Application to Computer Science, (J.Cuny, H.Ehrig, G.Engels, G.Rozenberg, eds.), Lecture Notes in Computer Science 1073, Springer-Verlag, Berlin, 1996, pp.211-224

[101] K.Skodinis, E.Wanke; Neighborhood-preserving confluent node replacement, Manuscript, November 1995

[102] Gy.Turán; On the complexity of graph grammars, Acta Cybernetica 6 (1983), 271-280

[103] W.Vogler; Recognizing edge replacement graph languages in cubic time, in [37], 676-687

[104] W.Vogler; On hyperedge replacement and BNLC graph grammars, Discrete Appl. Math. 46 (1993), 253-273

[105] E.Wanke; Algorithms for graph problems on BNLC structured graphs, Inform. and Comput. 94 (1991), 93-122

[106] E.Wanke; The complexity of connectivity problems on context-free graph languages, J. of Comp. Syst. Sci. 49 (1994), 57-82

[107] E.Welzl; Encoding graphs by derivations and implications for the theory of graph grammars, Proc. ICALP '84 (J.Paredaens,ed.), Lecture Notes in Computer Science 172, Springer-Verlag, Berlin, 1984, pp.503-513

[108] E.Welzl; On the set of all subgraphs of the graphs in a boundary NLC graph language, in *The Book of L* (G.Rozenberg, A.Salomaa, eds.), Springer-Verlag, Berlin, 1986, pp.445-459

[109] E.Welzl; Boundary NLC and partition controlled graph grammars, in [50], 593-609

Chapter 2

HYPEREDGE REPLACEMENT GRAPH GRAMMARS

F. DREWES,H.-J. KREOWSKI
Universität Bremen, Fachbereich 3, D-28334 Bremen (Germany)
E-mail: {drewes,kreo}@informatik.uni-bremen.de

A. HABEL
Universität Hildesheim, D-31141 Hildesheim (Germany)
E-mail: habel@informatik.uni-hildesheim.de

In this survey the concept of hyperedge replacement is presented as an elementary approach to graph and hypergraph generation. In particular, hyperedge replacement graph grammars are discussed as a (hyper)graph-grammatical counterpart to context-free string grammars. To cover a large part of the theory of hyperedge replacement, structural properties and decision problems, including the membership problem, are addressed.

Contents

2.1 Introduction

Graph grammars have been developed as an extension of the concept of formal grammars on strings to grammars on graphs. Among string grammars, context-free grammars have proved extremely useful in practical applications and powerful enough to generate a wide spectrum of interesting formal languages. It is therefore not surprising that analogous notions have been developed also for graph grammars. Corresponding to the different graph-grammar formalisms in the literature, and to the differing opinions about what the term "context-free" means, a number of different types of graph grammars have been given this attribute (cf. Feder [1], Pavlidis [2], Della Vigna and Ghezzi [3], Janssens and Rozenberg [4], Slisenko [5], Bauderon and Courcelle [6], Habel and Kreowski [7,8], Montanari and Rossi [9], Lautemann [10], Engelfriet [11,12], and Lengauer and Wanke [13]). Among the most general, there are the C-edNCE graph grammars surveyed in Chapter 1 and the hyperedge replacement grammars dealt with in this chapter.

Hyperedge replacement is an elementary approach of graph and hypergraph rewriting. It was introduced in the early seventies by Feder [1] and Pavlidis [2] (under other names) and has been intensively studied since the late seventies (starting with the special case of edge replacement) by Bauderon, Courcelle, and Engelfriet [6,14,15,16,17], Engelfriet, Heyker, Leih, and Rozenberg [18,19,20,21,22,23], Drewes, Habel, Kreowski, Lautemann, and Vogler [24,25,7,26,10,27,28,29,30,31,32,33,34], Lengauer and Wanke [35,36,13], and others.

A *hyperedge* is an atomic item with a fixed number of tentacles, called the *type* of the hyperedge. It can be attached to any kind of structure coming with a set of nodes by attaching each of its tentacles to a node. The hyperedge controls the sequence of these *attachment nodes* and can play the role of a place holder, which may be replaced with some other structure eventually. In such a *hyperedge replacement*, the hyperedge is removed, and the replacing structure R is embedded into the original structure. For this purpose, the considered structures are equipped with sequences of *external nodes*. R is then glued to the remainder of the original structure by fusing each external node with the corresponding attachment node. For this, the number of external nodes in R is required to equal the type of the replaced hyperedge. Thus, referring to the distinction between *gluing* and *connecting* approaches discussed in the introduction of Chapter 1, hyperedge replacement belongs to the gluing approaches.

The replacement of a hyperedge by some structure can be iterated if the original structure or the replacing one is equipped with further hyperedges. If the hyperedges are labelled we may define *productions*. They consist of a label as left-hand side and a replacing structure as right-hand side. Hyperedge replacement grammars mainly consist of a start structure and a finite set of such productions. If a hyperedge labelled with the left-hand side of a production is replaced with the right-hand side this is called a *direct derivation*. If such a grammar, besides the set of productions and the start structure, provides a specification of terminal structures, it can generate a language: the set of terminal structures derivable from the start structure.

The aim of this survey is to present the theory of hyperedge replacement graph grammars. To keep the technicalities as simple as possible, we deal with hypergraphs that consist of sets of nodes and sets of hyperedges as described above, rather than with hybrid objects where some underlying structures are equipped with hyperedges in addition. Hypergraphs are general and flexible enough to cover many interesting cases. In particular, the ordinary directed graphs are special hypergraphs because hyperedges of type 2 are just directed edges. Clearly, undirected graphs can also be handled as a special case.

If a hyperedge is replaced its context is not affected. Therefore, hyperedge replacement provides a context-free type of rewriting (as long as no additional application conditions are employed). This is the main reason for the fact that several structural results for hyperedge replacement grammars and languages are quite similar to the properties of context-free string grammars and languages and that many interesting problems turn out to be decidable. In this survey we try to cover some typical parts of the theory of hyperedge replacement. In particular, we consider structural properties and decision problems. It turns out that the generative power is increasing with the type of hyperedges involved in the derivation process even if one wants to generate only graphs or even string graphs. The generative power of hyperedge replacement grammars generating string graphs can be characterized in terms of tree-walking transducers.

Concerning decision problems we discuss two types of questions. First, one can ask whether the hypergraphs generated by a given hyperedge replacement grammar satisfy a property π of interest. This question—and related ones—are of course interesting for any kind of language generating device since the languages are usually infinite sets. In the case of hyperedge replacement grammars one can find a large class of such properties π for which the mentioned question is decidable (closely related results are discussed in Section 5.7 of Chapter 5). The second decision problem we are going to address is the classical membership problem: Does a given hypergraph belong to the considered language?

As the reader probably expects, the problem is decidable for hyperedge replacement languages. However, there is an important difference to context-free string grammars: The membership problem turns out to be NP-complete. Only restricted subclasses lead to polynomial membership algorithms.

Two further, equally important parts of the theory are left out here. They can be found in other chapters of this volume. The relation of hyperedge replacement and node replacement is discussed in Sections 1.3.3 and 1.3.4 of Chapter 1 and in Section 5.6 of Chapter 5. Furthermore, monadic second-order logic on graphs, its relation with hyperedge replacement, and its use in connection with hyperedge replacement is presented in Chapter 5 (see in particular Sections 5.6–5.8 of that chapter).

The introduction above should have made clear that the chapter presents an overview of the *theory* of hyperedge replacement graph grammars rather than of their practical use. However, it must be pointed out that hyperedge replacement is not only a concept of mathematical beauty, but is useful also from a practical point of view. This is no surprise at all. Graphs are used successfully in all branches of computer science, and context-free generation mechanisms of formal languages are not less important. This opens a wide area of potential applications. Just to mention some examples, hyperedge replacement can be used to model and support the design of VLSI circuits (see, for example, [13,37] by Lengauer and Wanke), and it has been used in a generator for diagram editors to describe the context-free part of the syntax of diagrams (see [38,39,40] by Minas and Viehstaedt). An early application to the generation of semi-structured control-flow diagrams was given by Farrow, Kennedy, and Zucconi [41] (see also Example 2.1). In general, wherever classes of graphs or hypergraphs are to be specified, generated, manipulated, etc. it seems a good idea to distinguish between context-free aspects and non-context-free ones, and to describe the former by a context-free rewriting mechanism like hyperedge or node replacement (see also Chapter 1).

The survey is organized in the following way. Section 2.2 provides the basic notions and notations of hyperedge replacement grammars. In Section 2.3, the context-freeness lemma is presented stating that derivations in hyperedge replacement grammars do not interfere with each other as long as they concern different hyperedges. This is the key result to the major part of the theory of hyperedge replacement as discussed in this chapter. Some structural properties (including a fixed-point theorem, a pumping lemma and a Parikh theorem) are presented in Section 2.4, while the generative power is considered in Section 2.5. Sections 2.6 and 2.7 are devoted to decision problems. While in Section 2.7 the membership problem is discussed, Section 2.6 concerns the decidability of properties of hypergraphs and hypergraph languages. Each of the main sections ends with bibliographic notes and hints to further results.

2.2 Hyperedge replacement grammars

In this section, we introduce hyperedge replacement grammars as hypergraph manipulating and hypergraph-language generating devices. A simple example may perhaps be useful to illustrate the idea. Let us consider directed graphs with multiple edges having two distinguished nodes *begin* and *end*. We can replace an edge e of a graph G with another graph G' by removing e from G, adding G' disjointly, and fusing the *begin*-node of G' with the source of e and the *end*-node of G' with the target of e. The resulting graph is denoted by $G[e/G']$. An example is given in Figure 2.1.

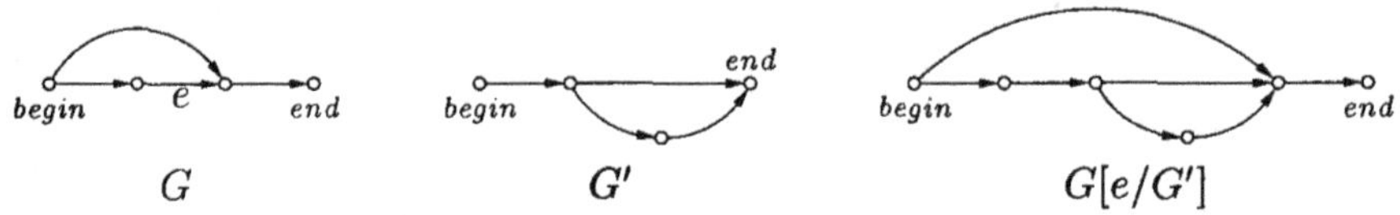

Figure 2.1: The replacement of an edge.

Note that $G[e/G']$ keeps the *begin*- and *end*-nodes of G.

Now, let us allow in addition to label edges with arbitrary symbols. Then we can build productions whose left-hand sides are labels and whose right-hand sides are graphs. A production $S ::= G'$ is applied to a graph G by choosing an edge e labelled with S and replacing it with G', which yields a *direct derivation* $G \Longrightarrow G[e/G']$. The productions shown in Figure 2.2, for instance, generate the set of *series-parallel graphs* (which describe a simple type of concurrent processes) if we start with a single, S-labelled edge and apply productions until no S-labelled edge is left. A sample derivation beginning with a slightly larger graph is shown in Figure 2.3. Here, $\Longrightarrow^*$ denotes the transitive and reflexive closure of $\Longrightarrow$, as usual.

The notion of replacement discussed so far is commonly called *edge replacement*, for obvious reasons. It is a special case of the notion of hyperedge replace-

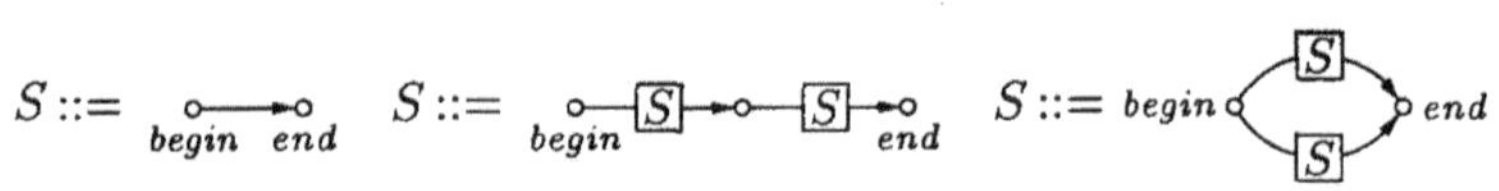

Figure 2.2: Productions to generate series-parallel graphs.

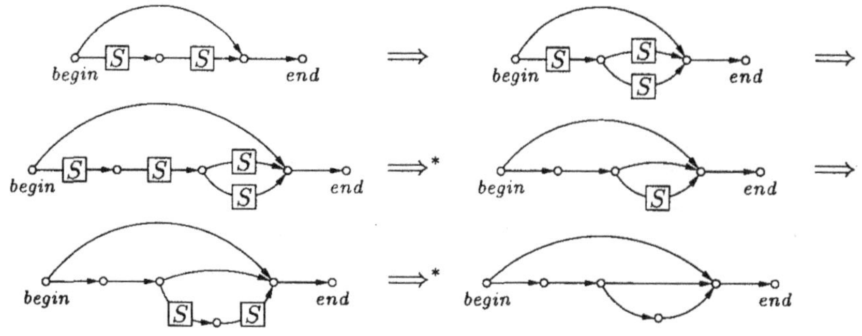

Figure 2.3: A derivation using the productions of Figure 2.2.

ment presented in this chapter. In the remainder of this section the approach will be defined formally. The first paragraph provides the basic notions concerning hyperedges and hypergraphs. In the introduced approach, a hyperedge is an atomic item with a label and an ordered set of tentacles. A set of nodes together with a collection of such hyperedges (usually with varying numbers of tentacles) forms a hypergraph if each tentacle is attached to a node. Directed and labelled graphs, undirected and unlabelled graphs as well as strings can be considered as special cases of hypergraphs. For technical reasons, we assume that the labels are typed by non-negative integers in such a way that, for each hyperedge, the type of its label and the number of its tentacles coincide.

Moreover, a hypergraph is equipped with a sequence of external nodes, which is used to construct the replacement of hyperedges by hypergraphs. The external nodes correspond to the nodes *begin* and *end* in the example above. To make the construction simpler, we assume in addition that the attachment nodes of each hyperedge on the one hand and the external nodes on the other hand are pairwise distinct. It must be pointed out, however, that this restriction is no vital one. If it is dropped one can prove normal-form results yielding systems of the type considered here.

Similar to the example presented above, the replacement of some hyperedges of a hypergraph by other hypergraphs yields an expanded hypergraph by removing the chosen hyperedges, adding the replacing hypergraphs, and fusing their external nodes with the corresponding attachment nodes of the replaced hyperedges. This notion of hyperedge replacement yields the basic steps in the derivation process of a hyperedge replacement grammar, where the pair of the label of each replaced hyperedge and the replacing hypergraph form a production of the grammar.

2.2.1 Hypergraphs

By $\mathbb{N}$ we denote the set of all natural numbers, including 0. For a set A, A^* denotes the set of all strings over A, including the empty string λ; $A^+ = A^* - \{\lambda\}$ denotes the set of all strings over A, except the empty string λ. For $w \in A^*$, $|w|$ denotes the *length* of w, $[w]$ denotes the set of all symbols occurring in w, and $w(i)$ denotes the i-th symbol in w, for $1 \leq i \leq |w|$. The free symbolwise extension $f^*: A^* \to B^*$ of a mapping $f: A \to B$ is defined by $f^*(a_1 \cdots a_k) = f(a_1) \cdots f(a_k)$ for all $k \in \mathbb{N}$ and $a_i \in A$ $(i = 1, \ldots, k)$.

In the following, let C be an arbitrary, but fixed set of *labels* and let $type: C \to \mathbb{N}$ be a *typing* function for C. A (hyperedge-labelled, multi-pointed) *hypergraph* H over C is a tuple (V, E, att, lab, ext) where V is a finite set of *nodes*, E is a finite set of *hyperedges*, $att: E \to V^*$ is a mapping assigning a sequence of pairwise distinct *attachment nodes* $att(e)$ to each $e \in E$, $lab: E \to C$ is a mapping that *labels* each hyperedge such that $type(lab(e)) = |att(e)|$, and $ext \in V^*$ is a sequence of pairwise distinct *external nodes*.

The components of a hypergraph H may be denoted by V_H, E_H, att, lab_H, ext_H, respectively. Furthermore, given a set $X \subseteq C$ of labels we denote by E_H^X the set $\{e \in E_H \mid lab_H(e) \in X\}$ of hyperedges of H with labels in X. The number of nodes plus the number of hyperedges of H is called the *size* of H, denoted by $|H|$. The class of all hypergraphs over C is denoted by $\mathcal{H}_C$.

If the hypergraph in question is understood, we say that $e \in E$ is an *m-edge* for some $m \in \mathbb{N}$ and m is its *type*, denoted by $type(e)$, if $type(lab(e)) = m$. In order to avoid confusion we may also write $type_H(e)$ if H is the hypergraph referred to. $H \in \mathcal{H}_C$ is an *n-hypergraph* for some $n \in \mathbb{N}$ and n its *type*, denoted by $type(H)$, if $|ext_H| = n$.

The sequence of external nodes may be empty so that ordinary hypergraphs (without external nodes) may be seen as 0-hypergraphs and in this way as special cases of hypergraphs with external nodes. An n-hypergraph H over C is considered as a (directed) *n-graph* if all hyperedges of H are 2-edges. The first node of the attachment nodes of a 2-edge corresponds to the source and the second one to the target. The two external nodes of a 2-graph G are denoted by $begin_G$ and end_G in this order.

As a convention, an *unlabelled* hyperedge is a hyperedge labelled with a special label $\sqcup$ which we do not draw in figures. Subject to this convention, unlabelled graphs and hypergraphs turn out to be special cases of the sort of hypergraphs defined above.

A graph $G = (\{v_0, v_1, \ldots, v_n\}, \{e_1, \ldots, e_n\}, att, lab, v_0 v_n)$ over C is called a *string graph* if $v_0, v_1, \ldots, v_n$ are pairwise distinct and $att(e_i) = v_{i-1} v_i$ for

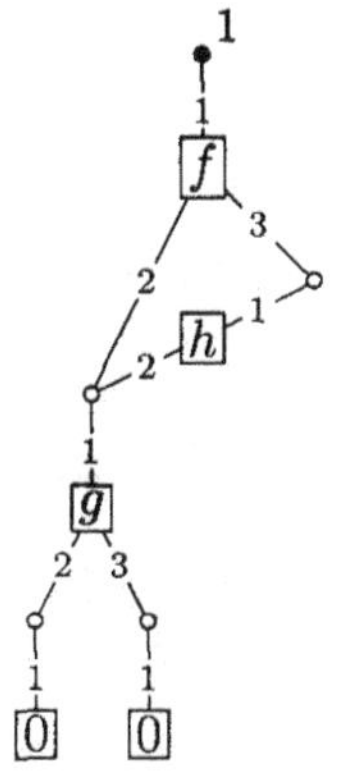

Figure 2.4: A hypergraph denoting the expression $f(g(0,0), h(g(0,0)))$.

$i = 1, \ldots, n$. If $w = lab(e_1) \cdots lab(e_n)$, then G is called the *string graph induced by* w and is denoted by $w^\bullet$. Such a string graph provides a unique graph representation of the string $w \in C^+$.

In drawings of hypergraphs, a dot (•) represents an external node, internal nodes are drawn as circles, and a box depicts a hyperedge with attachment nodes, where the label is inscribed in the box, and the i-th tentacle is attached to the i-th attachment node ($i = 1, \ldots, m$). In other words, the graphical representation makes use of the one-to-one correspondence between hypergraphs and bipartite graphs. As an example, see the 1-hypergraph in Figure 2.4, which represents a functional expression with sharing. A 2-edge may also be drawn as an arrow pointing from its first attached node to the second.

An m-hypergraph H with m nodes (that is, all nodes are external) and a single hyperedge e is said to be a *handle*[a] if $att_H(e) = ext_H$. If $lab_H(e) = A$, then H is said to be the *handle* induced by A and is denoted by $A^\bullet$.

Let $H, H' \in \mathcal{H}_C$. Then H is a *sub-hypergraph* of H', denoted by $H \subseteq H'$, if $V_H \subseteq V_{H'}$, $E_H \subseteq E_{H'}$, $att_H(e) = att_{H'}(e)$, and $lab_H(e) = lab_{H'}(e)$ for all $e \in E_H$. Note that nothing is assumed about the relation of the external nodes. H and H' are *isomorphic* if there is a pair $h = (h_V, h_E)$ of bijective mappings $h_V\colon V_H \to V_{H'}$ and $h_E\colon E_H \to E_{H'}$ with $h_V^*(att_H(e)) = att_{H'}(h_E(e))$ and $lab_H(e) = lab_{H'}(h_E(e))$ for all $e \in E_H$ as well as $h_V^*(ext_H) = ext_{H'}$.

[a]Note that this notion of *handles* differs from the one used in Section 1.4.3 of Chapter 1.

2.2.2 Hyperedge replacement

Let $H \in \mathcal{H}_C$ be a hypergraph, $B \subseteq E_H$ be a set of hyperedges to be replaced. Let $repl\colon B \to \mathcal{H}_C$ be a mapping with $type(repl(e)) = type(e)$ for all $e \in B$. Then the *replacement* of B in H by *repl* yields the hypergraph $H[repl]$ obtained by removing B from E_H, adding the nodes and hyperedges of $repl(e)$ for each $e \in B$ disjointly and fusing the i-th external node of $repl(e)$ with the i-th attachment node of e for each $e \in B$ and $i = 1, \ldots, type(e)$. All hyperedges keep their labels and attachment nodes; the external nodes of $H[repl]$ are those of H. If $B = \{e_1, \ldots, e_n\}$ and $repl(e_i) = R_i$ for $i = 1, \ldots, n$, then we also write $H[e_1/R_1, \ldots, e_n/R_n]$ instead of $H[repl]$.

If, for each $e \in B$, the replacing hypergraph $repl(e)$ consists of the attachment nodes as external nodes and nothing else, the replacement of B in H removes B and adds nothing. Hence, the result may be denoted by $H - B$ in this case. The replacement of some hyperedges is illustrated in Figure 2.5.

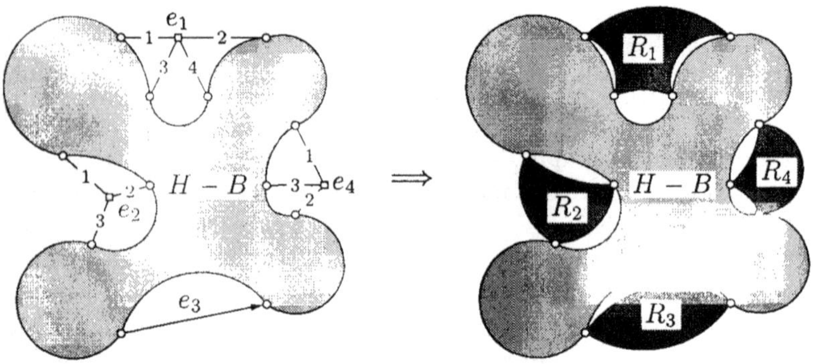

Figure 2.5: The replacement of (unlabelled) hyperedges $e_1, \ldots, e_4$ with $R_1, \ldots, R_4$.

Note that the result of a hyperedge replacement is defined only up to isomorphism. Therefore, to keep the technicalities as simple as possible, we usually do not distinguish between isomorphic copies of a hypergraph.

Hyperedge replacement enjoys some nice properties well-known from other rule-based formalisms. First of all, we have a sequentialization and parallelization property. It does not matter whether we replace some hyperedges of a hypergraph one after another, or simultaneously. The second property is confluence. Hyperedges of a hypergraph can be replaced in any order, without affecting the result. (In fact, this follows already from the sequentialization and parallelization property. If we replace hyperedges simultaneously there is no order among them at all.) The last and maybe most important property is associativity. If a hyperedge is replaced and afterwards a hyperedge of the new part is replaced with a third hypergraph, the same is obtained by first replacing the latter hyperedge and then replacing the first one with the result. These properties are stated formally below.

Sequentialization and parallelization. Let H be a hypergraph with pairwise distinct $e_1, \ldots, e_n \in E_H$ and let H_i be a hypergraph with $type(H_i) = type_H(e_i)$ for $i = 1, \ldots, n$. Then

$$H[e_1/H_1, \ldots, e_n/H_n] = H[e_1/H_1] \cdots [e_n/H_n].$$

Confluence. Let H be a hypergraph with distinct $e_1, e_2 \in E_H$ and let H_i be a hypergraph with $type(H_i) = type_H(e_i)$ for $i \in \{1, 2\}$. Then

$$H[e_1/H_1][e_2/H_2] = H[e_2/H_2][e_1/H_1].$$

Associativity. Let H, H_1, H_2 be hypergraphs with $e_1 \in E_H$ and $e_2 \in E_{H_1}$, such that $type_H(e_1) = type(H_1)$ and $type_{H_1}(e_2) = type(H_2)$. Then

$$H[e_1/H_1][e_2/H_2] = H[e_1/H_1[e_2/H_2]].$$

2.2.3 Hyperedge replacement derivations, grammars, and languages

Let $N \subseteq C$ be a set of *nonterminals*. A *production* over N is an ordered pair $p = (A, R)$ with $A \in N$, $R \in \mathcal{H}_C$ and $type(A) = type(R)$. A is called the *left-hand side* of p and is denoted by $lhs(p)$; R is called the *right-hand side* and is denoted by $rhs(p)$.

Let $H \in \mathcal{H}_C$ and let P be a set of productions. Let $e \in E_H$ and $(lab_H(e), R) \in P$. Then H *directly derives* $H' = H[e/R]$. In this case, we write $H \Longrightarrow_P H'$, or just $H \Longrightarrow H'$ if P is clear from the context, and call this a *direct derivation*. Examples of direct derivations have already been discussed in the beginning of this section (see Figure 2.3).

A sequence of direct derivations $H_0 \Longrightarrow \cdots \Longrightarrow H_k$ is called a *derivation of length k* from H_0 to H_k and is denoted by $H \overset{*}{\Longrightarrow}_P H'$ or $H_0 \Longrightarrow^* H_k$. If the length of the derivation matters, we write $H_0 \Longrightarrow^k H_k$. Additionally, if H_0' is isomorphic to H_0, we speak of a derivation from H_0 to H_0' of length 0.

Using the concepts of productions and derivations, hyperedge replacement grammars and languages can be introduced in a straightforward way. This is done below.

Definition 2.2.1 (***hyperedge replacement grammar***)

A *hyperedge replacement grammar* is a system $HRG = (N, T, P, S)$ where $N \subseteq C$ is a set of *nonterminals*, $T \subseteq C$ with $T \cap N = \emptyset$ is a set of *terminals*, P is a finite set of *productions* over N and $S \in N$ is the *start symbol*.

The *hypergraph language* $L(HRG)$ *generated by* HRG is $L_S(HRG)$, where for all $A \in N$, $L_A(HRG)$ consists of all hypergraphs in $\mathcal{H}_T$ derivable from $A^\bullet$ by applying productions of P:

$$L_A(HRG) = \{H \in \mathcal{H}_T \mid A^\bullet \overset{*}{\underset{P}{\Longrightarrow}} H\}.$$

We denote the class of all hyperedge replacement grammars by $\mathcal{HRG}$ and the class of all *hyperedge replacement languages* by $\mathcal{HRL}$. A hyperedge replacement grammar $HRG = (N, T, P, S)$ is said to be of *order* r (for some $r \in \mathbb{N}$) if for all $(A, R) \in P$, $type(R) \leq r$. A hyperedge replacement language L is of *order* r (for some $r \in \mathbb{N}$) if there is a hyperedge replacement grammar HRG of order r with $L(HRG) = L$. The classes of all hyperedge replacement grammars and languages of order r are denoted by $\mathcal{HRG}_r$ and $\mathcal{HRL}_r$, respectively. A hyperedge replacement grammar $ERG = (N, T, P, S)$ such that all right-hand sides of productions in P are graphs is also called an *edge replacement grammar*. Note that, if given such an edge replacement grammar, one may always assume without loss of generality that all nonterminal labels except perhaps the start symbol have type 2. The class of all edge replacement grammars is denoted by $\mathcal{ERG}$.

Even if one wants to generate graph languages (or string-graph languages) rather than hypergraph languages, one may use nonterminal hyperedges because the generative power of hyperedge replacement grammars increases with their order (see Section 2.5). By definition of derivations the set $L(HRG)$ is closed under isomorphisms. Moreover, $L(HRG)$ is homogeneous, that is, all its hypergraphs are of the same type. Therefore, non-homogeneous languages and languages not closed under isomorphism cannot be generated by the grammars introduced above.

Example 2.1 (generation of semi-structured control-flow graphs)
Control-flow graphs of so-called semi-structured programs are useful for data flow analysis, as shown by Farrow, Kennedy, and Zucconi [41]. These control-flow graphs (seen as hypergraphs) are generated by the hyperedge replacement grammar $FLOW\text{-}GRAPHS = (\{C, D\}, \{c, d\}, P, C)$, where P contains the productions given in Figure 2.6 in a kind of Backus-Naur-Form. (The types of nonterminal labels are those of the corresponding right-hand sides.) A derivation deriving a multi-exit loop is given in Figure 2.7, where tentacle numbers are omitted. □

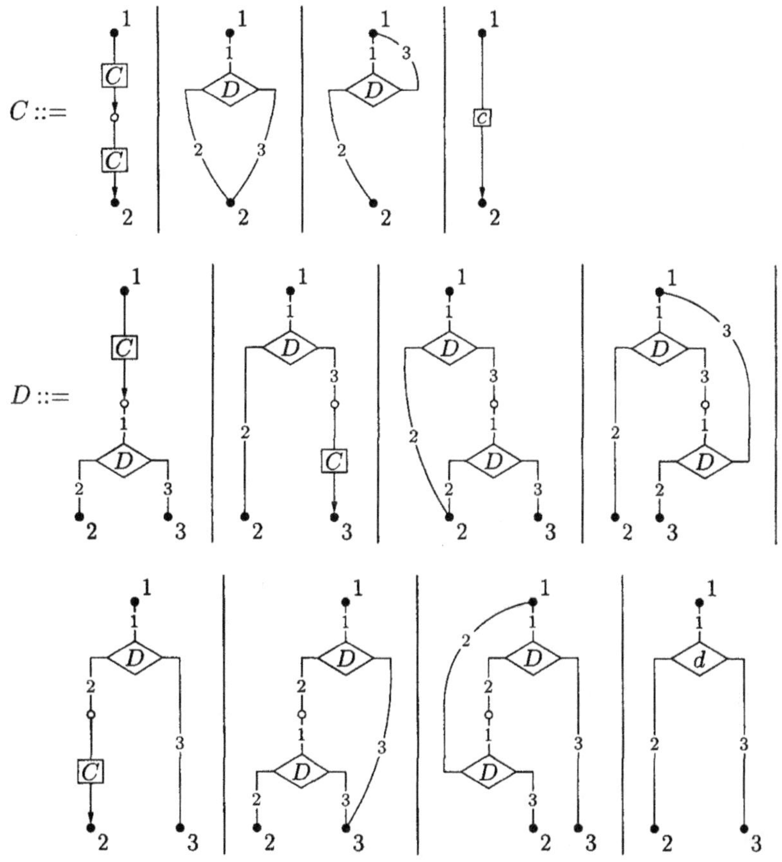

Figure 2.6: Productions of the hyperedge replacement grammar *FLOW-GRAPHS*.

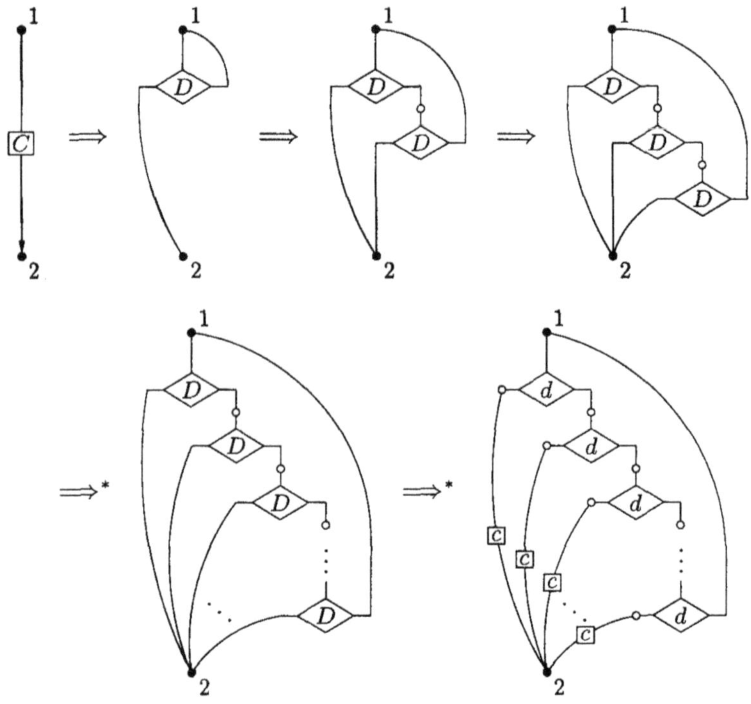

Figure 2.7: A derivation in the hyperedge replacement grammar *FLOW-GRAPHS*.

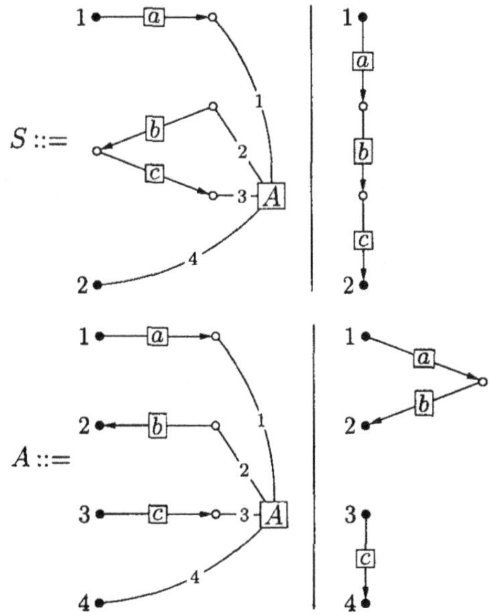

Figure 2.8: Productions of the hyperedge replacement grammar $A^nB^nC^n$.

Example 2.2 (generation of a string-graph language)
Let us consider the hyperedge replacement grammar $A^nB^nC^n$ = $(\{S, A\}, \{a, b, c\}, P, S)$ where P consists of the productions depicted in Figure 2.8. Beginning with $S^\bullet$, the application of the second production yields the string graph $(abc)^\bullet$. By applying the first production, then applying the third production $n - 1$ times, followed by an application of the fourth production, we obtain the derivation in Figure 2.9. Furthermore, the only hypergraphs in $L(A^nB^nC^n)$ are string graphs of the form $(a^nb^nc^n)^\bullet$ for $n \geq 1$. Thus, $L(A^nB^nC^n) = \{(a^nb^nc^n)^\bullet \mid n \geq 1\}$. □

2.2.4 Bibliographic notes

The kind of introduction chosen here is similar to the one in [42]. It must be noted, however, that the notion of direct derivations chosen here is a purely sequential one. A direct derivation replaces only one hyperedge. Instead, one can

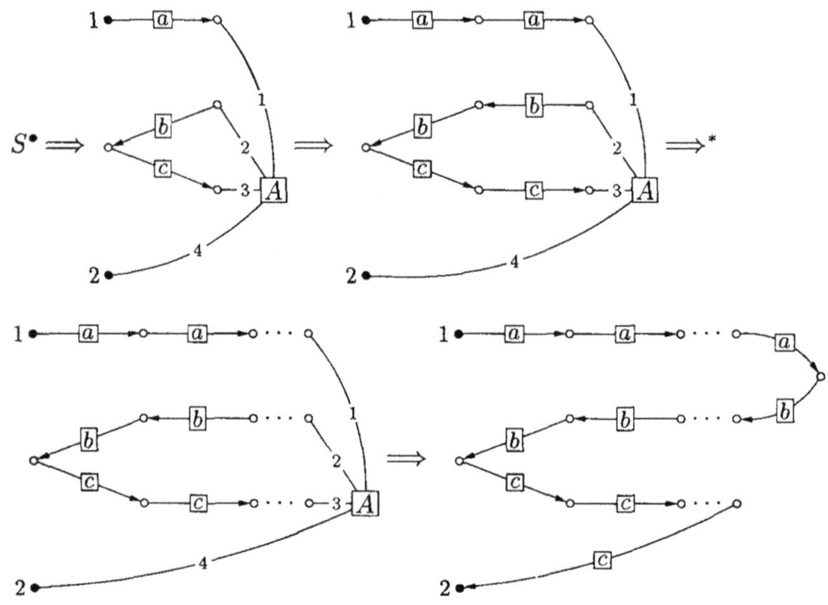

Figure 2.9: A derivation in $A^nB^nC^n$.

as well use a notion of parallel direct derivations, where arbitrarily many hyperedges can be replaced in one step. More precisely, we could say $H \Longrightarrow_P H[repl]$ if $repl: B \to \mathcal{H}_C$ is a mapping with $B \subseteq E_H$ and $(lab_H(e), repl(e)) \in P$ for all $e \in B$. Using the sequentialization property of hyperedge replacement quoted above, it is clear that every derivation consisting of this type of parallel direct derivations can be transformed into a derivation using only direct derivations as defined here. In particular, both notions of direct derivations lead to hyperedge replacement grammars of equal generative power.

The hyperedge replacement approach presented here is described on a set-theoretical level. Since hyperedge replacement may be seen as a special case of hypergraph replacement in the double-pushout approach (see [43]) and the double-pushout approach has an algebraic description (see Chapter 3), hyperedge replacement has an algebraic description, too. The relation between hyperedge replacement languages, logical definability of hypergraph languages, and recognizability has been intensively studied by Courcelle (see Chapter 5).

Courcelle, Engelfriet, and Rozenberg [44] studied the notion of separated handle-rewriting hypergraph grammars (see also Section 1.4.3 of Chapter 1). These grammars combine the rewriting mechanisms of vertex- and hyperedge replacement graph grammars. If we restrict our attention to the generation of graphs, handle rewriting grammars are as powerful as the C-edNCE graph grammars discussed in Chapter 1 and are thus more powerful than hyperedge replacement grammars. Looking at the more general situation where hypergraphs are generated, hyperedge replacement and handle-rewriting graph grammars turn out to be incomparable, though.

In [23] Engelfriet shows how regular tree grammars can be used to generate graphs, by generating expressions that denote graphs. Top-down and bottom-up tree transducers can then be used as a tool for proving properties of such graph generating tree grammars.

As mentioned in the introduction, the idea of hyperedge replacement can also be applied to structures other than hypergraphs. Considering pictures (that is, subsets of $\mathbb{R}^d$ for some $d \in \mathbb{N}$) as underlying structures Habel and Kreowski [45] introduced the so-called *collage grammars.* These allow to generate and to study (approximations of) fractal images and other sorts of pictures (see the work by Dassow, Drewes, Habel, Kreowski, and Taubenberger [46,47,48,49]).

2.3 A context-freeness lemma

In this section, we present a context-freeness lemma for hyperedge replacement grammars and languages. Furthermore, we employ the lemma in order to introduce derivation trees as a helpful and intuitive representation of derivations.

2.3.1 Context freeness

Most of the results presented in this survey (as, for example, the fixed-point theorem and the pumping lemma) are mainly based on the context-free nature of hyperedge replacement. Suppose we are given any notion of replacement allowing to replace primitive components of certain objects with other such objects. In the string case the objects would be strings and the primitive components would be the individual letters. In the case of hyperedge replacement we have hypergraphs as objects and hyperedges as primitive components. Intuitively, context-freeness means that in a derivation of an object O' from and object O we can consider the primitive components of O separately to derive from them the corresponding parts of O'. More precisely, O derives O' if and only if each nonterminal component x derives an object $repl(x)$ such that O' is the object obtained by replacing each nonterminal x of O with $repl(x)$.

In the case of type-2 string grammars we can derive a string w from $A_1 \cdots A_n$ if and only if $w = \mathit{repl}(A_1) \cdots \mathit{repl}(A_n)$, where each $\mathit{repl}(A_i)$ is a string derivable from A_i. In the case of hyperedge replacement grammars the situation is similar, though not as completely obvious. Intuitively, during a derivation that derives a hypergraph H from a hypergraph R the nonterminal hyperedges of R are turned step by step into larger hypergraphs. Thus, each nonterminal hyperedge of R gives rise to a particular sub-hypergraph of H. The important point is that the applicability of a production depends only on the existence of a hyperedge with the required label, and nothing but this hyperedge is affected by the application of the production. Therefore, if we are given a nonterminal hyperedge $e \in E_R^N$ we may simply forget about the context of e in R, keeping only the handle $\mathit{lab}_R(e)^\bullet$, and restrict the original derivation accordingly. In this way we get for every $e \in E_R^N$ a sub-derivation $\mathit{lab}_R(e)^\bullet \Longrightarrow^* \mathit{repl}(e)$ such that $R[\mathit{repl}] = H$. Clearly, the converse is also true: If we are given the hypergraphs $\mathit{repl}(e)$ with $\mathit{lab}_R(e)^\bullet \Longrightarrow^* \mathit{repl}(e)$ ($e \in E_R^N$) then the hypergraph $R[\mathit{repl}]$ is derivable from R. This property justifies to call hyperedge replacement grammars context-free.

Below, we state the result in a recursive version especially suitable for inductive proofs. If a derivation starts in a handle $A^\bullet$ it must necessarily have the form $A^\bullet \Longrightarrow R \Longrightarrow^* H$, where (A, R) is a production. The remainder $R \Longrightarrow^* H$ can now be decomposed as described above.

For the rest of this chapter let us employ the following assumption.

General assumption. Let N, T be two disjoint subsets of the alphabet C.

Theorem 2.3.1 (context-freeness lemma)
Let P be a set of productions over N. Let $H \in \mathcal{H}_C$, $A \in N$, and $k \in \mathbb{N}$. Then there is a derivation $A^\bullet \Longrightarrow^{k+1} H$ if and only if there is a production $(A, R) \in P$ and a mapping $\mathit{repl}\colon E_R^N \to \mathcal{H}_C$ with $H = R[\mathit{repl}]$, such that $\mathit{lab}_R(e)^\bullet \Longrightarrow^{k(e)} \mathit{repl}(e)$ for all $e \in E_R^N$ and $\sum_{e \in E_R^N} k(e) = k$.

Proof
We shall prove this result in detail because of its central character. By definition of direct derivations, if a derivation $A^\bullet \Longrightarrow^{k+1} H$ exists it must have the form $A^\bullet \Longrightarrow R \Longrightarrow^k H$ for some $(A, R) \in P$. Hence, the proof is finished if we can show the following.

> Let G, G' be hypergraphs. Then we have $G \Longrightarrow^k G'$ if and only if there is some $\mathit{repl}\colon E_G^N \to \mathcal{H}_C$ with $G' = G[\mathit{repl}]$, such that $\mathit{lab}_G(e)^\bullet \Longrightarrow^{k(e)} \mathit{repl}(e)$ for all $e \in E_G^N$ and $\sum_{e \in E_G^N} k(e) = k$.

We proceed by induction on k. For $k = 0$ both directions are trivial, so let us assume $k > 0$. Proving first the *only-if* direction, we must have $G \Longrightarrow G_0 \Longrightarrow^{k-1} G'$ where $G_0 = G[e_0/R_0]$ for some $e_0 \in E_G^N$ and some $(lab_G(e_0), R_0) \in P$. The induction hypothesis yields a mapping $repl_0 \colon E_{G_0}^N \to \mathcal{H}_C$ with $G' = G_0[repl_0]$, such that $lab_{G_0}(e)^\bullet \Longrightarrow^{k_0(e)} repl_0(e)$ for all $e \in E_{G_0}^N$, where the sum of the $k_0(e)$ is $k-1$. Now, let $repl(e) = repl_0(e)$ for $e \in E_G^N \setminus \{e_0\}$ and $repl(e_0) = R_0[repl'_0]$, where $repl'_0$ is the restriction of $repl_0$ to $E_{R_0}^N$. Then, if $repl''_0$ is the restriction of $repl_0$ to $E_G \setminus \{e_0\}$ we get $G' = G_0[repl_0] = G[e_0/R_0][repl_0] = G[repl''_0][e_0/R_0][repl'_0] = G[repl''_0][e_0/repl(e_0)] = G[repl]$, as required.

For the *if* direction, let $G' = G[repl]$ with $lab_G(e)^\bullet \Longrightarrow^{k(e)} repl(e)$ for all $e \in E_G^N$ and $\sum_{e \in E_G^N} k(e) = k$. We may assume without loss of generality that $k(e) > 0$ for all $e \in E_G^N$ (otherwise, consider the restriction of *repl* to those hyperedges). Suppose $E_G^N = \{e_1, \dots, e_n\}$. Then there are hypergraphs G_i $(i = 1, \dots, n)$ such that $lab_G(e_i)^\bullet \Longrightarrow G_i \Longrightarrow^{k(e_i)-1} repl(e_i)$ for $i = 1, \dots, n$. Making use of the (already proved) *only-if* direction there are $repl_i \colon E_{G_i}^N \to \mathcal{H}_C$ for $i = 1, \dots, n$ such that $G_i[repl_i] = repl(e_i)$, $lab_{G_i}(e)^\bullet \Longrightarrow^{k_i(e)} repl_i(e)$ for all $e \in E_{G_i}^N$, and $\sum_{e \in E_{G_i}^N} k_i(e) = k(e_i) - 1$.

Let $G_0 = G[e_1/G_1, \dots, e_n/G_n]$ and $repl_0(e) = repl_i(e)$ for all $e \in E_{G_i}^N$ $(i = 1, \dots, n)$. Then we get $G' = G[repl] = G[e_1/G_1[repl_1], \dots, e_n/G_n[repl_n]] = G[e_1/G_1, \dots, e_n/G_n][repl_0] = G_0[repl_0]$. By the induction hypothesis this yields a derivation $G_0 \Longrightarrow^{k-n} G'$ and by construction of G_0 we have $G \Longrightarrow^n G_0$, which completes the proof. □

The possibility of decomposing a derivation as stated in the context-freeness lemma may be illustrated as in Figure 2.10. The context-freeness lemma can also be formulated as a characterization of the languages generated from handles, which yields an alternative method of deriving hypergraphs.

Corollary 2.3.2
Let $HRG = (N, T, P, S) \in \mathcal{HRG}$. *For all* $A \in N$ *let* $rhs(A) = \{R \in \mathcal{H}_T \mid (A, R) \in P\}$. *Then*

$$L_A(HRG) = \bigcup_{R \in rhs(A)} \{R[repl \colon E_R^N \to \mathcal{H}_T] \mid repl(e) \in L_{lab_R(e)}(HRG) \text{ for } e \in E_R^N\}.$$

Proof
Apply the context-freeness lemma to terminal graphs which are the elements of the languages $L_A(HRG)$ for $A \in N$. □

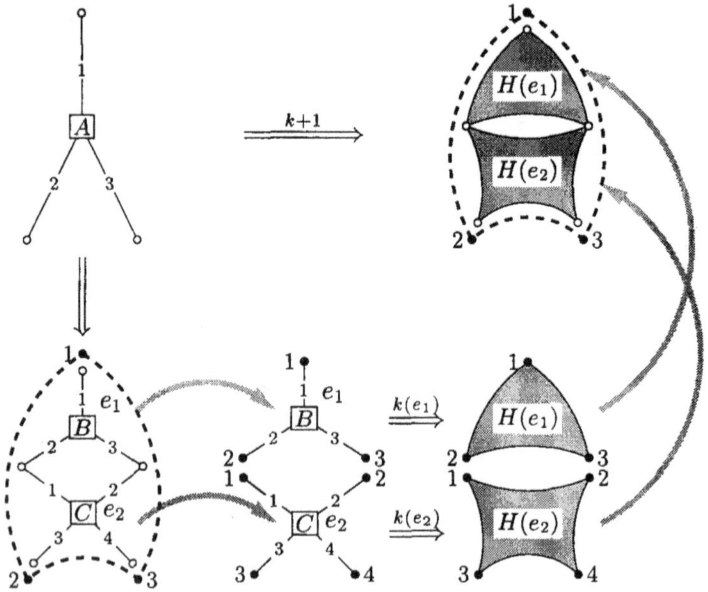

Figure 2.10: Decomposition of a derivation according to the context-freeness lemma

2.3.2 Derivation trees

The context-freeness lemma allows us to introduce the concept of derivation trees as a convenient representation of derivations. Consider some derivation $A^\bullet \Longrightarrow^* H$. If $k = 0$ then $H = A^\bullet$ and we may represent the derivation by a one-node tree whose only node is the label A. Otherwise, the context-freeness lemma states that the derivation has the form $A^\bullet \Longrightarrow R \Longrightarrow^* R[repl]$, where (A, R) is a production and $lab_R(e)^\bullet \Longrightarrow^* repl(e)$ for all $e \in E_R^N$. Thus, the derivation gives rise to a tree whose root is the production (A, R) and whose subtrees are the trees obtained recursively from the derivations $lab_R(e)^\bullet \Longrightarrow^* repl(e)$ for $e \in E_R^N$. Thus, derivation trees represent derivations up to some reordering of direct derivation steps (which does not affect the result of a derivation, as we know). The resulting hypergraph of a derivation tree is then given by $R[repl]$, where R is the right-hand side of the production in its root and $repl(e)$ is recursively obtained as the result of the subtree corresponding to $e \in E_R^N$.

Definition 2.3.3 (*derivation tree*)
The set $TREE(P)$ of *derivation trees over* P is recursively defined as follows.

- $N \subseteq TREE(P)$ with $root(A) = A$ and $result(A) = A^\bullet$ for $A \in N$.
- For every production $(A, R) \in P$ and every mapping $branch\colon E_R^N \to TREE(P)$ such that we have $type(e) = type(root(branch(e)))$ for all $e \in E_R^N$, the triple $t = (A, R, branch)$ is in $TREE(P)$.

  Furthermore, we let $root(t) = A$ and $result(t) = R[repl]$ where, for all $e \in E_R^N$, $repl(e) = result(branch(e))$.

One should notice that a derivation tree contains nonterminal labels $A \in N$ as subtrees if and only if its result is not terminal. The theorem below states the expected correspondence between derivations and derivation trees.

Theorem 2.3.4
Let P be a set of productions over N, let $A \in N$ and $H \in \mathcal{H}_T$. Then there is a derivation $A^\bullet \Longrightarrow^* H$ if and only if there is a derivation tree t over P with $root(t) = A$ and $result(t) = H$.

Proof
For the first direction, suppose t is a derivation tree over P with $root(t) = A$ and $result(t) = H$. The proof is by induction. If $t = A$ then $root(t)^\bullet = A^\bullet \Longrightarrow^0 A^\bullet = result(t)$. If $t = (A, R, branch)$, using the induction hypothesis we get a derivation $root(branch(e))^\bullet \Longrightarrow^* result(branch(e))$ for all $e \in E_R^N$. By the context-freeness lemma, this yields a derivation $root(t)^\bullet = A^\bullet \Longrightarrow^* R[repl] = result(t)$ with $repl(e) = result(branch(e))$ for all $e \in E_R^N$.

The proof for the other direction can be done in a similar way by induction on the length of derivations using the other direction of the context-freeness lemma. □

An interesting consequence of Theorem 2.3.4 is obtained by considering the root and the result of a derivation tree over P as a new production. Formally, let $P^* = \{(root(t), result(t)) \mid t \in TREE(P)\}$. Then, $A^\bullet \Longrightarrow_{P^*} H$ if and only if $A^\bullet \Longrightarrow_P^* H$, by Theorem 2.3.4. By induction on the length of derivations this yields, for every label A and all hypergraphs H,

$$A \Longrightarrow_P^* H \text{ if and only if } A \Longrightarrow_{P^*}^* H.$$

This property will be used in Section 2.7 in order to obtain a cubic membership algorithm for a certain class of edge replacement languages.

2.3.3 Bibliographic notes

To emphasize their context-freeness, hyperedge replacement grammars are sometimes called context-free hypergraph grammars (see, for example, [26,18] and Chapter 1 of this handbook). The context-freeness lemma was first formulated and proved for edge replacement grammars (see [7]); a formulation for hyperedge replacement grammars can be found in [26,42]. In [50] Courcelle presents an axiomatic definition of context-free rewriting. The definition requires that a context-free notion of replacement be associative and confluent. As mentioned in Section 2.2 hyperedge replacement indeed satisfies these requirements. Thus, hyperedge replacement is a context-free rewriting mechanism in Courcelle's sense. In fact, though in a somewhat different way, Theorem 2.3.1 also expresses nothing else than associativity and confluence of hyperedge replacement. It says that the order in which productions are applied is not important (on one side of the equivalence there is no order) and that we can first build the results of the sub-derivations and afterwards replace the hyperedges of the initial hypergraph with these results instead of doing it step by step (which is associativity).

In the literature, some suggestions concerning non-context-free extensions of hyperedge replacement are encountered. The parallel mode of rewriting as known from L-systems with tables may be employed. This was studied by Ehrig and Tischer [51] for edge replacement and by Kreowski [52] for hyperedge replacement. David, Drewes, and Kreowski [53] combined this with a rendezvous mechanism. Here, nodes of different right-hand sides added in a parallel step can be fused in a certain way whenever the replaced hyperedges are neighbours. This allows to overcome most of the restrictions. Besides the parallel case and other application conditions, Kreowski [54] discusses hyperedge replacement as a means to specify infinite (hyper)graphs by infinite derivations, which were first investigated by Bauderon [55,56].

Derivation trees as they are defined here are closely related to rule trees as introduced and investigated by Kreowski [25].

2.4 Structural properties

In this section, we discuss some structural properties of hyperedge replacement languages. They demonstrate that several results well-known for context-free string languages (like the fixed-point theorem, the pumping lemma, and Parikh's theorem) can be generalized to hyperedge replacement languages.

2.4.1 A fixed-point theorem

We start with a fixed-point theorem for hyperedge replacement languages generalizing Ginsburg's and Rice's fixed-point theorem for context-free string languages [57]. It is shown that hyperedge replacement languages are the least fixed points of their generating productions (considered as a system of language equations).

In the following, for every set C of labels let us denote by $\mathcal{HL}_C$ the set of all languages of hypergraphs over C, that is, $\mathcal{HL}_C$ is the powerset of $\mathcal{H}_C$.

Definition 2.4.1 (***system of equations*** **and** ***fixed-point***)
Let $N, T \subseteq C$ be sets of labels with $N \cap T = \emptyset$. A *system of equations* over N is a mapping $EQ: N \to \mathcal{HL}_{N \cup T}$. A mapping $L: N \to \mathcal{HL}_T$ is a *fixed-point* of EQ if, for all $A \in N$,

$$L(A) = \bigcup_{R \in EQ(A)} \{R[repl: E_R^N \to \mathcal{H}_T] \mid repl(e) \in L(lab_R(e)) \text{ for } e \in E_R^N\}.$$

$L: N \to \mathcal{HL}_T$ is a *least fixed-point* of EQ if L is a fixed-point of EQ and $L(A) \subseteq L'(A)$ for all $A \in N$ and all fixed-points L' of EQ.

Remarks:

1. A system of equations $EQ: N \to \mathcal{HL}_{N \cup T}$ over N may be denoted by $A = EQ(A)$ for $A \in N$ to emphasize the name. Actually, it is nothing else than a representation of a set P of productions over N. For P, there is an *associated system of equations* $EQ_P: N \to \mathcal{HL}_{N \cup T}$ defined as $EQ_P(A) = \{R \mid (A, R) \in P\}$. Conversely, a system of equations $EQ: N \to \mathcal{HL}_{N \cup T}$ over N yields a set of productions $P(EQ) = \{(A, R) \mid R \in EQ(A)\}$.

2. The *language family* $L: N \to \mathcal{HL}_T$ *generated by* a hyperedge replacement grammar HRG is given by $L(A) = L_A(HRG)$ for $A \in N$. ◻

According to Corollary 2.3.2, the language family generated by a hyperedge replacement grammar is a fixed-point of the system of equations associated with the set of productions of the grammar. Even better, the following holds.

Theorem 2.4.2 (fixed-point theorem)
Let $HRG = (N, T, P, S)$ be a hyperedge replacement grammar, $EQ_P: N \to \mathcal{HL}_{N \cup T}$ the system of equations associated with P, and $L: N \to \mathcal{HL}_T$ the language family generated by HRG. Then L is the least fixed-point of EQ_P.

Proof
Let $L': N \to \mathcal{HL}_T$ be a fixed-point of EQ. We have to show $L(A) \subseteq L'(A)$ for all $A \in N$, that is, $H \in L'(A)$ for all $H \in \mathcal{H}_T$ with $A^\bullet \Longrightarrow_P^* H$. This can be shown by induction on the length of derivations. $A^\bullet \Longrightarrow_P^1 H$ implies $(A, H) \in P$ with $H \in \mathcal{H}_T$, so that $H[empty] = H$ is defined for the empty mapping $empty: \emptyset \to \mathcal{H}_T$. Since L' is a fixed-point of EQ, $H = H[empty] \in L'(A)$.

Consider now $A^\bullet \Longrightarrow_P^{k+1} H$. Due to the context-freeness lemma, there are $(A, R) \in P$, and derivations $lab_R(e)^\bullet \Longrightarrow^{k(e)} repl(e)$ for $e \in E_R^N$ with $k(e) \leq k$ and $H = R[repl]$. By the induction hypothesis, $repl(e) \in L'(lab_R(e))$. Therefore, $H = R[repl] \in L'(A)$ because L' is a fixed-point of EQ. □

2.4.2 A pumping lemma

We now present a pumping lemma for hyperedge replacement languages. It says that each sufficiently large hypergraph belonging to a hyperedge replacement language can be decomposed into three hypergraphs *FIRST*, *LINK*, and *LAST*, so that a suitable composition of *FIRST*, k samples of *LINK* for each natural number k, and *LAST* yields also a member of the language. This theorem generalizes the well-known pumping lemma for context-free string languages (see, for instance, [58,59]). As in the string case, the pumping lemma can be used to show that certain languages are no hyperedge replacement languages.

A hypergraph H is *substantial* if $V_H \neq [ext_H]$ or $|E_H| > 1$. A hyperedge replacement grammar each of whose productions has a substantial right-hand side is *growing*. A growing hyperedge replacement grammar generates only substantial hypergraphs. Note that all except a finite number of hypergraphs in a hyperedge replacement language L are substantial, since $\{|ext_H| \mid H \in L\}$ is finite. Using a straightforward extension of the proof for the well-known result saying that one can eliminate empty and chain productions from context-free Chomsky grammars it is not hard to prove the following normal-form result, which is used later on.

Lemma 2.4.3
For every hyperedge replacement grammar HRG we can effectively construct a growing hyperedge replacement grammar HRG′ of the same order satisfying $L(HRG') = \{H \in L(HRG) \mid H \text{ substantial}\}$. □

Definition 2.4.4 (*composition* and *iterated composition*)

1. Let $X \in C$. Then $H \in \mathcal{H}_C$ is said to be X*-handled* if there is a unique $type(X)$-edge $e \in E_H$ with label X. In this case, e is called the X*-handle* of H; the hypergraph without the X-handle, $H - \{e\}$, is denoted by $H^\ominus$.

2. Let $H \in \mathcal{H}_C$ be a hypergraph with X-handle e and let $H' \in \mathcal{H}_C$ be a $type(X)$-hypergraph. Then the hypergraph $H[e/H']$ is called the *composition* of H and H' with respect to e and is abbreviated by $H \otimes H'$.

3. Let $H \in \mathcal{H}_C$ be an X-handled $type(X)$-hypergraph. Then for $k \in \mathbb{N}$, $H^k \in \mathcal{H}_C$ is recursively defined by $H^0 = X^\bullet$ and $H^{i+1} = H \otimes H^i$ for $i \geq 0$.

Note that, in the definition above, H^i is X-handled, thus H^{i+1} is well-defined.

Theorem 2.4.5 (pumping lemma)
Let L be some hyperedge replacement language of order r (for some $r \in \mathbb{N}$). Then there exist constants p and q such that the following is true: For every hypergraph H in L with $|H| > p$ there are an X-handled hypergraph $FIRST$, a substantial X-handled $type(X)$-hypergraph $LINK$, and a $type(X)$-hypergraph $LAST$ for some $X \in C$ with $H = FIRST \otimes LINK \otimes LAST$, $|LINK \otimes LAST| \leq q$, and $type(LINK) \leq r$, such that $FIRST \otimes LINK^k \otimes LAST \in L$ for all $k \in \mathbb{N}$. Furthermore, for every hypergraph H in L with $|V_H| > p$ we can choose $LINK$ in such a way that $V_{LINK} \setminus [ext_{LINK}] \neq \emptyset$.

Proof
Let $HRG = (N, T, P, S) \in \mathcal{HRG}_r$ with $L(HRG) = L$ and n the number of nonterminals. Since there are only finitely many non-substantial hypergraphs in $L(HRG)$ we may assume HRG is growing. Let max be the size of the largest right-hand side of HRG. Let $t \in TREE(P)$ with $root(t) = S$ and $H = result(t) \in \mathcal{H}_T$. If $|H| > max^n$, then t contains a path from the root to a leaf longer than n such that one of the nonterminals, say X, occurs twice. In other words, t has a subtree t' with root X which has a proper subtree t'' with root X. Choose $LAST = result(t'')$, $LINK = result(t' - t'')$ and $FIRST = result(t - t')$ where $t' - t''$ is obtained from t' by removing the subtree t'' and $t - t'$ by removing t' from t. Then, $FIRST$ and $LINK$ are X-handled, $LINK$ and $LAST$ are $type(X)$-hypergraphs, and $H = FIRST \otimes LINK \otimes LAST$. Since HRG is growing, $LINK$ must be substantial. As in the case of the pumping lemma for context-free string languages one can now show $FIRST \otimes LINK^k \otimes LAST \in L$ for all $k \in \mathbb{N}$.

If we even have $|V_H| > p$ we may choose $t' - t''$ in such a way that $result(t' - t'')$ has at least one internal node. This is because, in this case it suffices to consider the nodes of the derivation tree of which the corresponding right-hand sides have internal nodes. This yields the second statement of the theorem and thus finishes the proof. □

The composition of H using $FIRST$, $LINK$, and $LAST$ can be depicted as shown in Figure 2.11. The pumped hypergraphs have the shape depicted in Figure 2.12. Note that $FIRST^\ominus$, $LINK^\ominus$, and $LAST$ are not necessarily connected hypergraphs (see Example 2.3 below).

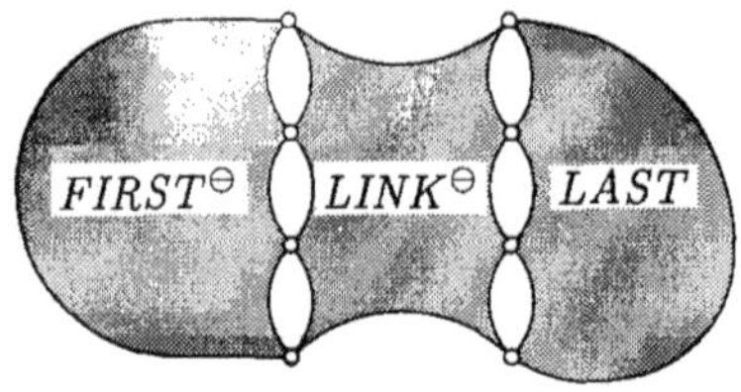

Figure 2.11: The result of composing *FIRST*, *LINK*, and *LAST*.

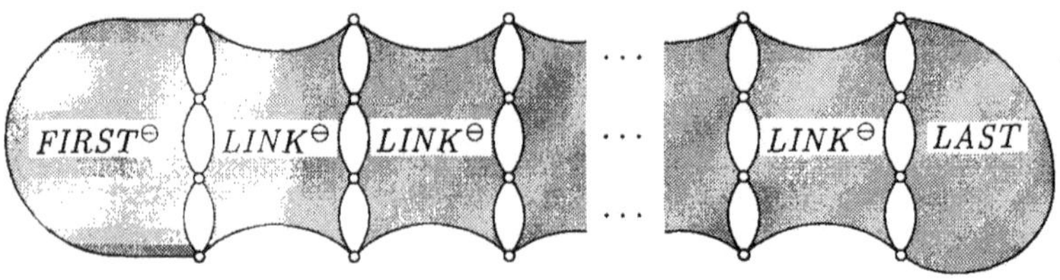

Figure 2.12: The hypergraph obtained by composing $FIRST^{\ominus}$, k samples of $LINK^{\ominus}$, and *LAST*.

As in the case of context-free string grammars, the pumping lemma can be used to show that the *finiteness problem* is decidable for hyperedge replacement languages. Another helpful consequence of the pumping lemma is the linear growth within hyperedge replacement languages. More explicitly, let L be an infinite hyperedge replacement language. Then the pumping of a large enough member of L yields an infinite sequence of hypergraphs in L, say $H_0, H_1, H_2, \ldots$, and constants $c, d \in \mathbb{N}$ with $c + d > 0$ such that $|V_{H_{i+1}}| = |V_{H_i}| + c$ and $|E_{H_{i+1}}| = |E_{H_i}| + d$, for all $i \geq 0$.

Example 2.3
First, we want to illustrate the pumping property for the string-graph language $L = \{(a^n b^n c^n)^\bullet \mid n \geq 1\}$ that can be generated by a hyperedge replacement grammar of order 4, as shown in Example 2.2. For $n \geq 3$, the string graph $(a^n b^n c^n)^\bullet$ can be decomposed as indicated in Figure 2.13 for the case $n = 4$. *FIRST* is a 2-hypergraph containing the two external nodes of the string graph as external nodes. It consists of three chains of edges of length $n - 2$, an a-chain, a b-chain, and a c-chain, where the a-chain is attached to the first external node and the c-chain is attached to the end of the b-chain. Moreover, it may be seen as 4-handled, where the first pair of tentacles is attached to the end of the a-chain and the beginning of the b-chain and the second pair of tentacles is attached to the end of the c-chain and the second external node. It is composed with the *LINK*-component with respect to this 4-handle. The

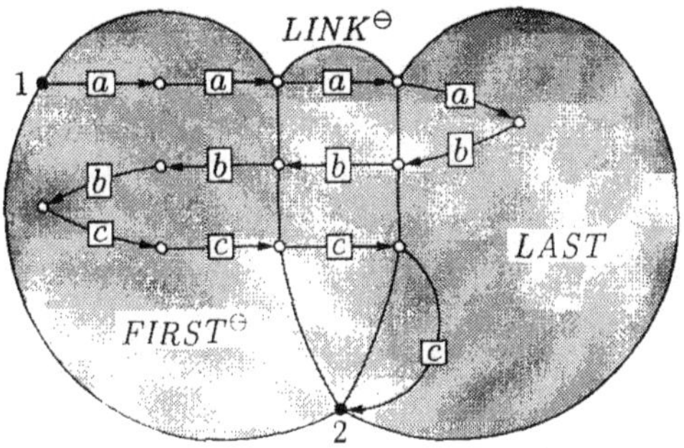

Figure 2.13: Decomposition of $(a^4b^4c^4)^\bullet$.

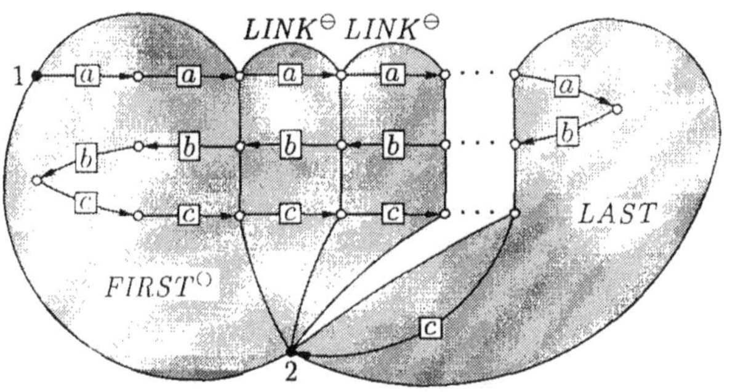

Figure 2.14: Pumping of $(a^4b^4c^4)^\bullet$.

LINK-component possesses a 4-handle which is attached to the target of the *a*-edge, the source of the *b*-edge, the target of the *c*-edge and the second external node. Note that neither $FIRST^\ominus$ (*FIRST* without the 4-handle) nor $LINK^\ominus$ (*LINK* without the 4-handle) nor *LAST* is connected. Moreover, *LINK* is non-trivial.

The pumped 2-hypergraphs have the shape shown in Figure 2.14. Obviously, each resulting 2-hypergraph is a string graph of the form $(a^nb^nc^n)^\bullet$ for some $n \geq 1$. □

One of the main uses of results like the pumping lemma is to prove that specific hypergraph languages are no hyperedge replacement languages. There is, for example, no hyperedge replacement language of unbounded connectivity. To see this, assume such a language is generated by a hyperedge replacement grammar of order r. Then, a sufficiently large graph of connectivity greater than r could be written as $FIRST \otimes LINK \otimes LAST$, where $|LINK \otimes LAST| \leq q$ for some fixed q, $type(LINK) \leq r$, and $LINK$ contains at least one internal node. But since q is fixed we may assume that $FIRST$ contains more than r nodes, so $FIRST \otimes LINK \otimes LAST$ is of connectivity at most r because removing the external nodes of $LINK$ from the graph yields at least two components.

As another example, the string-graph language $\{(a^{n^2})^\bullet \mid n \geq 1\}$ cannot be generated by a hyperedge replacement grammar because the growth of the number of edges is not linear. The pumping lemma can also be used to show that the string-graph language $\{(a_1^n \cdots a_{2k}^n)^\bullet \mid n \in \mathbb{N}\}$ cannot be generated by any hyperedge replacement grammar of order less than $2k$, for $k \geq 1$. This will be done in Section 2.5 (see Example 2.5).

2.4.3 Parikh's theorem

As a third structural result, we present the hyperedge replacement version of Parikh's theorem [60] which relates hyperedge replacement languages to semilinear sets. The proof makes use of Parikh's theorem for context-free string languages by mapping hyperedge replacement grammars to context-free string grammars. In the following, for $x = (x_1, \ldots, x_n)$ and $y = (y_1, \ldots, y_n)$ in $\mathbb{N}^n$ and $c \in \mathbb{N}$, let us define $x+y = (x_1+y_1, \ldots, x_n+y_n)$, $x-y = (x_1-y_1, \ldots, x_n-y_n)$, and $cx = (cx_1, \ldots, cx_n)$.

Definition 2.4.6 (*Parikh mapping* and *semilinear set*)

1. Let $T = \{a_1, \ldots, a_n\}$ be an alphabet and $\psi\colon \mathcal{H}_T \to \mathbb{N}^n$ be the mapping given by $\psi(H) = (\#_{a_1}(H), \ldots, \#_{a_n}(H))$, where $\#_{a_i}(H)$ denotes the number of a_i-labelled hyperedges in $H \in \mathcal{H}_T$. Then ψ is called a *Parikh mapping*. For every language $L \subseteq \mathcal{H}_T$, $\psi(L)$ denotes the set $\psi(L) = \{\psi(H) \mid H \in L\}$.

2. A set $S \subseteq \mathbb{N}^n$ is *linear* if S is of the form

$$S = \{x_0 + \sum_{i=1}^{k} c_i x_i \mid c_1, \ldots, c_k \in \mathbb{N}\},$$

   where $k \geq 1$ and $x_1, \ldots, x_k \in \mathbb{N}^n$.

3. $S \subseteq \mathbb{N}^n$ is *semilinear* if it is the union of a finite number of linear sets.

Theorem 2.4.7 (Parikh's theorem)
For all hyperedge replacement languages L and every Parikh mapping ψ, the set $\psi(L)$ is semilinear.

Proof
Let L be a hyperedge replacement language and $HRG = (N, T, P, S)$ be a hyperedge replacement grammar with $L = L(HRG)$. We construct a context-free string grammar G as follows: Let $str: \mathcal{H}_{N \cup T} \to (N \cup T)^*$ be the mapping assigning the string $str(H) = lab_H(e_1) \cdots lab_H(e_n)$ to a hypergraph H with $E_H = \{e_1, \ldots, e_n\}$ (ordered in some arbitrary way) and let $G = (N, T, str(P), S)$ with $str(P) = \{(A, str(R)) \mid (A, R) \in P\}$. Then $\psi(L(HRG)) = \psi_{str}(L(G))$ where ψ_{str} denotes the usual Parikh mapping for string languages. By Parikh's theorem for context-free string languages, the set $\psi_{str}(L(G))$ is semilinear. Consequently, the set $\psi(L(HRG)) = \psi(L)$ is semilinear. □

It is perhaps interesting to notice that one may attach a 1-edge with a special label to each internal node of a right-hand side of a production. Then counting the hyperedges with the special label means counting the number of (internal) nodes. Consequently, counting the number of nodes in hypergraphs of a hyperedge replacement language yields a semilinear set, too. As a second remark, it is not hard to see that there are several languages L (such as the set of all graphs) which have the semilinearity property but are no hyperedge replacement languages.

2.4.4 Bibliographic notes

The fixed-point theorem for hyperedge replacement languages was formulated for edge replacement in [7] by Habel and Kreowski and for hyperedge replacement by Bauderon and Courcelle in [6] (in a slightly different form). The reader should compare the fixed-point theorem with the way in which Courcelle defines *HR sets of hypergraphs* in Section 5.6 of Chapter 5. HR sets of hypergraphs are just the least fixed-points of systems of equations. However, the allowed equations do not make use of hyperedge replacement. They are built upon a more primitive set of operations, which nevertheless give rise to the same least fixed-points. Thus, HR sets of hypergraphs are hyperedge replacement languages and vice versa.

The pumping lemma is based on the pumping lemma for edge replacement languages given in Kreowski [24]. A hyperedge replacement version was first given in [26] and proved in [42]. Moreover, a number of consequences of the pumping lemma for hyperedge replacement languages can be found in [42]. In particular, it is shown that for hyperedge replacement languages of simple graphs the clique size and the minimum degree is bounded.

The extension of Parikh's theorem to hyperedge replacement languages was first published in [42]. Extensions of Parikh's theorem are also discussed in Section 5.7.3 of Chapter 5.

2.5 Generative power

In this section, we discuss the generative power of hyperedge replacement grammars, dependent on their order. By definition, $\mathcal{HRL}_0 \subsetneqq \mathcal{HRL}_1 \subsetneqq \mathcal{HRL}_2 \subsetneqq \cdots$ since there cannot occur any hypergraph of type k in a language $L \in \mathcal{HRL}_{k'}$, for $k > k'$. In the following the generative power of graph and of string-graph generating hyperedge replacement grammars is studied.

2.5.1 Graph-generating hyperedge replacement grammars

We first show that the generative power of hyperedge replacement grammars depends on their order, that is, on the maximum number of tentacles involved in the replacement of hyperedges, even if only graph-generating grammars are considered. For example the set of all (partial) k-trees can easily be shown to be in $\mathcal{HRL}_k$. Using the pumping lemma it turns out that this language is not in $\mathcal{HRL}_{k-1}$. In other words, the family $(\mathcal{HRL}_k)_{k\in\mathbb{N}}$ forms an infinite hierarchy of classes of hypergraph languages, that remains proper if restricted to graph languages because k-trees are graphs.

Example 2.4 (k-tree and partial k-tree)
As usual, let us call two nodes of a graph *adjacent* if they are connected by an edge. A *k-clique* of a graph is a subset of k pairwise adjacent nodes of this graph. For $k \geq 1$, the set $kTREE$ of all *k-trees* (see Rose [61]) is recursively defined as follows:

- Every complete graph with k nodes is a k-tree on k-nodes.
- Given a k-tree H on n nodes and a k-clique $G \subseteq H$, a k-tree on $n + 1$ nodes is obtained when a new $(n + 1)$-th node is made adjacent to each node of G.

A *partial* k-tree is an undirected graph that is obtained from a k-tree by removing an arbitrary number of edges. The set of all partial k-trees is denoted by $kTREE^{\mathrm{p}}$. For arbitrary, but fixed positive k, one can easily find a hyperedge replacement grammar HRG_{kTREE} of order k generating the set of all k-trees. The grammar simulates the recursive construction of a k-tree: There is one production that produces a new node *new* and $k+1$ hyperedges indicating the old k-clique and the k newly created k-cliques. The second production allows

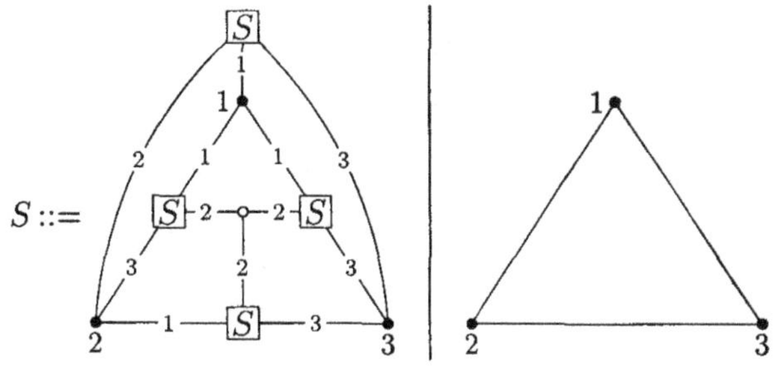

Figure 2.15: Productions to generate 3-trees.

to replace a hyperedge by a k-clique. For $k = 3$ this yields productions as shown in Figure 2.15. Thus, for $k \geq 1$, the set $kTREE$ can be generated by a grammar of order k. Suppose $kTREE$ is a language of order $k-1$. Then, except perhaps a finite number, all elements in $kTREE$ are of connectivity $\leq k-1$ due to the pumping lemma. On the other hand, each k-tree with at least $k+1$ nodes is of connectivity k, a contradiction.

Furthermore, for $k \geq 2$, the addition of a production removing an ordinary edge yields a hyperedge replacement grammar HRG_{kTREE^p} of order k generating the set of all partial k-trees. Hence, by the same arguments as above, for $k \geq 2$ the set $pkTREE$ of all partial k-trees can be generated by a grammar of order k, but not by a grammar of order $k-1$. □

Let $\mathcal{L}_{GRAPH}$ be the class of all graph languages. As an immediate consequence of Example 2.4, we get the following result, saying that the classes $\mathcal{GL}_k = \mathcal{HRL}_k \cap \mathcal{L}_{GRAPH}$ ($k \in \mathbb{N}$) of hyperedge replacement graph languages of order k form a proper hierarchy.

Theorem 2.5.1 (hierarchy theorem 1)
$\mathcal{GL}_k \subsetneq \mathcal{GL}_{k+1}$ for all $k \in \mathbb{N}$. □

2.5.2 *String-generating hyperedge replacement grammars*

Let us now turn to the classes of string-graph languages generated by hyperedge replacement grammars of order k. We denote by $\mathcal{SL}_k$ the set of all hyperedge replacement languages of order k which solely consist of string graphs. Compared with the situation encountered above, these classes turn out to behave slightly different. We show that the family $(\mathcal{SL}_{2k})_{k\geq 1}$ forms an infinite hierarchy that starts with the class of context-free string languages (represented as string-graph languages), but $\mathcal{SL}_{2k+1} = \mathcal{SL}_{2k}$ for all $k \in \mathbb{N}$.

It is quite obvious that, for a given string language L, $L^\bullet$ is in $\mathcal{SL}_2$ if and only if L is context-free. If L is context-free, just choose the set of all productions $(A, w^\bullet)$ for which (A, w) is a production in the given context-free grammar to build an edge replacement grammar yielding $L^\bullet$. For the other direction, using standard techniques one can remove productions that just delete an edge. Afterwards, it is not hard to see that every right-hand side of a production applied in a derivation yielding a string graph must be a string graph, and the productions of this kind obviously induce the needed context-free productions for the string case.

In the following, we present an example of a string-graph language that can be generated by a grammar of order $2k$, but not by a grammar of smaller order. As a consequence, we get a proper language hierarchy.

Example 2.5
The string-graph language $L_k = \{(a_1^n \cdots a_{2k}^n)^\bullet \mid n \in \mathbb{N}\}$ can be generated by the hyperedge replacement grammar of order $2k$, for every $k \geq 1$. The construction is a straightforward generalization of the one presented in Example 2.2.

Thus, L_k is a string-graph language of order $2k$. Making use of the pumping lemma, it can be shown that L_k is no string-graph language of order $2k-1$. To see this, assume L_k is a language of order $2k-1$. Let *FIRST*, *LINK*, and *LAST* be as in the pumping lemma, for some sufficiently large member of L_k. Then *LINK* has type less than $2k$ and is substantial. The latter implies $E_{LINK^\ominus} \neq \emptyset$ because $|V_H| = |E_H| + 1$ for all $H \in L_k$.

Since $FIRST \otimes LINK \otimes LAST$ is a string graph the connected components of $LINK^\ominus$ must be paths whose first and last nodes are in $[ext_{LINK}] \cup [att_{LINK}(e)]$, where e is the X-handle of *LINK*. Hence, the number of connected components of $LINK^\ominus$ which contain at least one edge is at most $k-1$. Moreover, none of these connected components can contain edges with different labels, since otherwise pumping obviously yields graphs not in L_k. But then $LINK^\ominus$ lacks at least one of the labels, so $FIRST \otimes LINK \otimes LAST \in L$ implies $FIRST \otimes LAST \notin L$ because $E_{LINK^\ominus} \neq \emptyset$. □

A similar reasoning as in the previous example shows that the string-graph language $W^k = \{(w^k)^\bullet \mid w \in T^+\}$ is of the same type. It can be generated by a hyperedge replacement grammar of order $2k$ but not by any one of order $2k-1$. The example proves the second hierarchy theorem.

Theorem 2.5.2 (hierarchy theorem 2)
For all $k \in \mathbb{N}$, $\mathcal{SL}_{2k} \subsetneq \mathcal{SL}_{2(k+1)}$. □

The obvious question is, of course, whether one can also prove $\mathcal{SL}_k \subsetneq \mathcal{SL}_{k+1}$. More general, which string-graph languages are generated by hyperedge replacement grammars of order k? In the following, we want to give a characterization by Engelfriet and Heyker [19] of these languages by means of *deterministic tree-walking transducers*, a class of finite automata operating on trees that was introduced by Aho and Ullman. These transducers act on regular sets of terms (that we may also consider as sets of ordered trees over some ranked alphabet). A regular set of terms is generated by a *regular tree grammar* (see [62]) $R = (\Sigma, \Gamma, \Psi, \gamma_0)$, where Σ is a ranked alphabet, Γ is a set of nonterminals with $\gamma_0 \in \Gamma$ (the *start symbol*), and Ψ is a finite set of productions of the form $\gamma \to f(\gamma_1, \ldots, \gamma_n)$, for $\gamma, \gamma_1, \ldots, \gamma_n \in \Gamma$ and $f \in \Sigma$ of rank n. The language $L(R)$ of terms generated by R is defined in the obvious way. It contains all terms over Σ that can be derived from γ_0 by considering the productions as term rewrite rules. The set $occ(t)$ of *occurrences* of a term t over Σ is defined as usual: If $t = f(t_1, \ldots, t_n)$ then $occ(t) = \{\lambda\} \cup \{io \mid 1 \le i \le n \text{ and } o \in occ(t_i)\}$. We denote by $o(t)$ the symbol at occurrence $o \in occ(t)$ and by t/o the subterm of t whose root is the occurrence o in t. The parent occurrence of oi ($i \in \mathbb{N}$) is o and the *parent occurrence* of λ is $\perp$.

Definition 2.5.3 (*deterministic tree walking transducer*)
A *deterministic tree walking transducer* (dtwt) is a tuple $M = (Q, R, \Delta, \delta, q_0, F)$, where Q is a finite set of states with $q_0 \in Q$ (the initial state) and $F \subseteq Q$ (the set of final states), $R = (\Sigma, \Gamma, \Psi, \gamma_0)$ a regular tree grammar, Δ the output alphabet, and $\delta \colon Q \times \Sigma \to Q \times D \times \Delta^*$ the transition function with $D = \{stay, up\} \cup \{down(i) \mid i \ge 1\}$.

A *configuration* of M is a tuple (q, t, o, w) with $q \in Q$, $t \in L(R)$, $o \in occ(t) \cup \{\perp\}$, and $w \in \Delta^*$. A step of M turns (q, t, o, w) into (q', t, o', ww') if we have $\delta(q, o(t)) = (q', d, w')$ and either

- $d = up$ and o' is the parent occurrence of o, or
- $d = stay$ and $o' = o$, or
- $o(t)$ is of rank $\ge i$, $d = down(i)$, and $o' = oi$.

If M, starting with configuration $(q_0, t, \lambda, \lambda)$, finally reaches a configuration $(q, t, \perp, w)$ with $q \in F$ then the computation is *successful* and yields the output w. $L(M)$ denotes the language of all outputs of M, and $\mathcal{DTWT}^+$ denotes the class of all dtwt's M with $\lambda \notin L(M)$.

A property of dtwt's that will turn out to be related to the order of hyperedge replacement grammars is the so-called crossing number. Consider a successful run of M on t. Every move from an occurrence o to an occurrence o' or from o' to o, where o' is the parent occurrence of o, is a *crossing* of the edge between o and o'. If $k \in \mathbb{N}$ is such that for no successful run on any input tree, an edge is crossed more than $2k$ times then M is said to be k-crossing. Note that every dtwt is k-crossing for some $k \in \mathbb{N}$, for if a computation crosses an edge more than $|Q|$ times, a node is visited twice with the same state, so the computation cannot end. Let us denote by $\mathcal{DTWT}_k^+$ the set of all k-crossing $M \in \mathcal{DTWT}^+$. Then, the following can be shown.

Lemma 2.5.4
For all $k \in \mathbb{N}$, $\mathcal{SL}_{2k+1} \subseteq L(\mathcal{DTWT}_k^+)$.

Proof
We roughly sketch the basic idea underlying the construction. One first proves that a given hyperedge replacement grammar $HRG = (N, T, P, S)$ generating a string-graph language can be modified (without changing the order), such that for every $A \in N$ there is some $out_A \in \{0,1\}^{type(A)}$ such that for all $H \in L_A(HRG)$ the out-degree of $ext_H(i)$ is $out_A(i)$, for $i = 1, \ldots, type(A)$. Furthermore, one can ensure by a straightforward construction that for every right-hand side R in HRG and all distinct $e, e' \in E_R^N$ we have $lab_R(e) \neq lab_R(e')$. Then one can construct a dtwt $M = (Q, R, T, \delta, q_0, F) \in \mathcal{DTWT}_k^+$ with $L(M) = L(HRG)$, as follows.

R is designed so that, roughly speaking, $L(R)$ is the set of derivation trees of HRG. The nonterminals of R are those of HRG, the start symbol is S, and the alphabet is P, where every $(A, H) \in P$ has rank $|E_H^N|$. For every production $p = (A, H) \in P$, choose an arbitrary (but fixed) order $e_1, \ldots, e_n$ on E_H^N, and let R contain the production $A \to p(lab_H(e_1), \ldots, lab_H(e_n))$. Now, M is constructed to walk on the trees $t \in L(R)$. Every node of a right-hand side of a production in P is a state of M. If M is at occurrence o of $t \in L(R)$ and $o(t) = (A, H)$, then the current state is a node v of H. M works by searching for the occurrence in t generating the terminal edge e whose first attached node is the image of v in the generated string graph. If there is some $e \in E_H^T$ with $att_H(e, 1) = v$, the occurrence has been found. M follows e by changing its state to $att_H(e, 2)$, outputs $lab_H(e)$, and stays at the same occurrence of t. If there is no such edge, but there is $e_i \in E_H^N$ and $j \in \{1, \ldots, type(lab_H(e_i))\}$ with $att_H(e_i, j) = v$ and $out_{lab_H(e_i)}(j) = 1$, then M moves to the i-th subtree by a $down(i)$-move and proceeds in state $ext_{H'}(j)$ (where H' is the right-hand side of the new occurrence). If neither one of these two possibilities holds, the edge sought is not generated within the current subtree, so v must be an external node $ext_H(i)$ of H (since a string graph is generated). Then M performs an *up*-move and—if H' is the right-hand side of the occurrence reached thereby—assumes state $att_{H'}(e,i)$, where e is the (unique) A-labelled hyperedge of H'.

In order to implement the *down*- and *up*-moves correctly we have to use some auxiliary states that remember i and (A,i), respectively, since we do not know H' in advance. Since M enters and leaves the occurrences of an input tree in states which are external nodes of the right-hand sides, and none of these external nodes is used twice it is not hard to see that M is k-crossing. □

Lemma 2.5.5
For all $k \in \mathbb{N}$, $L(\mathcal{DTWT}_k^+) \subseteq \mathcal{SL}_{2k}$.

Proof
The construction is based on the following ideas. First, one shows that it suffices to consider transducers M without *stay*-moves that produce exactly one output symbol in each step. (The latter is due to the fact that for $L_1^\bullet \in \mathcal{SL}_k$ the image L_2 of L_1 under a string homomorphism satisfies $L_2^\bullet \in \mathcal{SL}_k$, provided $\lambda \notin L_2$.) Now, if M satisfies these requirements and is k-crossing, every subtree t/o of an input tree t is visited at most k times, where a visit to t/o is considered to start at occurrence o at the time it is entered from its parent occurrence o' and ends at o' at the time o' is reached the next time. During each visit, a (nonempty) substring of the output string is produced. The visit—and thus the string produced —- is composed of the visits to the immediate subtrees, which are connected by the steps where M is at occurrence o, leaving it down- or upwards.
To exploit these observations, one designs a hyperedge replacement grammar *HRG* whose derivation trees are more or less the trees in $L(R)$, such that the result of a derivation tree corresponding to t/o is the disjoint union of the $l \leq k$ substrings generated by the l visits of M to t/o. The $2l$ end nodes of the strings are the external nodes, and the substrings resulting from the different visits are generated in parallel. Consistency can be ensured by guessing (using the nonterminal labels) the states M is in when the i-th visit begins and ends, respectively. The productions of *HRG* corresponding to a rule $\gamma_0 \to f(\gamma_1, \ldots, \gamma_n)$ of R are of the form (X_0, H), where H has nonterminal hyperedges $e_1, \ldots, e_n$ labelled with $X_1, \ldots, X_n$. Every label X_i is of the form $\langle \gamma_i, \mathit{begin}_i(1)\mathit{end}_i(1) \cdots \mathit{begin}_i(l_i)\mathit{end}_i(l_i)\rangle$, where $l_i \leq k$ and $\mathit{begin}_i(j), \mathit{end}_i(j)$ are the guessed states for the start and the end of the i-th visit to the tree generated by γ_i. The type of X_i is $2l_i$ and $V_H = \sum_{i=0}^{n}\{\mathit{begin}_i(1), \mathit{end}_i(1), \ldots, \mathit{begin}_i(l_i), \mathit{end}_i(l_i)\}$, where we define $att_H(e_i) = \mathit{begin}_i(1)\mathit{end}_i(1) \cdots \mathit{begin}_i(l_i)\mathit{end}_i(l_i)$ for $i = 1, \ldots, n$. In addition, H contains edges labelled with the output symbols M generates when leaving

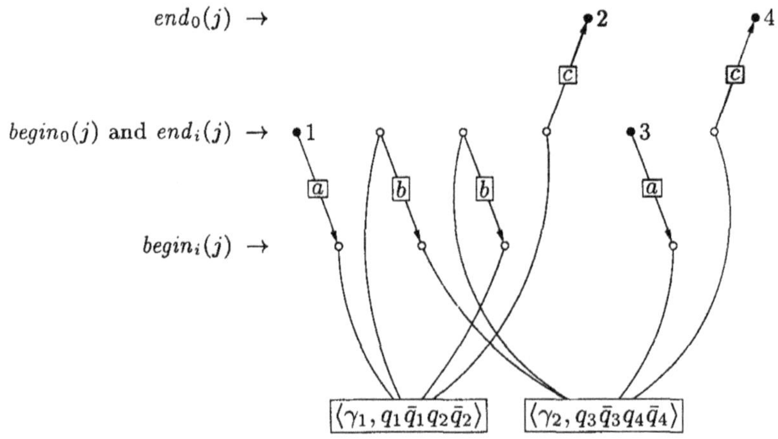

Figure 2.16: A Right-hand side in the grammar constructed to simulate a dtwt. The tentacles of hyperedges are to be numbered from left to right.

the node f, that is, the root of the subtree generated by γ_0, down- or upwards. As an example, see the hypergraph in Figure 2.16. For this to be a consistent right-hand side obtained from $\gamma_0 \to f(\gamma_1, \gamma_2)$ we must have, for instance, $\delta(\bar{q}_1, f) = (q_3, down(2), b)$. □

Lemmas 2.5.4 and 2.5.5 have a number of interesting consequences, of which we mention only two. First of all, it follows that, indeed, the string generating power of hyperedge replacement grammars increases only every second step.

Theorem 2.5.6
For all $k \in \mathbb{N}$, $\mathcal{SL}_{2k} = L(\mathcal{DTWT}_k^+) = \mathcal{SL}_{2k+1}$. □

As mentioned above, every dtwt is k-crossing for some $k \in \mathbb{N}$, so we get the following characterization of the class of all hyperedge replacement string-graph languages.

Theorem 2.5.7
$\mathcal{SL} = L(\mathcal{DTWT}^+)$. □

2.5.3 Further results and bibliographic notes

Most of the results concerning the generative power are formulated in [26] and proved in [42].

The notion of tree-walking transducers was invented by Aho and Ullman ([63]; see also [64]). Lemmas 2.5.4 and 2.5.5 and their consequences are taken from a paper by Engelfriet and Heyker [19], where the results are proved in detail. (Figure 2.16 is also from [19], although slightly modified.) Engelfriet and Heyker mention a number of additional consequences, including results for linear hyperedge replacement grammars that generate string-graph languages. Roughly speaking, one obtains the same results as presented here, by looking at two-way deterministic generalized sequential machines instead of dtwt's. (The former can be seen as dtwt's whose second component is a linear regular tree grammar.) From results known for dtwt's Engelfriet and Heyker also obtain the corollary that the class $\mathcal{SL}$, its linear counterpart, and $\mathcal{SL}_{2k}$ are substitution-closed full AFLs (see [65]).

Asking similar questions for term-generating rather than string-generating hyperedge replacement grammars Engelfriet and Heyker show in [20] that hyperedge replacement grammars have the same term-generating power as attribute grammars. In this context one should also mention the work by Engelfriet and Vogler [66]. They define the so-called tree-to-graph-to-tree transducers, a sort of tree transducers making use of hyperedge replacement in order to construct the computed output trees. The main result of [66] states that these transducers—in the deterministic total case—have the same power as macro tree transducers (for the later see [67]).

Quite a lot is also known about the generative power of hyperedge replacement grammars compared with other graph generating types of grammars. A comparison of hyperedge replacement grammars and boundary edNCE graph grammars is established by Engelfriet and Rozenberg in [18]. It is shown that hyperedge replacement grammars and boundary graph grammars of bounded nonterminal degree have the same power, both for generating sets of graphs and for generating sets of hypergraphs. Arbitrary boundary graph grammars have more generating power than hyperedge replacement grammars, but they have the same hypergraph generating power (subject to a representation of hypergraphs as bipartite graphs).

In [68], Engelfriet and Heyker compare the power of hyperedge replacement grammars and separated handle hypergraph grammars (or S-HH grammars) to generate hypergraph languages of bounded degree. It is shown that every S-HH language of bounded degree can be generated by a (separated) hyperedge replacement grammar. This implies that those two types of grammars generate the same class of graph languages of bounded degree. In the general case, incomparable classes of hypergraph languages are generated.

In [22], Engelfriet, Heyker, and Leih show that hyperedge replacement graph languages of bounded degree are generated by apex graph grammars.

In [16], Courcelle separates vertex replacement from hyperedge replacement: A set of graphs generated by vertex replacement is a hyperedge replacement language if and only if its graphs do not contain arbitrarily large complete bipartite graphs $K_{n,n}$ as subgraphs if and only if its graphs have a number of edges that is linearly bounded in terms of the number of vertices. These properties can be shown to be decidable.

In [17], Courcelle and Engelfriet give a grammar independent characterization of hyperedge replacement languages (see also Section 1.4.2 of Chapter 1). These are exactly the images of the recognizable sets of finite trees under certain graph transformations definable in monadic second-order logic. Several results follow saying that sets of graphs defined by vertex-replacement satisfying additional conditions (like bounded degree, planarity, bounded tree-width, or closure under subgraphs or minors) are hyperedge replacement languages.

2.6 Decision problems

A hyperedge replacement grammar specifies a hypergraph language. Unfortunately, the derivation process never produces more than a finite subset of the language explicitly (and even this may consume much time before significant members of the language occur). Hence, one may wonder what the hyperedge replacement grammar can tell about the generated language. As a matter of fact, the context-freeness lemma leads the way. Given a hyperedge replacement grammar and an arbitrary terminal hypergraph H with derivation $A^\bullet \Longrightarrow R \Longrightarrow^* H$, we get a decomposition of H into smaller components which are derivable from the handles of the hyperedges in R. This can be employed if one wants to reason about certain graph-theoretic properties or numerical functions on (hyper)graphs. The first leads to the notion of *compatible properties*, the second to *compatible functions*.

2.6.1 Compatible properties

If a graph-theoretic property can be tested for each H generated by a hyperedge replacement grammar by testing the property (or related properties) for the components and composing the results to a result for H, it is called *compatible*. It can be shown that compatibility implies decidability of the following questions:

- Is there a hypergraph in the generated language having the property?
- Do all hypergraphs in the generated language (except perhaps for a finite number) have the property?
- Are there only a finite number of hypergraphs in the generated language having the property?

These decision problems are reduced to the emptiness problem and to the finiteness problem by a filter theorem that shows that the intersection of a hyperedge replacement language with the set of hypergraphs that satisfy a compatible property is a hyperedge replacement language, too.

As an illustrating example, let us consider the property that two external nodes of a hypergraph are connected by a path. Let $H \in \mathcal{H}_C$ and $v, v' \in V_H$. A vv'-*path* (in H) is an alternating sequence $v_0 e_1 v_1 \cdots e_n v_n$ with $v_0, \ldots, v_n \in V_H$, $e_1, \ldots, e_n \in E_H$, $v = v_0$, and $v' = v_n$ such that, for $i = 1, \ldots, n$, v_{i-1} and v_i are attachment nodes of e_i. If there is such a path, H is called vv'-*connected*.

Let $HRG = (N, T, P, S) \in \mathcal{HRG}$, and consider $A^\bullet \Longrightarrow R \Longrightarrow^* H$ with $(A, R) \in P$ and $H \in \mathcal{H}_T$ as well as $lab_R(e)^\bullet \Longrightarrow^* H(e)$ for $e \in E_R^N$ given by the context-freeness lemma. Hence we can assume $H = R[repl]$.

If H is vv'-connected, there is a path $p = v_0 e_1 v_1 \cdots e_n v_n$ connecting v and v'. Then either $e_i \in E_R^T$ or $e_i \in E_{repl(e)}$ for some $e \in E_R^N$ $(i = 1, \ldots, n)$. Replace now, for all $e \in E_R^N$, each longest subpath $v_{i(e)} e_{i(e)+1} \cdots e_{j(e)} v_{j(e)}$ of p in $repl(e)$ where $i(e) < j(e)$ with $v_{i(e)}\, e\, v_{j(e)}$. This yields a vv'-path in R such that $repl(e)$ is $v_{i(e)} v_{j(e)}$-connected for each $e \in E_R^N$ on this path. Conversely, if one has such paths in R and in the components $repl(e)$, one gets obviously a vv'-path in H.

This means that the vv'-connectedness of H can be checked by looking for a vv'-path in R and checking corresponding connectedness properties for some components of H. If v and v' are external nodes of H, one obtains a terminating recursive procedure because the components have shorter derivations than H. Moreover, connectedness needs to be considered for only a finite number of pairs of natural numbers because the external nodes of a hypergraph are ordered and their number is bounded by the order of the given grammar if the involved hypergraph is derived from a handle.

Definition 2.6.1 (*compatible predicate*)
Let $\mathcal{C} \subseteq \mathcal{HRG}$ and let I be a (possibly infinite) index set, such that there is an effective procedure constructing for every $HRG = (N, T, P, S) \in \mathcal{C}$ a finite subset I_{HRG} of I, together with some distinguished index $i(type(S)) \in$

I_{HRG}. Let $PROP'$ be a decidable predicate defined on triples (R, ass, i), with $R \in \mathcal{H}_C$, $ass\colon E_R^N \to I$ a mapping, and $i \in I$, and let $PROP(H,i) = PROP'(H, empty, i)$ for $H \in \mathcal{H}_T$ and $i \in I$.

Then $PROP$ is $(\mathcal{C}, PROP')$*-compatible* if for every $HRG = (N,T,P,S) \in \mathcal{C}$, all derivations $A^\bullet \Longrightarrow R \Longrightarrow^* H$ with $H \in \mathcal{H}_T$, and all $i \in I_{HRG}$ we have that $PROP(H,i)$ holds if and only if there is a mapping $ass\colon E_R^N \to I_{HRG}$ such that $PROP'(R, ass, i)$ holds and $PROP(H(e), ass(e))$ holds for all $e \in E_R^N$.

The unary predicate $PROP_0$ that holds on $H \in \mathcal{H}_T$ with $type(H) = k$ if and only if $PROP(H, i(k))$ holds, is called $\mathcal{C}$*-compatible.*

It is not difficult to see that compatible predicates are closed under boolean operations. In addition to the properties of hyperedge replacement languages discussed in Section 2.4, one can now show one more structural property in connection with compatible properties: The set of all hypergraphs from a hyperedge replacement language satisfying a compatible property is again a hyperedge replacement language.

Theorem 2.6.2 (filter theorem)
Let $PROP_0$ be a $\mathcal{C}$-compatible predicate for some $\mathcal{C} \subseteq \mathcal{HRG}$. For every $HRG \in \mathcal{C}$ there is a hyperedge replacement grammar HRG_{PROP_0} such that $L(HRG_{PROP_0}) = \{H \in L(HRG) \mid PROP_0(H)\}$.

Proof
Let $HRG = (N,T,P,S)$ and let I_{HRG}, $PROP'$, and $i(type(S))$ be as in Definition 2.6.1. Then we construct a hyperedge replacement grammar $HRG' = (N', T, P', S')$ as follows.

- $N' = N \times I_{HRG}$;
- P' is the set of all pairs $((A,i),(R, ass))$ such that $(A,R) \in P$, $ass(e) \in I_{HRG}$ for $e \in E_R^N$, $i \in I_{HRG}$, and $PROP'(R, ass, i)$ holds, where (R, ass) denotes the hypergraph $(V_R, E_R, att_R, lab, ext_R)$ with $lab(e) = lab_R(e)$ if $lab_R(e) \in T$ and $lab(e) = (lab_R(e), ass(e))$ if $lab_R(e) \in N$;
- $S' = (S, type(S))$.

It remains to show $L(HRG') = \{H \in L(HRG) \mid PROP(H, type(S))\}$, which can be done by induction on the length of derivations in HRG and HRG' respectively, in a straightforward way. □

The filter theorem can be used to get several decidability results.

Theorem 2.6.3 (decidability of compatible properties)
Let $PROP_0$ be $\mathcal{C}$-compatible with respect to some class $\mathcal{C}$ of hyperedge replacement grammars. Then for all $HRG \in \mathcal{C}$, it is decidable whether

1. *$PROP_0$ holds for some $H \in L(HRG)$;*
2. *$PROP_0$ holds for all $H \in L(HRG)$;*
3. *$PROP_0$ holds for no $H \in L(HRG)$ except perhaps a finite number;*
4. *$PROP_0$ holds for all $H \in L(HRG)$ except perhaps a finite number.*

Proof
By the filter theorem, for every hyperedge replacement grammar $HRG \in \mathcal{C}$, we can effectively construct hyperedge replacement grammars HRG_{PROP_0} and $HRG_{\neg PROP_0}$ generating the sets

$$L(HRG_{PROP_0}) = \{H \in L(HRG) \mid PROP_0(H)\}$$

and

$$L(HRG_{\neg PROP_0}) = \{H \in L(HRG) \mid \neg PROP_0(H)\},$$

respectively. Now $PROP_0$ holds for some hypergraph $H \in L(HRG)$ if and only if $L(HRG_{PROP_0})$ is not empty. $PROP_0$ holds for a finite number of hypergraphs $H \in L(HRG)$ if and only if $L(HRG_{PROP_0})$ is finite. $PROP_0$ holds for all $H \in L(HRG)$ if and only if $L(HRG_{\neg PROP_0})$ is empty. And $PROP_0$ holds for all $H \in L(HRG)$ except perhaps a finite number if and only if $L(HRG_{\neg PROP_0})$ is finite. As a consequence of the pumping lemma, it was noted above that the finiteness problem for hyperedge replacement languages is decidable. Furthermore, the emptiness problem is decidable using the same proof as in the string case. Altogether, these facts yield the claimed results. □

2.6.2 *Compatible functions*

We now turn to the discussion of compatible functions, which generalize compatible predicates. A function on hypergraphs is said to be compatible with the derivation process of hyperedge replacement grammars if it can be computed in the way a compatible predicate can be tested. We restrict the discussion to a certain type of compatible functions that are composed of minima, maxima, sums, and products on natural numbers. They induce compatible predicates of the form: the function value of a graph exceeds a given fixed integer, or the function value does not exceed a fixed integer. Consequently, we get the corresponding decidability results for these predicates as a corollary.

Given a function on hypergraphs (like the size, the maximum degree, the number of components, etc.) and considering the set of values for a hypergraph language, one may wonder whether this set is finite or not. The question is whether the language is bounded with respect to the given function. It can be shown that this boundedness problem is decidable for a class of hyperedge replacement grammars if the function is compatible and composed of sums, products and maxima of natural numbers.

In order to discuss this result, it is appropriate to enrich $\mathbb{N}$ by a special value $\diamond$. This additional element is useful because sometimes the functions one would like to consider have no sensible integer value for some arguments. For example, if we are interested in computing the shortest path between two external nodes of a hypergraph we have to take into account that there is perhaps no path at all between these nodes. Therefore, let $\mathbb{N}^\diamond = \mathbb{N} + \{\diamond\}$ with the following properties for every index set I and $n, n_i \in \mathbb{N}^\diamond$ for $i \in I$, where $I' = \{i \in I \mid n_i \neq \diamond\}$:

- $\diamond \leq n$,
- $\sum_{i \in I} n_i = \diamond$ and $\prod_{i \in I} n_i = \diamond$ if and only if $n_j = \diamond$ for some $j \in I$,
- $\min_{i \in I} n_i = \min_{i \in I'} n_i$ and $\max_{i \in I} n_i = \max_{i \in I'} n_i$, and
- $\min_{i \in I} n_i = \diamond$ and $\max_{i \in I} n_i = \diamond$ for $I = \emptyset$.

Now, the notion of a compatible function is defined as follows.

Definition 2.6.4 (*compatible function*)

1. Let $\mathcal{C} \subseteq \mathcal{HRG}$ and let I be a (possibly infinite) index set, such that there is an effective procedure which constructs for every $HRG = (N, T, P, S) \in \mathcal{C}$ a finite subset I_{HRG} of I and a distinguished index $i(type(S)) \in I_{HRG}$. Let VAL be a set of values, f' be a function on triples (R, ass, i) with $R \in \mathcal{H}_C$, $ass\colon E_R^N \times I \to VAL$, and $i \in I$, and let $f(H, i) = f'(H, empty, i)$ for $H \in \mathcal{H}_T$ and $i \in I$.

   Then f is *$(\mathcal{C}, f')$-compatible* if for all $HRG = (N, T, P, S) \in \mathcal{C}$, all derivations $A^\bullet \Longrightarrow R \Longrightarrow^* H$ with $H \in \mathcal{H}_T$, and all $i \in I_{HRG}$, $f(H, i) = f'(R, ass, i)$, where $ass\colon E_R^N \times I \to VAL$ is given by $ass(e, j) = f(H(e), j)$ for $e \in E_R^N$ and $j \in I_{HRG}$.

   The function $f_0\colon \mathcal{H}_T \to VAL$ given by $f_0(H) = f(H, i(k))$ for all $H \in \mathcal{H}_T$ with $type(H) = k$ is called *$\mathcal{C}$-compatible*.

2. A function $f\colon \mathcal{H}_T \times I \to \mathbb{N}^\circ$ is said to be *$(\mathcal{C}, \min, \max, +, \cdot)$-compatible* if there exists a function f' such that f is $(\mathcal{C}, f')$-compatible and for each right-hand side R of some production in $\mathcal{C}$ and each $i \in I$, $f'(R, -, i)$ corresponds to an expression formed with variables $ass(e, j)$ ($e \in E_R^N$, $j \in I$) and constants from $\mathbb{N}$ by addition, multiplication, minimum, and maximum. The function f is *$(\mathcal{C}, \max, +, \cdot)$-compatible* if the operation min does not occur.

   The function $f_0\colon \mathcal{H}_T \to \mathbb{N}^\circ$ given by $f_0(H) = f(H, i(k))$ for all $H \in \mathcal{H}_T$ with $type(H) = k$ is $(\mathcal{C}, \min, \max, +, \cdot)$-*compatible*, or *$(\mathcal{C}, \max, +, \cdot)$-compatible*, respectively.

Remark:
Let $f_0\colon \mathcal{H}_T \to \mathbb{N}^\circ$ be a $(\mathcal{C}, \min, \max, +, \cdot)$-compatible function, let $n \in \mathbb{N}^\circ$, and let $\bowtie$ be one of the binary predicates $\leq$, $<$, $=$, $\geq$, and $>$. Then it is not difficult to see that the predicate $PROP(f_0, \bowtie)$ given for $H \in \mathcal{H}_T$ by $PROP(f_0, \bowtie)(H) \iff f_0(H) \bowtie n$ is $\mathcal{C}$-compatible. This together with the decidability results of the previous section yields the solution of some particular decision problems. For all $HRG \in \mathcal{C}$ it is decidable whether

- $f_0(H) \bowtie n$ holds for some $H \in L(HRG)$;
- $f_0(H) \bowtie n$ holds for only a finite number of hypergraphs $H \in L(HRG)$.
- $f_0(H) \bowtie n$ holds for all $H \in L(HRG)$.
- $f_0(H) \bowtie n$ holds for all $H \in L(HRG)$ except a finite number. □

As an example, let us have a look at the function that yields the number of simple paths of a 2-graph connecting the two external nodes. Given a graph G, a path is called *simple* if each node appears only once. Let $PATH_G$ denote the set of all simple $begin_G end_G$-paths of G and let $numpath(G) = |PATH_G|$ be its cardinality. Using the context-freeness lemma, if $A^\bullet \Longrightarrow R \Longrightarrow^* G$ we get $PATH_G = \bigcup_{p \in PATH_R} \{replace(p, path) \mid path\colon E_R^N \to PATH_{G(e)}\}$, where $replace(p, path)$ denotes the path obtained from p by replacing all edges $e \in E_R^N$ on p by the corresponding paths $path(e)$. This obviously yields

$$numpath(G) = \sum_{p \in PATH_R} \prod_{e \in E_R^N \text{ on } p} numpath(G(e)).$$

These observations can be reformulated in terms of compatible functions. Let $\mathcal{C}$ be the class of all edge replacement grammars, $I = \{np\}$, $VAL = \mathbb{N}^\circ$, and

f' be the function given by $f'(R, ass, np) = \sum_{p \in PATH_R} \prod_{e \in E_R^N \text{ on } p} ass(e, np)$. Then we have $f(G, np) = f'(G, empty, np) = \sum_{p \in PATH_G} 1 = numpath(G)$ and $f(G, np) = f'(R, ass, np)$ with $ass(e, np) = f(G(e), np)$ $(e \in E_R)$, so f turns out to be $(\mathcal{ERG}, f')$-compatible. Consequently, the function $numpath = f(-, np)$ is $(\mathcal{ERG}, \max, +, \cdot)$-compatible.

Other compatible functions are the number of nodes, the size, the density of a hypergraph, that is, the ratio of the number of edges and the number of nodes, the minimum-path length (of paths connecting external nodes), the maximum-simple-path length (of paths connecting external nodes), the number of simple cycles, the minimum-cycle length, the maximum-simple-cycle length, the minimum degree, the maximum degree and the number of components.

Theorem 2.6.5 (meta-theorem for boundedness problems)
Let f_0 be a $(\mathcal{C}, \max, +, \cdot)$-compatible function for a class $\mathcal{C}$ of hyperedge replacement grammars. Then, for all $HRG \in \mathcal{C}$, it is decidable whether there is a natural number $n \in \mathbb{N}$ such that $f_0(H) \leq n$ for all $H \in L(HRG)$.

Proof
Let f, f' be the required functions, so that f is $(\mathcal{C}, f')$-compatible and $f_0(H) = f(H, i(type(H)))$ for $H \in \mathcal{H}_T$. Let $HRG = (N, T, P, S) \in \mathcal{C}$. Then one can show that f_0 is unbounded on $L(HRG)$ if and only if there are $A \in N$, $j \in I_{HRG}$ and A-handled hypergraphs X and Y with $X_0, Y_0 \in \mathcal{H}_T$ such that the following hold:

1. $A^\bullet \Longrightarrow^* H$ for some $H \in \mathcal{H}_T$.
2. $S^\bullet \Longrightarrow^* Y$ such that $f(H, j) \leq f(Y \otimes H, i_0) = f_0(Y \otimes H)$ for all $H \in \mathcal{H}_T$ with $A^\bullet \Longrightarrow^* H$.
3. $A^\bullet \Longrightarrow^* X$ such that $f(H, j) < f(X \otimes H, j)$ for all $H \in \mathcal{H}_T$ that satisfy $A^\bullet \Longrightarrow^* H$.

To check for such A, j, X and Y, one has to inspect the finite number of derivations that start in handles and are of length up to the number of nonterminals. The somewhat tedious technical details are omitted. □

2.6.3 Further results and bibliographic notes

Results which are closely related to those discussed in this section are surveyed in Section 5.7 of Chapter 5 (see also the remarks on inductive predicates and inductively computable functions below). The results about compatibility are taken from [27,30] by Habel, Kreowski, and Vogler. It must be mentioned, however, that a finite index set was used in these papers. For the case of compatible predicates an infinite index set was first used by Habel, Kreowski, and Lautemann in [34].

The filter theorem is published in [28,42]. A corresponding result can be found in [15]. The statements (1) and (2) of the decidability result are published in [27]. In that paper, a direct proof is given which is based on the idea of constructing the set of handles from which a hypergraph with the desired property can be derived and checking whether the start handle belongs to the constructed set. The statements (3) and (4) of the decidability result are presented in [28,42]. In [31] Lautemann shows how these and several related results in connection with tree decompositions and bounded tree-width can by systematized by the use of finite tree automata and well-known characterizations of recognizability. Some of the methods for finite tree automata are carried over to deal with certain tree automata with infinite state sets. In this way it is shown that graph properties defined by formulae of monadic second order logic with arithmetic can be decided efficiently for graphs of bounded tree-width, a result which was first shown (in somewhat more general form) by Arnborg, Lagergren, and Seese [69].

In [13], Lengauer and Wanke consider efficient ways of analyzing graph languages generated by so-called cellular graph grammars. Cellular graph grammars are in fact hyperedge replacement graph grammars, defined in a slightly different way. In particular, they generate the same class of graph languages. A characteristic of graph properties called finiteness is defined, and combinatorial algorithms are presented for deciding whether a graph language generated by a given cellular graph grammar contains a graph with a given finite graph property. Structural parameters are introduced that bound the complexity of the decision procedure and special cases for which the decision can be made in polynomial time are discussed. The results provide explicit and efficient combinatorial algorithms.

In [15], Courcelle introduces the notion of a recognizable set of graphs. Every set of graphs definable in monadic second-order logic is recognizable, but not vice versa. It is shown that the monadic second-order theory of hyperedge replacement languages is decidable. The notion of F-inductive predicates studied in [15] is closely related to the concept of compatible properties (see also Section 5.7.2 of Chapter 5).

In [34], Habel, Kreowski, and Lautemann compare compatible, finite, and inductive graph properties and show that the three notions are essentially equivalent. Consequently, three lines of investigation in the theory of hyperedge replacement—including the one discussed here—merge into one.

In [70], Wanke and Wiegers investigate the decidability of the bandwidth problem on linear hyperedge replacement languages. In particular, they show the following. Let $\mathcal{HRG}_{\text{lin}}$ denote the class of all linear hyperedge replacement grammars generating graphs. Then it is undecidable whether an instance grammar $HRG \in \mathcal{HRG}_{\text{lin}}$ generates a graph G having bandwidth k, for any fixed integer $k \geq 3$. The result implies that the bandwidth-k property for $k \geq 3$ is not $\mathcal{HRG}_{\text{lin}}$-compatible.

Theorem 2.6.5 is published in [30]. Related work is done in [71] by Courcelle and Mosbah, who investigate monadic second-order evaluations on tree-decomposable graphs and come up with a general method to translate these evaluations of graph expressions over suitable semirings. Their method allows the derivation of polynomial algorithms for a large number of problems on families of graphs, and especially on graphs definable by hyperedge replacement. The notion of inductively computable functions introduced by Courcelle and Mosbah ([71], see also Section 5.7.3 of Chapter 5) is a generalization of the concept of inductive predicates which corresponds closely to the notion of compatible functions.

Boundedness problems for hyperedge replacement grammars correspond closely to boundedness problems for finite tree automata with cost functions over a suitable semiring, investigated by Seidl in [72]. Cost functions for tree automata are mappings from transitions to polynomials over some semiring or, in the so-called k-dimensional case, to k-tuples of polynomials. (The dimensions correspond to the different indices used in the definition of a compatible function.) Four semirings are considered, namely the semiring $\mathbf{N}$ over $\mathbb{N}$ with addition and multiplication, the semiring $\mathbf{A}$ over $\mathbb{N} \cup \{\infty\}$ with maximum and addition, the semiring $\mathbf{T}$ over $\mathbb{N} \cup \{-\infty\}$ with minimum and addition, and the semiring $\mathbf{F}$ over the finite subsets of $\mathbb{N}$ with union and addition. It turns out that for the semirings $\mathbf{N}$ and $\mathbf{A}$, it is decidable in polynomial time (if the dimension k is fixed) whether or not the costs of accepting computations is bounded; for $\mathbf{F}$, it is decidable in polynomial time whether or not the cardinalities of occurring cost sets are bounded. In all three cases, explicit upper bounds are derived. Moreover, for the semiring $\mathbf{T}$, the decidability of boundedness is proved, but a polynomial-time algorithm is obtained only in the case that the degrees of occurring polymonials are at most 1.

In [37], Wanke considers another type of decidability problems on hyperedge replacement grammars, called integer subgraph problems. The main idea is to transform the decidability problem on hyperedge replacement languages concerning function values of graph$\times$subgraph pairs to a decidability problem over semilinear sets.

Finally, one should mention an approach to generalize the idea of compatible functions that was invented by Engelfriet and Drewes [23,73,49,74,75]. The approach is based on the observation that the definition of a compatible function can be considered as a tree transduction. The input trees are derivation trees of the hypergraphs in question and the output trees are expressions using maximum, addition, and multiplication (viewed as trees), that denote the computed numbers. Both the derivation trees (denoting hypergraphs) and the expressions (denoting natural numbers) may be viewed as terms in appropriate algebras (over hypergraphs and natural numbers, respectively). Thus, the idea is to consider such computations by tree transductions in general. A mapping $f\colon A \to B$ between two algebras A and B is said to be computed by a tree transducer if every term denoting an element a of A is transformed into a term denoting $f(a)$ in B.

2.7 The membership problem

In this section the possibilities for (efficient) algorithmic solutions to the *membership problem* for hyperedge replacement languages are discussed. We consider membership with respect to a given hyperedge replacement grammar HRG, so the membership problem is defined as follows.

Given: A hyperedge replacement grammar HRG.
Instance: A hypergraph H.
Question: Is H in the language generated by HRG?

Algorithms that solve the membership problem are called membership algorithms. In many cases, one would also wish to *parse* hypergraphs with respect to a given grammar, that is, to find a derivation tree for a given hypergraph, if possible. For technical convenience, we concentrate on the membership problem here, but it is easy to see that both the presented algorithms can be extended in a straightforward way to yield solutions to the parsing problem as efficiently as they solve the membership problem.

2.7.1 NP-completeness

It is not hard to see that the membership problem is in NP for every hyperedge replacement grammar HRG, because for every such grammar there is a linear function f such that, for every hypergraph H of size n, if $H \in L(HRG)$ then H has a derivation of length at most $f(n)$. Thus we can simply use nondeterminism to guess an appropriate derivation and test whether the generated hypergraph is isomorphic to H (again using nondeterminism).

Clearly, one would like to have more efficient algorithms than the one just sketched, and in particular deterministic ones. It turns out, however, that the complexity of the membership problem marks one of the few areas where the results for hyperedge replacement grammars differ significantly from what we know about context-free Chomsky grammars. Whereas, for the latter we have the well-known membership algorithm by Cocke, Kasami, and Younger, which runs in cubic time, hyperedge replacement grammars are able to generate NP-complete graph languages. Thus, one can expect that there is no generally applicable membership algorithm for hyperedge replacement languages. Intuitively, the reason for this is that, in a string, symbols that result from one and the same nonterminal form a substring, and there are only $O(n^2)$ many substrings of a given string of length n. In contrast, from a nonterminal hyperedge a somewhat disconnected subgraph may be generated in a hyperedge replacement grammar, so that, in general, there are exponentially many possible choices for a subgraph to be generated from a given hyperedge. There are simply too many combinations to be tested, which leads to the two NP-completeness results presented below. For this, let us call a hyperedge replacement grammar *linear* if none of its right-hand sides has more than one nonterminal hyperedge. The *node degree* of a graph is the maximal number of edges attached to a node of this graph, and the node degree of a graph languages is the maximum node degree of graphs in that language.

Theorem 2.7.1 (NP-completeness 1)
There is a linear edge replacement grammar that generates an NP-complete graph language of degree 2.

Proof
We reduce the NP-complete *Hamiltonian path problem* (see Garey, Johnson [76], problem GT39, p. 199) to the membership problem for a particular edge replacement language generated by a linear grammar. In fact, there is a slightly easier way to prove Theorem 2.7.1 using the NP-complete problem 3-PARTITION (see [77]). However, we are going to use a modified version of the proof in order to obtain another theorem below. As we do not know how this can be done using 3-PARTITION the Hamiltonian path problem is employed here.

Let us first define the Hamiltonian path problem. We use a slightly more general version on undirected hypergraphs. An undirected, unlabelled hypergraph is a triple $h = (V_h, E_h, att_h)$, where V_h and E_h are the finite sets of vertices and hyperedges, respectively, and att_h is a mapping assigning to every hyperedge $e \in E_h$ the set $att_h(e) \subseteq V_h$ of attached nodes. Now, the Hamiltonian path problem can be formulated as follows.

Instance: An undirected, unlabelled hypergraph h.
Question: Is there a Hamiltonian path in h, that is, is there an alternating sequence $v_0e_1v_1 \dots e_nv_n$ of pairwise distinct nodes and hyperedges of h such that $V_h = \{v_0, \dots, v_n\}$ and $v_{i-1}, v_i \in att_h(e_i)$ for all i, $1 \leq i < n$?

Clearly, the generalization to hypergraphs does not affect the NP-hardness of the problem. Let us define $\mathcal{S}$ to be the set of all finite sets $S = \{s_1, \dots, s_n\}$ such that the s_i $(1 \leq i \leq n)$ are strings of equal length of the form u_i-u_i, where $u_i \in \{0,1\}^*$. We say that u_i-u_i is *word adjacent* to u_j-u_j if $u_i(l) = u_j(l)$ for some l, $1 \leq l \leq |u_i|$.

Every $S \in \mathcal{S}$ defines an undirected, unlabelled hypergraph $h(S)$. If S is as above, with $|s_i| = 2m+1$ for $i = 1, \dots, n$, the set of nodes of $h(S)$ is S and the set of hyperedges is $\{1, \dots, m\}$. Node s_i is in the set of attachments of hyperedge j if $u_i(j) = 1$. By definition, two nodes are adjacent in $h(S)$ if the corresponding strings are word adjacent, so $h(S)$ has a Hamiltonian path if and only if there is a sequence $i_1, \dots, i_n$ with $\{i_1, \dots, i_n\} = \{1, \dots, n\}$ such that, for $i = 1, \dots, n-1$, s_i and s_{i+1} are word adjacent. Clearly, every undirected, unlabelled hypergraph h is represented by some $S \in \mathcal{S}$. Furthermore, given any sensible representation of h we can compute some $S_h \in \mathcal{S}$ with $h = h(S_h)$ in polynomial time.

We are going to construct a linear edge replacement grammar such that each of the generated graphs is a string graph, except that some of the edges are missing. Thus, such a graph represents a multiset of strings. The individual strings of the multisets generated have the form w-w', where $w, w' \in \{0,1\}^*$. The multisets themselves will be of the form

$$\begin{array}{l}\{w_0\text{-}w_{11}1w_{12},\\ \quad w'_{11}1w'_{12}\text{-}w_{21}1w_{22},\\ \qquad w'_{21}1w'_{22}\text{-}w_{31}1w_{32},\\ \qquad\qquad \ddots\\ \qquad\qquad\quad w'_{n1}1w'_{n2}\text{-}w_1\}\end{array}$$

where each w_{i1} is generated along with the corresponding w'_{i1}, and each w_{i2} is generated together with the corresponding w'_{i2}. This is used to ensure that w_{ij} and w'_{ij} are of equal length, so that the digit 1 between w_{i1} and w_{i2} appears at the same position as the one between w'_{i1} and w'_{i2}. Thus, if such a multiset S is in $\mathcal{S}$ it defines a hypergraph with a Hamiltonian path passing the nodes represented by the strings above one after the other.

Consider the grammar $ERG = (\{S, A, B, C, D\}, \{\text{-}, 0, 1\}, P, S)$, where all terminals and nonterminals are of type 2, except for S, which is of type 0, and where P consists of the productions given in Figure 2.17. (In Figure 2.17, a

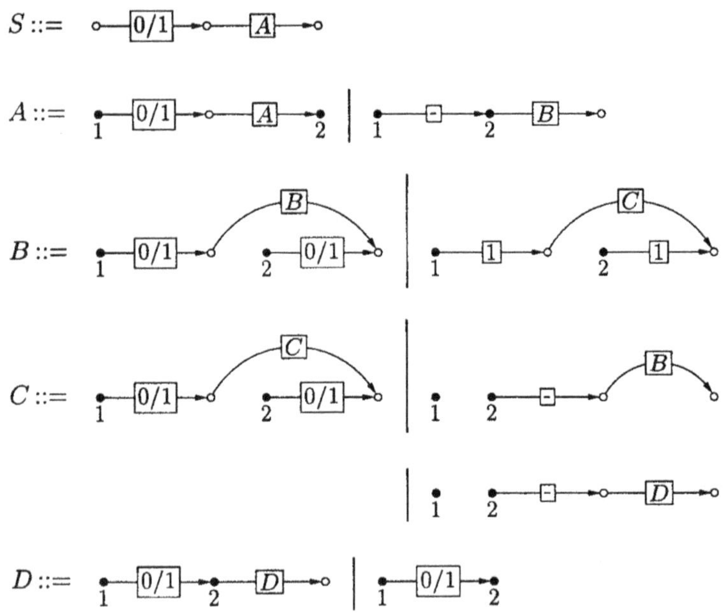

Figure 2.17: Productions of *ERG* in the proof of Theorem 2.7.1.

right-hand side some of whose edges are labelled with "0/1" stands for all right-hand sides in which the respective edges are labelled with 0 or 1.) If we interpret the graphs in $L(ERG)$ as multisets of strings in the obvious way, this yields a set $\mathcal{S}(ERG)$ whose elements are these multisets. It is not too hard to see that, as explained above, all these multisets are of the form

$$\{w_0\text{-}w_{11}1w_{12},\ w'_{11}1w'_{12}\text{-}w_{21}1w_{22},\ w'_{21}1w'_{22}\text{-}w_{31}1w_{32},\ \ldots,\ w'_{n1}1w'_{n2}\text{-}w_1\}.$$

The w_{i1} and w'_{i1} are generated using the nonterminal B, and the w_{i2} and w'_{i2} result from a nonterminal C. As a consequence, w_{ij} and w'_{ij} are indeed of equal length, so that the 1 between w_{i1} and w_{i2} appears in the same position as the one between w'_{i1} and w'_{i2}. Therefore, one can show that the following holds:

> $\mathcal{S}(ERG) \cap \mathcal{S}$ is the set of all $\{s_1, \ldots, s_n\} \in \mathcal{S}$ for which there is an ordering $i_1, \ldots, i_n$ of $1, \ldots, n$ such that s_{i_j} is word adjacent to $s_{i_{j+1}}$ for all i, $1 \leq i < n$.

Clearly, for all $S \in \mathcal{S}$ some graph $H(S)$ representing S in the same way as above can be constructed in polynomial time. Now, $H(S) \in L(ERG)$ if and only if $h(S)$ has a Hamiltonian path. We have thus reduced the Hamiltonian path problem to the membership problem for $L(ERG)$, as required. □

If we drop the requirement of bounded degree in Theorem 2.7.1 we can even generate an NP-complete language of connected graphs, as one can see by an easy modification of the grammar used.

Theorem 2.7.2 (NP-completeness 2)
There is an edge replacement grammar that generates an NP-complete graph language of connected graphs.

Proof
We construct an edge replacement grammar ERG' that generates for every graph $H \in L(ERG)$, where ERG is the edge replacement grammar in the proof of Theorem 2.7.1, a graph obtained from H as follows. First, a new node v_0 is added. Then, every edge e labelled - is removed and its two attached nodes are identified, yielding a new node v_e. Finally, for each v_e a new unlabelled edge with attachment $v_0 v_e$ is added. The correspondence between the graphs in $L(ERG)$ and their counterparts in $L(ERG')$ constructed this way is a bijection that can be computed in polynomial time. Therefore, it remains to give an appropriate grammar ERG'. We may choose for this the grammar $ERG' = (\{S, A, B, C, D\}, \{0, 1\}, P', S)$, where the types of symbols are as in ERG and the productions are the ones given in Figure 2.18. In order to understand how the grammar works, the reader should notice that, applying in a first phase only the productions with left-hand sides S, A, and B yields a hypergraph as in Figure 2.19. Now, it is easy to see that the remaining productions yield just the type of graphs aimed at. □

2.7.2 Two polynomial algorithms

In view of Theorems 2.7.1 and 2.7.2 there is not much hope that one can find an efficient solution to the membership problem for arbitrary hyperedge replacement grammars (and not even for linear edge replacement grammars), since this would require P=NP. It is therefore natural to look for restrictions that can be imposed on the languages or grammars to allow for efficient membership algorithms. The proofs of Theorem 2.7.1 and 2.7.2 give a hint. It seems that, in the first case, the high complexity of the generated language is caused by the fact that the graphs are disconnected, whereas in the the second case the potentially unbounded degree of nodes seems to be responsible. In fact, the

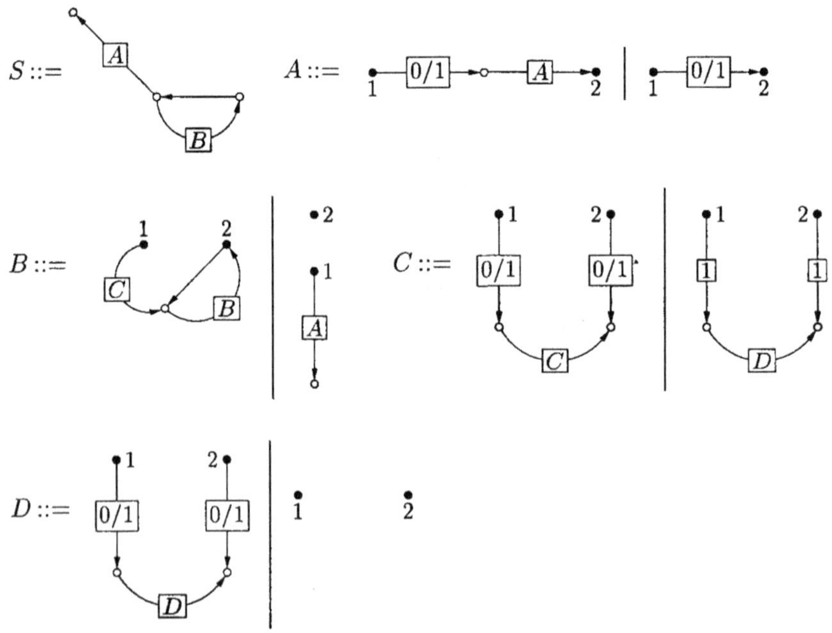

Figure 2.18: The productions of ERG' in the proof of Theorem 2.7.2.

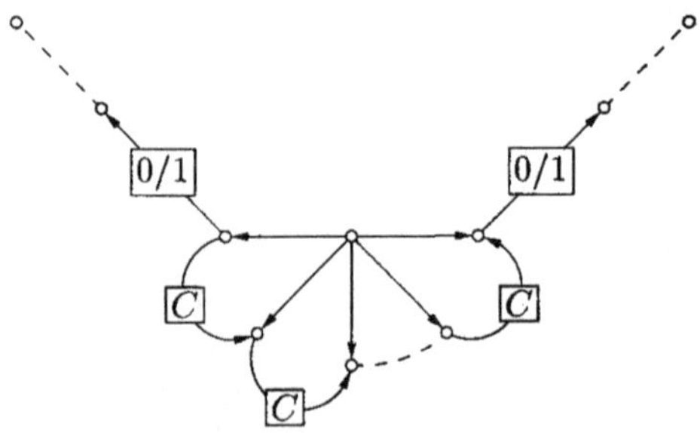

Figure 2.19: An intermediate graph generated by the productions in Figure 2.18.

two cases do not differ too much: The unbounded degree in the second case is nothing else than a hidden disconnectedness that reappears if a single node is deleted. Later on, this observation will lead to the notion of k-separability, which is needed to formulate a condition under which the first membership algorithm we will discuss runs in polynomial time. As opposed to the string case both algorithms we are going to present rely on the fact that the grammar considered is supposed to be fixed. It is easy to see that the running time of these algorithms becomes exponential if the grammar is made part of the input.

In the following, we use the notation $H = H_0<e_1/H_1, \dots, e_n/H_n>$ to express the fact that $H = H_0[e_1/H_1, \dots, e_n/H_n]$, where $V_H = V_{H_0} \cup \cdots \cup V_{H_n}$, $E_H = E_{H_0} \cup \cdots \cup E_{H_n}$, $ext_H = ext_{H_0}$, and $lab_H(e) = lab_{H_i}(e)$, $att_H(e) = att_{H_i}(e)$ for all $e \in E_H \cap E_{H_i}$, $0 \leq i \leq n$. In other words, $H_0<\cdots>$ is hyperedge replacement on *concrete* hypergraphs without the possibility to take isomorphic copies (and is hence not always defined). The notion of X-candidates is central to the first algorithm we are going to present.

Definition 2.7.3 (*X-candidate*)
Let H be hypergraph and let $X \in V_H^*$. A hypergraph $H' \subseteq H$ is an *X-candidate* of H if $ext_{H'} = X$ and for every internal node v of H' and every hyperedge $e \in E_H$ with $v \in [att_H(e)]$ we have $e \in E_{H'}$.

By definition of hyperedge replacement, every hyperedge of a hypergraph $H_0<e_1/H_1, \dots, e_n/H_n>$ that is attached to an internal node of H_i for some i, $1 \leq i \leq n$, is a hyperedge of H_i. This can be used to prove that, in $H_0<e_1/H_1, \dots, e_n/H_n>$, every H_i ($1 \leq i \leq n$) is an $att_{H_0}(e_i)$-candidate of H for $i = 1, \dots, n$. This fact restricts the number of sub-hypergraphs to be tested recursively in the membership algorithm below. We formulate the algorithm for growing hyperedge replacement grammars. By Lemma 2.4.3 this means no loss of generality.

Algorithm 2.7.4
Given: *A growing hyperedge replacement grammar $HRG = (N, T, P, S)$.*
Input: *A hypergraph $\bar{H}$ over T.*
Output: *$Derive(S, \bar{H})$, where $Derive(A, H) =$*

(1) *if $A^\bullet \Longrightarrow R$ with $V_R \subseteq V_H$ and $E_R^T \subseteq E_H$*
  (where $\{e_1, \dots, e_n\} = E_R^N$)
(2) *such that there are substantial $att_R(e_i)$-candidates H_i of $\bar{H}$*
  ($i = 1, \dots, n$) with
(3) *1. $H = R<e_1/H_1, \dots, e_n/H_n>$*
(4) *2. for $i = 1, \dots, n$ we have $Derive(lab_R(e_i), H_i)$*
(5) *then return* **true** *else return* **false** $\Box$

Theorem 2.7.5
For all hypergraphs $\bar{H} \in \mathcal{H}_T$ Algorithm 2.7.4 terminates and yields **true** *if $\bar{H} \in L(HRG)$, and* **false** *otherwise.*

Proof
Since HRG is growing all hypergraphs that can be derived from $A^\bullet$ are substantial. Therefore, Algorithm 2.7.4 is nothing else than an algorithmic formulation of the context-freeness lemma. Since the hypergraph R chosen in step (1) and the $att_R(e_i)$-candidates H_i of step (2) are substantial we have $|H_i| < |H|$ for $i = 1, \ldots, n$, so the algorithm must eventually terminate. □

Since there is only a finite set of productions in HRG there are only polynomially many possibilities to consider in step (1), and the equality in step (3) can also be tested in polynomial time (note that the test is for equality, not isomorphy). Hence, there is only a single point that causes an exponential running time of the algorithm: the number of X-candidates to be tested. Therefore, we aim at a condition to be imposed on $L(HRG)$ that implies a polynomial upper bound on the number of X-candidates. Let us denote by $H|_V$, where H is a hypergraph and $V \subseteq V_H$, the hypergraph $(V, E_H, att, lab_H, \lambda)$, where $att(e)$ is the restriction of $att_H(e)$ to V, for all $e \in E$. A *connected component* of H is a maximal connected sub-hypergraph of H. (Note that a 0-edge is a connected component on its own.)

Definition 2.7.6 (*k-separability*)
For $k \in \mathbb{N}$ the k-separability $k\text{-}sep(H)$ of a hypergraph H is the maximum number of connected components of $H|_V$, where V ranges over all subsets of V_H of size at most k. For every language L of hypergraphs, $k\text{-}sep_L\colon \mathbb{N} \to \mathbb{N}$ is defined by

$$k\text{-}sep_L(n) = \max\{k\text{-}sep(H) \mid H \in L \text{ and } |H| \leq n\}.$$

The following theorem says that logarithmic k-separability of $L(HRG)$ (where HRG is of order k) results in a polynomial upper bound on the running time of Algorithm 2.7.4.

Theorem 2.7.7
Let $HRG \in \mathcal{HRG}_k$ be growing. If $k\text{-}sep_{L(HRG)} \in O(\log n)$ then Algorithm 2.7.4 can be implemented to run in polynomial time in $|\bar{H}|$.

Proof
Let $\bar{H} \in \mathcal{H}_T$ and $X \in V_{\bar{H}}^*$ with $|X| \leq k$. By definition of X-candidates, if a node of a connected component G of $\bar{H}|_{[X]}$ occurs in an X-candidate H

of $\bar{H}$, then $G \subseteq H$. Since there are only a logarithmic number of connected components in $\bar{H}|_{[X]}$ this yields a polynomial bound on the number of X-candidates in $\bar{H}$ if $\bar{H} \in L(HRG)$. Since the number of sequences $X \in V_{\bar{H}}^*$ with $|X| \leq k$ is bounded by $|V_{\bar{H}}|^k$ this yields a polynomial bound on the number of X-candidates to be considered in the algorithm (that is, if there are more X-candidates in the input graph it can be rejected immediately).

We are now able to make use of the well-known idea underlying the membership algorithm for context-free Chomsky grammars by Cocke, Kasami, and Younger. In an initial phase, we compute the set of all X-candidates of $\bar{H}$, where $|X| \leq k$. This can be done in polynomial time since it essentially suffices to determine all combinations of connected components of the hypergraphs $\bar{H}|_V$, where $|V| \leq k$. Now, every call to *Derive* with parameters A and H is performed only once and the result is stored in a list which is looked up when repeated calls occur. As a consequence, the algorithm runs in polynomial time in the number of X-components, which is polynomial in $|\bar{H}|$, as required. □

The major drawback of Algorithm 2.7.4 is that, even if $k\text{-}sep_{L(HRG)}$ is a constant, the running time is bounded by a polynomial whose degree may vary with *HRG*. In view of step (1) of the algorithm, for instance, the degree depends on the size of right-hand sides used in the grammar. We now want to discuss a membership algorithm for edge replacement grammars that always runs in cubic time. It applies to all input graphs that are *almost 2-connected*, a notion to be defined below. Again, the algorithm exploits the idea by Cocke, Kasami, and Younger.

From now on, let us consider some arbitrary (but fixed) edge replacement grammar (N, T, P, S) with $type(A) = 2$ for all $A \in N \cup T$. (In particular, every graph is assumed to be of type 2 in the following, without mentioning.)

Definition 2.7.8 (*almost k-connected hypergraph*)
A hypergraph H is *almost k-connected* for some $k \in \mathbb{N}$ if

- $|V_H| \geq k$ and
- for all hypergraphs H_1, H_2 and every hyperedge $e \in E_{H_1}$, if we have $H = H_1[e/H_2]$ and $V_{H_1} \setminus [att_{H_1}(e)] \neq \emptyset \neq V_{H_2} \setminus [ext_{H_2}]$ then $type(H_2) \geq k$.

Intuitively, almost k-connectedness means that a graph cannot be decomposed into non-trivial parts using hypergraphs of type less than k. The graph H shown in Figure 2.20, for example, is almost 2-connected, but H' is not.
A substantial graph H (of type 2) is a *bond* if $V_H = ext_H$, that is, if H consists of at least two edges between the external nodes. A substantial string graph is

Figure 2.20: H is almost 2-connected while H' is not.

said to be a *chain*. H is called a *block* if it is almost 3-connected, $|V_H| > 3$, and $[att_H(e)] \neq [ext_H]$ for all $e \in E_H$. Notice that bonds, chains, and blocks are substantial and that these three classes of graphs are mutually disjoint. The smallest bonds and chains are those with two edges. We have the following theorem about almost 2-connected graphs, that we state without proof.

Lemma 2.7.9

1. *Let $H = H_0\langle e/H_1\rangle$ for substantial graphs H_0 and H_1. Then H is almost 2-connected if and only if both H_0 and H_1 are almost 2-connected.*

2. *Every substantial 2-connected graph H can be written as $H_0\langle e/H_1\rangle$, for substantial graphs H_0, H_1, unless H is a bond or chain with two edges, or a block.* □

In the following, let us reserve a label τ to be used as a special one in some constructions, that is, we assume $\tau \notin N \cup T$. For the algorithm we are going to explain the following notions are of basic importance.

Definition 2.7.10 (*total and collapsed split tree*)
Let H be a hypergraph over T.

1. A *total split tree* of H is a derivation tree for H over P_t, where P_t is the set of all productions over $\{\tau\}$ whose right-hand sides are bonds and chains with two edges, and blocks.

2. Let P_c be the set of all productions over $\{\tau\}$ whose right-hand sides are bonds, chains, or blocks. A *collapsed split tree* of H is a derivation tree for H over P_c such that no edge in this derivation tree connects two bonds or two chains.

By definition, the result of a (total or collapsed) split tree t is a substantial, almost 2-connected graph. Since the left-hand side of productions in a split tree is always τ we may consider a split tree as a pair $(H, branch)$ rather than as a triple $(\tau, H, branch)$. Using Lemma 2.7.9 it is not hard to see that every substantial, almost 2-connected graph has some total split tree. From a total split tree one can obtain a collapsed one as follows: As long as a subtree $(H, branch)$ exists with $branch(e) = (H', branch')$ for some $e \in E_H^N$, where both H and H' are bonds or both are chains, replace the subtree by $(H[e/H'], branch'')$, where

$$branch''(e') = \begin{cases} branch(e') & \text{if } e' \in E_N^H - \{e\} \\ branch'(e') & \text{if } e' \in E_N^{H'}. \end{cases}$$

Then $H[e/H']$ is a bond (chain, respectively), so the procedure eventually leads to a collapsed split tree. As a consequence, every substantial, almost 2-connected graph has a total as well as a collapsed split tree, and for every (total or collapsed) split tree t the graph $result(t)$ is substantial and almost 2-connected.

The algorithm we are going to present makes use of a result by MacLane [78] saying that collapsed split trees are unique in a certain sense. It is easy to see that collapsed split trees cannot be really unique. This is because, if H and H' are graphs with $e \in E_H$, and we let G and G' be obtained from them by reversing e in H and interchanging the external nodes of H' we obviously obtain $H[e/H'] = G[e/G']$. Hence, for split trees we must allow to reverse nonterminal edges if at the same time the external nodes of the graphs that replace them are interchanged. This is stated more precisely below.

Definition 2.7.11 ***(similar split trees)***

Let $t = (H, branch)$ and $t' = (H', branch')$ be split trees with $E_H^N = \{e_1, \ldots, e_n\}$ and, for $i = 1, \ldots, n$, $branch(e_i) = (H_i, branch_i)$. Then t and t' are *similar* if there is some $I \subseteq \{1, \ldots, n\}$ such that the following hold.

1. The graph obtained from H by reversing all e_i with $i \in I$ is isomorphic to H' via some isomorphism h.

2. For all $i \in I$, $(H_i', branch_i)$ and $branch(h_E(e_i))$ are similar, where H_i' is obtained from H_i by reversing ext_{H_i} if $i \in I$ and $H_i' = H_i$ otherwise.

As one can easily show by induction, similarity of t and t' implies that $result(t)$ and $result(t')$ are isomorphic. The algorithm we are developing is based on the following result by MacLane [78].

Fact 2.7.12 (uniqueness of collapsed split trees)

All collapsed split trees of a substantial, almost 2-connected graph are similar. □

Using Fact 2.7.12 we want to prove the following.

Theorem 2.7.13
Let ERG be an edge replacement grammar. Then there is a cubic algorithm that takes as input a graph H and decides whether H is an almost 2-connected member of $L(ERG)$. In particular, if all graphs in $L(ERG)$ are almost 2-connected, the algorithm decides whether $H \in L(ERG)$. □

We are now going to develop the means to prove Theorem 2.7.13. Let us first assume some edge replacement grammar $ERG = (N, T, P, S)$ as in the theorem is given. By Lemma 2.4.3 it may be assumed that ERG is growing. We modify ERG as follows.

(1) Remove all productions whose right-hand side is not almost 2-connected. By Lemma 2.7.9 we are allowed to do so because this modification does not affect the set of almost 2-connected graphs in the language generated.

(2) As long as there is a production (A, R) such that $H = R[e/H']$ for some substantial H and H', choose a new nonterminal label A' for e in H and replace (A, R) by the two productions (A, H) and (A', H'). Obviously, such a modification does not influence the generated language. By Lemma 2.7.9 the right-hand sides of the grammar obtained are bonds and chains with two edges, and blocks.

(3) Complete the grammar as follows: For every label A (nonterminal as well as terminal ones), add a new label $\bar{A}$, and add for all productions (A, R) in the grammar obtained in step (2), all productions (X, H), where $X \in \{A, \bar{A}\}$ and H is obtained from R by reversing some edges while changing their label from $lab_R(e)$ to $\overline{lab_R(e)}$, and reversing ext_R if $X = \bar{A}$.

Let the edge replacement grammar resulting from steps (1)–(3) be given by $ERG_0 = (N_0, T_0, P_0, S)$. Then it follows by a straightforward induction that the set of graphs over T generated by ERG_0 coincides with the one of the grammar obtained by steps (1) and (2), that is, $L(ERG_0) = L(ERG)$. For every derivation tree t over P_0^* (see page 115 for the definition of P_0^*) let us denote by $unlabel(t)$ the derivation tree we obtain by replacing all nonterminal labels with τ. We have the following lemma.

Lemma 2.7.14
Let $H \in \mathcal{H}_{T_0}$ and $A \in N_0$. If $A^\bullet \Longrightarrow^*_{P_0} H$ and t is a collapsed split tree of H, then there is a derivation tree t' over P_0^* with root A such that $unlabel(t') = t$.

Proof
Let t_0 be a derivation tree for H in ERG_0. As remarked above, by modification

(2) $unlabel(t_0)$ is a total split tree for every derivation tree t_0 over P_0. Hence there is some derivation tree t_0' over P_0^* such that $unlabel(t_0')$ is a collapsed split tree. (We can construct t_0' from t_0 in the same way a collapsed split tree can be constructed from a total one; see the paragraph after Definition 2.7.10.) By Fact 2.7.12 the given collapsed split tree t and $unlabel(t_0')$ are similar. By modification (3) this means there is a derivation tree t' over P_0^* with $unlabel(t') = t$. □

We are now able to sketch the construction of the cubic algorithm we aimed at, thereby proving Theorem 2.7.13.

Proof of Theorem 2.7.13
To find out whether an input graph H is in $L(ERG_0)$, where ERG_0 is constructed as above, proceed as follows. First, compute a collapsed split tree t of H. By a result proved by Hopcroft and Tarjan [79] this can be done in linear time. By Lemma 2.7.14, if $H \in L(ERG_0)$ (and, of course, only then) we can exchange every τ in t by an appropriate nonterminal label to obtain a derivation tree over P_0^* for H whose root is S. To find out whether this is possible, for all subtrees t' of t we compute by a bottom-up approach the set $poslab(t')$ of all $A \in N_0$ such that $A^\bullet \Longrightarrow_{P_0}^* result(t')$. This can be done efficiently, as follows. Suppose $t' = (R, branch)$ with $branch(e) = t_e$ for $e \in E_R^N$, and we have already computed $poslab(t_e)$ for all $e \in E_R^N$. Then, we have to decide whether $A^\bullet \Longrightarrow_{P_0}^* R_1$ for some R_1 obtained from R by labelling every edge labelled with τ in R with one of the labels in $poslab(t_e)$. Since R is either a bond, a chain, or a block there are three possible cases to consider.

1. R is a bond.
   Construct R' from R by giving a hyperedge $e \in E_R$ the label $\{lab_R(e)\}$ if $lab_R(e) \in T_0$ and the label $poslab(t_e)$ if $lab_R(e) = \tau$. Let R'' be the graph obtained from R' by reversing all edges e with $att_{R'}(e) = end_{R'}\, begin_{R'}$, while exchanging $lab_{R'}(e)$ with $\{\bar{A} \mid A \in lab_{R'}(e)\}$. By the completion performed in modification (3) and Theorem 2.4.7 it is not hard to see that $A \in poslab(t')$ if and only if $\psi(R'')$ is in some fixed semilinear set associated with A. (To be able to apply Theorem 2.4.7, add productions $(A, X^\bullet)$, where $X \subseteq N_0$ and $A \in X$, or $A \in T_0$ and $X = \{A\}$. Then we have to decide whether $A^\bullet \Longrightarrow_{P_0}^* R''$.) Due to results by Fischer, Meyer, and Rosenberg [80] membership in a semilinear set can be decided in linear time.

2. R is a chain.
   Then, since in a derivation $A^\bullet \Longrightarrow_{P_0}^* R_1$ we cannot use any production with a substantial right-hand side which is not a chain, we can use

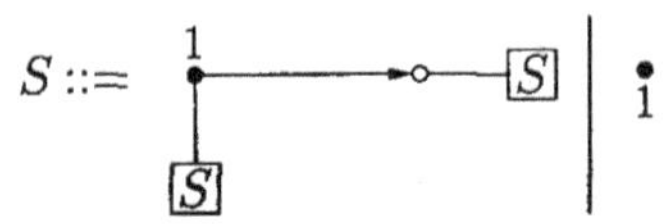

Figure 2.21: A hyperedge replacement grammar with componentwise derivations.

the well-known algorithm by Cocke, Kasami, and Younger to find out *poslab*(t'). (Note that we do not have to consider all the possible R_1 separately, because the CKY-algorithm works with sets of nonterminals like our *poslab*(t_e) provide.) Since the CKY-algorithm runs in cubic time this step takes cubic time in the size of R.

3. R is a block.
   Here, we only have to test a finite number of possibilities since R must be isomorphic to a right-hand side in ERG_0, up to the labelling. Hence, this case can be handled in constant time.

Altogether, we need at most a cubic number of steps (in $|H|$) in order to find *poslab*(t), and accept H if $S \in$ *poslab*(t). □

2.7.3 Further results and bibliographic notes

The proof of NP-completeness given in the beginning of this section was found by Lange and Welzl [81], who presented it for *string grammars with disconnecting* rather than for edge replacement, as we did here.

Algorithm 2.7.4 is due to Lautemann [29]. The paper contains a second variant of the algorithm that does not rely on logarithmic k-separability. Instead, it is required that the considered hyperedge replacement grammar has *componentwise derivations*. Roughly speaking, this means that an X-candidate is derivable from $A^\bullet$ if and only if each of the connected components this X-candidate consists of is derivable from $A^\bullet$. As an example, one may consider the grammar $HRG = (\{S\}, \{\sqcup\}, P, S)$ (where "$\sqcup$" means "unlabelled") whose productions are the two given in Figure 2.21. $L(HRG)$ is the set of all (rooted, directed) trees, and a graph as in Figure 2.22 on the left is derivable from $S^\bullet$ if and only if each of its components (shown on the right) can be derived from $S^\bullet$. Under the condition that the grammar satisfies this requirement, we may look at the (linear number of) individual connected components rather than at all X-candidates, which yields a polynomial algorithm. As Lautemann remarks in

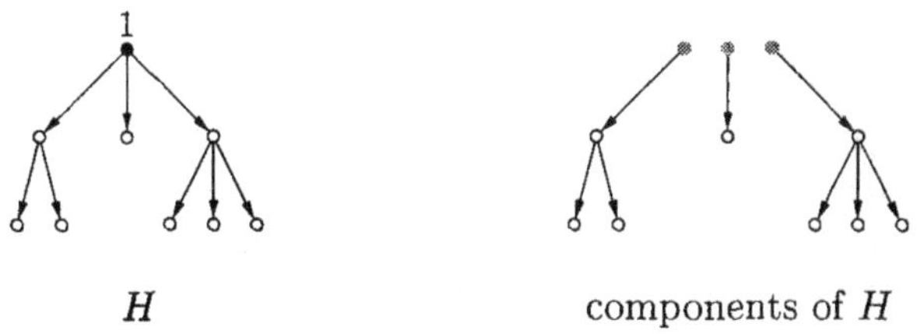

Figure 2.22: A graph generated by HRG and its components.

his paper, more sophisticated notions of componentwise derivations, that still lead to polynomial algorithms, may more or less obviously be thought of.

The cubic membership algorithm was presented by Vogler in [82]. As shown by Drewes [33], Theorem 2.7.13 and the results used to prove it can be generalized to hyperedge replacement grammars of order k generating almost k-connected k-hypergraphs, where a k-hypergraph is a hypergraph all of whose hyperedges are of type k. This yields the case treated here by choosing $k = 2$. It may be interesting to notice that the generalized version of the algorithm is still cubic; the exponent does not depend on the order of the grammar.

As mentioned, the generalization works only for k-hypergraphs. It is therefore natural to wonder whether this restriction can be avoided. Is there still a polynomial algorithm if a grammar of order k generates, for instance, a language of almost k-connected graphs? Unfortunately, it is likely that this question has to be answered negatively, since it was shown by Drewes [32] that this case is again NP-complete for all $k > 2$. (For $k = 2$, we have Theorem 2.7.13.) Intuitively, this is caused by the fact that, if we replace every hyperedge in a k-connected k-hypergraph by, say, a clique on the k attached nodes, we retain k-connectedness, but loose the information which of the resulting edges belong together, so that a unique reconstruction of the hypergraph is not possible.

2.8 Conclusion

In this survey, we have outlined the theory of hyperedge replacement as a grammatical device for the generation of hypergraph, graph, and string languages with an emphasis on the sequential mode of rewriting. Hyperedge replacement (without application conditions) has a context-free nature which is the key for an attractive mathematical theory including structural properties and decidability results.

Although several interesting (hyper)graph languages occurring in computer science and graph theory can be generated by hyperedge replacement grammars, their generative power is bound to be restricted. As we saw in Section 2.4 there is, for instance, no way to generate a language of unbounded connectivity. As mentioned in the bibliographic notes of Section 2.3, extensions can be found in the literature, which make it possible to overcome one or the other deficiency. However, one cannot expect to get additional power for free. Normally, extensions lack some of the useful properties of hyperedge replacement, or their rewriting mechanisms are more complicated and therefore harder to reason about. In many cases hyperedge replacement seems to be a good compromise between the wish to have a nicely developed mathematical theory and the demand for reasonable generative power.

Acknowledgement

We thank Grzegorz Rozenberg and the anonymous referee for helpful comments on a draft of this chapter. The work presented here has been supported by ESPRIT Basic Research Working Group 7183 (COMPUGRAPH II).

References

1. Jerome Feder. Plex languages. *Information Sciences*, 3:225–241, 1971.
2. Theodosios Pavlidis. Linear and context-free graph grammars. *Journal of the ACM*, 19(1):11–23, 1972.
3. Pierluigi Della Vigna and Carlo Ghezzi. Context-free graph grammars. *Information and Control*, 37:207–233, 1978.
4. Dirk Janssens and Grzegorz Rozenberg. Restrictions, extensions and variations of NLC grammars. *Information Sciences*, 20:217–244, 1980.
5. A.O. Slisenko. Context-free graph grammars as a tool for describing polynomial-time subclasses of hard problems. *Information Processing Letters*, 14:52–56, 1982.
6. Michel Bauderon and Bruno Courcelle. Graph expressions and graph rewriting. *Mathematical Systems Theory*, 20:83–127, 1987.
7. Annegret Habel and Hans-Jörg Kreowski. Characteristics of graph languages generated by edge replacement. *Theoretical Computer Science*, 51:81–115, 1987.
8. Annegret Habel and Hans-Jörg Kreowski. May we introduce to you: Hyperedge replacement. In H. Ehrig, M. Nagl, G. Rozenberg, and A. Rosenfeld, editors, *Graph-Grammars and Their Application to Computer Science*, volume 291 of *Lecture Notes in Computer Science*, pages 15–26, 1987.

9. Ugo Montanari and Francesca Rossi. An efficient algorithm for the solution of hierarchical networks of constraints. In H. Ehrig, M. Nagl, G. Rozenberg, and A. Rosenfeld, editors, *Graph-Grammars and Their Application to Computer Science*, volume 291 of *Lecture Notes in Computer Science*, pages 440–457, 1987.
10. Clemens Lautemann. Efficient algorithms on context-free graph languages. In T. Lepistö and A. Salomaa, editors, *Automata, Languages and Programming*, volume 317 of *Lecture Notes in Computer Science*, pages 362–378, 1988.
11. Joost Engelfriet. Context-free NCE graph grammars. In J. Csirik, J. Demetrovics, and F. Gécseg, editors, *Fundamentals of Computation Theory*, volume 380 of *Lecture Notes in Computer Science*, pages 148–161, 1989.
12. Joost Engelfriet. A characterization of context-free NCE graph languages by monadic second-order logic on trees. In H. Ehrig, H.-J. Kreowski, and G. Rozenberg, editors, *Graph Grammars and Their Application to Computer Science*, volume 532 of *Lecture Notes in Computer Science*, pages 311–327, 1991.
13. Thomas Lengauer and Egon Wanke. Efficient decision procedures for graph properties on context-free graph languages. *Journal of the ACM*, 40:368–393, 1993.
14. Bruno Courcelle. Graph rewriting: An algebraic and logical approach. In J. van Leeuwen, editor, *Handbook of Theoretical Computer Science*, volume Vol. B., pages 193–242. Elsevier, Amsterdam, 1990.
15. Bruno Courcelle. The monadic second-order logic of graphs I: Recognizable sets of finite graphs. *Information and Computation*, 85:12–75, 1990.
16. Bruno Courcelle. Context-free graph grammars: Separating vertex replacement from hyperedge replacement. In Z. Ésik, editor, *Fundamentals of Computation Theory*, volume 710 of *Lecture Notes in Computer Science*, pages 181–193, 1993.
17. Bruno Courcelle and Joost Engelfriet. A logical characterization of the sets of hypergraphs defined by hyperedge replacement systems. *Mathematical Systems Theory*, 28:515–552, 1995.
18. Joost Engelfriet and Grzegorz Rozenberg. A comparison of boundary graph grammars and context-free hypergraph grammars. *Information and Computation*, 84:163–206, 1990.
19. Joost Engelfriet and Linda Heyker. The string generating power of context-free hypergraph grammars. *Journal of Computer and System Sciences*, 43:328–360, 1991.
20. Joost Engelfriet and Linda Heyker. Context-free hypergraph grammars have the same term-generating power as attribute grammars. *Acta Informatica*, 29:161–210, 1992.

21. Joost Engelfriet. A Greibach normal form for context-free graph grammars. In W. Kuich, editor, *Automata, Languages and Programming*, volume 623 of *Lecture Notes in Computer Science*, pages 138–149, 1992.
22. Joost Engelfriet, Linda Heyker, and George Leih. Context-free graph languages of bounded degree are generated by apex graph grammars. *Acta Informatica*, 31:341–378, 1994.
23. Joost Engelfriet. Graph grammars and tree transducers. In *CAAP'94*, volume 787 of *Lecture Notes in Computer Science*, pages 15–36, 1994.
24. Hans-Jörg Kreowski. A pumping lemma for context-free graph languages. In V. Claus, H. Ehrig, and G. Rozenberg, editors, *Graph-Grammars and Their Application to Computer Science and Biology*, volume 73 of *Lecture Notes in Computer Science*, pages 270–283, 1979.
25. Hans-Jörg Kreowski. Rule trees represent derivations in edge replacement systems. In G. Rozenberg and A. Salomaa, editors, *The Book of L*, pages 217–232. Springer-Verlag, Berlin, 1986.
26. Annegret Habel and Hans-Jörg Kreowski. Some structural aspects of hypergraph languages generated by hyperedge replacement. In F. Brandenburg, G. Vidal-Naquet, and M. Wirsing, editors, *STACS 87*, volume 247 of *Lecture Notes in Computer Science*, pages 207–219, 1987.
27. Annegret Habel, Hans-Jörg Kreowski, and Walter Vogler. Metatheorems for decision problems on hyperedge replacement graph languages. *Acta Informatica*, 26:657–677, 1989.
28. Annegret Habel and Hans-Jörg Kreowski. Filtering hyperedge-replacement languages through compatible properties. In M. Nagl, editor, *Graph-Theoretic Concepts in Computer Science*, volume 411 of *Lecture Notes in Computer Science*, pages 107–120, 1990.
29. Clemens Lautemann. The complexity of graph languages generated by hyperedge replacement. *Acta Informatica*, 27:399–421, 1990.
30. Annegret Habel, Hans-Jörg Kreowski, and Walter Vogler. Decidable boundedness problems for sets of graphs generated by hyperedge replacement. *Theoretical Computer Science*, 89:33–62, 1991.
31. Clemens Lautemann. Tree automata, tree decomposition, and hyperedge replacement. In H. Ehrig, H.-J. Kreowski, and G. Rozenberg, editors, *Graph Grammars and Their Application to Computer Science*, volume 532 of *Lecture Notes in Computer Science*, pages 520–537, 1991.
32. Frank Drewes. NP-completeness of k-connected hyperedge-replacement languages of order k. *Information Processing Letters*, 45:89–94, 1993.
33. Frank Drewes. Recognising k-connected hypergraphs in cubic time. *Theoretical Computer Science*, 109:83–122, 1993.
34. Annegret Habel, Hans-Jörg Kreowski, and Clemens Lautemann. A comparison of compatible, finite, and inductive graph properties. *Theoretical Computer Science*, 110:145–168, 1993.

35. Thomas Lengauer and Egon Wanke. Efficient analysis of graph properties on context-free graph languages. In T. Lepistö and A. Salomaa, editors, *Automata, Languages and Programming*, volume 317 of *Lecture Notes in Computer Science*, pages 379–393, 1988.
36. Thomas Lengauer and Egon Wanke. Efficient solution of connectivity problems on hierarchically defined graphs. *SIAM Journal on Computing*, 17:1063–1080, 1988.
37. Egon Wanke. On the decidability of certain integer subgraph problems on context-free graph languages. *Information and Computation*, 113:26–49, 1994.
38. Mark Minas and Gerhard Viehstaedt. Specification of diagram editors providing layout adjustment with minimal change. In *Proc. IEEE Symp. on Visual Languages (VL'93)*, pages 324–329. IEEE Comp. Society Press, 1993.
39. Mark Minas and Gerhard Viehstaedt. *DiaGen*: A generator for diagram editors providing direct manipulation and execution of diagrams adjustment with minimal change. In *Proc. IEEE Symp. on Visual Languages (VL'95)*. IEEE Comp. Society Press, 1995.
40. Gerhard Viehstaedt. A generator for diagram editors. Doctoral Dissertation, University of Erlangen, Germany, 1995.
41. R. Farrow, K. Kennedy, and L. Zucconi. Graph grammars and global program data flow analysis. In *Proc. 17th Ann. IEEE Symp. on Found. of Comp. Sci.*, pages 42–56, Houston, 1976.
42. Annegret Habel. *Hyperedge Replacement: Grammars and Languages*, volume 643 of *Lecture Notes in Computer Science*. Springer-Verlag, Berlin, 1992.
43. Annegret Habel. Hypergraph grammars: Transformational and algorithmic aspects. *Journal of Information Processing and Cybernetics EIK*, 28:241–277, 1992.
44. Bruno Courcelle, Joost Engelfriet, and Grzegorz Rozenberg. Handle-rewriting hypergraph grammars. *Journal of Computer and System Sciences*, 46:218–270, 1993.
45. Annegret Habel and Hans-Jörg Kreowski. Collage grammars. In H. Ehrig, H.-J. Kreowski, and G. Rozenberg, editors, *Graph Grammars and Their Application to Computer Science*, volume 532 of *Lecture Notes in Computer Science*, pages 411–429, 1991.
46. Frank Drewes, Annegret Habel, Hans-Jörg Kreowski, and Stefan Taubenberger. Generating self-affine fractals by collage grammars. *Theoretical Computer Science*, 145:159–187, 1995.

47. Jürgen Dassow, Annegret Habel, and Stefan Taubenberger. Chain-code pictures and collages generated by hyperedge replacement. In H. Ehrig, H.-J. Kreowski, and G. Rozenberg, editors, *Proc. Fifth Intl. Workshop on Graph Grammars and Their Application to Computer Science*, volume 1073 of *Lecture Notes in Computer Science*, pages 412–427, Springer, 1996.
48. Frank Drewes and Hans-Jörg Kreowski. (Un)decidability of properties of pictures generated by collage grammars. *Fundamenta Informaticae* 25, 295–325, 1996.
49. Frank Drewes. Language theoretic and algorithmic properties of *d*-dimensional collages and patterns in a grid. *Journal of Computer and System Sciences*, 1996. To appear.
50. Bruno Courcelle. An axiomatic definition of context-free rewriting and its application to NLC graph grammars. *Theoretical Computer Science*, 55:141–181, 1987.
51. Hartmut Ehrig and Karl Wilhelm Tischer. Graph grammars and applications to specialization and evolution in biology. *Journal of Computer and System Sciences*, 11:212–236, 1975.
52. Hans-Jörg Kreowski. Parallel hyperedge replacement. In G. Rozenberg and A. Salomaa, editors, *Lindenmayer Systems*, pages 271–282. Springer-Verlag, Berlin Heidelberg New York, 1992.
53. Gnanamalar David, Frank Drewes, and Hans-Jörg Kreowski. Hyperedge replacement with rendezvous. In P. Jouannaud, editor, *Proc. Theory and Practice of Software Development*, volume 668 of *Lecture Notes in Computer Science*, pages 167–181, 1993.
54. Hans-Jörg Kreowski. Five facets of hyperedge replacement beyond context-freeness. In Z. Ésik, editor, *Fundamentals of Computation Theory*, volume 710 of *Lecture Notes in Computer Science*, pages 69–86, 1993.
55. Michel Bauderon. Infinite hypergraphs I. Basic properties. *Theoretical Computer Science*, 82:177–214, 1991.
56. Michel Bauderon. Infinite hypergraphs II. Systems of recursive equations. *Theoretical Computer Science*, 103:165–190, 1992.
57. Seymour Ginsburg and Gordon Rice. Two families of languages related to ALGOL. *Journal of the ACM*, 9:350–371, 1962.
58. Y. Bar-Hillel, M. Perles, and E. Shamir. On formal properties of simple phrase-structure grammars. *Zeitschrift für Phonetik, Sprachwissenschaft und Kommunikationsforschung*, 14:143–177, 1961.
59. John E. Hopcroft and Jeffrey D. Ullman. *Formal Languages and Their Relation to Automata.* Addison-Wesley, Reading, Mass., 1969.

60. R.J. Parikh. On context-free languages. *Journal of the ACM*, 13:570–581, 1966.
61. Donald J. Rose. On simple characterizations of k-trees. *Discrete Mathematics*, 7:317–322, 1974.
62. Ferenc Gécseg and Magnus Steinby. *Tree Automata.* Akadémiai Kiadó, Budapest, 1984.
63. Alfred V. Aho and Jeffrey D. Ullman. Translations on a context free grammar. *Information and Control*, 19:439–475, 1971.
64. Joost Engelfriet, Grzegorz Rozenberg, and Giora Slutzki. Tree transducers, L systems, and two-way machines. *Journal of Computer and System Sciences*, 20:150–202, 1980.
65. John E. Hopcroft and Jeffrey D. Ullman. *Introduction to Automata Theory, Languages and Computation.* Addison–Wesley, Reading, Massachusetts, 1979.
66. Joost Engelfriet and Heiko Vogler. The translation power of top-down tree-to-graph transducers. *Journal of Computer and System Sciences*, 49:258–305, 1994.
67. Joost Engelfriet and Heiko Vogler. Macro tree transducers. *Journal of Computer and System Sciences*, 31:71–146, 1985.
68. Joost Engelfriet and Linda Heyker. Hypergraph languages of bounded degree. *Journal of Computer and System Sciences*, 48:58–89, 1994.
69. Stefan Arnborg, Jens Lagergren, and Detlef Seese. Problems easy for tree-decomposable graphs. *Journal of Algorithms*, 12:308–340, 1991.
70. Egon Wanke and Manfred Wiegers. Undecidability of the bandwidth problem on linear graph languages. *Information Processing Letters*, 33:193–197, 1989.
71. Bruno Courcelle and Mohamed Mosbah. Monadic second-order evaluations on tree-decomposable graphs. *Theoretical Computer Science*, 109:49–82, 1993.
72. Helmut Seidl. Finite tree automata with cost functions. *Theoretical Computer Science*, 126:113–142, 1994.
73. Frank Drewes. A lower bound on the growth of functions computed by tree transductions. *Fundamenta Informaticae*, 1996. To appear; short version in Lecture Notes in Computer Science 787 (CAAP 94).
74. Frank Drewes. The use of tree transducers to compute translations between graph algebras. In *Proc. Fifth Intl. Workshop on Graph Grammars and Their Application to Computer Science*, volume 1073 of *Lecture Notes in Computer Science*, pages 196–210, Springer, 1996.
75. Frank Drewes. Computation by tree transductions. Doctoral Dissertation, University of Bremen, Germany, 1996.

76. Michael R. Garey and David S. Johnson. *Computers and Intractability. A Guide to the Theory of NP-Completeness.* W.H. Freeman and Company, New York, 1979.
77. Ijsbrand Jan Aalbersberg, Andrzej Ehrenfeucht, and Grzegorz Rozenberg. On the membership problem for regular DNLC grammars. *Discrete Applied Mathematics*, 13:79–85, 1986.
78. Saunders MacLane. A structural characterization of planar combinatorial graphs. *Duke Mathematical Journal*, 3:460–472, 1937.
79. John E. Hopcroft and Robert E. Tarjan. Dividing a graph into triconnected components. *SIAM Journal on Computing*, 2(3):135–158, 1973.
80. Patrick C. Fischer, Albert R. Meyer, and Arnold L. Rosenberg. Counter machines and counter languages. *Mathematical Systems Theory*, 2:265–283, 1968.
81. Klaus-Jörn Lange and Emo Welzl. String grammars with disconnecting or a basic root of the difficulty in graph grammar parsing. *Discrete Applied Mathematics*, 16:17–30, 1987.
82. Walter Vogler. Recognizing edge replacement graph languages in cubic time. In H. Ehrig, H.-J. Kreowski, and G. Rozenberg, editors, *Graph Grammars and Their Application to Computer Science*, volume 532 of *Lecture Notes in Computer Science*, pages 676–687, 1991.

Chapter 3

ALGEBRAIC APPROACHES TO GRAPH TRANSFORMATION

PART I: BASIC CONCEPTS AND DOUBLE PUSHOUT APPROACH

A. CORRADINI, U. MONTANARI, F. ROSSI
Dipartimento di Informatica, Corso Italia 40, I-56125 Pisa, Italy

H. EHRIG, R. HECKEL, M. LÖWE
Technische Universität Berlin, Fachbereich 13 Informatik, Franklinstraße 28/29, D-10587 Berlin, Germany

The algebraic approaches to graph transformation are based on the concept of gluing of graphs, modelled by pushouts in suitable categories of graphs and graph morphisms. This allows one not only to give an explicit algebraic or set theoretical description of the constructions, but also to use concepts and results from category theory in order to build up a rich theory and to give elegant proofs even in complex situations. In this chapter we start with an overwiev of the basic notions common to the two algebraic approaches, the ***double-pushout (DPO) approach*** and the ***single-pushout (SPO) approach***; next we present the classical theory and some recent development of the double-pushout approach. The next chapter is devoted instead to the single-pushout approach, and it is closed by a comparison between the two approaches.

Contents

3.1 Introduction

The algebraic approach to graph grammars has been invented at the Technical University of Berlin in the early seventies by H. Ehrig, M. Pfender and H.J. Schneider in order to generalize Chomsky grammars from strings to graphs [1]. The main idea was to generalize the concatenation of strings to a gluing construction for graphs. This allowed to formulate a graph rewriting step by two gluing constructions. The approach was called "algebraic" because graphs are considered as special kinds of algebras and the gluing for graphs is defined by an "algebraic construction", called *pushout*, in the category of graphs and total graph morphisms. This basic idea allows one to apply general results from algebra and category theory in the algebraic theory of graph grammars. In addition to the idea to generalize Chomsky grammars from strings to graphs, it was B. K. Rosen's idea to use graph grammars as a generalization of term rewriting systems where terms are represented by specific kinds of trees. In fact, graphs are suitable to model terms with shared subterms, which are no longer trees, and therefore term rewriting with shared subterms can be modelled by graph rewriting.

Graph grammars in general provide an intuitive description for the manipulation of graphs and graphical structures as they occur in programming language semantics, data bases, operating systems, rule based systems and various kinds of software and distributed systems. In most of these cases graph grammars allow one to give a formal graphical specification which can be executed as far as corresponding graph grammar tools are available. Moreover, theoretical results for graph grammars are useful for analysis, correctness and consistency proofs of such systems. The algebraic approach [2,3,4,5] has been worked out for several years and provides interesting results for parallelism analysis [6,7], efficient evaluation of functional expressions [8,9] and logic programs [10], synchronization mechanisms [11], distributed systems [11,12,13,5], object-oriented systems [14], applied software systems [15], implementation of abstract data types [16] and context-free hyperedge replacement [17] (see also Chapter 3).

Historically, the first of the algebraic approaches to graph transformation is the so-called *double-pushout (DPO) approach* introduced in [1], which owes its name to the basic algebraic construction used to define a direct derivation step: This is modelled indeed by two gluing diagrams (i.e., pushouts) in the category of graphs and total graph morphisms. More recently a second algebraic approach has been proposed in [5], which defines a basic derivation step as a *single* pushout in the category of graphs and *partial* graph morphisms:

Hence the name of *single-pushout (SPO) approach*.[a] This chapter includes an informal description of the basic concepts and results common to the DPO and SPO approaches, followed by a more detailed, formal presentation of the DPO approach. On the other hand, the SPO approach is the main topic of the next chapter, which is closed by a formal comparison of the two approaches.

More precisely, in Section 3.2 many relevant concepts and results of the algebraic approaches to graph transformation are introduced in the form of problems stated in quite an abstract way: This provides an outline that will be followed (to a great extent) along the presentation of the two approaches, each of which will provide its own answers to such problems. The stated problems concern independence and parallelism of direct derivations, embedding of derivations, amalgamation, and distribution. The last part of the section shortly recalls some other issues addressd in the literature of the algebraic approaches, including computation based semantics, application conditions, structuring of grammars, consistency conditions, and generalization of the algebraic approaches to other structures. Some of these topics will be presented in detail for just one of the two approaches, namely computation based semantics (in particular the so-called *models of computation*) for the DPO, and application conditions for the SPO approach; for the other topics relevant references are indicated only.

Section 3.3 starts the presentation of the DPO approach by introducing the very basic notions, including the category of graphs we shall use and the definitions of production, direct derivation, and derivation; also the fundamental notion of pushout and the gluing conditions (i.e., the conditions that ensures the applicability of a production at a given match) are discussed in some depth. Section 3.4 presents the main results of the theory of parallelism for the DPO approach, including the notions of parallel and sequential independence, the Local Church Rosser and the Parallelism theorems, and the synthesis and analysis constructions. This section on the one hand answers the problems about independence and parallelism raised in Section 3.2.2 and, on the other hand, it provides some basic notions exploited in the following section.

Section 3.5 presents various models of computations for graph grammars on different levels of abstraction, i.e., various categories having graphs as objects and graph derivations as arrows. All such models are obtained from the most concrete one by imposing suitable equivalences on graphs and derivations. In particular, in the most abstract model, called the *abstract truly-concurrent model*, all isomorphic graphs are identified, as well as all derivations that are

[a] Other definitions of graph rewriting using a single pushout in a category of partial graph morphisms exist: Their relationship with the SPO approach is discussed in Section 4.2.2.

related by the classical shift-equivalence (which considers as equal all derivations which differ for the order of independent direct derivations only) or by equivalence $\equiv_3$, which relates derivations that are isomorphic and satisfy a further technical (but fundamental) condition. Finally Section 3.6 summarizes the main results concerning embedding of derivations, amalgamation and distribution for the DPO approach, providing answers for the remaining problems raised in Section 3.2.

Two appendices close the chapter. Appendix A (Section 3.8) introduces the definition of binary coproducts in a category, and shows that coproducts cannot be assumed to be commutative (in a strict way) unless the category is a preorder. This fact justifies the way parallel productions are defined and manipulated in Section 3.4, which may appear more complex than the corresponding definitions in previous papers in the literature. Appendix B (Section 3.9) presents the proof of the main technical result of Section 3.5. That proof is based on the analysis and synthesis constructions which are reported as well, not only for completeness, but also because they demonstrate a typical example of reasoning that is used when results in the algebraic approach to graph transformations are proven on an abstract level.

A formal, detailed comparison of the DPO and SPO approaches is presented in Section 4.6, where it is also shown that the DPO approach can be embedded (via a suitable translation) into the SPO approach.

All the categorical notions used in this chapter are explicitly introduced; nevertheless, some knowledge of the basic concepts of category theory would be helpful. Standard references are [18,19].

How to read the chapters on the algebraic approaches.

Readers interested in just one of the algebraic approaches are suggested to read either this chapter (for the DPO approach), or Section 3.2 and then all Chapter 4 excluded Section 4.6 (for the SPO approach; a few references to Section 3.3 will have to be followed).

Alternatively, some parts may be skipped, depending on the specific interests of the reader. All readers (and in particular newcomers) are encouraged to read Section 3.2 first, where they can get a precise, although informal idea of the main topics addressed in the theory of the algebraic approaches. Such topics are then presented with all technical details for the DPO approach in Section 3.3, in the first part of Section 3.4 (till Theorem 3.4.6), and in Section 3.6. For the SPO approach, they are presented instead in Section 4.2 and 4.3. After these sections, the reader interested in a careful comparison of the two approaches can read Section 4.6.

3.2 Overview of the Algebraic Approaches

The purpose of this section is to provide an overview of the two algebraic approaches to graph transformation, the double-pushout (DPO) and the single-pushout (SPO) approach. Taking a problem-oriented point of view allows us to discuss the main questions raised (and solutions proposed) by both approaches quite informally using the same conceptual terminology. Explicit technical constructions and results are then presented for the DPO approach in the remaining sections of this chapter, and for the SPO approach in the next chapter. A detailed comparison of the two approaches can be found in Section 4.6.

Among the various approaches to graph transformation, the algebraic approaches are characterized by the use of categorical notions for the very basic definitions of graph transformation rules (or *productions*, as they are usually called), of matches (i.e., of occurrences of the left-hand side of a production in a graph), and of rule applications (called *direct derivations*). The main advantage is clearly that techniques borrowed from category theory can be used for proving properties and results about graph transformation, and such proofs are often be simpler than corresponding proofs formulated without categorical notions. In addition, such proofs are to a great extent independent from the structure of the objects that are rewritten. Therefore the large body of results and constructions of the algebraic approaches can be extended quite easily to cover the rewriting of arbitrary structures (satisfying certain properties), as shown for example by the theory of High-Level Replacement Systems [20,21].

In this section, we first introduce informally the basic notions of graph, production and derivation for the DPO and SPO approaches in Section 3.2.1. We address questions concerning the independence of direct derivations and their parallel application in Section 3.2.2; the embedding of derivations in larger contexts and the related notion of derived production in Section 3.2.3; and the relationship between the amalgamation of productions along a common subpart (a sort of synchronization) and the distributed application of several productions to a graph in Section 3.2.4. Finally Section 3.2.5 shortly recalls some other issues addressed in the literature of the algebraic approaches, including computation based semantics, application conditions, structuring of grammars, consistency conditions, and generalization of the algebraic approaches to other structures.

3.2.1 Graphs, Productions and Derivations

The basic idea of all graph transformation approaches is to consider a *production* $p : L \rightsquigarrow R$, where graphs L and R are called the left- and the right-hand side, respectively, as a finite, schematic description of a potentially infinite set of *direct derivations*. If the *match* m fixes an occurrence of L in a *given graph* G, then $G \stackrel{p,m}{\Longrightarrow} H$ denotes the direct derivation where p is applied to G leading to a *derived graph* H. Intuitively, H is obtained by replacing the occurrence of L in G by R. The essential questions distinguishing the particular graph transformation approaches are:

1. What is a "graph"?
2. How can we match L with a subgraph of G?
3. How can we define the replacement of L by R in G?

In the algebraic approaches, a graph is considered as a two sorted *algebra* where the sets of vertices V and edges E are the carriers, while source $s : E \to V$ and target $t : E \to V$ are two unary operations. Moreover we have label functions $lv : V \to L_V$ and $le : E \to L_E$, where L_V and L_E are fixed label alphabets for vertices and edges, respectively. Since edges are objects on their own right, this allows for multiple (parallel) edges with the same label.

Each graph production $p : L \rightsquigarrow R$ defines a partial correspondence between elements of its left- and of its right-hand side, determining which nodes and edges have to be preserved by an application of p, which have to be deleted, and which must be created. To see how this works, consider the production $p_1 : L_1 \rightsquigarrow R_1$ which is applied on the left of Figure 3.1 to a graph G_1, leading to the direct derivation (1). The production assumes three vertices on the left-hand side L_1, and leaves behind two vertices, connected by an edge, on the right-hand side R_1. The numbers written near the vertices and edges define a partial correspondence between the elements of L_1 and R_1: Two elements with the same number are meant to represent the same object before and after the application of the production. In order to apply p_1 to the given graph G_1, we have to find an occurrence of the left-hand side L_1 in G_1, called match in the following. A match $m : L \to G$ for a production p is a graph homomorphism, mapping nodes and edges of L to G, in such a way that the graphical structure and the labels are preserved. The match $m_1 : L_1 \to G_1$ of the direct derivation (1) maps each element of L_1 to the element of G_1 carrying the same number. Applying production p_1 to graph G_1 at match m_1 we have to delete every object from G_1 which matches an element of L_1 that has no corresponding element in R_1, i.e., vertex 2 of G_1 is deleted. Symmetrically, we add to G_1 each element of R_1 that has no corresponding element in L_1, i.e., edge 5 is inserted. All remaining elements of G_1 are preserved, thus, roughly spoken, the derived graph H_1 is constructed as $G_1 - (L_1 - R_1) \cup (R_1 - L_1)$.

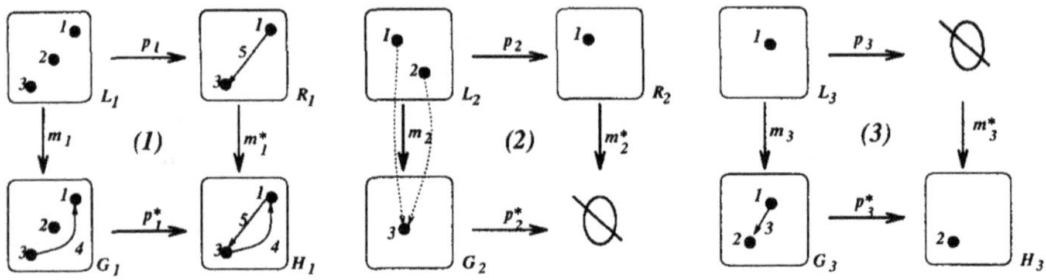

Figure 3.1: Direct derivations.

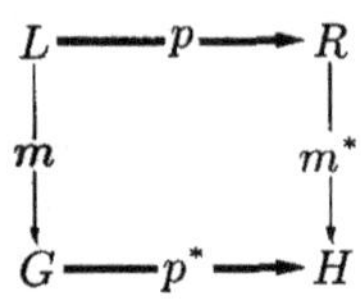

Figure 3.2: Example and schematic representation of direct derivation $G \stackrel{p,m}{\Longrightarrow} H$.

The application of a production can be seen as an embedding into a context, which is the part of the given graph G that is not part of the match, i.e., the edge 4 in our example. Hence, at the bottom of the direct derivation, the *co-production* $p_1^* : G_1 \rightsquigarrow H_1$ relates the given and the derived graphs, keeping track of all elements that are preserved by the direct derivation. Symmetrically, the *co-match* $m^*{}_1$, which is a graph homomorphism, too, maps the right-hand side R_1 of the production to its occurrence in the derived graph H_1. A direct derivation from G to H resulting from an application of a production p at a match m is schematically represented as in Figure 3.2, and denoted by $d = (G \stackrel{p,m}{\Longrightarrow} H)$.

In general L does not have to be isomorphic to its image $m(L)$ in G, i.e., elements of L may be identified by m. This is not always unproblematic, as we may see in the direct derivation (2) of Figure 3.1. A production p_2, assuming two vertices and deleting one of them, is applied to a graph G_2 containing only one vertex, i.e., vertices 1 and 2 of L_2 are both mapped to the vertex 3 of G_2. Thus, the production p_2 specifies both the deletion and the preservation of 3: We say that the match m_2 contains a conflict. There are three possible ways to solve this conflict: Vertex 3 of G_2 may be preserved, deleted, or the application of the production p_2 at this match may be forbidden. In the derivation (2) of Figure 3.1 the second alternative has been choosen, i.e., the vertex is deleted.

Another difficulty may occur if a vertex shall be deleted which is connected to an edge that is not part of the match, as shown in direct derivation (3) of Figure 3.1: Deleting vertex 1 of G_3, as specified by production p_3, would leave behind the edge 3 without a source vertex, i.e., the result would no longer be a graph. Again there are two ways to avoid such dangling edges. We may either delete the edge 3 together with its source node, as in derivation (3) of Figure 3.1, or forbid the application of p_3.

The basic difference between the DPO and the SPO approach lies in the way they handle the above problematic situations. In the DPO approach, rewriting is not allowed in these cases (i.e., productions p_2 and p_3 are not applicable to matches m_2 and m_3, respectively). On the contrary, in the SPO approach such direct derivations are allowed, and deletion has priority over preservation; thus (2) and (3) of Figure 3.1 are legal SPO direct derivations.

In both algebraic approaches direct derivations are modeled by gluing constructions of graphs, that are formally characterized as pushouts in suitable categories having graphs as objects, and (total or partial) graph homomorphisms as arrows. A production in the DPO approach is given by a pair $L \xleftarrow{l} K \xrightarrow{r} R$ of graph homomorphisms from a common *interface graph* K, and a direct derivation consists of two gluing diagrams of graphs and total graph morphisms, as (1) and (2) in the left diagram below. The *context graph* D is obtained from the given graph G by deleting all elements of G which have a pre-image in L, but none in K. Via diagram (1) this deletion is described as an inverse gluing operation, while the second gluing diagram (2) models the actual insertion into H of all elements of R that do not have a pre-image in K. In order to avoid problematic situations like (2) and (3) in Figure 3.1, in the DPO approach the match m must satisfy an application condition, called the *gluing condition*. This condition consists of two parts. To ensure that D will have no dangling edges, the *dangling condition* requires that if p specifies the deletion of a vertex of G, then it must specify also the deletion of all edges of G incident to that node. On the other hand the *identification condition* requires that every element of G that should be deleted by the application of p has only one pre-image in L. The gluing condition ensures that the application of p to G deletes exactly what is specified by the production.

$$\begin{array}{ccccc} L & \xleftarrow{l} & K & \xrightarrow{r} & R \\ \downarrow m & (1) & \downarrow d & (2) & \downarrow m^* \\ G & \xleftarrow{l^*} & D & \xrightarrow{r^*} & H \end{array} \qquad\qquad \begin{array}{ccc} L & \xrightarrow{p} & R \\ \downarrow m & (3) & \downarrow m^* \\ G & \xrightarrow{p^*} & H \end{array}$$

In the SPO approach a production p is a partial graph homomorphism. A direct derivation is given by a single gluing diagram, as (3) in the right diagram above, where the match m is required to be total. No gluing condition has to be satisfied for applying a production, and this may result in side effects as shown in Figure 3.1. In fact, on the one hand, the deletion of a node of G automatically causes the deletion of all incident edges, even if this is not specified by the production. On the other hand conflicts between deletion and preservation of nodes and edges are solved in favor of deletion, and in this case the co-match m^* becomes a partial homomorphism, because elements of the right-hand side that should have been preserved but are deleted because of some conflict do not have an image in H.

Direct derivations in the SPO approach are more expressive than DPO derivations, in the sense that they may model effects that can not be obtained in the more restricted DPO approach. A formal correspondence between direct derivations in the two approaches is presented in Section 4.6.

A *graph grammar* $\mathcal{G}$ consists of a set of productions P and a *start graph* G_0. A sequence of direct derivations $\rho = (G_0 \overset{p_1}{\Longrightarrow} G_1 \overset{p_2}{\Longrightarrow} \dots \overset{p_n}{\Longrightarrow} G_n)$ constitutes a *derivation* of the grammar, also denoted by $G_0 \Longrightarrow^* G_n$. The *language* $\mathcal{L}(\mathcal{G})$ generated by the grammar $\mathcal{G}$ is the set of all graphs G_n such that $G_0 \Longrightarrow^* G_n$ is a derivation of the grammar.

3.2.2 Independence and Parallelism

Parallel computations can be described in two different ways. If we stick to a sequential model, two parallel processes have to be modeled by interleaving arbitrarily their atomic actions, very similarly to the way multitasking is implemented on a single processor system. On the contrary, explicit parallelism means to have one processor per process, which allows the actions to take place simultaneously.

Interleaving

Considering the interleaving approach first, two actions are concurrent, i.e., potentially in parallel, if they may be performed in any order with the same result. In terms of graph transformations, the question whether two actions (direct derivations) are concurrent or not can be asked from two different points of view.

1. Assume that a given graph represents a certain system state. The next evolution step of this state is obtained by the application of a production

at a certain match, where the production and the match are chosen non-deterministically from a set of possible alternatives. Clearly, each choice we make leads to a distinct derivation sequence, but the question remains, whether two of these sequences indeed model different computations or if we have only chosen one of two equivalent interleavings. (Here, two derivation sequences from some graph G_0 to G_n are equivalent if both insert and delete exactly the same nodes and edges.) In other words, given two alternative direct derivations $H_1 \overset{p_1}{\Longleftarrow} G \overset{p_2}{\Longrightarrow} H_2$ we ask ourselves if there are direct derivations $H_1 \overset{p_2}{\Longrightarrow} X \overset{p_1}{\Longleftarrow} H_2$, showing that the two given direct derivations are not mutually exclusive, but each of them can instead be postponed after the application of the other, yielding the same result.

2. Given a derivation, intuitively two consecutive direct derivations $G \overset{p_1}{\Longrightarrow} H_1 \overset{p_2}{\Longrightarrow} X$ are concurrent if they can be performed in a different order, as in $G \overset{p_2}{\Longrightarrow} H_2 \overset{p_1}{\Longrightarrow} X$, without changing the result. The existence of two different orderings ensures that there is no causal dependency between these applications of p_1 and p_2.

Summarizing, we can say that two alternative derivations are concurrent if they are not mutually exclusive while two consecutive derivations are concurrent if they are not causally dependent. To stress the symmetry between these two conditions they are called *parallel* and *sequential independence*, respectively, in the more formal statement of this problem below. Originally investigated in view of the local confluence property of (term) rewriting systems, it is called the Local Church-Rosser Problem.

Problem 3.2.1 (Local Church Rosser)

1. *Find a condition, called* parallel independence, *such that two alternative direct derivations $H_1 \overset{p_1}{\Longleftarrow} G \overset{p_2}{\Longrightarrow} H_2$ are parallel independent iff there are direct derivations $H_1 \overset{p_2}{\Longrightarrow} X$ and $H_2 \overset{p_1}{\Longrightarrow} X$ such that $G \overset{p_1}{\Longrightarrow} H_1 \overset{p_2}{\Longrightarrow} X$ and $G \overset{p_2}{\Longrightarrow} H_2 \overset{p_1}{\Longrightarrow} X$ are equivalent.*

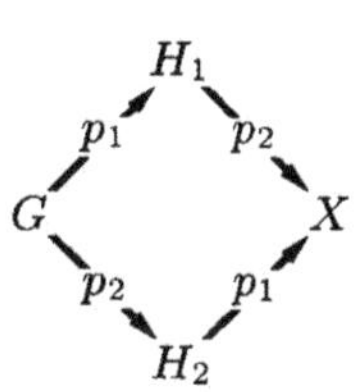

2. *Find a condition, called* sequential independence, *such that a derivation* $G \overset{p_1}{\Longrightarrow} H_1 \overset{p_2}{\Longrightarrow} X$ *is sequentially independent iff there is an equivalent derivation* $G \overset{p_2}{\Longrightarrow} H_2 \overset{p_1}{\Longrightarrow} X$. ⧉

More concretely, two alternative direct derivations are parallel independent if their matches do only overlap in items that are preserved by both derivations, that is, if none of the two deletes any item that is also accessed by the other one. If an item is preserved by one of the direct derivations, say $G \overset{p_1}{\Longrightarrow} H_1$, but deleted by the other one, the second one can be delayed after the first; we say that $G \overset{p_2}{\Longrightarrow} H_2$ is *weakly parallel independent* of $G \overset{p_1}{\Longrightarrow} H_1$. *Parallel independence* can then be defined as mutual weakly parallel independence.

Two consecutive direct derivations are sequentially independent if (*a*) the match of the second one does not depend on elements generated by the first one, and (*b*) the second derivation does not delete an item that has been accessed by the first. The first part of this condition ensures that p_2 may already be applied before p_1, leading to an alternative direct derivation $G \overset{p_2}{\Longrightarrow} H_2$. We say that $H_1 \overset{p_2}{\Longrightarrow} X$ is *weakly sequentially independent* of $G \overset{p_1}{\Longrightarrow} H_1$. From (*b*) we can show that the latter is weakly parallel independent of $G \overset{p_2}{\Longrightarrow} H_2$ in the above sense. So two consecutive direct derivations $G \overset{p_1}{\Longrightarrow} H_1 \overset{p_2}{\Longrightarrow} X$ are *sequentially independent* if $H_1 \overset{p_2}{\Longrightarrow} X$ is weakly sequentially independent of $G \overset{p_1}{\Longrightarrow} H_1$ and the latter is weakly *parallel* independent of $G \overset{p_2}{\Longrightarrow} H_2$.

Explicit Parallelism

Parallel computations can be represented more directly by parallel application of productions. *Truly* parallel application of productions essentially requires to abstract from any possible application order, which implies that no intermediate graph is generated. In other words, a (true) parallel application may be modeled by the application of a single production, the *parallel production*.

Given productions $p_1 : L_1 \rightsquigarrow R_1$ and $p_2 : L_2 \rightsquigarrow R_2$, their parallel composition is denoted by $p_1 + p_2 : L_1 + L_2 \rightsquigarrow R_1 + R_2$. Intuitively, $p_1 + p_2$ is just the disjoint union of p_1 and p_2. Direct derivations like $G \overset{p_1+p_2}{\Longrightarrow} X$, using a parallel production $p_1 + p_2$, are referred to as *parallel direct derivations*.

Now we have seen two different ways to model parallel computations by graph transformation, either using interleaving sequences or truly parallel derivations. How are they related? This question can be asked from two different points of view. Given a parallel direct derivation $G \overset{p_1+p_2}{\Longrightarrow} X$ we may look for a sequentialization, say $G \overset{p_1}{\Longrightarrow} H_1 \overset{p_2}{\Longrightarrow} X$. On the other hand, starting from the latter sequence we can try to put the two consecutive direct derivations in parallel.

The problem of finding necessary and sufficient conditions for these sequentialization and parallelization constructions, also called *analysis* and *synthesis* in the literature, is stated in the following way.

Problem 3.2.2 (Parallelism)

1. *Find* a sequentialization condition *such that a parallel direct derivation* $G \stackrel{p_1+p_2}{\Longrightarrow} X$ *satisfies this condition iff there is an equivalent sequential derivation* $G \stackrel{p_1}{\Longrightarrow} H_1 \stackrel{p_2}{\Longrightarrow} X$ *(or* $G \stackrel{p_2}{\Longrightarrow} H_2 \stackrel{p_1}{\Longrightarrow} X$*, respectively).*

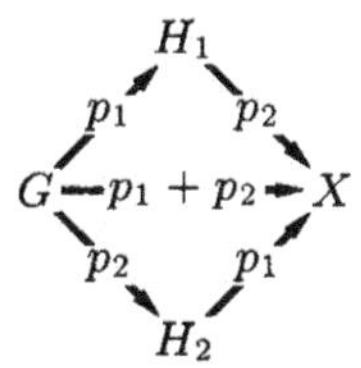

2. *Find* a parallelization condition *such that a sequential derivation* $G \stackrel{p_1}{\Longrightarrow} H_1 \stackrel{p_2}{\Longrightarrow} X$ *(or* $G \stackrel{p_2}{\Longrightarrow} H_2 \stackrel{p_1}{\Longrightarrow} X$*) satisfies this condition iff there is an equivalent parallel direct derivation* $G \stackrel{p_1+p_2}{\Longrightarrow} X$. □

The solution to this problem is closely related to the notions of sequential and parallel independence. In the double pushout approach it turns out that each parallel direct derivation $G \stackrel{p_1+p_2}{\Longrightarrow} X$ may be sequentialized to derivations $G \stackrel{p_1}{\Longrightarrow} H_1 \stackrel{p_2}{\Longrightarrow} X$ and $G \stackrel{p_2}{\Longrightarrow} H_2 \stackrel{p_1}{\Longrightarrow} X$, which are in turn sequentially independent; indeed, the sequentialization condition is already part of the gluing condition for the application of the parallel production. Vice versa, consecutive direct derivations can be put in parallel if they are sequentially independent, thus the parallelization condition requires sequential independence of the derivation. Hence in the double pushout approach the two ways of modeling concurrency are in fact equivalent, in the sense that they allow one to represent exactly the same amount of parallelism.

In the single pushout setting, however, a parallel direct derivation $G \stackrel{p_1+p_2}{\Longrightarrow} X$ can be sequentialized to $G \stackrel{p_1}{\Longrightarrow} H_1 \stackrel{p_2}{\Longrightarrow} X$ if $G \stackrel{p_2}{\Longrightarrow} H_2$ is weakly parallel independent of $G \stackrel{p_1}{\Longrightarrow} H_1$, where the two alternative direct derivations from G are obtained by restricting the match of p_1+p_2 into G to the left-hand side of p_1 and p_2, respectively. Since for a given parallel direct derivation this condition may not hold, it turns out that in this setting parallel direct derivations can express a higher degree of concurrency with respect to derivation sequences, because there are parallel direct derivations that do not have an equivalent interleaving sequence. On the other hand, $G \overset{p_1}{\Longrightarrow} H_1 \overset{p_2}{\Longrightarrow} X$ can be put in parallel if the second direct derivation is weakly sequentially independent of the first one.

3.2.3 Embedding of Derivations and Derived Productions

Systems are usually not monolithic, but consist of a number of subsystems which may be specified and implemented separately. After defining each subsystem over its own local state space, they have to be joined together in order to build the whole system. Then the local computations of each subsystem have to be embedded into the computations of the enclosing system.

Speaking about derivations, $\rho = (G_0 \overset{p_1}{\Longrightarrow} \cdots \overset{p_n}{\Longrightarrow} G_n)$ is embedded into a derivation $\delta = (X_0 \overset{p_1}{\Longrightarrow} \cdots \overset{p_n}{\Longrightarrow} X_n)$ if both have the same length n and each graph G_i of ρ is embedded into the corresponding graph X_i of δ in a compatible way. The embedding is represented by a family of injections $(G_i \overset{e_i}{\longrightarrow} X_i)_{i\in\{0,\dots n\}}$, denoted by $\rho \overset{e}{\longrightarrow} \delta$. If it exists, it is uniquely determined by the first injection $e_0 : G_0 \to X_0$. Given ρ and e_0, the following problem asks under which conditions there is a derivation δ such that e_0 induces an embedding e of ρ into δ.

Problem 3.2.3 (Embedding)
Find an embedding condition *such that, for a given derivation $\rho = (G_0 \overset{p_1}{\Longrightarrow} \cdots \overset{p_n}{\Longrightarrow} G_n)$, an injection $e_0 : G_0 \to X_0$ satisfies this condition iff there is a derivation $\delta = (X_0 \overset{p_1}{\Longrightarrow} \cdots \overset{p_n}{\Longrightarrow} X_n)$ and an embedding $e : \rho \to \delta$.* □

$$
\begin{array}{ccccccc}
G_0 & \xrightarrow{p_1} & G_1 & \xrightarrow{p_1} & \cdots & \xrightarrow{p_n} & G_n \\
\downarrow{\scriptstyle e_0} & & \downarrow{\scriptstyle e_1} & & \cdots & & \downarrow{\scriptstyle e_n} \\
X_0 & \xrightarrow{p_1} & X_1 & \xrightarrow{p_1} & \cdots & \xrightarrow{p_n} & X_n
\end{array}
$$

Let us consider the simpler problem of embedding a direct derivation $G \overset{p}{\Longrightarrow} H$ along an injection $e_0 : G \to X$ first. According to our general introduction, the direct derivation of Figure 3.2 defines a co-production $p^* : G \rightsquigarrow H$, relating the given and the derived graphs. The injection $e_0 : G \to X$ provides us with a match for this *directly derived production* p^*. The embedding condition for $G \overset{p}{\Longrightarrow} H$ is now equivalent to the applicability of the directly derived production p^* at the match e_0.

This idea can be generalized to derivations of arbitrary length: Given $\rho = (G_0 \overset{p_1}{\Longrightarrow} \cdots \overset{p_n}{\Longrightarrow} G_n)$, we have to define a *derived production* $p^* : G_0 \rightsquigarrow G_n$ that is applicable at an injective match $G_0 \xrightarrow{e_0} X_0$ if and only if e_0 induces an embedding e of ρ into a derivation δ.

Problem 3.2.4 (Derived Production)
Let $\rho = (G_0 \overset{p_1}{\Longrightarrow} \cdots \overset{p_n}{\Longrightarrow} G_n)$ *be a derivation. Find a* derived production $p^* : G_0 \rightsquigarrow G_n$ *such that for each injection* $G_0 \xrightarrow{e_0} X_0$ *there is a direct derivation* $X_0 \overset{p^*}{\Longrightarrow} X_n$ *at* e_0 *iff* e_0 *induces an embedding* $e : \rho \to \delta$ *of* ρ *into a derivation* $\delta = (X_0 \overset{p_1}{\Longrightarrow} \cdots \overset{p_n}{\Longrightarrow} X_n)$. □

In order to obtain this derived production we will introduce a *sequential composition* $p_1;p_2$ of two productions $p_1 : L_1 \rightsquigarrow R_1$ and $p_2 : L_2 \rightsquigarrow R_2$ with $R_1 = L_2$. Then we define the *derived production* p^* for a given derivation sequence $G_0 \overset{p_1}{\Longrightarrow} \cdots \overset{p_n}{\Longrightarrow} G_n$ by $p^* = p_1^*;\ldots;p_n^*$.

The embedding condition of $G_0 \overset{p_1}{\Longrightarrow} \cdots \overset{p_n}{\Longrightarrow} G_n$ for e_0 is equivalent to the applicability of the derived production p^* at this match. In the DPO approach this means that e_0 has to satisfy the gluing condition for p^*, which in this case reduces to the dangling condition, since the identification condition is ensured by the injectivity of e_0. In the SPO approach, instead, the embedding is always possible since SPO derivations do not need any application conditions.

There are other interesting applications for derived productions and derivations. First they can be used to shortcut derivation sequences. Thereby they reduce the number of intermediate graphs to be constructed and fix the interaction of the productions applied in the sequence. Similar to a database transaction we obtain an "all-or-nothing" semantics: The single actions described by the direct derivations are either performed together (the derived production is applied) or they leave the current state unchanged (the derived production is not applicable). Due to the absence of intermediate states the transaction may not be interrupted by other actions.

In parallel and concurrent systems intermediate states correspond to synchronization points. A parallel derivation sequence, as introduced in Section 3.2.2, may be seen as a synchronous computation: Several independent processors perform a direct derivation each and wait for the next clock tick to establish a global intermediate state. Then they proceed with the next application. The derived production forgets about these synchronization points. Therefore it may perform even sequentially dependent derivations in a single step.

The representation of derivations by derived productions, however, is not at all minimal. Given a derivation δ, each derivation ρ that can be embedded into

δ defines a derived production which can simulate the effect of ρ. If we look for a more compact representation, we can take all those derived productions corresponding to derivations ρ which are minimal w.r.t. the embedding relation. We speak of *minimal derived productions.* Intuitively, the minimal derived production of a derivation ρ does not contain any context but is constructed from the elements of the productions of ρ, only.

3.2.4 Amalgamation and Distribution

According to Subsection 3.2.2, truly parallel computation steps can be modeled by the application of a parallel production, constructed as the disjoint union of two elementary productions. This provides us with a notion of parallel composition of productions and derivations. In many specification and programming languages for concurrent and distributed systems, however, such a parallel composition is equipped with some synchronization mechanism in order to allow for cooperation of the actions that are put in parallel. If two graph productions shall not work concurrently but cooperatively, they have to synchronize their applications w.r.t. commonly accessed elements in the graph: A common subproduction specifies the shared effect of the two productions. This general idea is used in this subsection in two different ways. First we consider systems with global states, represented as graphs in the usual way. In this setting the synchronized application of two subproduction-related productions may be described by the application of a so-called amalgamated production, i.e., the gluing of the elementary productions w.r.t. the common subproduction. In the second part we consider distributed states, where each production is applied to one local state, such that synchronized local derivations define a distributed derivation.

Amalgamation

The synchronization of two productions p_1 and p_2 is expressed by a *subproduction* p_0 which is somehow embedded into p_1 and p_2. The *synchronized productions* are then denoted by $p_1 \overset{g^1}{\leftarrow} p_0 \overset{g^2}{\rightarrow} p_2$. They may be glued along p_0, leading to the *amalgamated production* $p_1 \oplus_{p_0} p_2$. A direct derivation $G \overset{p_1 \oplus_{p_0} p_2}{\Longrightarrow} H$ using $p_1 \oplus_{p_0} p_2$ is called *amalgamated derivation.* It may be seen as the simultaneous application of p_1 and p_2 where the common action described by the subproduction p_0 is performed only once. If the subproduction p_0 is empty, the amalgamated production $p_1 \oplus_{p_0} p_2$ specializes to the parallel production $p_1 + p_2$.

Distribution

Like an amalgamated production is somehow distributed over two synchronized productions, a graph representing a certain system state can be splitted into local graphs, too. A *distributed graph* $DG = (G_1 \overset{g_1}{\leftarrow} G_0 \overset{g_2}{\rightarrow} G_2)$ consists of two *local graphs* G_1 and G_2 that share a common *interface graph* G_0, embedded into G_1 and G_2 by graph morphisms g_1 and g_2, respectively. Gluing G_1 and G_2 along G_0 yields the global graph $\oplus DG = G_1 \oplus_{G_0} G_2$ of DG. Then DG is also called a *splitting* of $\oplus DG$. The embeddings g_1, g_2 of the interface into the local graphs may be either total or partial; accordingly we call DG a *total* or a *partial splitting*. A partial splitting models a distributed state where the local states have been updated in an inconsistent way, which may result from a loss of synchronization of the local updates. Imagine, for example, a vertex v that is shared by the two local graphs, i.e., which is in G_1, G_2, and in the interface G_0. If this vertex is deleted from the local graph G_1 but not from the interface G_0, the embedding $g_1 : G_0 \rightarrow G_1$ becomes undefined for the vertex v, i.e., the two local states have inconsistent information about their shared substate. Then, the global graph $\oplus DG$ of this partial splitting does not contain the vertex v, but only those elements where the local informations are consistent with each other.

Distributed graphs can be transformed by synchronized productions: Given *local direct derivations* $d_i = (G_i \overset{p_i}{\Longrightarrow} H_i)$ for $i \in \{0, 1, 2\}$ which are compatible with the splitting $G_1 \overset{g_1}{\leftarrow} G_0 \overset{g_2}{\rightarrow} G_2$, d_0 becomes a subderivation of d_1 and d_2. In this case, the *(direct) distributed derivation* $d_1||_{d_0}d_2 : DG \Longrightarrow DH$ transforms the given distributed graph DG into the new one $DH = (H_1 \overset{h_1}{\leftarrow} H_0 \overset{h_2}{\rightarrow} H_2)$. The distributed derivation is *synchronous* if both DG and DH are total splittings.

Now, since synchronized rules can be amalgamated and distributed graphs may be glued together, it seems natural to simulate a distributed derivation by an amalgamated one. Vice versa, assuming a splitting of the global graph, we may want to distribute a given amalgamated derivation. The corresponding conditions, enabling these constructions, are called *amalgamation* and *distribution condition*, respectively.

Problem 3.2.5 (Distribution)
Let $DG = (G_1 \overset{g_1}{\leftarrow} G_0 \overset{g_2}{\rightarrow} G_2)$ *be a distributed graph and* $p_1 \overset{in^1}{\leftarrow} p_0 \overset{in^2}{\rightarrow} p_2$ *be synchronized productions.*

1. *Find an* amalgamation condition *such that a distributed derivation* $d_1||_{d_0}d_2 : DG \Longrightarrow DH$ *with* $d_i = (G_i \overset{p_i}{\Longrightarrow} H_i)$ *satisfies this condition iff there is an equivalent amalgamated derivation* $\oplus DG \overset{p_1 \oplus_{p_0} p_2}{\Longrightarrow} \oplus DH$.

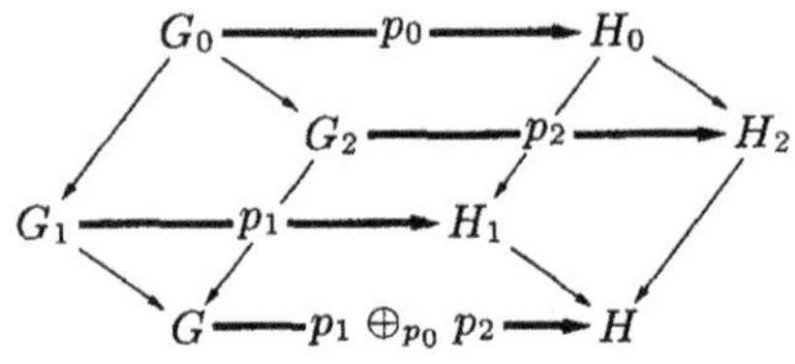

2. *Find a* distribution condition *such that an amalgamated derivation* $G \overset{p_1 \oplus_{p_0} p_2}{\Longrightarrow} H$ *satisfies this condition iff there is an equivalent distributed derivation* $d_1||_{d_0}d_2 : DG \Longrightarrow DH$ *with* $G = \oplus DG$, $H = \oplus DH$ *and* $d_i = (G_i \overset{p_i}{\Longrightarrow} H_i)$. □

Simulating a distributed derivation by an amalgamated one means to observe some local activities from a global point of view. Since for asynchronous systems such a global view does not always exist, it is not too surprising that the amalgamation construction in 1. is only defined if the given distributed graph is a total splitting, i.e., represents a consistent distributed state. In order to distribute an amalgamated derivation, the match of the amalgamated production has to be splitted as well. Such a splitting may not exist if the matches of the elementary productions are not in their corresponding local graphs. If the distributed derivation shall be synchronous, we need an additional condition, which can be seen as a distributed dangling condition. It takes care that nothing is deleted in any local component which is in the image of some interface, unless its preimage is deleted as well.

3.2.5 Further Problems and Results

This subsection is an overview of some further issues, problems and results addressed in the literature of the algebraic approaches.

Semantics

If a graph grammar is considered from the point of view of formal languages, its usual semantics is the class of graphs derivable from the start graph using the productions of the grammar, its *generated language*. Using graph grammars for specification of systems, however, we are not only interested in the generated graphs, corresponding to the states of the system, but also in the derivations, i.e., the systems computations. There are several proposals for semantics of graph grammars that represent graphs and derivations at different levels of abstraction.

- A *model of computation* of a graph grammar is a category having graphs as objects and derivations as arrows. Different models of computations can be defined for a grammar, by imposing different equivalence relations on graphs and derivations. The main problem in this framework is to find an adequate notion of equivalence on graphs and derivations, that provides representation independence as well as abstraction of the order of independent (concurrent) derivation steps. Models of computations for grammars in the DPO are introduced in Section 3.5.

- Processes are well-accepted as a truly concurrent semantics for Petri nets (see for example [22]), and such a semantics has been defined for both algebraic approaches to graph grammars as well, using slightly different terminologies. In a *graph process* (DPO) or *concurrent derivation* (SPO), the graphs of a derivation are no longer represented explicitly. Instead a so-called *type graph* (DPO) or *core graph* (SPO) is constructed by collecting all nodes and edges of the derivation. Moreover, the linear ordering of the direct derivations is replaced by a partial order representing the causal dependencies among the production applications, or *actions* (SPO). DPO processes and SPO concurrent derivations are introduced in [23] and [24,25], respectively.

- An *event structure* [26] consists of a collection of *events* together with two relations "$\leq$" and "#" modeling *causal dependency* and *mutual exclusion*, respectively. Event structures are widely accepted as abstract semantic domains for systems that exhibit concurrency and nondeterminism. Such a truly concurrent semantics is more abstract than the one based on processes, where it is still possible to reconstruct the graphs belonging to a derivation (from the type or core graph), while this is not possible from the event structure. Event structure semantics have been proposed for the double-pushout approach in [27,28,29].

Control

In order to specify systems with a meaningful behavior it is important that the application of productions can be controlled somehow.

- *Application conditions* restrict the applicability of individual productions by describing the admissible matches. In the SPO approach they are considered in some detail in Section 4.4.

- Imperative control structures like sequential composition, iteration, nondeterministic choice, etc, lead to the notions of programmed graph transformation and transactions. Such concepts are discussed, for example, in Chapter 7, while in the algebraic approaches they have been introduced in [30].

Structuring

Systems usually consists of a number of subsystems, which are specified separately and have some meaning in their own right. Structuring mechanisms for graph transformation systems based on colimits in the *category of graph grammars* are investigated in [31,32], and are shown to be compatible with the functorial semantics introduced in [33].

Analysis

A main advantage of formal specification methods is the possibility of formal reasoning about properties of the specified systems.

- *Consistency conditions* describe properties of all derived graphs. The problem of specifying these conditions and of analyzing them w.r.t. a given grammar has been considered in [34,27] in the SPO approach.
- Since graph grammars generalize term rewriting systems, *rewriting properties* as, for example, confluence, termination, etc., are interesting here as well. Corresponding results for (DPO) hypergraph rewriting can be found in [36,37], while confluence of critical pairs in the SPO approach has been studied in [38].

More general structures

Different application areas of graph transformation require different notions of graphs. Fortunately, most of the results of the algebraic approaches are obtained on a categorical level, that is, independently of the specific definition of graph. Main parts of the theory have already been generalized to arbitrary categories of structures in high-level replacement (HLR) systems (mainly in the DPO approach [21], see [6] for corresponding concepts in the SPO approach). The SPO approach has been developed for graph structures from the very beginning [40,5], which have recently been extended to generalized graph structures in [16].

3.3 Graph Transformation Based on the DPO Construction

Following the outline of Section 3.2.1, in this section we introduce the formal definition of the very basic concepts of the double-pushout approach to graph transformation. This theory has been started in [1]; comprehensive tutorials can be found in [2,3,4].

First of all, let us introduce the class of graphs that we shall consider in the rest of this chapter (and also in the next one). These are labeled, directed multigraphs, i.e., graphs where edges and nodes are labelled over two different sets of labels, and between two nodes many parallel edges, even with the same label, are allowed for. A fundamental fact is that graphs and graph morphisms form a category since this allows one to formulate most of the definitions, constructions and results in pure categorical terms. Only seldom some specific properties of the category of graphs are exploited, usually to present the concrete, set-theoretical counterpart of some abstract categorical construction.

Definition 3.3.1 (labeled graphs)
Given two fixed alphabets Ω_V and Ω_E for node and edge labels, respectively, a *(labeled) graph (over* (Ω_V, Ω_E)*)* is a tuple $G = \langle G_V, G_E, s^G, t^G, lv^G, le^G \rangle$, where G_V is a set of *vertices* (or *nodes*), G_E is a set of *edges* (or *arcs*), $s^G, t^G : G_E \to G_V$ are the *source* and *target* functions, and $lv^G : G_V \to \Omega_V$ and $le^G : G_E \to \Omega_E$ are the *node* and the *edge labeling* functions, respectively.

A *graph morphism* $f : G \to G'$ is a pair $f = \langle f_V : G_V \to G'_V, f_E : G_E \to G'_E \rangle$ of functions which preserve sources, targets, and labels, i.e., which satisfies $f_V \circ t^G = t^{G'} \circ f_E$, $f_V \circ s^G = s^{G'} \circ f_E$, $lv^{G'} \circ f_V = lv^G$, and $le^{G'} \circ f_E = le^G$. A graph morphism f is an *isomorphism* if both f_V and f_E are bijections. If there exists an isomorphism from graph G to graph H, then we write $G \cong H$; moreover, $[G]$ denotes the isomorphism class of G, i.e., $[G] = \{H \mid H \cong G\}$. An *automorphism* of graph G is an isomorphism $\phi : G \to G$; it is *non-trivial* if $\phi \neq id_G$.

The category having labeled graphs as objects and graph morphisms as arrow is called **Graph**.

Within this chapter we shall often call an isomorphism class of graphs *abstract graph*, like $[G]$; correspondingly, the objects of **Graph** will be called sometimes *concrete* graphs.

Example 3.1 (client-server systems)
As a running example, we will use a simple graph grammar which models the evolution of client-server systems. The (labeled) graphs are intended to represent possible configurations containing servers and clients (represented by nodes labeled by S and C, respectively), which can be in various states, indicated by edges. A loop on a server labeled by *idle* indicates that the server is ready to accept requests from clients; a loop on a client labeled by *job* means that the client is performing some internal activity, while a loop labeled by *req* means that the client issued a request. An edge from a client to a server

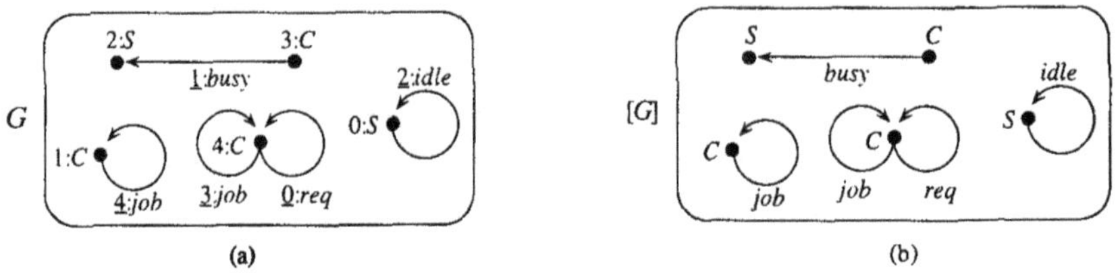

Figure 3.3: (a) A sample labeled graph and (b) its isomorphism class.

labeled by *busy* models the situation where the server is processing a request issued by the client.

In our running example, the alphabets of labels contain only the labels just mentioned, thus $\Omega_V = \{C, S\}$ and $\Omega_E = \{\textit{idle, job, req, busy}\}$. Figure 3.3(a) shows the graphical representation of graph $G = \langle G_V, G_E, s^G, t^G, lv^G, le^G \rangle$ with $G_V = \{0, 1, 2, 3, 4\}$, $G_E = \{\underline{0}, \underline{1}, \underline{2}, \underline{3}, \underline{4}\}$. Edges are drawn in the usual way as arrows from the source to the target, and the label of each node or edge is written after its identity, separated by a colon. We shall use natural numbers to denote nodes and underlined numbers to denote edges. Graph G represents a system with three clients and two servers, whose states are specified by the depicted edges.
Figure 3.3(b) shows the graphical representation of the isomorphism class of graphs $[G]$: it is obtained from G by deleting all natural numbers, i.e., by forgetting the identity of nodes and edges. Their labels are kept because isomorphisms must preserve labels. □

A production in the double-pushout approach is usually defined as a *span*, i.e., a pair of graph morphisms with common source. In the formal definition that follows, a production is defined instead as a structure $p : (L \stackrel{l}{\leftarrow} K \stackrel{r}{\rightarrow} R)$, where l and r are the usual two morphisms, and the additional component p is the production name. The name of a production plays no role when a production is applied to a graph (see Definition 3.3.5 of direct derivation below), but it is relevant in certain transformations of derivations (as in the *analysis* construction, Proposition 3.4.8) and when relating different derivations (as in Definition 3.5.10); in fact, since the name will often be used to encode the construction that yields the production, it allows one to distinguish, for example, between two productions obtained in completely unrelated ways, but that happen to have the same associated span. Thus on the one hand we insist that the name is a relevant part of a productions, but, on the other hand, when applying a production to a graph the name is sometimes disregarded.

Definition 3.3.2 (graph productions, graph grammars)
A *graph production* $p : (L \stackrel{l}{\leftarrow} K \stackrel{r}{\rightarrow} R)$ is composed of a *production name* p and a pair of injective graph morphisms $l : K \rightarrow L$ and $r : K \rightarrow R$. The graphs L, K, and R are called the *left-hand side* (lhs), the *interface*, and the *right-hand side* (rhs) of p, respectively. Two productions $p : (L \stackrel{l}{\leftarrow} K \stackrel{r}{\rightarrow} R)$ and $p' : (L' \stackrel{l'}{\leftarrow} K' \stackrel{r'}{\rightarrow} R')$ are *span-isomorphic* if there are three isomorphisms $L \stackrel{\phi_L}{\longrightarrow} L'$, $K \stackrel{\phi_K}{\longrightarrow} K'$ and $R \stackrel{\phi_R}{\longrightarrow} R'$ that make the two resulting squares commutative, i.e., $\phi_L \circ l = l' \circ \phi_K$ and $\phi_R \circ r = r' \circ \phi_K$. If no confusion is possible, we will sometimes make reference to a production $p : (L \stackrel{l}{\leftarrow} K \stackrel{r}{\rightarrow} R)$ simply as p, or also as $L \stackrel{l}{\leftarrow} K \stackrel{r}{\rightarrow} R$.

A *graph grammar* $\mathcal{G}$ is a pair $\mathcal{G} = \langle (p : L \stackrel{l}{\leftarrow} K \stackrel{r}{\rightarrow} R)_{p \in P}, G_0 \rangle$ where the first component is a family of productions indexed by production names in P, and G_0 is the *start graph.* [b]

Suppose (without loss of generality) that in a production $p : (L \stackrel{l}{\leftarrow} K \stackrel{r}{\rightarrow} R)$ morphisms l and r are inclusions. Then, intuitively, the production specifies that all the items in K have to be preserved (in fact, they are both in L and in R); that the items of L which are not in K have to be deleted, and that the items in R which are not in K have to be created.

Example 3.2 (graph grammar for client-server systems)
The graph grammar which models the evolution of client-server systems depicted as labeled graphs (as in Example 3.1) includes three productions, named *REQ*, *SER*, and *REL*, respectively, which are presented in Figure 3.4. Production *REQ* models the issuing of a request by a client. After producing the request, the client continues its internal activity (*job*), while the request is served asynchronously; thus a request can be issued at any time, even if other requests are pending or if the client is being served by a server. Production *SER* connects a client that issued a request with an idle server through a *busy* edge, modeling the beginning of the service. Production *REL* (for *release*) disconnect the client from the server (modeling the end of the service), restoring the *idle* state of the server.
It is worth stressing that the three productions neither delete nor create nodes, which means that the number of clients and servers is invariant. The interface of each production is a subgraph of both the left- and the right-hand sides, thus

[b] Usually the definition of a grammar also includes a set of terminal labels, used to identify the graphs belonging to the generated language. Since we are not directly interested in graph languages, we omit them.

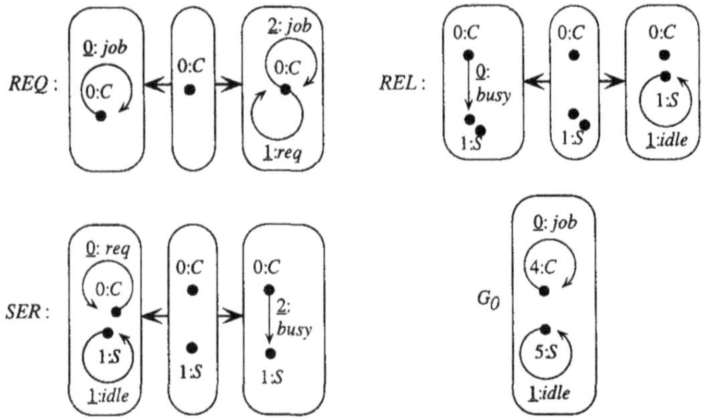

Figure 3.4: Productions and start graph of the grammar modeling a client-server systems.

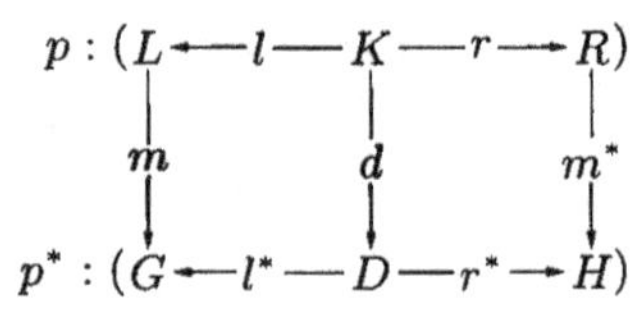

Figure 3.5: Direct derivation as double-pushout construction.

the morphisms are just inclusions. Note moreover that the natural numbers used as identities of nodes and edges are chosen arbitrarily.

Formally, the grammar is called $\mathcal{C}$-$\mathcal{S}$ (for *client-server*), and it is defined as $\mathcal{C}\text{-}\mathcal{S} = \langle\{REQ, SER, REL\}, G_0\rangle$, where G_0 is the start graph depicted in Figure 3.4. □

Given a production $p : (L \xleftarrow{l} K \xrightarrow{r} R)$ and a graph G, one can try to apply p to G if there is an occurrence of L in G, i.e., a graph morphism, called match, $m : L \to G$. If p is applicable to match m yielding a graph H, then we obtain a direct derivation $G \stackrel{p,m}{\Longrightarrow} H$. Such a direct derivation is a "double-pushout construction", the construction that characterizes the algebraic approach we are presenting, and it is shown in Figure 3.5.

Intuitively, regarding graphs as distributed states of a system, a *pushout* is a sort of "generalized union" that specifies how to merge together two states having a common substate [42]. For example, if the right square of Figure 3.5 is a pushout in **Graph**, then graph H is obtained by gluing together graphs

R and D along the common subgraph K, and r^* and m^* are the resulting injections.

Therefore the double-pushout construction can be interpreted as follows. In order to apply the production p to G, we first need to find an occurrence of its left-hand side L in G, i.e., a match $m : L \to G$. Next, we have to delete from G all the (images of) items of L which are not in the interface graph K; in other words, we have to find a graph D and morphisms d and l^* such that the resulting square is a pushout: The context graph D is therefore required to be a *pushout complement object* of $\langle l, m\rangle$. Finally, we have to embed the right-hand side R into D, to model the addition to D of the items to be created, i.e., those that are in R but not in K: This embedding is expressed by the right pushout.

In order to introduce this construction formally we first have to define pushouts and pushout complements. It is worth stressing that the advantage of describing such operation via a categorical construction instead of using standard set-theoretical notions is two-fold. On the one hand, in a set-theoretical framework one needs to handle explicitly the possible clashing of names used in the two states (in the non-shared parts): this is given for free in the categorical construction, at the price of determining the result only up to isomorphism. On the other hand, category theory provides handy techniques, based for example on diagram chasing, for describing and relating complex constructions, that it would be quite cumbersome to handle using standard set theory.

Definition 3.3.3 (pushout [18] and pushout complement [2])
Given a category $\mathbf{C}$ and two arrows $b : A \to B$, $c : A \to C$ of $\mathbf{C}$, a triple $\langle D, g : B \to D, f : C \to D\rangle$ as in the diagram below is called a *pushout* of $\langle b, c\rangle$ if

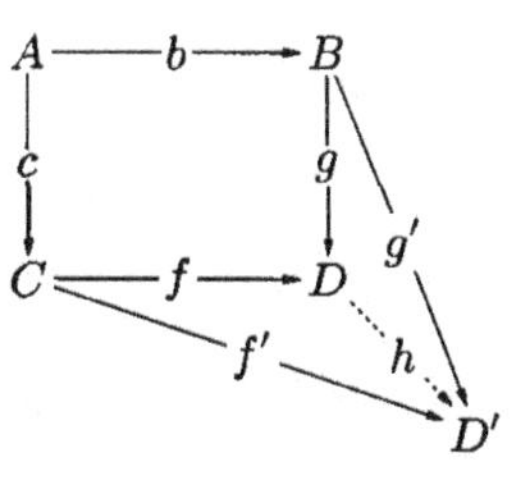

[Commutativity] $g \circ b = f \circ c$, and

[Universal Property] for all objects D' and arrows $g' : B \to D'$ and $f' : C \to D'$, with $g' \circ b = f' \circ c$, there exists a unique arrow $h : D \to D'$ such that $h \circ g = g'$ and $h \circ f = f'$.

In this situation, D is called a *pushout object* of $\langle b, c\rangle$. Moreover, given arrows $b : A \to B$ and $g : B \to D$, a *pushout complement* of $\langle b, g\rangle$ is a triple $\langle C, c : A \to C, f : C \to D\rangle$ such that $\langle D, g, f\rangle$ is a pushout of $\langle b, c\rangle$. In this case C is called a *pushout complement object* of $\langle b, g\rangle$.

Example 3.3 (Pushout in Set and in Graph)
The paradigmatic example of category is **Set**, i.e., the category having sets as objects and total functions as arrows. It is easy to show that the pushout of two arrows in **Set** always exists, and that it is characterized (up to isomorphism) as follows. If $b : A \to B$ and $c : A \to C$ are two functions, then the pushout object of $\langle b, c\rangle$ is the set $D = (B + C)_{\approx}$, i.e., the quotient set of the disjoint union of B and C, modulo the equivalence relation "$\approx$", which is the least equivalence relation such that for all $a \in A$, $b(a) \approx c(a)$, together with the two functions $g : B \to D$ and $f : C \to D$ that map each element to its equivalence class.

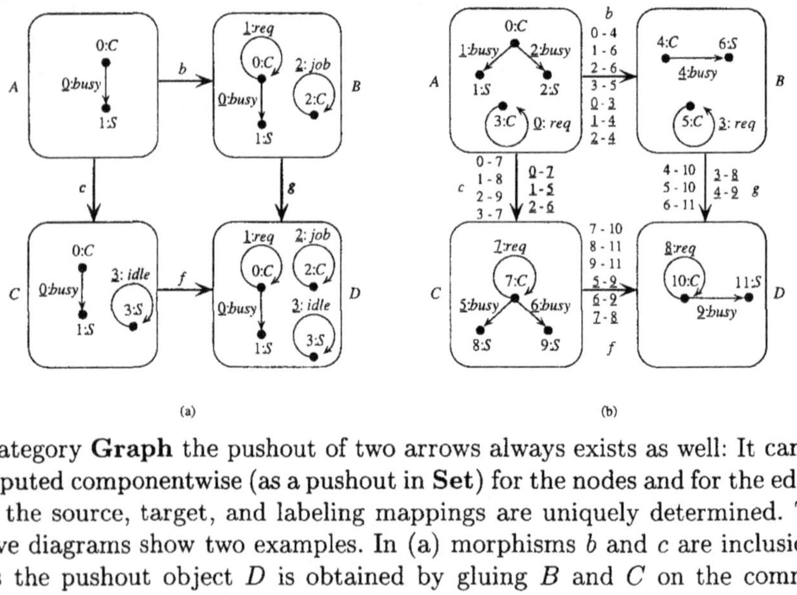

(a) (b)

In category **Graph** the pushout of two arrows always exists as well: It can be computed componentwise (as a pushout in **Set**) for the nodes and for the edges, and the source, target, and labeling mappings are uniquely determined. The above diagrams show two examples. In (a) morphisms b and c are inclusions, thus the pushout object D is obtained by gluing B and C on the common subgraph A. Diagram (b), where all morphisms are specified by listing the pairs *argument-result* (e.g., $b_V(0) = 4$ and $c_E(\underline{1}) = \underline{5}$), shows instead a pushout in **Graph** of two non-injective morphisms b and c. □

In any category, if the pushout object of two arrows exists then it is unique up to a unique isomorphism because of the universal property (see Fact 3.8.2 for the corresponding statement for coproducts). On the contrary, since the pushout complement object is not characterized directly by a universal property, for a pair of arrows there may not be any coproduct, or there can exist many pushout complement objects which are not isomorphic. For example, consider the diagrams below in the category of sets and functions. For the left diagram, there exist no set and functions which can close the square making it a pushout, while in the right diagram there exist two pushout complement objects C and C' which are not isomorphic.

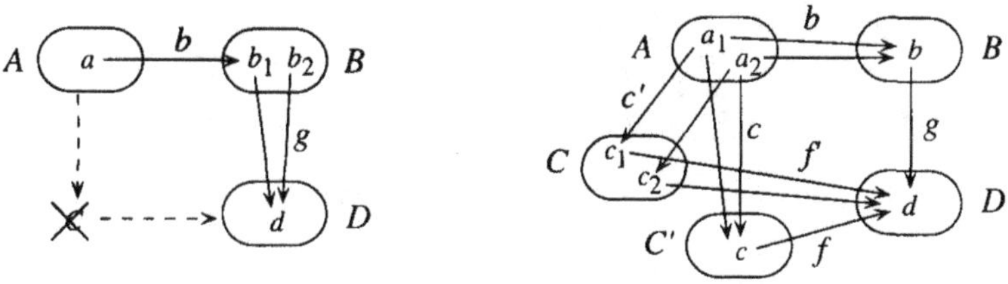

The characterization of sufficient conditions for the existence and uniqueness of the pushout complement of two arrows is a central topic in the algebraic theory of graph grammars, because, as we shall see below, in the DPO approach it allows one to check for the applicability of a production to a given match. As shown for example in [2], in category **Graph**, the pushout complement object of two morphisms $\langle b, g\rangle$ exists iff the *gluing condition* is satisfied; moreover, it is unique if b is injective.

Proposition 3.3.4 (existence of pushout complements)
Let $b : A \to B$ and $g : B \to D$ be two morphisms in **Graph**. *Then there exists a pushout complement $\langle C, c : A \to C, f : C \to D\rangle$ of $\langle b, g\rangle$ if and only if the following conditions are satisfied:*

[***Dangling condition***] *No edge $e \in D_E - g_E(B_E)$ is incident to any node in $g_V(B_V - b_V(A_V))$;*

[***Identification condition***] *There is no $x, y \in B_V \cup B_E$ such that $x \neq y$, $g(x) = g(y)$ and $y \notin b(A_V \cup A_E)$.*

In this case we say that $\langle b, g\rangle$ satisfy the gluing condition *(or g satisfies the* gluing condition *with respect to b). If moreover morphism b is injective,then the pushout complement is unique up to isomorphism, i.e., if $\langle C, c, f\rangle$ and $\langle C', c', f'\rangle$ are two pushout complements of $\langle b, g\rangle$, then there is an isomorphism $\phi : C \to C'$ such that $\phi \circ c = c'$ and $f' \circ \phi = f$.* □

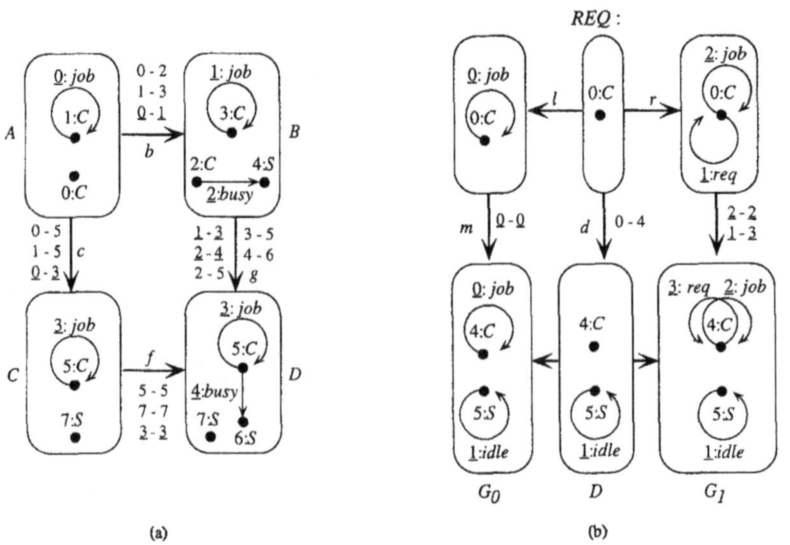

Figure 3.6: (a) Pushout complement in **Graph**. (b) A direct derivation.

Example 3.4
Morphisms $\langle b, g\rangle$ of Figure 3.6 (a) satisfy the dangling condition, because the only edge incident to node $6 : S$ (which is in $g_V(B_V - b_V(A_V))$) is $\underline{4} : busy$, which is not in $D_E - g_E(B_E)$. They also satisfy the identification condition because the only items identified by g, nodes $2 : C$ and $3 : C$, are in the image of b. Since b is injective, a pushout complement object of $\langle b, g\rangle$ is given by $C = D - g(B - b(A))$, i.e., it is the subgraph of D obtained by removing all items that are in the image of g but not in the image of $g \circ b$. Morphism $C \xrightarrow{f} D$ is the inclusion, and $A \xrightarrow{c} C$ is the (codomain) restriction of g to C. □

Definition 3.3.5 (direct derivation)
Given a graph G, a graph production $p : (L \xleftarrow{l} K \xrightarrow{r} R)$, and a *match* $m : L \to G$, a *direct derivation from G to H using p (based on m)* exists if and only if the diagram in Figure 3.5 can be constructed, where both squares are required to be pushouts in **Graph**. In this case, D is called the *context* graph, and we write $G \stackrel{p,m}{\Longrightarrow} H$, or also $G \stackrel{p}{\Longrightarrow} H$; only seldom we shall write $G \stackrel{\langle p,m,d,m^*,l^*,r^*\rangle}{\Longrightarrow} H$, indicating explicitly all the morphisms of the double-pushout.

Example 3.5
Figure 3.6 (b) shows the direct derivation $G_0 \stackrel{REQ}{\Longrightarrow} G_1$. All horizontal morphisms are inclusions. The vertical morphisms are instead depicted explicitly by listing enough pairs *argument-result* (for example, the morphism from the right-hand side of the production to graph G_1, denoted by $\{\underline{2}\text{-}\underline{2}, \underline{1}\text{-}\underline{3}\}$, maps edges $\underline{2}$ and $\underline{1}$ to edges $\underline{2}$ and $\underline{3}$, respectively, and therefore node 0 to node 4). Morphisms $\langle l, m\rangle$ obviously satisfy the dangling condition because no node is deleted by *REQ*, and also the identification condition because m is injective. Graph D is the pushout complement object of $\langle l, m\rangle$, and G_1 is the pushout object of $\langle r, d\rangle$. □

Definition 3.3.6 (sequential derivations)
A *sequential derivation (over $\mathcal{G}$)* is either a graph G (called an *identity derivation*, and denoted by $G : G \Rightarrow^* G$), or a sequence of direct derivations $\rho = \{G_{i-1} \stackrel{p_i}{\Longrightarrow} G_i\}_{i\in\{1,\dots,n\}}$ such that p_i is a production of $\mathcal{G}$ for all $i \in \{1, \dots, n\}$. In the last case, the derivation is written $\rho : G_0 \Rightarrow^*_{\mathcal{G}} G_n$ or simply $\rho : G_0 \Rightarrow^* G_n$. If $\rho : G \Rightarrow^* H$ is a (possibly identity) derivation, then graphs G and H are called the *starting* and the *ending graph* of ρ, and will be denoted by $\sigma(\rho)$ and $\tau(\rho)$, respectively. The *length* of a sequential derivation ρ is the number of direct derivations in ρ, if it is not identity, and 0 otherwise. The *sequential composition* of two derivations ρ and ρ' is defined if and only if $\tau(\rho) = \sigma(\rho')$; in this case it is denoted $\rho\,;\,\rho' : \sigma(\rho) \Rightarrow^* \tau(\rho')$, and it is obtained by identifying $\tau(\rho)$ with $\sigma(\rho')$.

The identity derivation $G : G \Rightarrow^* G$ is introduced just for technical reasons. By the definition of sequential composition, it follows that for each derivation $\rho : G \Rightarrow^* H$ we have $G\,;\,\rho = \rho = \rho\,;\,H$.

Example 3.6 (derivation)
Figure 3.7 shows a sequential derivation using grammar $\mathcal{C}$-$\mathcal{S}$ and starting from the start graph G_0. The derivation models the situation where a request is issued by the client, and while it is handled by the server, a new request is issued. □

3.4 Independence and Parallelism in the DPO approach

According to the overview of Section 3.2.1 we have to define two notions of independence on direct derivations formalizing the following concepts:

- two alternative direct derivations $H_1 \stackrel{p_1,m_1}{\Longleftarrow} G \stackrel{p_2,m_2}{\Longrightarrow} H_2$ are not in conflict (parallel independence), and

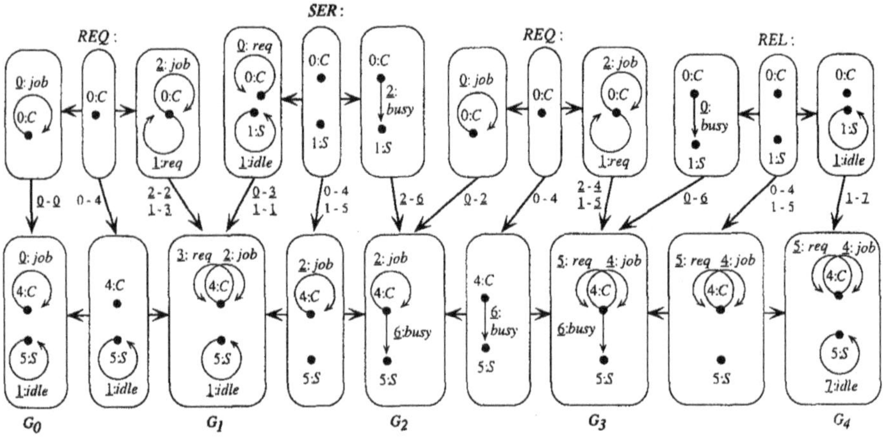

Figure 3.7: A sequential derivation in grammar C-S starting from graph G_0.

- two consecutive direct derivations $G \overset{p_1,m_1}{\Longrightarrow} H_1 \overset{p_2,m_2'}{\Longrightarrow} X$ are not causally dependent (sequential independence).

Intuitively, two alternative direct derivations are *parallel independent* (of each other), if each of them can still be applied after the other one has been performed. This means that neither $G \overset{p_1,m_1}{\Longrightarrow} H_1$ nor $G \overset{p_2,m_2}{\Longrightarrow} H_2$ can delete elements of G which are also needed by the other direct derivation. In other words, the overlapping of the left-hand sides of p_1 and p_2 in G must be included in the intersection of the corresponding interface graphs K_1 and K_2.

Definition 3.4.1 (parallel independence)
Let $G \overset{p_1,m_1}{\Longrightarrow} H_1$ and $G \overset{p_2,m_2}{\Longrightarrow} H_2$ be two direct derivations from the same graph G, as in Figure 3.8. They are *parallel independent* if $m_1(L_1) \cap m_2(L_2) \subseteq m_1(l_1(K_1)) \cap m_2(l_2(K_2))$.
This property can be formulated in categorical terms in the following way: There exist two graph morphisms $L_1 \xrightarrow{k_2} D_2$ and $L_2 \xrightarrow{k_1} D_1$ such that $l_2^* \circ k_2 = m_1$ and $l_1^* \circ k_1 = m_2$.

Two consecutive direct derivations $G \overset{p_1,m_1}{\Longrightarrow} H_1 \overset{p_2,m_2}{\Longrightarrow} X$ are *sequentially independent* if they may be swapped, i.e., if p_2 can be applied to G, and p_1 to the resulting graph. Therefore p_2 at match m_2 cannot delete anything that has been explicitly preserved by the application of p_1 at match m_1, and, moreover, it cannot use (neither consuming nor preserving it) any element generated by

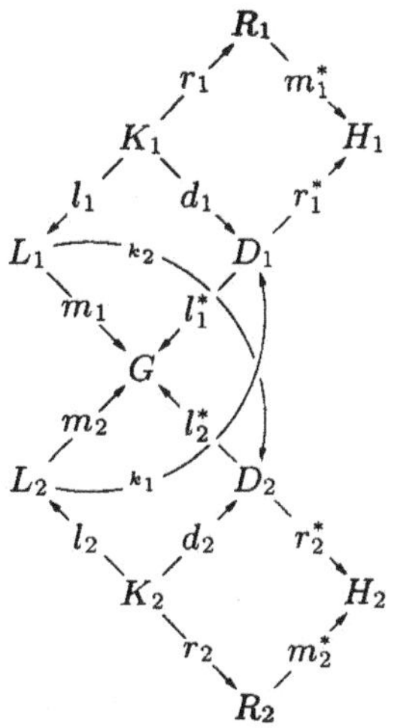

Figure 3.8: Parallel independence of direct derivations

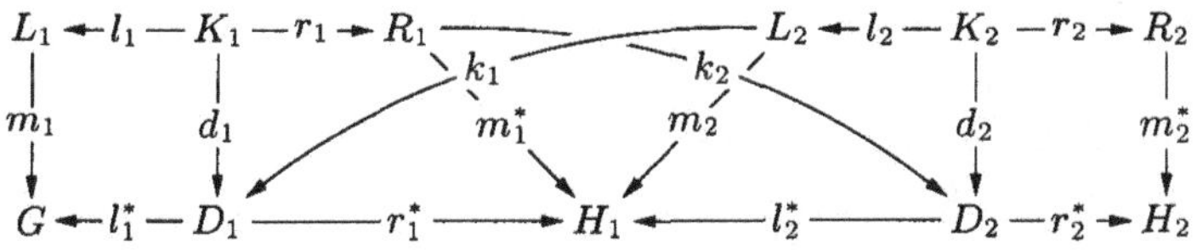

Figure 3.9: Sequential independent derivation.

p_1; this implies that the overlapping of R_1 and L_2 in H_1 must be included in the intersection of the interface graphs K_1 and K_2.

Definition 3.4.2 (sequential independence)
Given a two-step derivation $G \overset{p_1,m_1}{\Longrightarrow} H_1 \overset{p_2,m_2}{\Longrightarrow} H_2$ (as in Figure 3.9), it is *sequential independent* iff $m_1^*(R_1) \cap m_2(L_2) \subseteq m_1^*(r_1(K_1)) \cap m_2(l_2(K_2))$.
This property can be formulated also in categorical terms in the following way: There exist two graph morphisms $R_1 \xrightarrow{k_2} D_2$ and $L_2 \xrightarrow{k_1} D_1$ such that $l_2^* \circ k_2 = m_1^*$ and $r_1^* \circ k_1 = m_2$.

Example 3.7 (sequential independence)
Consider the derivation of Figure 3.7. The first two direct derivations are *sequential dependent*; in fact, the edge $\underline{3}$:*req* of graph G_1 is in the image of both the right-hand side of the first production and the left-hand side of the second one, but it is in the context of neither the first nor the second direct derivation: Thus the required morphisms for sequential independence do not exist. On the

contrary, in the same figure both the derivation from G_1 to G_3 and that from G_2 to G_4 are sequential independent, and the required morphisms are obvious. □

The following theorem says that these two definitions of parallel and sequential independence indeed solve the Local Church-Rosser Problem of Section 3.2.2 in the DPO approach.

Theorem 3.4.3 (Local Church Rosser)

1. *Let* $G \overset{p_1,m_1}{\Longrightarrow} H_1$ *and* $G \overset{p_2,m_2}{\Longrightarrow} H_2$ *be two parallel independent direct derivations, as in Figure 3.8, and let* $m_2' = r_1^* \circ k_1$*, where arrow* $L_2 \overset{k_1}{\longrightarrow} D_1$ *exists by parallel independence. Then morphisms* $\langle l_2, m_2' \rangle$ *satisfy the gluing condition, and thus production* p_2 *can be applied to match* m_2'*. Moreover, derivation* $G \overset{p_1,m_1}{\Longrightarrow} H_1 \overset{p_2,m_2'}{\Longrightarrow} X$ *is sequential independent.*
2. *Let* $G \overset{p_1,m_1}{\Longrightarrow} H_1 \overset{p_2,m_2}{\Longrightarrow} H_2$ *be a sequential independent derivation as in Figure 3.9, and let* $m_2' = l_1^* \circ k_1$*, where arrow* $L_2 \overset{k_1}{\longrightarrow} D_1$ *exists by sequential independence. Then morphisms* $\langle l_2, m_2' \rangle$ *satisfy the gluing condition, and thus production* p_2 *can be applied to match* m_2'*. Moreover, direct derivations* $G \overset{p_1,m_1}{\Longrightarrow} H_1$ *and* $G \overset{p_2,m_2'}{\Longrightarrow} Y$ *are parallel independent.* □

The proof of the Local Church Rosser theorem is postponed, because it follows easily from the proof of the Parallelism Theorem presented later.

The notion of direct derivation introduced in Definition 3.3.5 is intrinsically sequential: it models the application of a single production of a grammar $\mathcal{G}$ to a given graph. The categorical framework provides an easy definition of the parallel application of more than one production to a graph. The following definition of parallel production is slightly more complex than analogous definitions previously presented in the literature, which only used a binary, commutative and associative operator "+" on productions. Such a complexity is justified by the result reported in Section 3.8, where it is shown that the coproduct cannot be assumed to be commutative (in a strict way), unless the category is a preorder.

Definition 3.4.4 (parallel productions)
Given a graph grammar $\mathcal{G}$, a *parallel production (over* $\mathcal{G}$*)* has the form $\langle (p_1, in^1), \ldots, (p_k, in^k) \rangle : (L \overset{l}{\leftarrow} K \overset{r}{\rightarrow} R)$ (see Figure 3.10), where $k \geq 0$, $p_i : (L_i \overset{l_i}{\leftarrow} K_i \overset{r_i}{\rightarrow} R_i)$ is a production of $\mathcal{G}$ for each $i \in \{1, \ldots, k\}$, L is a

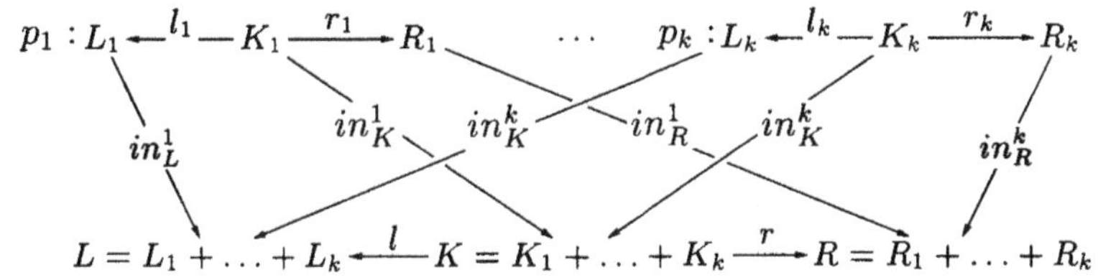

Figure 3.10: The parallel production $\langle (p_1, in^1), \ldots, (p_k, in^k)\rangle : (L \overset{l}{\leftarrow} K \overset{r}{\rightarrow} R)$.

coproduct object[c] of the graphs in $\langle L_1, \ldots, L_k\rangle$, and similarly R and K are coproduct objects of $\langle R_1, \ldots, R_k\rangle$ and $\langle K_1, \ldots, K_k\rangle$, respectively. Moreover, l and r are uniquely determined by the families of arrows $\{l_i\}_{i\leq k}$ and $\{r_i\}_{i\leq k}$, respectively (they are defined in a way similar to morphism $f + g$ of Definition 3.8.3). Finally, for each $i \in \{1, \ldots, k\}$, in^i denotes the triple of injections $\langle in^i_L : L_i \rightarrow L, in^i_K : K_i \rightarrow K, in^i_R : R_i \rightarrow R\rangle$. Note that all the component productions are recorded in the name of a parallel production.

A parallel production like the one above is *proper* if $k > 1$; the *empty* production is the (only) parallel production with $k = 0$, having the empty graph $\emptyset$ as left- and right-hand sides and as interface. Each production $p : (L \overset{l}{\leftarrow} K \overset{r}{\rightarrow} R)$ of $\mathcal{G}$ will be identified with the parallel production $\langle (p, \langle id_L, id_K, id_R\rangle)\rangle : (L \overset{l}{\leftarrow} K \overset{r}{\rightarrow} R)$.

Two parallel productions over $\mathcal{G}$, $p = \langle (p_1, in^1), \ldots, (p_k, in^k)\rangle : (L \overset{l}{\leftarrow} K \overset{r}{\rightarrow} R)$ and $q = \langle (q_1, \underline{in}^1), \ldots, (q_{k'}, \underline{in}^{k'})\rangle : (L' \overset{l'}{\leftarrow} K' \overset{r'}{\rightarrow} R')$, are *isomorphic via* Π if $k = k'$ and Π is a permutation of $\{1, \ldots, k\}$ such that and $p_i = q_{\Pi(i)}$ for each $i \in \{1, \ldots, k\}$; that is, the component productions of $\mathcal{G}$ are the same, up to a permutation. We say that p and q are *isomorphic* if there is a Π such that they are isomorphic via Π.

The *graph grammar with parallel productions* $\mathcal{G}^+$ generated by a grammar $\mathcal{G}$ has the same start graph of $\mathcal{G}$, and as productions all the parallel productions over $\mathcal{G}$. A *parallel direct derivation over* $\mathcal{G}$ is a direct derivation over $\mathcal{G}^+$; it is *proper* or *empty* if the applied parallel production is proper or empty, respectively. If $G \overset{\emptyset}{\Longrightarrow} H$ is an empty direct derivation, then morphisms l^* and r^* are isomorphisms (see Figure 3.5); morphism $r^* \circ l^{*-1} : G \rightarrow H$ is called the *isomorphism induced* by the empty direct derivation.

A *parallel derivation over* $\mathcal{G}$ is a sequential derivation over $\mathcal{G}^+$.

---

[c] Binary coproducts are introduced in Definition 3.8.1. The generalization to arbitrary coproducts is straightforward. The coproduct of a list of graphs is given by their disjoint union (the coproduct object) and by all the injections.

It is worth noting that two isomorphic parallel productions are, *a fortiori*, span-isomorphic (see Definition 3.3.2).

Fact 3.4.5 (isomorphic parallel productions are span-isomorphic)
Let $\langle(p_1, in^1), \ldots, (p_k, in^k)\rangle : (L \stackrel{l}{\leftarrow} K \stackrel{r}{\rightarrow} R)$ *and* $\langle(q_1, \underline{in}^1), \ldots, (q_k, \underline{in}^k)\rangle :$ $(L' \stackrel{l'}{\leftarrow} K' \stackrel{r'}{\rightarrow} R')$ *be two parallel productions isomorphic via* Π. *Then they are span-isomorphic.* □

Proof
By definition, we have that for each $X \in \{L, K, R\}$, $\langle X, in^1_X, \ldots, in^k_X\rangle$ and $\langle X', \underline{in}^1_X, \ldots, \underline{in}^k_X\rangle$ are two coproducts of the same objects, and thus there is a unique isomorphism $X \stackrel{\phi_X}{\longrightarrow} X'$ which commutes with the injections (see Fact 3.8.2). It can be shown easily that these isomorphisms satisfy $\phi_L \circ l = l' \circ \phi_K$ and $\phi_R \circ r = r' \circ \phi_K$. □

In the rest of the chapter we will assume, without loss of generality, that a "canonical" *choice of coproducts* is given once and for all, i.e., a mapping associating with every pair of graphs $\langle A, B\rangle$ a fixed coproduct $\langle A + B, in_1^{A+B} : A \to A + B, in_2^{A+B} : B \to A + B\rangle$ (see Definitions 3.8.1 and 3.8.3). Then for each pair of parallel productions $p : (L \stackrel{l}{\leftarrow} K \stackrel{r}{\rightarrow} R)$ and $p' : (L' \stackrel{l'}{\leftarrow} K' \stackrel{r'}{\rightarrow} R')$, where $p = \langle(p_1, in^1), \ldots, (p_k, in^k)\rangle$ and $p' = \langle(p'_1, in'^1), \ldots, (p'_{k'}, in'^{k'})\rangle$ we denote by $p+p'$ the parallel production $\langle(p_1, in \circ in^1), \ldots, (p_k, in \circ in^k), (p'_1, in' \circ in'^1), \ldots, (p'_{k'}, in' \circ in'^{k'})\rangle : (L + L' \stackrel{l+l'}{\leftarrow} K + K' \stackrel{r+r'}{\rightarrow} R + R')$, where '+' is the coproduct functor induced by the choice of coproducts (Definition 3.8.3), and in and in' are triples of canonical injections. Moreover, given a list $\langle p_1, \ldots, p_k\rangle$ of parallel productions of $\mathcal{G}$, by $p_1+p_2+\ldots+p_k$ we denote the parallel production $((\ldots(p_1+p_2)+\ldots)+p_k)$. Note that the + operator on productions defined in this way is in general neither associative nor commutative: such properties would depend on analogous properties of the choice of coproducts. Nevertheless, it is obvious that + is associative and commutative "up to isomorphism", in the sense that, for example, productions $p + p'$ and $p' + p$ are isomorphic. The canonical choice of coproducts is assumed here only to simplify the syntax of parallel productions: it does not imply that coproducts are unique (see also Proposition 3.8.6 and Corollary 3.8.7).

Example 3.8 (parallel direct derivation)
Figure 3.11 shows a parallel direct derivation based on the productions of Example 3.2. The upper part of the diagram shows the parallel production *SER + REQ*. The component productions *SER* and *REQ* and the corresponding

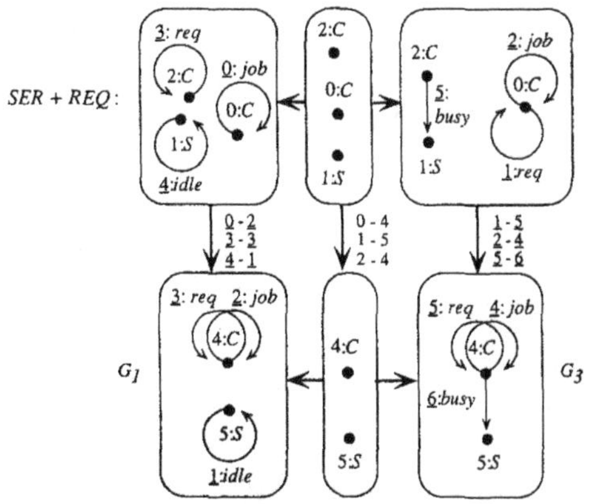

Figure 3.11: A parallel direct derivation via a non-injective match.

inclusions are not depicted explicitly in order to simplify the drawing; anyway, it is easy to recognize in the parallel production the isomorphic copies of *SER* and *REQ*.

The parallel production *SER* + *REQ* is applied to graph G_1 via a non-injective match. This direct derivation models the situation where a new request is issued by client 4 exactly when the service of an older request is started by server 5. □

Let us consider now Problem 3.2.2 of Section 3.2.2, asking under which conditions a parallel direct derivation can be sequentialized into the application of the component productions, and, viceversa, under which conditions the sequential application of two (or more) productions can be put in parallel. As the following Parallelism Theorem states, such conditions are nothing else than the parallel and sequential independence conditions already introduced. More precisely, it states that the *sequentialization condition* is already part of the gluing condition of the parallel production, that is, each parallel direct derivation may be sequentialized in an arbitrary order leading to sequentially independent derivations. Vice versa, the *parallelization condition* is the sequential independence of the derivation, since consecutive direct derivations can be put in parallel if they are sequentially independent.

Theorem 3.4.6 (Parallelism)
Given (possibly parallel) productions $p_1 : (L_1 \stackrel{l_1}{\longleftarrow} K_1 \stackrel{r_1}{\longrightarrow} R_1)$ *and* $p_2 : (L_2 \stackrel{l_2}{\longleftarrow} K_2 \stackrel{r_2}{\longrightarrow} R_2)$ *the following statements are equivalent:*

1. *There is a parallel direct derivation* $G \stackrel{p_1+p_2,m}{\Longrightarrow} X$

2. *There is a sequentially independent derivation* $G \stackrel{p_1,m_1}{\Longrightarrow} H_1 \stackrel{p_2,m'_2}{\Longrightarrow} X$.

□

The rest of this section is devoted to the proofs of the Parallelism and of the Local Church Rosser Theorems, and to the presentation of some technical results that will be used in the next section. First of all, let us clarify the relationship between a match of a parallel production and the induced matches of the component productions.

Proposition 3.4.7 (applicability of parallel productions)
Let $q = \langle (p_1, in^1), \ldots, (p_k, in^k) \rangle : (L \stackrel{l}{\leftarrow} K \stackrel{r}{\rightarrow} R)$ *be a parallel production (see Definition 3.4.4), and let* $L \stackrel{m}{\longrightarrow} G$ *be a match for it. For each* $i \in \{1, \ldots, k\}$, *let* $L_i \stackrel{m_i}{\longrightarrow} G$ *be the match of the i-th production in G induced by m, defined as* $m_i = m \circ in^i_L$.
Then $\langle l, m \rangle$ *satisfies the gluing condition, i.e., there is a parallel direct derivation* $G \stackrel{q,m}{\Longrightarrow} H$, *if and only if for all* $i \in \{1, \ldots, k\}$ $\langle l_i, m_i \rangle$ *satisfies the gluing condition, i.e.,* p_i *can be applied at match* m_i, *say with result* H_i, *and for each* $1 \leq i < j \leq k$, *direct derivations* $G \stackrel{p_i,m_i}{\Longrightarrow} H_i$ *and* $G \stackrel{p_j,m_j}{\Longrightarrow} H_j$ *are parallel independent.*

Proof outline
The parallel independence of the matches of the composing productions follows from the gluing conditions of the parallel direct derivations, and viceversa. □

Therefore there is a very tight relationship between parallel independence and parallel direct derivations: different matches of productions in a graph induce a match of the corresponding parallel production if and only if they are pairwise parallel independent. It is worth stressing here that this property does not hold in the SPO approach, where a parallel production can be applied even if the induced matches are not parallel independent (as discussed in Sections 4.3.1 and 4.6).

The next important results state that every parallel direct derivation can be decomposed in an arbitrary way as the sequential application of the component productions, and, conversely, that every sequential independent derivation can be transformed into a parallel direct derivation. These constructions are in general non-deterministic, and will be used in Section 3.5 to define suitable relations among derivations. The constructive proofs of Lemmas 3.4.8 and 3.4.9 are reported in Section 3.9 (see also [43,21]). It is worth stressing that since we do not assume commutativity of coproducts, the statements of the following lemmas are slightly different from the corresponding statements in the related literature.

Lemma 3.4.8 (analysis of parallel direct derivations)
Let $\rho = (G \stackrel{q}{\Longrightarrow} H)$ be a parallel direct derivation using the parallel production $q = p_1 + \ldots + p_k : (L \stackrel{l}{\leftarrow} K \stackrel{r}{\rightarrow} R)$. Then for each ordered partition $\langle I = \langle i_1, \ldots, i_n \rangle, J = \langle j_1, \ldots, j_m \rangle \rangle$ of $\{1, \ldots, k\}$ (i.e., $I \cup J = \{1, \ldots, k\}$ and $I \cap J = \emptyset$) there is a constructive way to obtain a sequential independent derivation $\rho' = (G \stackrel{q'}{\Longrightarrow} X \stackrel{q''}{\Longrightarrow} H)$, called an analysis *of ρ, where $q' = p_{i_1} + \ldots + p_{i_n}$, and $q'' = p_{j_1} + \ldots + p_{j_m}$. Such a construction is in general not deterministic (for example, graph X is determined only up to isomorphisms). If ρ and ρ' are as above, we shall write $\rho' \in ANAL_{I,J}(\rho)$ (or $\langle \rho', \rho \rangle \in ANAL_{I,J}$).* □

Therefore any parallel direct derivation can be transformed into a sequence of sequential independent direct derivations by repeated applications of the above construction.

Lemma 3.4.9 (synthesis of sequential independent derivations)
Let $\rho = (G \stackrel{q'}{\Longrightarrow} X \stackrel{q''}{\Longrightarrow} H)$ be a sequential independent derivation. Then there is a constructive way to obtain a parallel direct derivation $\rho' = (G \stackrel{q'+q''}{\Longrightarrow} H)$, called a synthesis *of ρ. Also this construction is in general not deterministic. If ρ and ρ' are as above, we shall write $\rho' \in SYNT(\rho)$.* □

Example 3.9 (analysis and synthesis)
Let ρ be the two-steps derivation from graph G_1 to graph G_3 depicted in Figure 3.7, and let ρ' be the parallel direct derivation of Figure 3.11. Since ρ is sequential independent, the synthesis construction can be applied to it, and one possible result is indeed ρ'; thus we have $\rho' \in SYNT(\rho)$. Conversely, the application of the analysis construction to ρ' can produce ρ, and we have $\rho \in ANAL_{\langle 1 \rangle, \langle 2 \rangle}(\rho')$. Also, through a different analysis of ρ' we may obtain the derivation $\rho_1 ; \rho_2$ of Figure 3.15 (a): In fact, we have $\rho_1 ; \rho_2 \in ANAL_{\langle 2 \rangle, \langle 1 \rangle}(\rho')$. □

It is worth stressing that the analysis and synthesis constructions are not inverse to each other in a strict sense, but they are so up to isomorphism. Isomorphisms of (direct) derivations and other equivalences among derivations will be discussed in depth in the next section.

We are now ready to present the outline of the proofs of the Parallelism and of the Local Church Rosser Theorems

Proof outline of Theorem 3.4.6 (Parallelism)
Lemma 3.4.8 provides a constructive proof that $1 \Rightarrow 2$, and, similarly, Lemma 3.4.9 provides a constructive proof that $2 \Rightarrow 1$. □

Proof outline of Theorem 3.4.3 (Local Church Rosser)
$G \stackrel{p_1,m_1}{\Longrightarrow} H_1$ and $G \stackrel{p_2,m_2}{\Longrightarrow} H_2$ are two parallel independent direct derivations, iff (by Proposition 3.4.7) there is a graph H such that $G \stackrel{p_1+p_2,[m_1,m_2]}{\Longrightarrow} H$ is a parallel direct derivation, iff (by Theorem 3.4.6) there is a sequentially independent derivation $G \stackrel{p_1,m_1}{\Longrightarrow} H_1 \stackrel{p_2,m_2'}{\Longrightarrow} X$. □

3.5 Models of Computation in the DPO Approach

In the previous sections we considered mainly definitions and properties (for the DPO approach) of *direct* derivations, i.e., of the basic graph rewriting steps. Since the derivations of a grammar are intended to model the computations of the system modeled by the grammar, it is certainly worth studying the operational behaviour of a grammar by developing a formal framework where its derivations can be described and analyzed. A *model of computation* for a grammar is a mathematical structure which contains relevant information concerning the potential behaviour of the grammar, that is, about all the possible computations (derivations) of the grammar. Since the basic operation on derivations is concatenation, it is quite obvious that *categories* are very suited as mathematical structures for this use. In fact, a model of computation for a given grammar is just a category having graphs as objects, and where every arrow is a derivation starting from the source graph and ending at the target graph. Then the categorical operation of sequential composition of arrows corresponds to the concatenation of derivations. Models of computation of this kind (i.e., categories of computations, possibly equipped with some additional algebraic structure) have been proposed (with various names) for many formalisms, like Phrase Structure Grammars [44], Petri nets [45,46], CCS [47], Logic Programming [48,49], and others. Although in this section we focus on the DPO approach, it is obvious that models of computation can be defined easily for the SPO approach as well, and also for all other approaches to graph transformation.

A *concrete* model of computation for a grammar is defined easily according to the above intuition: it has all concrete graphs as objects and all derivations as arrows. However, such a model often contains a lot of redundant information; in fact one is usually interested in observing the behaviour of a grammar from a more abstract perspective, ignoring some details which are considered as unessential. Consequently, more abstract models of computation are often of interest, where derivations (and graphs) differing only for insignificant aspects are identified. In general such abstract models can be defined by imposing an equivalence relation on derivations (and graphs) of the most concrete model: Such relation equates exactly those derivations (graphs) which cannot be distinguished by the chosen observation criterion. A very natural requirement for such an equivalence is that it is a congruence with respect to sequential composition. If such a condition is satisfied, one gets automatically the *abstract* model of computation corresponding to the chosen observation by taking the quotient category of the concrete model with respect to the equivalence relation.

In the following subsections, besides the concrete model for a graph grammar in the DPO approach, we consider two different observation mechanisms on graph derivations, and we define the corresponding abstract models. Such observation mechanisms are intended to capture *true concurrency* and *representation independence*, respectively. The equivalence we use for capturing true concurrency is a slight variation of the well-known *shift equivalence* [43,3], and it is presented in Section 3.5.1. On the other hand, the most natural equivalence on graph derivations intended to capture representation independence (i.e., the equivalence equating pairs of isomorphic derivations) fails to be a congruence with respect to sequential composition. Therefore it has to be refined carefully through the use of the so-called *standard isomorphisms* (first introduced in [50]). This will be the topic of Sections 3.5.2 and 3.5.3. The full proofs of the main results of Section 3.5.3 are given in Section 3.9.

Besides the models corresponding to these two equivalences, we shall introduce in Section 3.5.4 a third abstract model, called the *abstract, truly-concurrent model*, which satisfactorily composes the two equivalences, and we will relate it to the other models through suitable functors. In our view, such a model represents the behaviour of a graph grammar at a very reasonable level of abstraction, at which many classical results of the algebraic theory of grammars can be lifted. This model is used for example in [28] as a starting point for the definition of a truly-concurrent semantics for graph grammars based on Event Structures [26]. The present section is based mainly on the paper [51].

3.5.1 The Concrete and Truly Concurrent Models of Computation

As suggested above, the simplest, more concrete model of computation is just a category having graphs as objects and derivations as arrows. Thus the next definition just emphasizes the obvious fact that since derivations can be composed, they can be regarded as arrows of a category. All along this section, by derivation we mean a possibly parallel derivation.

Definition 3.5.1 (the concrete model of computation)
Given a graph grammar $\mathcal{G}$, the *concrete model of computation for* $\mathcal{G}$, denoted $\mathbf{Der}(\mathcal{G})$, is the category having all graphs as objects (thus it has the same objects as category **Graph**), and where $\rho : G \to H$ is an arrow iff ρ is a (parallel) derivation, $\sigma(\rho) = G$ and $\tau(\rho) = H$ (see Definitions 3.3.6 and 3.4.4). Arrow composition is defined as the sequential composition of derivations, and the identity of object G is the identity derivation $G : G \Rightarrow^* G$ (see Definition 3.3.6).

The following example shows why, in our view, the concrete model of computation for a grammar contains too much information.

Example 3.10 (concrete model of computation for grammar $\mathcal{C}$-$\mathcal{S}$)
Let us have a look at the structure of category $\mathbf{Der}(\mathcal{C}\text{-}\mathcal{S})$. As objects, it contains all labeled graphs introduced in Example 3.1. Thus, for example, for each non-empty graph G it contains infinitely many distinct objects isomorphic to G.

As arrows are concerned, consider for example the graphs G_1 and G_3 of Figure 3.7. Between these two graphs we have one arrow for the two-step derivation of Figure 3.7 which applies first SER and then REQ, and infinitely many additional distinct arrows, one for each isomorphic copy of that derivation, which can be obtained by changing arbitrarily the identity of nodes and edges in all the graphs of the derivation, except the starting and the ending ones. Moreover, between G_1 and G_3 there are infinitely many distinct arrows representing the parallel direct derivation of Figure 3.11 and all its isomorphic copies. There are also infinitely many arrows having G_1 both as source and as target: among them there is the identity derivation $G_1 : G_1 \Rightarrow^* G_1$, and some of the empty direct derivations mentioned at the end of Definition 3.4.4. □

Therefore many arrows of the concrete model should be considered as not distinguishable, when looking at derivations from a more abstract perspective. Postponing to the next section the discussion about how isomorphic derivations can be identified, we introduce here an equivalence intended to capture true concurrency, yielding the *truly-concurrent model of computation*.

According to the Parallelism Theorem (Theorem 3.4.6), two productions can be applied at parallel independent matches in a graph either at the same time, or one after the other in any order, producing the same resulting graph. An observation mechanism which does not distinguish two derivations where independent rewriting steps are performed in different order is considered to observe the *concurrent* behaviour of a grammar, instead of the *sequential* one. The corresponding equivalence is said to capture "true concurrency". Such an equivalence has been deeply studied in the literature, where it is called *shift equivalence* [43,2,52,15], and it is based on the analysis and synthesis constructions (Lemmas 3.4.8 and 3.4.9): It equates two derivations if they are related by a finite number of applications of analysis and synthesis. The name "shift equivalence" refers to the basic operation of "shifting" a production one step towards the left, if it is sequential independent from the previous direct derivation: In fact, Proposition 3.5.5 below will show that the analysis and synthesis constructions can be simulated by repeated shifts.

Definition 3.5.2 (shift equivalence)
If ρ and ρ' are two derivations such that $\rho' \in ANAL_{I,J}(\rho)$ (see Lemma 3.4.8), $\rho_1 = \rho_2 ; \rho ; \rho_3$, $\rho_1' = \rho_2 ; \rho' ; \rho_3$, and ρ_2 has length $n-1$, then we will write $\rho_1' \in ANAL^n_{I,J}(\rho_1)$, indicating that ρ_1' is an analysis of ρ_1 at step n. Similarly, if $\rho' \in SYNT(\rho)$ (see Lemma 3.4.9), $\rho_1 = \rho_2 ; \rho ; \rho_3$, $\rho_1' = \rho_2 ; \rho' ; \rho_3$ and ρ_2 has length $n-1$, then we will write $\rho_1' \in SYNT^n(\rho_1)$, indicating that ρ_1' is obtained from ρ_1 via a synthesis construction at step n.

Now, let $ANAL$ be the union of all analysis relations (i.e., $ANAL = \bigcup\{ANAL^n_{I,J} \mid n \in \mathbb{N} \wedge I, J \in \mathbb{N}^*\}$), and similarly let $SYNT$ be the union of all synthesis relations ($SYNT = \bigcup\{SYNT^n \mid n \in \mathbb{N}\}$). Furthermore, let $SHIFT^0_0$ be the relation defined as $\langle\rho_1, \rho_2\rangle \in SHIFT^0_0$ iff ρ_1 is an identity derivation $G : G \Rightarrow^* G$ and $\rho_2 : G \stackrel{\emptyset}{\Longrightarrow} G$ is an empty direct derivation such that the induced isomorphism is the identity $id_G : G \to G$ (see Definition 3.4.4).

The smallest equivalence relation on derivations containing $ANAL \cup SYNT \cup SHIFT^0_0$ is called *shift equivalence* and it is denoted by $\equiv_{sh}$. If ρ is a derivation, by $[\rho]_{sh}$ we denote the equivalence class containing all derivations shift-equivalent to ρ.

From the last definition it follows immediately that $\rho \equiv_{sh} \rho'$ implies $\sigma(\rho) = \sigma(\rho')$ and $\tau(\rho) = \tau(\rho')$, because synthesis and analysis do not affect the extremes of a derivation. The truly-concurrent model of computation for a graph grammar is, like the concrete model of Definition 3.5.1, a category having concrete graphs as objects; however, its arrows are equivalence classes of derivations modulo the shift equivalence. There is an obvious functor relating the concrete model and the truly-concurrent one.

Definition 3.5.3 (the truly-concurrent model of computation)
Given a graph grammar $\mathcal{G}$, the *truly-concurrent model of computation for* $\mathcal{G}$, denoted $\mathbf{Der}(\mathcal{G})/_{sh}$, is the category having graphs as objects, and as arrows equivalence classes of derivations with respect to the shift equivalence. More precisely, $[\rho]_{sh} : G \to H$ is an arrow of $\mathbf{Der}(\mathcal{G})/_{sh}$ iff $\sigma(\rho) = G$ and $\tau(\rho) =$

H. The composition of arrows is defined as $[\rho]_{sh} ; [\rho']_{sh} = [\rho ; \rho']_{sh}$, and the identity of G is the equivalence class $[G]_{sh}$, containing the identity derivation G.

We denote by $[\_]_{sh}$ the obvious functor from $\mathbf{Der}(\mathcal{G})$ to $\mathbf{Der}(\mathcal{G})/_{sh}$ which is the identity on objects, and maps each concrete derivation to its shift equivalence class.

Example 3.11 (truly-concurrent model of computation for $\mathcal{C}$-$\mathcal{S}$) Consider the concrete model for $\mathcal{C}$-$\mathcal{S}$ described in Example 3.10. Category $\mathbf{Der}(\mathcal{C}\text{-}\mathcal{S})/_{sh}$ has the same objects but much fewer arrows. For example, since the two-steps derivation $G_1 \Rightarrow_{SER} G_2 \Rightarrow_{REQ} G_3$ of Figure 3.11 and the parallel direct derivation $G_1 \Rightarrow_{SER+REQ} G_3$ of Figure 3.11 are shift-equivalent, they are represented by the same arrow between graphs G_1 and G_3; moreover, since the analysis and synthesis constructions are non-deterministic, it turns out that also all the isomorphic copies of the mentioned derivations are shift-equivalent to them. Therefore there is only one arrow from G_1 to G_3 representing the application of productions *SER* and *REQ* in any order, as one would expect in an abstract model.

However, since the objects of the category are concrete graphs, looking at the arrows starting from G_1 one may still find infinitely many arrows corresponding to the application of the same productions *SER* and *REQ*, but ending at different isomorphic copies of G_3. This fact shows that the truly-concurrent model still contains too many arrows, as it may manifests *unbounded non-determinism*, like in the case just described. □

Since arrows of the truly-concurrent model are equivalence classes of derivations, it is natural to ask if they contain a "standard representative", i.e., a derivation which can be characterized as the only one in the equivalence class which enjoys a certain property. Canonical derivations provide a partial answer to this question. In the thesis of Hans-Jörg Kreowski ([43]) it is shown that any graph derivation can be transformed into a shift-equivalent *canonical* derivation, where each production is applied as early as possible. In general, since the analysis and synthesis constructions are not deterministic, from a given derivation many distinct canonical derivations can be obtained: nevertheless, all of them are isomorphic. This is formalized by a result in [43] stating that every derivation has, up to isomorphism, a unique shift-equivalent canonical derivation.

Given a derivation, a shift-equivalent canonical derivation can be obtained by repeatedly "shifting" the application of the individual productions as far as possible towards the beginning of the derivation. It is worth stressing that, unlike the original definition, we explicitly consider here the deletion of empty direct derivations.

Definition 3.5.4 (shift relation, canonical derivations)
Let us write $\rho' \in ANAL_i^n(\rho)$ if $\rho' \in ANAL_{I,J}^n(\rho)$, where $I = \langle i \rangle$ and $J = \langle 1, \ldots, i-1, i+1, \ldots, k \rangle$; i.e., when ρ' is obtained from derivation ρ by anticipating the application of the i-th production of the n-th direct derivation.

Now, for each pair of natural numbers $\langle n, j \rangle$, both greater than 0, the relation $SHIFT_j^n$ is defined as $SHIFT_j^n = SYNT^{n-1} \circ ANAL_j^n$ ($\circ$ is the standard composition of relations). Moreover, for each $n > 0$, we write $\langle \rho_1, \rho_2 \rangle \in SHIFT_0^n$ iff ρ_1 has length n, $\rho_1 = \rho\,;\, G \stackrel{q}{\Longrightarrow} H$, $\rho_2 = \rho\,;\, G \stackrel{q}{\Longrightarrow} X \stackrel{\emptyset}{\Longrightarrow} H$, and the double pushout $G \stackrel{q}{\Longrightarrow} H$ is obtained from $G \stackrel{q}{\Longrightarrow} X$ by composing the right-hand side pushout with the isomorphism from X to H induced by the empty direct derivation $X \stackrel{\emptyset}{\Longrightarrow} H$. Finally, let $SHIFT_0^0$ be as in Definition 3.5.2.

Relation $<_{sh}$ is defined as the union of relations $SHIFT_j^n$ for each $n,\, j \in \mathbb{N}$. The transitive and reflexive closure of $<_{sh}$, denoted $\leq_{sh}$, is called the *shift relation*. A derivation is *canonical* iff it is minimal with respect to $\leq_{sh}$.

Thus we have that $\rho_1 <_{sh} \rho_2$ iff ρ_1 is obtained from ρ_2 either by removing the last direct derivation if it is empty (if ρ_1 is the identity, then an additional condition must be satisfied), or by moving the application of a single production one step towards left. It is worth stressing here that for each $n \in \mathbb{N}$, relation $SHIFT_0^n$ is deterministic, in the sense that $\langle \rho_1, \rho_2 \rangle, \langle \rho_1', \rho_2 \rangle \in SHIFT_0^n$ implies $\rho_1 = \rho_1'$.

Example 3.12 (canonical derivation)
Let δ_1 be the derivation depicted in Figure 3.7. Figures 3.12 and 3.13 show two derivations, δ_2 and δ_4, both shift-equivalent to δ_1, and where δ_4 is canonical. Let us explain how δ_4 can be obtained from δ_1 through repeated applications of the shift construction. First, note that we have $\langle \delta_2, \delta_1 \rangle \in SHIFT_1^3$, i.e., δ_2 is obtained from δ_1 by moving the (second) application of REQ one step towards left. More formally, by Definition 3.5.4 we have $SHIFT_1^3 = SYNT^2 \circ ANAL_1^3$; thus δ_2 is obtained applying first an analysis construction to direct derivation $G_2 \Rightarrow_{REQ} G_3$, obtaining the two-step derivation $G_2 \Rightarrow_{REQ} G_3 \Rightarrow_{\emptyset} G_3$, and then by applying the synthesis construction to the sequential independent derivation $G_1 \Rightarrow_{SER} G_2 \Rightarrow_{REQ} G_3$, obtaining the direct derivation $G_1 \Rightarrow_{SER+REL} G_3$. In particular, note that the application of the empty derivation in δ_2 is generated by the analysis construction (see the corresponding proof in Section 3.9).

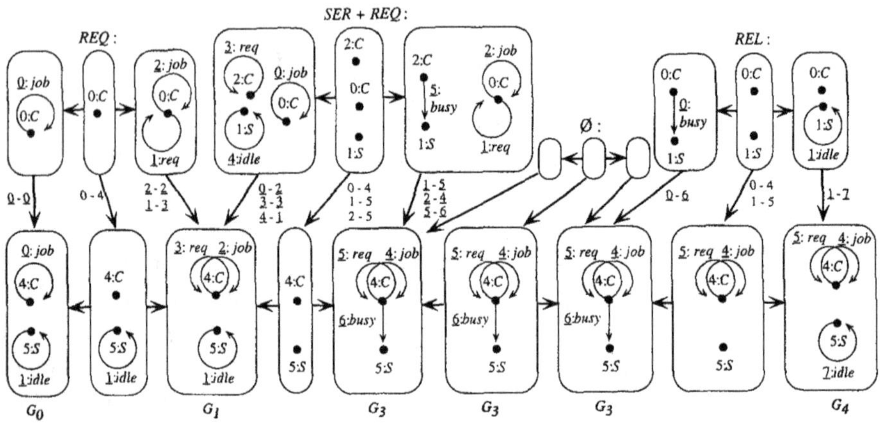

Figure 3.12: Derivation δ_2, shift-equivalent to derivation δ_1 of Figure 3.7. More precisely, $\langle \delta_2, \delta_1 \rangle \in SHIFT^3_1$.

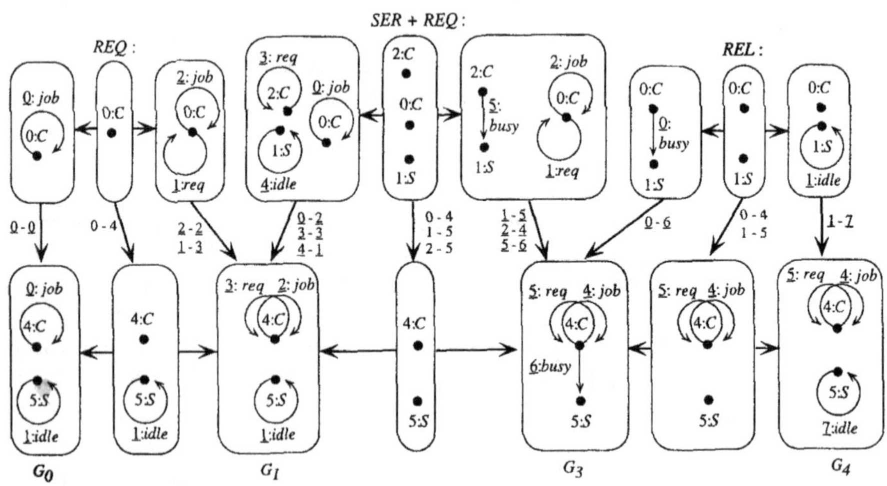

Figure 3.13: The canonical derivation δ_4, shift-equivalent to derivation δ_1 of Figure 3.7.

Next, we can apply the shift construction to the last part of δ_2, because the derivation $G_3 \Rightarrow_\emptyset G_3 \Rightarrow_{REL} G_4$ is clearly sequential independent. In this way we get a new derivation $\delta_3 \in SHIFT^4_1(\delta_2)$, which is obtained by moving one step earlier the application of *REL*; derivation δ_3 is not depicted, but it is obtained easily by adding an empty direct derivation $G_4 \Rightarrow_\emptyset G_4$ to δ_4. Now, since the last direct derivation of δ_3 is empty, by removing it we get the last derivation $\delta_4 \in SHIFT^3_0(\delta_3)$. Finally, the fact that δ_4 is canonical can be shown easily by checking that every production application in it deletes at least one item (edge) generated by the previous direct derivation. Thus the shift construction cannot be applied anymore. □

The next proposition on the one hand justifies the name of the equivalence introduced in Definition 3.5.2, on the other hand it guarantees the existence of a shift-equivalent canonical derivation for any given derivation. This fact will be used in the next section.

Proposition 3.5.5 (properties of the shift relation)

1. *The smallest equivalence relation containing the $\leq_{sh}$ relation is exactly relation $\equiv_{sh}$ introduced in Definition 3.5.2.*

2. *Relation $\leq_{sh}$ is well-founded, i.e., there is no infinite chain $\rho_1, \rho_2, \ldots$ of derivations such that $\rho_{i+1} <_{sh} \rho_i$ for all $i \in \mathbb{N}$. Thus for each derivation ρ there exists a canonical (i.e., minimal with respect to $\leq_{sh}$) derivation ρ' such that $\rho' \leq_{sh} \rho$.*

Proof outline

For the first point, denote by $\approx_{sh}$ the smallest equivalence relation containing $\leq_{sh}$. Now $\approx_{sh} \subseteq \equiv_{sh}$ is obvious by the definition. Conversely, one can prove that $\equiv_{sh} \subseteq \approx_{sh}$ by showing that for each $n \in \mathbb{N}$ and $I, J \in \mathbb{N}^*$, $ANAL^n_{I,J} \subseteq \approx_{sh}$ and $SYNT^n \subseteq \approx_{sh}$; in other words, if two derivations are related by a single analysis (or synthesis) construction, then they can be related by a sequence of basic shift constructions. For synthesis, if $\rho' \in SYNT(\rho)$, and $\rho = G \overset{q}{\Longrightarrow} X \overset{p_1+\ldots+p_k}{\Longrightarrow} H$, then ρ' can be obtained from ρ by anticipating the production applications in $X \overset{p_1+\ldots+p_k}{\Longrightarrow} H$ one at a time, and deleting the remaining empty direct derivation. For analysis the solution is a bit more tricky, because relation $ANAL_{I,J}$ allows one to reorder all the productions in the parallel direct derivation. This effect can be simulated via shifts by moving individual productions back and forth until they reach the desired position.

For the second point, see Lemma 4.6 and Theorem 5.3 of [43]. The corresponding proofs can be adjusted easily in order to match the slightly different definitions. □

3.5.2 Requirements for capturing representation independence.

Graph grammars are often introduced (in most of the approaches) as a formalism for specifying the evolution of certain systems. In this perspective, some sort of labeled graphs are used to represent system states, while graph transformation rules encode the basic system transformations as in the running example of this section.

Almost invariably, two isomorphic graphs are considered as representing the *same* system state. In fact, such a state is determined by the topological structure of the graph and by the labels only, while the true identity of nodes and edges is considered just as a representation detail. Therefore, in order to define a model of computation which represents faithfully the behaviour of a system, it is very natural to look for an equivalence which equates all isomorphic graphs. This is even more demanding in the algebraic approaches to graph grammars. In fact, since the pushout object of two arrows is unique only up to isomorphism, from the definition of direct derivation (Definition 3.3.5) it follows that given a graph G and a production p, if $G \stackrel{p}{\Longrightarrow} H$, then $G \stackrel{p}{\Longrightarrow} H'$ for each graph H' isomorphic to H. Thus the application of a production to a graph can produce infinitely many distinct results, which implies that both in the concrete and in the truly-concurrent models of computation there are infinitely many arrows starting from a graph and corresponding to distinct applications of the same production. This fact is highly counter-intuitive, because in the above situation one would expect a deterministic result, or, at most, a finite set of possible outcomes.

Indeed, in the classical theory of the algebraic approach to graph grammars, one often reasons (more or less explicitly) in terms of *abstract graphs*, i.e., of isomorphism classes of concrete graphs: With this choice, given as above a graph G and a production p with match m (satisfying the gluing condition), there is only one abstract graph $[H]$ such that $[G] \stackrel{p,m}{\Longrightarrow} [H]$, thus a direct derivation becomes in a sense deterministic.

Such a solution is satisfactory if one is interested just in the set of (abstract) graphs generated by a grammar. However, if one considers instead a kind of semantics which associates with each grammar all the possible derivations (like the models of computation we are interested in), even with abstract graphs one still has too much representation-dependent information. In fact, in the above situation there are still infinitely many distinct direct derivations (i.e., double-pushout diagrams) from $[G]$ to $[H]$ via p and m (as it follows from Definition 3.3.5). Like for graphs, one can find many distinct derivations which should not be considered distinct as system evolutions, because they differ just for the identity of nodes and edges in the involved graphs. Therefore it is natural to reason not only in terms of abstract graphs, but also in terms of ***abstract derivations***, i.e., suitable equivalence classes of derivations.

We shall say that an equivalence on graphs and graph derivations captures "representation independence" if it equates all isomorphic graphs (and not more), and if it equates derivations which differ for representation details. It is worth stressing here that the shift equivalence (Definition 3.5.2), which has been recognized to capture the essentials of the semantics of parallel graph derivations, is clearly not representation independent, because it is able to relate only derivations starting from and ending with the same concrete graphs.

In the following we first formalize the notion of "isomorphism of derivations" (which is the most natural candidate for the equivalence we are looking for), showing that the obvious definition based on the categorical notion of isomorphism of diagrams is not acceptable from a semantical point of view, because it would identify too many derivations. Next we will introduce some reasonable requirements for equivalences on derivations intended to capture representation independence, formalizing them as suitable algebraic properties. More precisely, we will look for an equivalence that is a congruence with respect to sequential composition, and that is compatible with the construction of canonical derivations, i.e., such that any pair of canonical derivations obtained from two equivalent derivations are equivalent as well. It is worth stressing that these requirements are motivated by the goal of defining an abstract, truly-concurrent model of computation for a grammar: If one would like to carry on to the abstract setting other results of the theory (for example those concerning concurrency and amalgamation (see Section 3.6.1 and [2]), it might be the case that additional requirements should be considered.

Let us start with the definition of isomorphism of derivations.

Definition 3.5.6 (isomorphism of derivations)
Let $\rho : G_0 \Rightarrow^* G_n$ and $\rho' : G'_0 \Rightarrow^* G'_n$ be two derivations having the same length, as depicted in Figure 3.14. They are *isomorphic* (written $\rho \equiv_0 \rho'$) if there exists a family of isomorphisms $\{\phi_{G_0} : G_0 \to G'_0, \phi_{X_i} : X_i \to X'_i\}$ with $X \in \{L, K, R, G, D\}$ and $i \in \{1, \dots, n\}$ between corresponding graphs of the two derivations, such that all those squares commute, which are obtained by composing two corresponding morphisms $X \to Y$ and $X' \to Y'$ of the two derivations and the isomorphisms ϕ_X and ϕ_Y relating the source and the target graphs.

It is not difficult to see that the equivalence induced by isomorphisms of derivations identifies too many derivations. In fact, as it is clear from the picture,

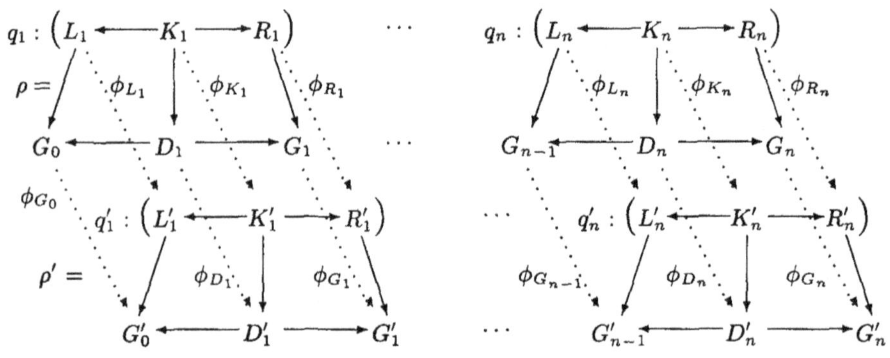

Figure 3.14: Isomorphism of derivations.

the productions in corresponding direct derivations only need to be span-isomorphic. Thus derivations which can hardly be considered as differing just for representation details may be identified. For example, if a grammar $\mathcal{G}$ contains two span-isomorphic productions p and p', two direct derivations using p and p' would be considered as equivalent. Even worst, two parallel productions $q = p_1 + \ldots + p_k$ and $q' = p'_1 + \ldots + p'_{k'}$ may happen to be span-isomorphic even if some of the involved sequential productions are distinct or if $k \neq k'$, i.e., if the number of applied productions is different.

SYNCHR :

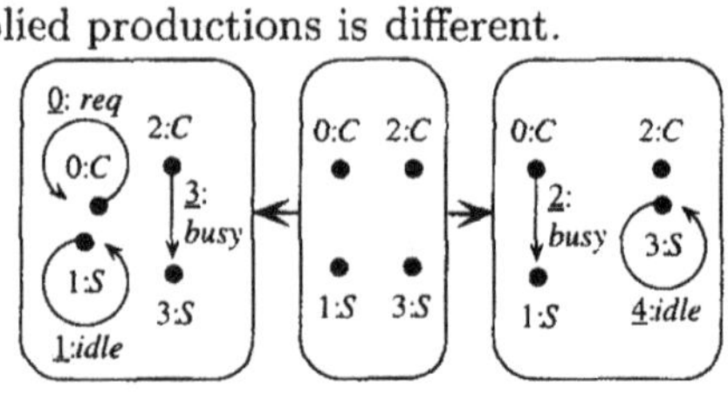

Example 3.13 (extending the graph grammar)
Suppose that grammar $\mathcal{C}\text{-}\mathcal{S}$ is extended with the above production *SYNCHR*, which is intended to model the situation where the service of a request starts exactly when another service terminates.

Such production happens to be span-isomorphic to the parallel production *SER* + *REL*, thus two direct derivations $G \overset{SYNCHR}{\Longrightarrow} H$ and $G \overset{SER+REL}{\Longrightarrow} H$ would be identified by equivalence $\equiv_0$. However, such direct derivations are "semantically" very different: the second one is a parallel direct derivations, and, using the analysis construction, it can be sequentialized in any order, while the other is a sequential direct derivation, which cannot be decomposed. □

In the next section we will propose other equivalences on derivations which require that the productions applied at each direct derivation are not only span-isomorphic, but also isomorphic: this is the main reason why in Definition 3.4.4 we defined a production as a pair including both a name and a span.

In the rest of this section, we establish some further requirements for a notion of equivalence on derivations which naturally captures representation independence. Throughout this section $\approx$ denotes an equivalence relation on derivations, and an *abstract* derivation, denoted $[\rho]$, is an equivalence class of derivations modulo $\approx$. The first, obvious requirement is that the notion of abstract derivation is consistent with the notion of abstract graph, i.e., any abstract derivation should start from an abstract graph and should end in an abstract graph.

Definition 3.5.7 (well-defined equivalences)
For each abstract derivation $[\rho]$, define $\sigma([\rho]) = \{\sigma(\rho') \mid \rho' \in [\rho]\}$, and similarly $\tau([\rho]) = \{\tau(\rho') \mid \rho' \in [\rho]\}$. Then equivalence $\approx$ is *well-defined* if for each derivation $\rho : G \Rightarrow^* H$ we have that $\sigma([\rho]) = [\sigma(\rho)]$, and $\tau([\rho]) = [\tau(\rho)]$.

If $\approx$ is well-defined, then for each derivation $\rho : G \Rightarrow^* H$ we can write $[\rho] : [G] \Rightarrow^* [H]$. Since our goal is to define abstract models of computation as quotient categories of concrete ones, the second requirement is that the equivalence is a congruence with respect to sequential composition.

Definition 3.5.8 (equivalence allowing for sequential composition)
Given two abstract derivations $[\rho]$ and $[\rho']$ such that $\tau([\rho]) = \sigma([\rho'])$, their *sequential composition* $[\rho] ; [\rho']$ is defined iff for all $\rho_1, \rho_2 \in [\rho]$ and $\rho'_1, \rho'_2 \in [\rho']$ such that $\tau(\rho_1) = \sigma(\rho'_1)$ and $\tau(\rho_2) = \sigma(\rho'_2)$, one has that $\rho_1 ; \rho'_1 \approx \rho_2 ; \rho'_2$. In this case $[\rho] ; [\rho']$ is, by definition, the abstract derivation $[\rho_1 ; \rho'_1]$.

An equivalence $\approx$ *allows for sequential composition* iff $[\rho] ; [\rho']$ is defined for all $[\rho]$ and $[\rho']$ such that $\tau([\rho]) = \sigma([\rho'])$.

The third requirement guarantees that we reobtain the result of uniqueness of canonical derivations in the abstract framework, i.e., that each abstract derivation has a unique shift-equivalent abstract canonical derivation.

Definition 3.5.9 (uniqueness of canonical derivations)
Equivalence $\approx$ *enjoys uniqueness of canonical derivations* iff for each pair of equivalent derivations $\rho \approx \rho'$ and for each pair $\langle \rho_c, \rho'_c \rangle$ of canonical derivations, $\rho \equiv_{sh} \rho_c$ and $\rho' \equiv_{sh} \rho'_c$ implies that $\rho_c \approx \rho'_c$.

3.5.3 Towards an equivalence for representation independence.

In this section we introduce two different equivalences on derivations, and analyze their properties with respect to the requirements established in Section 3.5.2. Actually, since by the definitions all the equivalences are clearly well-defined in the sense of Definition 3.5.7, we will focus only on the other requirements. The first equivalence we consider is obtained from equivalence $\equiv_0$ of Definition 3.5.6 by requiring that the productions applied at each (parallel) direct derivation are isomorphic and not just span-isomorphic.

Definition 3.5.10 (equivalence $\equiv_1$)
Let $\rho : G_0 \Rightarrow^* G_n$ and $\rho' : G'_0 \Rightarrow^* G'_n$ be two derivations such that $\rho \equiv_0 \rho'$ as in Figure 3.14. Then they are 1-*equivalent* (written $\rho \equiv_1 \rho'$) if

1. there exists a family of permutations $\langle \Pi_1, \ldots, \Pi_n \rangle$ such that for each $i \in \{1, \ldots, n\}$, productions q_i and q'_i are isomorphic via Π_i;
2. For each $i \in \{1, \ldots, n\}$, the family of isomorphisms $\{\phi_{L_i} : L_i \to L'_i, \phi_{K_i} : K_i \to K'_i, \phi_{R_i} : R_i \to R'_i\}$ that exists by Definition 3.5.6 is exactly the family of isomorphisms between corresponding graphs of q_i and q'_i which is induced by permutation Π_i, according to Fact 3.4.5.

We say that ρ *and* ρ' *are* 1-*equivalent via* $\langle \Pi_i \rangle_{i \leq n}$ if $\langle \Pi_i \rangle_{i \leq n}$ is the family of permutations of point 1 above. Given a derivation ρ, its equivalence class with respect to $\equiv_1$ is denoted by $[\rho]_1$, and is called a 1-*abstract derivation.*

Let us consider now the algebraic properties of equivalence $\equiv_1$. We prove below that $\equiv_1$ enjoys uniqueness of canonical derivations. This result is based on an important lemma which states that $\equiv_1$ behaves well with respect to analysis, synthesis, and shift of derivations. This is the main technical result of this section, and the whole Section 3.9 is devoted to its proof.

Lemma 3.5.11 (analysis, synthesis and shift preserve equivalence $\equiv_1$)
Let $\rho \equiv_1 \rho'$ via $\langle \Pi_i \rangle_{i \leq n}$, and let ϕ_{G_0} and ϕ_{G_n} be the isomorphisms between their starting and ending graphs (see Figure 3.14).

1. *If $\rho_1 \in ANAL^i_j(\rho)$ then there is at least one derivation $\rho_2 \in ANAL^i_{\Pi_i(j)}(\rho')$. Moreover, for each $\rho_2 \in ANAL^i_{\Pi_i(j)}(\rho')$, it holds that $\rho_2 \equiv_1 \rho_1$.*
2. *If $\rho_1 \in SYNT^i(\rho)$ then there is at least one derivation $\rho_2 \in SYNT^i(\rho')$. Moreover, for each $\rho_2 \in SYNT^i(\rho')$, it holds that $\rho_2 \equiv_1 \rho_1$.*

3. *If* $\rho_1 \in SHIFT^i_j(\rho)$ *then there is at least one derivation* $\rho_2 \in SHIFT^i_{\Pi_i(j)}(\rho')$. *Moreover, for each* $\rho_2 \in SHIFT^i_{\Pi_i(j)}(\rho')$, *it holds that* $\rho_2 \equiv_1 \rho_1$.

Furthermore, in all these three cases, the isomorphisms relating the starting and ending graphs of the 1-equivalent derivations ρ_2 *and* ρ_1 *are exactly isomorphisms* ϕ_{G_0} *and* ϕ_{G_n}. □

The observation that analysis, synthesis and shift preserve not only equivalence $\equiv_1$, but also the isomorphisms between the starting and ending graphs of equivalent derivations will be used later in Proposition 3.5.16.

Theorem 3.5.12 ($\equiv_1$ enjoys uniqueness of canonical derivations)
Let ρ, ρ' *be two derivations such that* $\rho \equiv_1 \rho'$, *and let* ρ_c, ρ'_c *be two canonical derivations such that* $\rho \equiv_{sh} \rho_c$ *and* $\rho' \equiv_{sh} \rho'_c$. *Then* $\rho_c \equiv_1 \rho'_c$.

Proof outline
Let $\sqsubset_{sh}$ be the relation on 1-abstract derivations defined as $[\rho_1]_1 \sqsubset_{sh} [\rho_2]_1$ if $\rho_1 <_{sh} \rho_2$, and let $\sqsubseteq_{sh}$ be its reflexive and transitive closure. Relation $\sqsubset_{sh}$ is clearly well-defined by point 3 of Lemma 3.5.11; moreover it is well-founded because so is $<_{sh}$, by Proposition 3.5.5.

Furthermore, $\sqsubset_{sh}$ enjoys the following local confluence property: if $[\rho_2]_1 \sqsubset_{sh} [\rho_1]_1$ and $[\rho_3]_1 \sqsubset_{sh} [\rho_1]_1$, then there is a $[\rho_4]_1$ such that $[\rho_4]_1 \sqsubseteq_{sh} [\rho_2]_1$ and $[\rho_4]_1 \sqsubseteq_{sh} [\rho_3]_1$. This can be proved by following the outline of Lemma 4.7 of [43] (where the case of concrete derivations is considered, and all critical pairs of relation $<_{sh}$ are analyzed), and by exploiting Lemma 3.5.11 for handling the equivalence classes.

Local confluence and well-foundedness (i.e., termination) of relation $\sqsubset_{sh}$ imply confluence, which implies the statement because canonical derivations are minimal with respect to $<_{sh}$. □

Despite the last result, equivalence $\equiv_1$ is not completely satisfactory, because it does not allow for sequential composition, as the following counterexample shows (a similar counterexample is discussed in [50]).

Example 3.14 ($\equiv_1$ does not allow for sequential composition)
Let ρ_1 ; ρ_2 and ρ'_2 be the derivations of Figure 3.15 (a) and (b), respectively. It is easy to check that $\rho_2 \equiv_1 \rho'_2$: In fact the productions applied are the same, and the family of isomorphisms $\Phi_2 = \{\phi_{G'_2} : G'_2 \to G'_2, \phi_{K_2} : K_2 \to K'_2, \phi_{G_3} : G_3 \to G'_3\}$ with $\phi_{G'_2}(\underline{3}) = \underline{5}$, $\phi_{G'_2}(\underline{5}) = \underline{3}$, $\phi_{K_2}(\underline{5}) = \underline{3}$, $\phi_{G_3}(\underline{5}) = \underline{3}$, and identities elsewhere, makes everything commute. Clearly, we also have

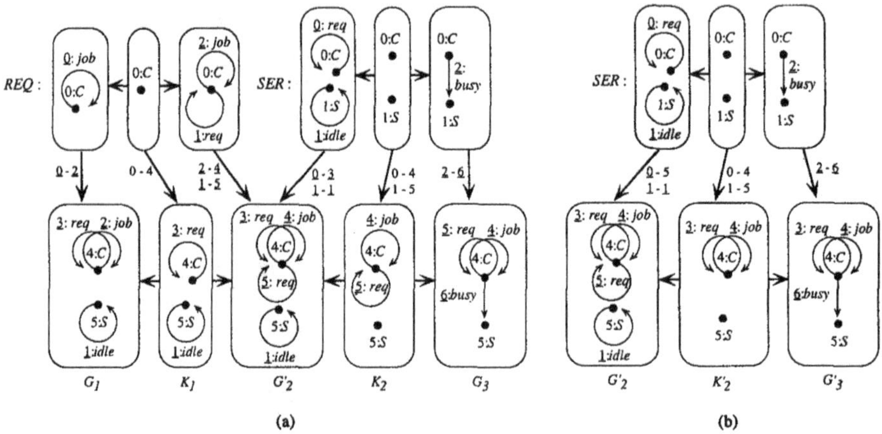

Figure 3.15: (a) Derivation $\rho_1 ; \rho_2$, and (b) direct derivation ρ'_2. Since $\tau(\rho_1) = G'_2 = \sigma(\rho'_2)$, $\rho_1 ; \rho'_2$ is well defined.

$\rho_1 \equiv_1 \rho_1$. using the family of isomorphisms $\Phi_1 = \{id_{G_1}, id_{K_1}, id_{G'_2}\}$. Moreover, $\rho_1 ; \rho'_2$ is defined because $\tau(\rho_1) = G'_2 = \sigma(\rho'_2)$, but derivations $\rho_1 ; \rho_2$ and $\rho_1 ; \rho'_2$ are not 1-equivalent, because families Φ_1 and Φ_2 do not agree on graph G'_2, and actually there is no family of isomorphisms between the corresponding graphs of the two derivations making the resulting diagram commutative.

Note that the two derivations $\rho_1 ; \rho_2$ and $\rho_1 ; \rho'_2$ of Figure 3.15 are "semantically" different (and thus they should be kept distinct by a reasonable abstract model): in fact, in the first derivation the first request issued by the client is served, while in the second derivation the second request is served.

Figure 3.16 shows an alternative, equivalent way of presenting the same counterexample. We used a graphical representation of 1-abstract direct derivations, where all the involved graphs are abstract (thus identities of nodes and edges have been dropped). The figure shows abstract derivations $[\rho_1]_1$ (left) and $[\rho_2]_1 = [\rho'_2]_1$ (right). In both abstract double-pushouts all morphisms are determined by the labels of nodes and edges, except those having $[G'_2]$ as target: Thus we added subscripts to the two edges of $[G'_2]$ labeled by *req*, indicating for each morphism which edge is in the codomain. Such subscripts only have a "local scope", in the following sense: if in $[\rho_2]_1$ (or $[\rho_1]_1$) we exchange 1 and 2 in the morphisms having abstract graph $[G'_2]$ as target, then the resulting abstract derivation is still $[\rho_2]_1$ (or $[\rho_1]_1$).

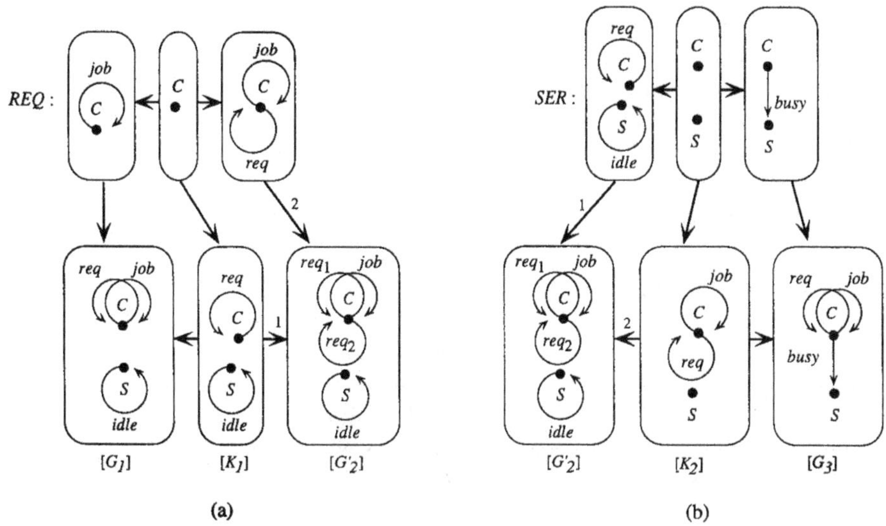

Figure 3.16: The abstract direct derivations corresponding to ρ_1 and ρ_2 of Figure 11 (a).

Now the fact that the sequential composition of $[\rho_1]_1$ and $[\rho_2]_1$ does not yield a well-defined 1-abstract derivation follows from the observation that there are two possible ways of matching $\tau([\rho_1]_1)$ with $\sigma([\rho_2]_1)$, yielding different results. $\square$

In [50] it is shown that the problem is due to the fact that in the categorical framework many relevant notions cannot be defined "up to isomorphism", not even arrow composition. The solution proposed there will be used in the next notion of equivalence. We first need to introduce "standard isomorphisms".

Definition 3.5.13 (standard isomorphisms)
A family s of *standard isomorphisms* in category **Graph** is a family of isomorphisms indexed by pairs of isomorphic graphs (i.e., $s = \{s(G, G') \mid G \cong G'\}$), satisfying the following conditions for each G, G' and $G'' \in |\mathbf{Graph}|$:

- $s(G, G') : G \to G'$;
- $s(G, G) = id_G$;
- $s(G'', G') \circ s(G, G'') = s(G, G')$.

In the equivalence we are going to introduce, denoted $\equiv_3$, we make a limited use of standard isomorphisms: We require that the starting and ending graphs of equivalent derivations are related by such isomorphisms (see also [51]). It is worth recalling that in [51] also another equivalence is introduced, denoted $\equiv_2$, that requires that all the isomorphisms of the form ϕ_{X_i} for $X \in \{G, D\}$ are standard: However, it is shown, through a counter-example, that such equivalence does not enjoy uniqueness of canonical derivations.

Definition 3.5.14 (Equivalence $\equiv_3$)
Let s be an arbitrary but fixed family of standard isomorphisms of category **Graph**, and let ρ, ρ' be two derivations. We say that ρ and ρ' are *3-equivalent* (written $\rho \equiv_3 \rho'$) iff they are 1-equivalent, and moreover the isomorphisms ϕ_{G_0} and ϕ_{G_n} relating their starting and ending graphs (see Figure 3.14) are standard. An equivalence class of derivations with respect to $\equiv_3$ is denoted by $[\rho]_3$, and is called a *3-abstract derivation.*

Now the fact that equivalence $\equiv_3$ allows for sequential composition follows easily by the above considerations.

Proposition 3.5.15 ($\equiv_3$ allows for sequential composition)
Let ρ_1, ρ_1', ρ_2 and ρ_2' be four derivations such that $\rho_1 \equiv_3 \rho_1'$, $\rho_2 \equiv_3 \rho_2'$, $\tau(\rho_1) = \sigma(\rho_2)$ and $\tau(\rho_1') = \sigma(\rho_2')$. Then $\rho_1 \,;\, \rho_2 \equiv_3 \rho_1' \,;\, \rho_2'$.

Proof
By hypothesis there are two families of isomorphisms making ρ_1 and ρ_2 3-equivalent to ρ_1' and ρ_2', respectively. The two families must agree on $\tau(\rho_1) = \sigma(\rho_2)$ (because there is only one standard isomorphism between a given pair of isomorphic graphs), and thus their union provides a family of isomorphisms which makes 3-equivalent $\rho_1 \,;\, \rho_2$ and $\rho_1' \,;\, \rho_2'$. □

Example 3.15 (sequential composition of 3-abstract derivations)
The counterexample of Example 3.14 does not apply to equivalence $\equiv_3$: In fact, looking at the direct derivations of Figure 3.15, we have that $\rho_2 \not\equiv_3 \rho_2'$, because one of the isomorphisms in Φ_2 (namely, $\phi_{G_2'}$) is certainly not standard (it is not the identity, see Definition 3.5.13).

Consider now the abstract direct derivations of Figure 3.16. It can be shown that they are a faithful representation of 3-abstract derivations, provided that we give the subscripts 1 and 2 in graph $[G_2']$ a "global scope" in the following sense: if we assume that the right abstract derivation is $[\rho_2]_3$, then by exchanging subscripts 1 and 2 in the morphisms having $[G_2']$ as target, we obtain abstract derivation $[\rho'_2]_3$, which is a different one.[d] Moreover, subscripts can now be used to concatenate abstract derivations: $[\rho_1]_3 ; [\rho_2]_3$ is obtained by matching the two copies of $[G'_2]$ in such a way that subscripts are preserved. □

The fact that the shift relation preserves the isomorphisms relating the starting and ending graphs of two equivalent derivations, as stated explicitly in Lemma 3.5.11, guarantees that equivalence $\equiv_3$ enjoys uniqueness of canonical derivations. Therefore equivalence $\equiv_3$ satisfies all the requirements of Section 3.5.2, as summarized in the next proposition.

Proposition 3.5.16 (Properties of equivalence $\equiv_3$)
Equivalence $\equiv_3$ is well-defined, it allows for sequential composition, and it enjoys uniqueness of canonical derivations.

Proof outline
For the uniqueness of canonical derivations, the same proof of Theorem 3.5.12 can be used. In fact, since the analysis, synthesis and shift relations preserve not only equivalence $\equiv_1$, but also the isomorphisms relating the starting and the ending graphs of 1-equivalent derivations (Lemma 3.5.11), it is immediate to show that they preserve equivalence $\equiv_3$ as well. Sequential composition is well-defined by Proposition 3.5.15. □

3.5.4 The abstract models of computation for a grammar.

Since equivalence $\equiv_3$ on derivations is a congruence with respect to sequential composition, and since all canonical derivations which are shift-equivalent to 3-equivalent derivations belong to the same 3-equivalence class, then this equivalence allows one to abstract out from the representation dependent aspects of derivations in a satisfactory way, defining the *abstract model of computation* for a grammar.

Definition 3.5.17 (abstract model of computation)
Given a grammar $\mathcal{G}$, its *abstract model of computation* **ADer**$(\mathcal{G})$ is defined as follows. The objects of **ADer**$(\mathcal{G})$ are abstract graphs, i.e., isomorphism classes of objects of **Graph**. Arrows of **ADer**$(\mathcal{G})$ are 3-abstract derivations, i.e., equivalence classes of (parallel) derivations of $\mathcal{G}$ with respect to equivalence $\equiv_3$. The source and target mappings are given by σ and τ, respectively; thus $[\rho]_3 : \sigma([\rho]_3) \to \tau([\rho]_3)$. Composition of arrows is given by the sequential composition of the corresponding 3-abstract derivations, and for each object

[d] Subscripts 1 and 2 can be replaced by any other marking able to distinguish the two *req* edges: the relevant fact is that once such a marking is fixed, by permuting the marks one gets a different abstract derivation.

$[G]$ the identity arrow is the 3-abstract derivation $[G : G \Rightarrow^* G]_3$ containing the identity derivation G.

Category $\mathbf{ADer}(\mathcal{G})$ is equipped with an equivalence relation $\equiv_{\underline{sh}}$ on arrows, defined as $[\rho]_3 \equiv_{\underline{sh}} [\rho']_3$ if $\rho \equiv_{sh} \rho'$.

Proposition 3.5.18 (category $\mathbf{ADer}(\mathcal{G})$ is well-defined)
Category $\mathbf{ADer}(\mathcal{G})$ is well-defined. Moreover, equivalence $\equiv_{\underline{sh}}$ is a congruence with respect to arrow compositions, i.e., if $[\rho_1]_3 \equiv_{\underline{sh}} [\rho_1']_3$ and $[\rho_2]_3 \equiv_{\underline{sh}} [\rho_2']_3$, then $[\rho_1]_3 ; [\rho_2]_3 \equiv_{\underline{sh}} [\rho_1']_3 ; [\rho_2']_3$ (if the compositions are defined).

Proof outline
The two facts follow easily from the properties of equivalence $\equiv_3$ summarized in Proposition 3.5.16 and from Theorem 3.5.12. □

Since $\equiv_{\underline{sh}}$ is a congruence with respect to arrow composition, it is safe to define the quotient category of $\mathbf{ADer}(\mathcal{G})$ with respect to $\equiv_{\underline{sh}}$: This yields the abstract, truly concurrent model of computation.

Definition 3.5.19 (abstract, truly-concurrent model of computation)
The *abstract, truly concurrent model of computation* for a grammar $\mathcal{G}$ is defined as category $\mathbf{ADer}(\mathcal{G})/_{\underline{sh}}$, having the same objects as $\mathbf{ADer}(\mathcal{G})$, and as arrows equivalence classes of arrows of $\mathbf{ADer}(\mathcal{G})$ with respect to equivalence $\equiv_{\underline{sh}}$.

The next definition and result show the relationship among the four models of computations for a grammar, introduced in Definitions 3.5.1, 3.5.3, 3.5.17, and 3.5.19, respectively, in terms of functors relating them.

$$\begin{array}{ccc}
\mathbf{Der}(\mathcal{G}) & \xrightarrow{[_]_{sh}} & \mathbf{Der}(\mathcal{G})/_{sh} \\
\Big\downarrow \mathcal{A} & & \Big\downarrow \underline{\mathcal{A}} \\
\mathbf{ADer}(\mathcal{G}) & \xrightarrow{[_]_{\underline{sh}}} & \mathbf{ADer}(\mathcal{G})/_{\underline{sh}}
\end{array}$$

Definition 3.5.20 (functors among models of computation)
Let $\mathcal{G}$ be a graph grammar. The various models of computation for $\mathcal{G}$ introduced in Definitions 3.5.1, 3.5.3, 3.5.17, and 3.5.19, are related by the four functors (as in the above diagram) defined as follows:

- Functor $[\_]_{sh} : \mathbf{Der}(\mathcal{G}) \to \mathbf{Der}(\mathcal{G})/_{sh}$ is the functor introduced in Definition 3.5.3 which is the identity on objects, and maps each concrete derivation to its shift equivalence class.

- Functor $[\_]_{\underline{sh}} : \mathbf{ADer}(\mathcal{G}) \to \mathbf{ADer}(\mathcal{G})/_{\underline{sh}}$ is the identity on objects, and maps each 3-abstract derivation to its equivalence class with respect to $\equiv_{\underline{sh}}$.

- Functor $\mathcal{A} : \mathbf{Der}(\mathcal{G}) \to \mathbf{ADer}(\mathcal{G})$ maps each object G to its isomorphism class $[G]$, and each concrete derivation ρ to its 3-abstract equivalence class $[\rho]_3$.

- Functor $\underline{\mathcal{A}} : \mathbf{Der}(\mathcal{G})/_{sh} \to \mathbf{ADer}(\mathcal{G})/_{\underline{sh}}$ maps each object G to its isomorphism class $[G]$, and each arrow $[\rho]_{sh}$ to the $\underline{sh}$-equivalence class of the corresponding 3-abstract derivation, i.e., to $[[\rho]_3]_{\underline{sh}}$.

Theorem 3.5.21 (functors among models of computation)
All the functors of Definition 3.5.20 are well-defined, and the diagram commutes. Moreover, functor $\underline{\mathcal{A}}$ is an equivalence of categories.

Proof outline
Well-definedness is obvious for functors $[\_]_{sh}$, $[\_]_{\underline{sh}}$ and $\mathcal{A}$. For $\underline{\mathcal{A}}$, we have to show that if $\rho \equiv_{sh} \rho'$, then $[\rho]_3 \equiv_{\underline{sh}} [\rho']_3$. This follows directly from the definition of $\equiv_{\underline{sh}}$. Also the commutativity of the diagram is immediate by the definitions.

To prove that $\underline{\mathcal{A}}$ is an equivalence of categories it suffices to show that it is full and faithful, and that each object $[G]$ of $\mathbf{ADer}(\mathcal{G})/_{\underline{sh}}$ is isomorphic to $\underline{\mathcal{A}}(H)$ for some object H of $\mathbf{Der}(\mathcal{G})_{sh}$. This last fact and fullness are ensured by surjectivity of $\underline{\mathcal{A}}$. For faithfulness, let $[\rho]_{sh}$ and $[\rho']_{sh}$ be two parallel arrows of $\mathbf{Der}(\mathcal{G})_{sh}$. We have to show that $\underline{\mathcal{A}}([\rho]_{sh}) = \underline{\mathcal{A}}([\rho']_{sh})$ implies $\rho \equiv_{sh} \rho'$.

By definition of $\underline{\mathcal{A}}$, $\underline{\mathcal{A}}([\rho]_{sh}) = \underline{\mathcal{A}}([\rho']_{sh})$ implies $[\rho]_3 \equiv_{\underline{sh}} [\rho']_3$. Then $\rho \equiv_{sh} \rho'$ follows by the general property that if $\rho \equiv_3 \rho'$, $\sigma(\rho) = \sigma(\rho')$, and $\tau(\rho) = \tau(\rho')$, then $\rho \equiv_{sh} \rho'$ (that is, equivalence $\equiv_3$ is contained in the shift equivalence, if restricted to derivations having the same starting and ending graphs). The last property can be proved by exploiting the fact that the shift equivalence allows one to add empty direct derivations at the end of a derivation, to move them inside the derivation, and to change every intermediate graph of a derivation with an isomorphic copy. □

Let us shortly comment on the relationship among the various models of computation. The concrete model $\mathbf{Der}(\mathcal{G})$ records all the possible derivations of grammar $\mathcal{G}$ without abstracting any detail; its definition is straightforward because the sequential composition of concrete derivations is well-understood, and it is obtained simply by the juxtaposition of the corresponding diagrams. Since the shift equivalence proposed in the literature relates only derivations between the same graphs, it can be used to define a congruence on the concrete derivations yielding the truly concurrent model of computation $\mathbf{Der}(\mathcal{G})/_{sh}$. However, this category is still too concrete, because it has all graphs as objects. Therefore from each graph there are still infinitely many "identical" derivations, differing only for the ending graph. In the other direction, we showed that if one wants to reason on derivations disregarding "representation dependent" details, then equivalence $\equiv_3$ can be used, because it allows for sequential composition; moreover, it is compatible with the shift equivalence.

Finally, thanks to the properties of $\equiv_3$, the abstract, truly concurrent model of computation $\mathbf{ADer}(\mathcal{G})/_{\underline{sh}}$ can be defined. This model seems to represent all the derivations of grammar $\mathcal{G}$ at a very reasonable level of abstraction, both independent of representation-dependent details and consistent with the shift equivalence. The fact that functor $\underline{\mathcal{A}}$ is an equivalence of categories essentially means that category $\mathbf{Der}(\mathcal{G})/_{sh}$ already has between each pair of objects the "right" structure, although this structure has, in general, infinitely many copies.

3.6 Embedding, Amalgamation and Distribution in the DPO approach

In this section we shortly recall some concepts and results of the DPO approach concerning the topics addressed in Sections 3.2.3 and 3.2.4.

3.6.1 *Embedding of Derivations and Derived Productions*

In Section 3.2.3 the Embedding Problem (Problem 3.2.3) asked for conditions ensuring that a derivation can be embedded into a larger context. Given derivations $\rho = (G_0 \stackrel{p_1,m_0}{\Longrightarrow} \cdots \stackrel{p_k,m_{k-1}}{\Longrightarrow} G_k)$ and $\delta = (X_0 \stackrel{p_1,n_0}{\Longrightarrow} \cdots \stackrel{p_k,n_{k-1}}{\Longrightarrow} X_k)$, an *embedding* of ρ into δ is a family of injections $e = (G_i \stackrel{e_i}{\longrightarrow} X_i)_{i\in\{1,\ldots,k\}}$ such that for all $i \in \{1, \ldots, k\}$

- $e_i \circ m_i = n_i$, and
- there is an injection $D_i \stackrel{c_i}{\longrightarrow} Y_i$ making the two resulting squares commutative (see the diagram below).

The embedding of ρ into δ is denoted by $\rho \stackrel{e}{\longrightarrow} \delta$. Given ρ and δ, the embedding e is completely determined by the first injection e_0, called the *embedding morphism* of e.

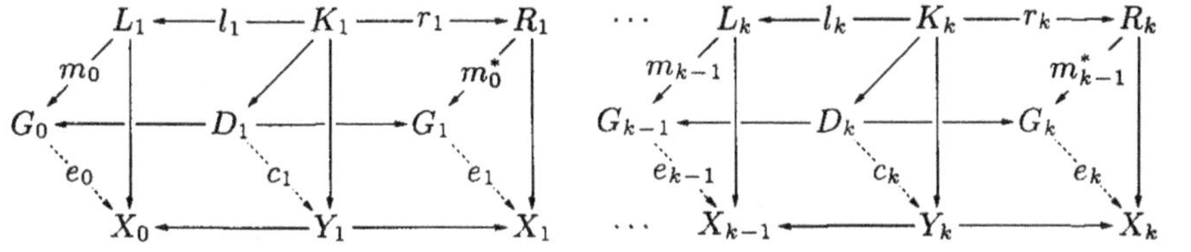

The applicability of a *derived production* simulating the effect of the derivation ρ shall ensure the existence of the embedding $\rho \xrightarrow{e} \delta$. First we consider the basic case of a direct derivation leading to the notion of directly derived production: Let $d = (G \overset{p,m}{\Longrightarrow} H)$ be a direct derivation. Then $\langle d \rangle : (G \xleftarrow{l^*} D \xrightarrow{r^*} H)$ denotes the *directly derived production* of d, where $G \xleftarrow{l^*} D \xrightarrow{r^*} H$ is the co-production of d, i.e., the bottom row of the double-pushout diagram (1)+(2) on the left of Figure 3.17. The production name $\langle d \rangle$ records the construction of the directly derived production by the direct derivation d. A direct derivation $X \overset{\langle d \rangle, e}{\Longrightarrow} Z$ using $\langle d \rangle$ is called a *directly derived derivation*. The following equivalence, relating directly derived derivations with derivations using the original production, is also known as vertical composition property of derivation diagrams.

Proposition 3.6.1 (Directly Derived Production)
Given the directly derived production $\langle d \rangle$ of a derivation $d = (G \overset{p,m}{\Longrightarrow} H)$ as in Figure 3.17, the following statements are equivalent:

1. *There is a directly derived derivation $X \overset{\langle d \rangle, e}{\Longrightarrow} Z$.*
2. *There is a direct derivation $X \overset{p, e \circ m}{\Longrightarrow} Z$.*

Proof outline
By composition and decomposition of the pushout diagrams (1) to (4) on the left of Figure 3.17. □

Now we want to extend this equivalence to derivations $\rho = (G_0 \overset{p_1, m_0}{\Longrightarrow} \cdots \overset{p_k, m_{k-1}}{\Longrightarrow} G_k)$ with $k > 1$. To this aim we introduce the notion of "sequential composition" of two productions, in order to obtain a single derived production $\langle \rho \rangle$ from the composition of the sequence of directly derived productions $\langle d_1 \rangle; \langle d_2 \rangle; \ldots; \langle d_k \rangle$. We speak of *sequential composition* because the application of such a composed production has the same effect of a sequential derivation based on the original productions.

Given two productions $p_1 : (L_1 \xleftarrow{l_1} K_1 \xrightarrow{r_1} R_1)$ and $p_2 : (L_2 \xleftarrow{l_2} K_2 \xrightarrow{r_2} R_2)$ with $R_1 = L_2$, the *sequentially composed production* $p_1; p_2 : (L_1 \xleftarrow{l_1 \circ l_2'} K \xrightarrow{r_2 \circ r_1'}$

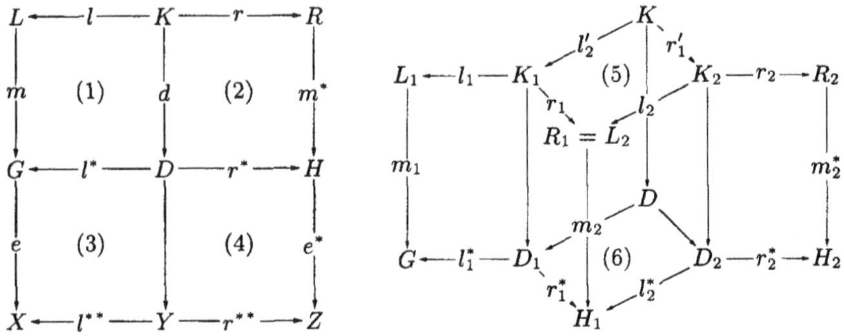

Figure 3.17: Horizontal and sequential composition of direct derivations.

R_2) is obtained as shown at the top of the right diagram of Figure 3.17, where (5) is a pullback diagram. Derivation $G \overset{p_1,m_1}{\Longrightarrow} H_1 \overset{p_2,m_2}{\Longrightarrow} H_2$, where the match m_2 of the second direct derivation is the co-match of the first, may now be realized in one step using the sequentially composed production $p_1;p_2$. Vice versa, each direct derivation $G \overset{p_1;p_2,m_1}{\Longrightarrow} H_2$ via $p_1;p_2$ can be decomposed in a derivation $G \overset{p_1,m_1}{\Longrightarrow} H_1 \overset{p_2,m_1^*}{\Longrightarrow} H_2$.

Proposition 3.6.2 (Sequential Composition)
Given two productions p_1 and p_2 and their sequential composition $p_1;p_2$ as above, the following statements are equivalent:

1. *There is a direct derivation $G \overset{p_1;p_2,m_1}{\Longrightarrow} H_2$.*

2. *There is a derivation $G \overset{p_1,m_1}{\Longrightarrow} H_1 \overset{p_2,m_1^*}{\Longrightarrow} H_2$ such that m_1^* is the co-match of m_1 w.r.t. p_1.*

Proof outline
By the 3-cube pushout-pullback-lemma [33]. □

Combining directly derived productions by sequential composition we now define derived productions for derivations of length greater than one: The *derived production* $\langle\rho\rangle$ of a derivation $\rho = (G_0 \overset{p_1,m_0}{\Longrightarrow} \cdots \overset{p_k,m_{k-1}}{\Longrightarrow} G_k)$ is defined as the sequential composition $\langle d_1\rangle;\ldots;\langle d_k\rangle$ of the directly derived productions $d_i = (G_{i-1} \overset{p_i,m_{i-1}}{\Longrightarrow} G_i)$ of ρ. A direct derivation $K \overset{\langle\rho\rangle,e}{\Longrightarrow} Y$ using $\langle\rho\rangle$ is called *derived derivation*.

Each derivation sequence may be shortcut by a corresponding derived derivation. Vice versa, each derived derivation corresponds to a derivation using the original productions. Hence, these notions solve the Derived Production Problem 3.2.4 of Section 3.2.3 in the DPO approach, which is formally stated by the theorem below.

Theorem 3.6.3 (Derived Productions)
Let ρ be a derivation and $\langle\rho\rangle$ its derived production as defined above. Then the following statements are equivalent:

1. *There is a derived derivation $X_0 \stackrel{\langle\rho\rangle,e_0}{\Longrightarrow} X_k$.*
2. *There is a derivation $\delta = (X_0 \stackrel{p_1,n_0}{\Longrightarrow} \cdots \stackrel{p_k,n_{k-1}}{\Longrightarrow} X_k)$ and an embedding $\rho \stackrel{e}{\longrightarrow} \delta$, where e_0 is the embedding morphism of e.*

Proof outline
Follows from Proposition 3.6.1, Proposition 3.6.2 by induction over the length k of the derivation ρ. □

Let us come back to Problem 3.2.3 of Section 3.2.3 asking for an embedding of a derivation into a larger context. This question is now answered using the equivalence above, that is, a derivation ρ may be embedded via an embedding morphism e_0 if and only if the corresponding derived production $\langle\rho\rangle$ is applicable at e_0.

Theorem 3.6.4 (Embedding)
Let $\rho = (G_0 \stackrel{p_1,m_0}{\Longrightarrow} \cdots \stackrel{p_k,m_{k-1}}{\Longrightarrow} G_k)$ be a derivation, $\langle\rho\rangle$ its derived production, and $G_0 \stackrel{e_0}{\longrightarrow} X_0$ an embedding morphism. Then there is a derivation $\delta = (X_0 \stackrel{p_1,e_0\circ m_0}{\Longrightarrow} \cdots \stackrel{p_k,e_{k-1}\circ m_{k-1}}{\Longrightarrow} X_k)$ and an embedding $\rho \stackrel{e}{\longrightarrow} \delta$ if and only if $\langle\rho\rangle$ is applicable at e_0, that is, if there is a derived derivation $X_0 \stackrel{\langle\rho\rangle,e_0}{\Longrightarrow} X_k$.

Proof outline
Theorem 3.6.4 follows from Theorem 3.6.3. □

The embedding theorem above differs from the classical one in [2], which considers not only the embedding of a derivation into a larger context (called "JOIN" in [2]), but also the reduction of the context of a derivation (called "CLIP"). In fact, Theorem 3.6.4 corresponds to the "JOIN" part of the classical embedding theorem, and the JOIN condition is just the applicability of the derived production.

As anticipated in Section 3.2.3 already, the construction of a derived production for a given derivation ρ is not only useful in order to solve the embedding problem but is interesting in its own right. A related concept has been introduced in [20] under the name "concurrent production", where two productions are composed w.r.t. a given *dependency relation* specifying an overlapping of the right-hand side of the first with the left-hand side of the second production. The resulting concurrent production enjoys similar properties as the derived production above, and is, moreover, minimal w.r.t. to the context included in the interface graph

3.6.2 Amalgamation and Distribution

Section 3.2.4 introduced two concepts modeling the synchronization of direct derivations, namely amalgamated and (synchronous) distributed derivations. In this subsection we formalize these two concepts in the DPO approach and show how they are related.

Amalgamation

Amalgamated derivations describe the synchronized application of productions by a so-called amalgamated production. The synchronization of two productions p_1 and p_2 is modeled by identifying a common subproduction p_0: Let $p_i : (L_i \overset{l_i}{\longleftarrow} K_i \overset{r_i}{\longrightarrow} R_i)$ be three productions for $k \in \{0,1,2\}$. Then production p_0 together with graph morphisms $in^1_L : L_0 \rightarrow L_1, in^1_K : K_0 \rightarrow K_1$ and $in^1_R : R_0 \rightarrow R_1$ is a *subproduction* of p_1 if the resulting squares (1) and (2) below commute. The embedding of p_0 into p_1 is denoted by $in^1 = \langle in^1_L, in^1_K, in^1_R \rangle : p_0 \rightarrow p_1$.

$$
\begin{array}{ccccc}
L_0 & \xleftarrow{l_0} & K_0 & \xrightarrow{r_0} & R_0 \\
\downarrow in^1_L & (1) & \downarrow in^1_K & (2) & \downarrow in^1_R \\
L_1 & \xleftarrow{l_1} & K_1 & \xrightarrow{r_1} & R_1
\end{array}
$$

Productions p_1 and p_2 are synchronized w.r.t. p_0, shortly denoted by $p_1 \overset{in^1}{\leftarrow} p_0 \overset{in^2}{\rightarrow} p_2$, if p_0 is a subproduction of both p_1 and p_2 with embeddings in^1 and in^2, respectively. The amalgamated production is obtained as the gluing of p_1 and p_2 w.r.t. p_0, formally described by pushout constructions: Given productions p_i for $i \in \{0,1,2\}$ as above, the *amalgamated production* is $p_1 \oplus_{p_0} p_2 : (L \overset{l}{\longleftarrow} K \overset{r}{\longrightarrow} R)$ in the left diagram below, where subdiagrams (1), (2) and (3) are pushouts, and l and r are induced by the universal property of (2) such that all subdiagrams commute. A direct derivation $G \overset{p,m}{\Longrightarrow} X$ using the amalgamated production $p = p_1 \oplus_{p_0} p_2$ is called *amalgamated derivation.*

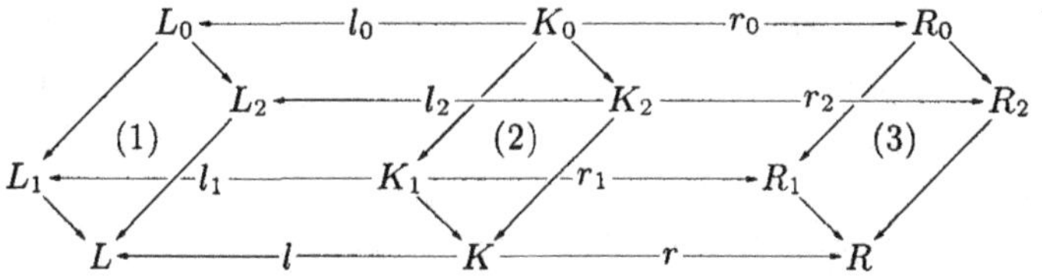

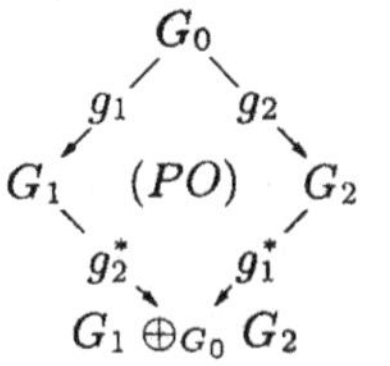

The concept of synchronizing derivations by amalgamation of productions is introduced in [11]. Based on a sequential decomposition of amalgamated productions into the common subproduction p_0 and the remainders of the elementary production $p_1 - p_0$ and $p_2 - p_0$, a sequential decomposition of amalgamated derivations is developed. Generalizing the Parallelism Theorem, the Amalgamation Theorem of [11] then establishes a connection between amalgamated derivations and derivations using the elementary productions p_1, p_2 and their remainders.

Distribution

The concept of distributed graphs allows to model the state of a system by interrelated local substates: A *distributed graph* $DG = (G_1 \overset{g_1}{\leftarrow} G_0 \overset{g_2}{\rightarrow} G_2)$ consists of two *local graphs* G_1 and G_2, an *interface graph* G_0, and graph morphisms g_1 and g_2 embedding the interface graph into the local graphs. The *global graph* $\oplus DG = G_1 \oplus_{G_0} G_2$ of DG is defined as the pushout object of g_1 and g_2 in the right diagram above. Given $G = \oplus DG$, the distributed graph is called a *splitting* of G. In the DPO approach we consider only total splittings, i.e., distributed graphs $G_1 \overset{g_1}{\leftarrow} G_0 \overset{g_2}{\rightarrow} G_2$ where g_1 and g_2 are total graph morphisms.

The local graphs can be transformed by local direct derivations. In order to form a distributed derivation, these direct derivations have to fulfill some compatibility conditions, which are formulated in terms of their matches [22,23]: Let DG be a distributed graph, $p_1 \overset{in^1}{\leftarrow} p_0 \overset{in^2}{\rightarrow} p_2$ synchronized productions, and $L_i \overset{m_i}{\longrightarrow} G_i$, $(i \in \{0,1,2\})$ be matches for p_i which are compatible with DG, i.e., $g_k \circ m_0 = m_k \circ in_L^k$ for $k \in \{1,2\}$. In this case, the family $(m_i)_{i \in \{0,1,2\}}$ is called a *distributed match* for $p_1 \overset{in^1}{\leftarrow} p_0 \overset{in^2}{\rightarrow} p_2$ in DG. It satisfies the *distributed gluing condition* if all m_i satisfy the local gluing condition of p_i for $i \in \{0,1,2\}$, and the *connection condition*, i.e., $g_k(G_0 - m_0(L_0 - l_0(K_0)) \subseteq G_k - m_k(L_k - l_k(K_k))$ for $k \in \{1,2\}$.

The connection condition states, that a distributed derivation is not allowed to delete, for example, a vertex v_1 of the local graph G_1 if a corresponding vertex v_0 in the interface graph G_0 (i.e., such that $g_1(v_0) = v_1$) is preserved. Given a distributed match as above, there are local direct derivations $d_i = (G_i \overset{p_i,m_i}{\Longrightarrow} H_i)$ for $i \in \{0,1,2\}$ and distributed graphs $DD = (D_1 \overset{c_1}{\leftarrow} D_0 \overset{c_2}{\rightarrow} D_2)$ and $DH = (H_1 \overset{h_1}{\leftarrow} H_0 \overset{h_2}{\rightarrow} H_2)$ such that $d_1||_{d_0}d_2 : DG \Longrightarrow DH$ forms a *distributed derivation*, as shown in the left diagram of Figure 3.18. The graph morphisms c_1 and c_2 of the distributed graph DD are defined by $c_1 = l_1^{*-1} \circ g_1 \circ l_0^*$ and $c_2 = l_2^{*-1} \circ g_2 \circ l_0^*$. The graph morphisms h_1 and h_2 of DH are induced by the universal property of the right-hand side pushout $\langle r_0^*, m_0^* \rangle$ of the direct derivation d_0.

Distributed derivations in the DPO approach are synchronous by definition, since only total splittings are considered. Given a total splitting $\oplus DG$, the connection condition ensures that also the distributed graphs DD and DH are total splittings [23].

The idea of modeling the distribution of states by a splitting a global graph into local subgraphs, related by an interface, has been developed in [12]. The derivation mechanism of [12], however, is more restricted than the one presented here because the local direct derivations are not allowed to change the interface graph G_0. This corresponds to distributed derivations (in the above sense) using a constant interface production p_0. In order to be able to change the interface graph, a global direct derivation had to be performed with the global graph $\oplus DG$ of the distributed graph DG, and two additional operations SPLIT and JOIN had to be introduced in order to switch between the distributed and the global representation. Distributed derivations with dynamic interfaces (like in this section) have been introduced in [19] in the SPO approach.

Of course, splitting a state into two substates with one interface is a very basic kind of a distribution. More general topologies are allowed in [23], where a distributed graph may be any diagram in the category ***Graph*** of graphs and total graph morphisms. In fact, the distribution concepts presented in this section are obtained by restricting distributed derivations in the sense of [23] to diagrams of the form $G_1 \overset{g_1}{\longleftarrow} G_0 \overset{g_2}{\longrightarrow} G_2$. The following distribution theorem is a special case of a corresponding theorem in [23]. It provides the solution to Problem 3.2.5 by relating amalgamated and distributed derivations.

Theorem 3.6.5 (Distribution)
Let DG be a distributed graph and $p_1 \overset{in^1}{\leftarrow} p_0 \overset{in^2}{\rightarrow} p_2$ be synchronized productions. Then the following statements are equivalent:

1. *There are an amalgamated derivation $\oplus DG \overset{p_1 \oplus_{p_0} p_2, m}{\Longrightarrow} H$, and a distributed match $(m_i)_{i \in \{0,1,2\}}$ such that*

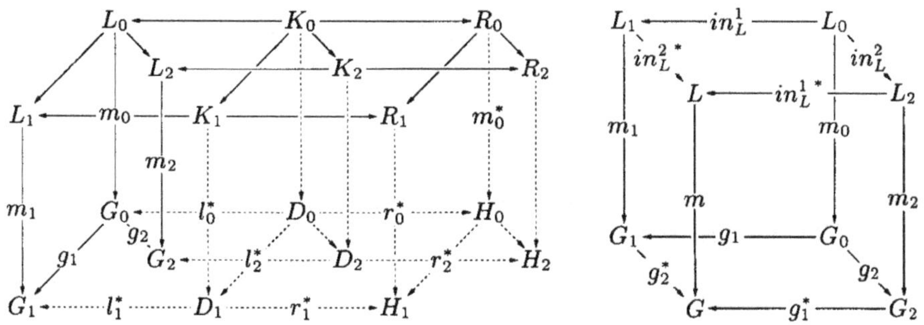

Figure 3.18: Distributed derivation, and applicability of subproduction p_i to local graph G_i.

(a) $g_2^* \circ m_1 = m \circ in_L^{2*}$ and $g_1^* \circ m_2 = m \circ in_L^{1*}$ (i.e., the front and left square in the right diagram of Figure 3.18 commute),

(b) $(m_i)_{i\in\{0,1,2\}}$ satisfies the connection condition, and

(c) for $i = 0,1,2$, $m_i|^{S_i} : L_i \to S_i$ satisfies the local gluing condition of p_i, where $S_1 = g_2^{*-1}(m(L))$, $S_2 = g_1^{*-1}(m(L))$, $S_0 = (g_2^* \circ g_1)^{-1}(m(L))$, and $m_i|^{S_i}$ denotes the codomain restriction of m_i to S_i.

2. There is a distributed graph DH with $\oplus DH = H$ and a distributed derivation $d_1||_{d_0}d_2 : DG \Longrightarrow DH$ with $d_i = (G_i \overset{p_i,m_k}{\Longrightarrow} H_i)$ for $i = 0,1,2$.

Proof outline
"1. $\Longrightarrow$ 2.": According to the definition of distributed derivations we have to show that there are local direct derivations $d_i = (G_i \overset{p_i,m_i}{\Longrightarrow} H_i)$ for $i = 0,1,2$. Then, the existence of the distributed derivation $d_1||_{d_0}d_2 : DG \Longrightarrow DH$ follows from the existence of a distributed match satisfying (a) and (b) of 1. Condition (c) covers the difference between the gluing conditions of the local derivations and that of the given amalgamated derivation, i.e., the gluing condition for d_i follows from the gluing condition of $\oplus DG \overset{p_1\oplus_{p_0}p_2,m}{\Longrightarrow} H$ and the *difference gluing condition* (c).
"2. $\Longrightarrow$ 1.": Given a distributed derivation $d_1||_{d_0}d_2 : DG \Longrightarrow DH$ as in Figure 3.18 on the left, the amalgamated derivation $\oplus DG \overset{p_1\oplus_{p_0}p_2,m}{\Longrightarrow} \oplus DH$ is obtained by constructing the global graphs $\oplus DG, \oplus DD$, and $\oplus DH$ as pushouts, and the morphisms of the co-production $\oplus DG \overset{l^*}{\longleftarrow} \oplus DD \overset{r^*}{\longrightarrow} \oplus DH$ as the universal morphisms induced by the pushout property of $\oplus DD$. The vertical morphisms $m : L \to \oplus DG$, $d : K \to \oplus DD$, and $m^* : R \to \oplus DH$ are induced

in a similar way by the universal property of the pushout objects L, K, and R of $\langle in_L^1, in_L^2\rangle$, $\langle in_K^1, in_K^2\rangle$, and $\langle in_R^1, in_R^2\rangle$. □

3.7 Conclusion

For a detailed comparison between the DPO and the SPO approach and for concluding remarks about the contents of the two parts we address the reader to Sections 4.6 and 4.7.

Acknowledgements

Most of the research results presented in this chapter have been developed within the ESPRIT Working Groups COMPUGRAPH (1989-1992) and COMPUGRAPH II (1992-1996).

3.8 Appendix A: On commutativity of coproducts

This appendix addresses a quite technical topic: it introduces the notion of binary coproduct in a category, and shows that coproducts cannot be assumed to be commutative (in a strict way) unless the category is a preorder. This fact justifies the way we defined parallel productions (Definition 3.4.4) and their manipulation in Section 3.4, which may appear more complex than the corresponding definitions in previous papers in the literature.

In any category, a coproduct is an instance of the more general concept of *colimit* [18]. In the category **Set** of sets and functions (as well as in **Graph**) the coproduct of two objects is their disjoint union. In the following discussion we stick to binary coproducts, but most statements can be adapted to any kind of limit or colimit construction.

Definition 3.8.1 (coproducts)
Let $\mathbf{C}$ be a category and A, B be two objects of $\mathbf{C}$. A *coproduct* of $\langle A, B\rangle$ is a triple $\langle A+B, in_1^{A+B}, in_2^{A+B}\rangle$ where $A+B$ is an object and $in_1^{A+B} : A \to A+B$, $in_2^{A+B} : B \to A+B$ are arrows of $\mathbf{C}$, such that the following universal property holds (see Figure 3.19(a)):

- for all pair of arrows $\langle f : A \to C, g : B \to C\rangle$ there exists a unique arrow $[f, g] : A + B \to C$ such that $[f, g] \circ in_1^{A+B} = f$ and $[f, g] \circ in_2^{A+B} = g$.

In this case $A + B$ is called a *coproduct object* of $\langle A, B\rangle$, and in_1^{A+B}, in_2^{A+B} are called the *injections*; arrow $[f, g]$ is called the *copairing* of f and g. One says that category $\mathbf{C}$ *has coproducts* if each pair of objects of $\mathbf{C}$ has a coproduct.

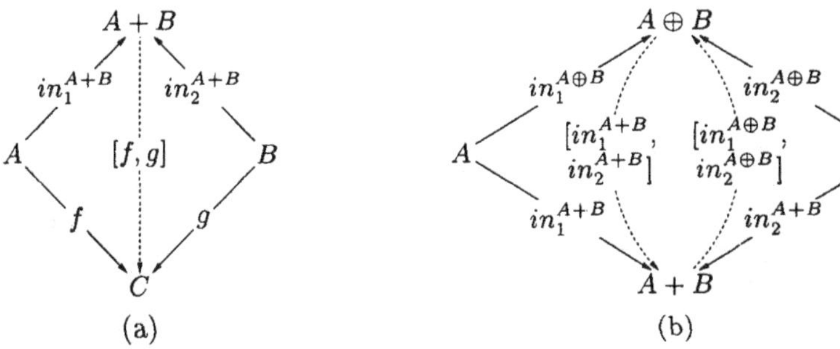

Figure 3.19: (a) Coproduct. (b) Isomorphisms between distinct coproduct objects.

In the following we assume that **C** has coproducts. From the definition it is evident that the coproduct of two objects is in general not unique. However, given any two coproducts of two objects they are not only isomorphic, but there exists *only one* isomorphism commuting with the injections. This fact holds for limits and colimits in general [18].

Fact 3.8.2 (uniqueness of coproducts up to isomorphism)
Let $\langle A+B, in_1^{A+B}, in_2^{A+B}\rangle$ *and* $\langle A\oplus B, in_1^{A\oplus B}, in_2^{A\oplus B}\rangle$ *be two coproducts of* $\langle A, B\rangle$ *in* **C**. *Then objects* $A+B$ *and* $A\oplus B$ *are isomorphic, and there exists only one isomorphism* $\phi : A+B \to A\oplus B$ *such that* $in_1^{A\oplus B} = \phi \circ in_1^{A+B}$ *and* $in_2^{A\oplus B} = \phi \circ in_2^{A+B}$. *As shown in Figure 3.19(b),* ϕ *is determined by the universal property of* $A+B$ *as* $\phi = [in_1^{A\oplus B}, in_2^{A\oplus B}]$, *and similarly for its inverse* $\phi^{-1} = [in_1^{A+B}, in_2^{A+B}]$. □

Often it is useful to regard the formal operator + used to denote coproduct objects as a real function, returning a specific object of **C**, and not merely specifying a class of isomorphic objects. This can be obtained by fixing a *choice of coproducts*, i.e., by fixing a specific coproduct for each pair of objects. Every choice of coproducts determines a coproduct functor as follows.

Definition 3.8.3 (coproduct functor)
Suppose that a *choice of coproducts* for **C** is given, i.e., a mapping associating with each pair of objects $\langle A, B\rangle$ a coproduct, say $\langle A+B, in_1^{A+B}, in_2^{A+B}\rangle$. This determines a bifunctor $+ : \mathbf{C}^2 \to \mathbf{C}$, called the *coproduct functor*, defined on objects as $+(A,B) = A+B$. On arrows, if $f : A \to C$ and $g : B \to D$ are in **C**, then $+(f,g)$ (more often written $f+g$) is uniquely determined, using the universal property of coproducts, as $f+g \stackrel{def}{=} [in_1^{C+D} \circ f, in_2^{C+D} \circ g] : A+B \to C+D$, as shown in Figure 3.20(a).

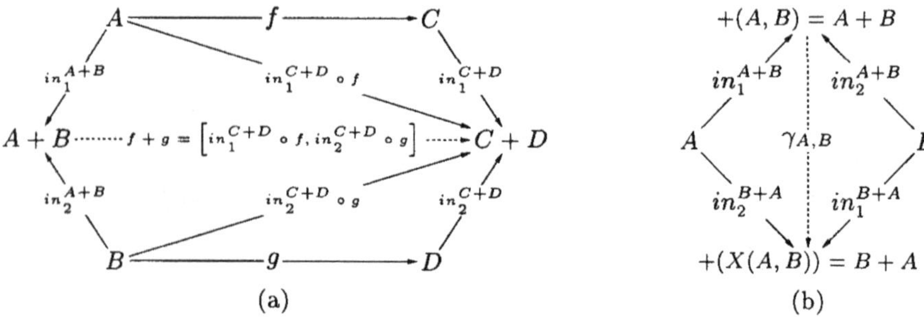

Figure 3.20: (a) Coproduct of arrows. (b) Commutativity up to natural isomorphism.

Coproducts are widely used in the algebraic theory of graph grammars, for example in the definition of parallel productions (see Definition 3.4.4). A natural question that arises in this framework is: Can one consider the coproduct functor as commutative? We show here that although every coproduct functor is commutative up to a natural isomorphism, in general it is not safe to assume that it is *strictly* commutative, i.e., commutative up to the *identity* natural transformation.

Fact 3.8.4 (commutativity up to natural isomorphism of coproducts) *Any coproduct functor* $+ : \mathbf{C}^2 \to \mathbf{C}$ *is commutative up to a natural isomorphism, in the following sense. Let* $X : \mathbf{C}^2 \to \mathbf{C}^2$ *be the functor that exchanges its arguments (i.e.,* $X(A, B) = \langle B, A \rangle$*). Then there exists a natural transformation*[e] $\gamma : + \Rightarrow + \circ X : \mathbf{C}^2 \to \mathbf{C}$*, such that* $\gamma_{A,B}$ *is an isomorphism for each* A, B*. The component* $\gamma_{A,B}$ *of the natural isomorphism* γ *is easily determined as in Figure 3.20(b), and the proof of the naturality of* γ *is routine.*

□

We are now ready to say when a coproduct functor is strictly commutative. Although the following definition may look very restrictive, it seems to be the only reasonable one in a categorical framework: compare it, for example, to the definition of *strict monoidal categories* in [18].

Definition 3.8.5 (strictly commutative coproduct functors)
A coproduct functor $+ : \mathbf{C}^2 \to \mathbf{C}$ is *strictly commutative* if the natural isomorphism $\gamma : + \Rightarrow + \circ X : \mathbf{C}^2 \to \mathbf{C}$ is the identity natural transformation. This

[e]Given two functors $F, G : \mathbf{C} \to \mathbf{D}$, a ***natural transformation*** $\gamma : F \Rightarrow G : \mathbf{C} \to \mathbf{D}$ is a family of arrows of $\mathbf{D}$ indexed by objects of $\mathbf{C}$, $\gamma = \{\gamma_A \mid A \in Obj(\mathbf{C})\}$, such that $\gamma_A : F(A) \to G(A)$, and satisfying the following ***naturality*** condition: for each arrow $f : A \to B$ of $\mathbf{C}$, $G(f) \circ \gamma_A = \gamma_B \circ F(f)$ (see [18]).

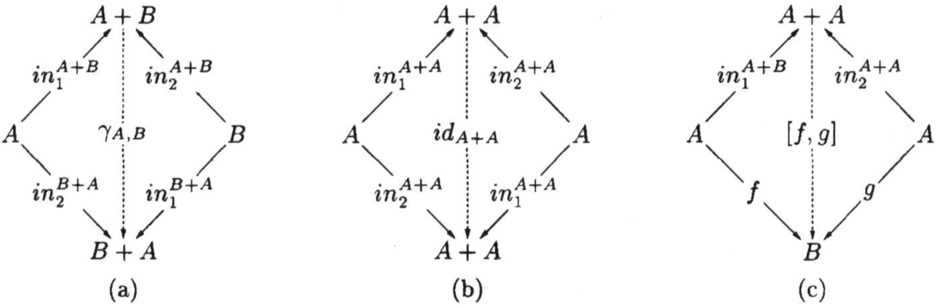

Figure 3.21: Strict commutativity of coproducts implies that all parallel arrows (like f and g) are identical.

means not only that $A + B = B + A$ (they are the same object), but also that $\gamma_{A,B} = id_{A+B}$ for each pair $\langle A, B \rangle$.

The following result shows that if a category **C** has a coproduct functor that is strictly commutative, then **C** is a preorder, i.e., there is at most one arrow between each pair of objects. Since none of the categories of graphs (or of other structures) considered in the algebraic theory of graph grammars is a preorder, this fact shows that one cannot assume, in general, that "coproducts are strictly commutative".

Proposition 3.8.6 (strictly commutative coproducts and preorders) *Let* **C** *be a category that has coproducts, and suppose that there exists a choice of coproducts for* **C** *such that the induced coproduct functor* $+ : \mathbf{C}^2 \to \mathbf{C}$ *is strictly commutative. Then* **C** *is a preorder, i.e., for each pair of arrows* $f, g : A \to B$*, it holds* $f = g$.

Proof
By the definition of strict commutativity, $\gamma_{A,B} = id_{A+B}$; thus the diagram in Figure 3.21(a) shows that the injections of A in $A + B$ and in $B + A$, i.e., in_1^{A+B} and in_2^{B+A}, are exactly the same arrow, because the two triangles must commute. Because of the same reason, if we consider the coproduct of A with itself, then it turns out that there is *only one injection* $in_1^{A+A} = in_2^{A+A}$ from A to $A + A$, as depicted in Figure 3.21(b). Suppose now that there are two arrows $f, g : A \to B$. By the universal property there is only one arrow $[f, g] : A + A \to B$ (as shown in Figure 3.21(c)) such that $[f, g] \circ in_1^{A+A} = f$ and $[f, g] \circ in_2^{A+A} = g$. Therefore, since $in_1^{A+A} = in_2^{A+A}$, we have that $f = g$, and **C** is a preorder. □

Other assumptions on the properties of coproducts or on the category **C** may yield similar results, as stated by the following corollary.

Corollary 3.8.7

1. *Say that a category* **C** *has* unique coproducts *if it has coproducts and for each pair of objects* $\langle A, B\rangle$ *there is only one triple* $\langle A+B, in_1^{A+B} : A \to A+B, in_2^{A+B} : B \to A+B\rangle$ *satisfying the universal property. If a category* **C** *has unique coproducts, then it is a preorder.*

2. *If* **C** *has coproducts and* **C** *has no non-trivial automorphisms (i.e., for each object* A *there is only one isomorphism from* A *to itself, the identity), then* **C** *is a preorder.* □

Both statements are easily proved by showing first that the hypotheses imply that there is only one injection from A to $A + A$, and then using the diagram of Figure 3.21(c).

Finally, it is worth stressing that the *uniqueness of coproduct objects* is not sufficient to deduce that a category is a preorder. More formally, if **C** has coproducts, say that it has *unique coproduct objects* if for each pair of objects $\langle A, B\rangle$ and for each pair of coproducts of $\langle A, B\rangle$, say $\langle C, in_1^C : A \to C, in_2^C : B \to C\rangle$ and $\langle D, in_1^D : A \to D, in_2^D : B \to D\rangle$, one has $C = D$. For example, the (skeletal) full sub-category of finite sets having as objects $\{\underline{n} \mid n \in \mathbb{N}\}$, where $\underline{n} = \{1, 2, \ldots, n\}$, has unique coproduct objects but it is not a preorder.

3.9 Appendix B: Proof of main results of Section 3.5

This appendix presents the proof of Lemma 3.5.11, which is the main technical result of Section 3.5.3. This proof is based on the constructive proofs of Lemmas 3.4.8 and 3.4.9 (i.e., the synthesis and analysis constructions) which are reported below for completeness, and also because they show in an instructive way the kind of reasoning that is used when proving results in the algebraic approach to graph transformations. To start with, we need the following well-known technical lemma [3].

Lemma 3.9.1 (Butterfly lemma)
Let a_1 *and* a_2 *be injective graph morphisms. Then the square (1) of Figure 3.22 is a pushout if and only if there are graphs* X_1, X_2 *and graph morphisms* r, s, v, w, t *and* u, *such that in the right part of the same figure diagrams (2), (3), and (4) are pushouts, (4) is a pullback, and all triangles commute.* □

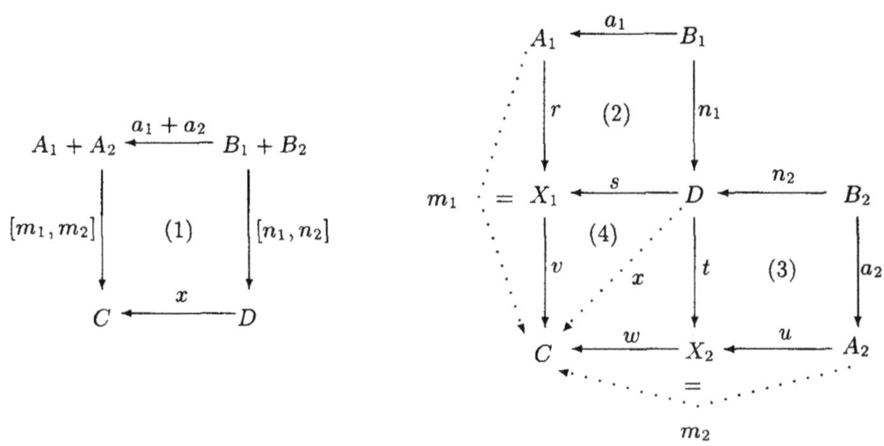

Figure 3.22: Butterfly Lemma.

Proof of Lemma 3.4.8 (analysis of parallel direct derivations)
Let $\rho = (G \stackrel{q}{\Longrightarrow} H)$ *be a parallel direct derivation using the parallel production* $q = p_1 + \ldots + p_k : (L \stackrel{l}{\leftarrow} K \stackrel{r}{\rightarrow} R)$. *Then for each ordered partition* $\langle I = \langle i_1, \ldots, i_n \rangle, J = \langle j_1, \ldots, j_m \rangle \rangle$ *of* $\{1, \ldots, k\}$ *(i.e.,* $I \cup J = \{1, \ldots, k\}$ *and* $I \cap J = \emptyset$*) there is a constructive way to obtain a sequential independent derivation* $\rho' = (G \stackrel{q'}{\Longrightarrow} X \stackrel{q''}{\Longrightarrow} H)$, *called an* analysis *of* ρ, *where* $q' = p_{i_1} + \ldots + p_{i_n}$, *and* $q'' = p_{j_1} + \ldots + p_{j_m}$.

If $I = \langle\rangle$ (or $J = \langle\rangle$) the statement follows by taking $X = G$ ($X = H$), $q'' = q$ ($q' = q$), and an empty direct derivation as q' (q''). Now let us assume that q is a proper parallel production, and that I and J are not empty. By the hypotheses on I and J, productions $q' + q''$ and q are clearly isomorphic via the permutation Π defined as $\Pi(x) = i_x$ if $1 \leq x \leq n$ and $\Pi(x) = j_{x-n}$ if $n < x \leq n+m$ (see Definition 3.4.4). Thus since the left diagram of Figure 3.23 is a double-pushout by hypothesis, the right diagram is a double-pushout as well, where we supposed that $q' : (L_1 \stackrel{l_1}{\leftarrow} K_1 \stackrel{r_1}{\rightarrow} R_1)$ and $q'' : (L_2 \stackrel{l_2}{\leftarrow} K_2 \stackrel{r_2}{\rightarrow} R_2)$. Therefore we can apply Lemma 3.9.1 to (1) and (2), obtaining the pushouts (3)–(5) and (6)–(8), respectively, shown in Figure 3.24. Now let (9) be the pushout of arrows l and u, where the pushout object is called X (left part of Figure 3.25). Then we can build the required derivation via q' and q'' as in the right part of Figure 3.25, where we exploited the well-known fact that a square consisting of two adjacent pushouts is again a pushout. Such derivation is sequential independent because it is easy to check that arrows $f : L_2 \rightarrow X_2$

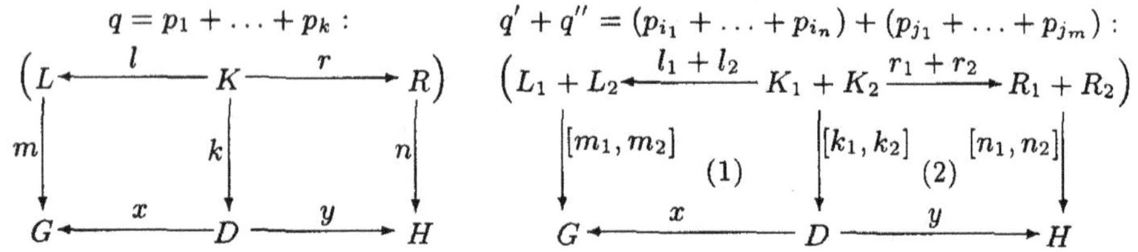

Figure 3.23: Splitting a parallel production.

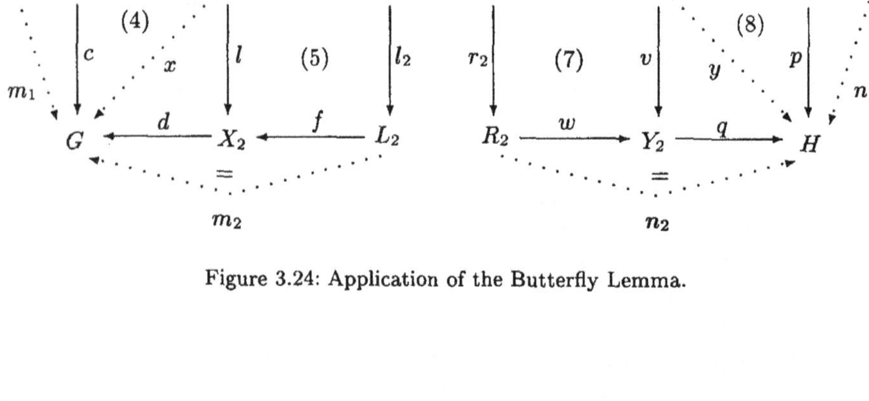

Figure 3.24: Application of the Butterfly Lemma.

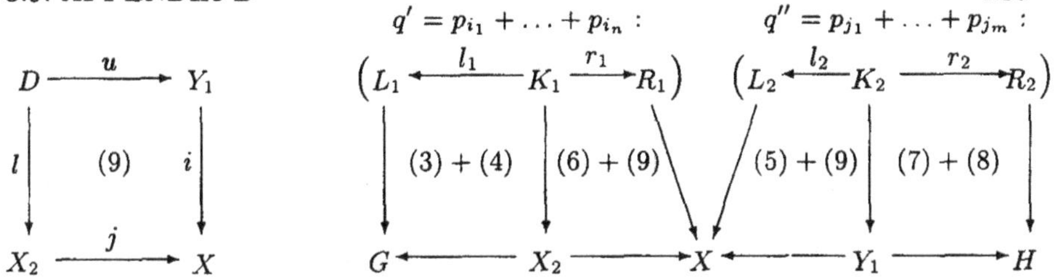

Figure 3.25: Building the sequential derivation.

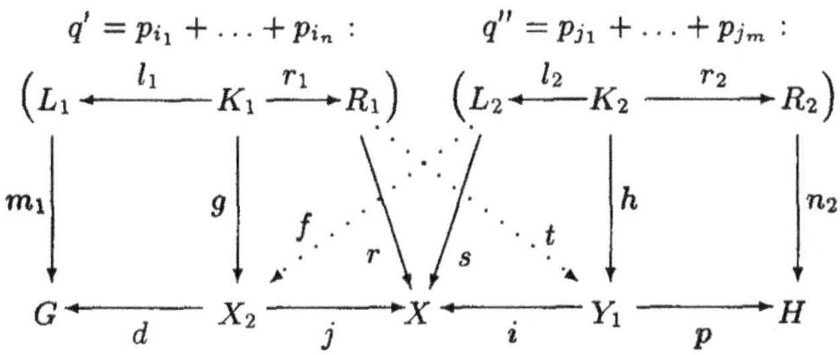

Figure 3.26: A sequential independent derivation.

and $t : R_1 \to Y_1$ satisfy the conditions of Definition 3.4.2. □

Proof of Lemma 3.4.9 (synthesis of sequential independent derivations)
Let $\rho = (G \stackrel{q'}{\Longrightarrow} X \stackrel{q''}{\Longrightarrow} H)$ *be a sequential independent derivation. Then there is a constructive way to obtain a parallel direct derivation* $\rho' = (G \stackrel{q'+q''}{\Longrightarrow} H)$, *called a* synthesis *of* ρ.

Let ρ be the sequential independent derivation of Figure 3.26. We have to show that the commutative diagrams of Figure 3.24 can be constructed, where squares from (3) to (8) have to be pushouts, and (4) and (8) need to be pullbacks. Then by two applications of the Butterfly Lemma we obtain the parallel production in the right part of Figure 3.23, as required.

The diagrams of Figure 3.24 are constructed as follows. Some of the graphs and of the morphisms are taken from the derivation ρ of Figure 3.26. The others are obtained constructively in the following way:

- $\langle D, k_2, l\rangle$ is taken as a pushout complement of $\langle l_2, f\rangle$. It is determined up to isomorphism because l_2 is injective. Thus square (5) is a pushout by construction, and l is injective because so is l_2 and in category **Graph** pushouts push out monos.

- Morphism $k_1 : K_1 \to D$ is determined by showing that morphism $g : K_1 \to X_2$ factorizes through l, i.e., that there exists a k_1 such that $l \circ k_1 = g$; then, since l is injective, k_1 is unique.

  We have to show that the image of K_1 in X_2 (i.e., $g(K_1)$) is contained in the image of D in X_2, i.e., in $l(D)$. Since X_2 is the pushout object of $\langle l_2, k_2 \rangle$, this is false only if $g(K_1)$ contains an item $z \in f(L_2)$ such that $z \notin f(l_2(K_2))$. But since X is the pushout object of $\langle l_2, h \rangle$ and $j \circ f = s$, this would mean that $j(z)$ is in $s(L_2)$ but not in $i(Y_1)$, which is absurd by the existence of a morphism $t : R_1 \to Y_1$ such that $i \circ t = r$.

- $\langle X_1, a, b \rangle$ is taken as a pushout of $\langle l_1, k_1 \rangle$. Thus (3) is a pushout by construction, and b is injective because so is l_1.

- Morphism x is defined as $x = d \circ l$.

- Morphism c is uniquely determined by condition $m_1 \circ l_1 = d \circ g = d \circ l \circ k_1 = x \circ k_1$, because (3) is a pushout.

- By a well-known decomposition property of pushouts, square (4) is a pushout because so are (3) and (3) + (4). Thus c is injective because so is l.

- Square (4) is a pullback, because any pushout in **Graph** made of injective morphisms is a pullback as well.

- Let us show that $j \circ l$ factorizes through i: this uniquely determines morphism u so that square (9) in Figure 3.24 commutes and u is injective, because so are l, j, and i; moreover, square (9) is a pushout because so are (5) and (5) + (9).

  To show that $j \circ l$ factorizes through i we prove that the image of D in X, $j(l(D))$, is contained in the image of Y_1 in X, $i(Y_1)$. Otherwise, since X is the pushout object of $\langle l_2, h \rangle$, there is an item $z = j(l(z''))$ such that $z \in s(L_2)$ and $z \notin s(l_2(K_2))$. Let $z' \in L_2$ be such that $s(z') = z$ (thus $z' \notin l_2(K_2)$); then we have $j(f(z')) = s(z') = z = j(l(z''))$, which implies $f(z') = l(z'')$ by injectivity of j. But since X_2 is the pushout object of $\langle l_2, k_2 \rangle$, this implies that $z' \in l_2(K_2)$, which is absurd.

- $\langle Y_2, w, v \rangle$ is taken as a pushout of $\langle k_2, r_2 \rangle$. Therefore (7) is a pushout and v is injective because so is r_2.

- Morphism y is defined as $y = p \circ u$.

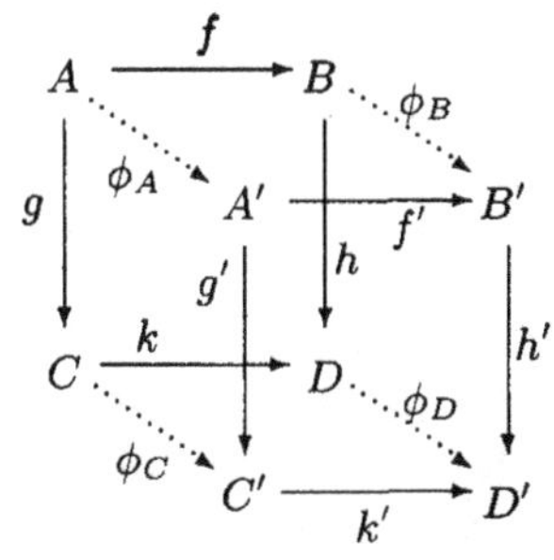

Figure 3.27: Isomorphic pushouts and pushout complements.

- Morphism q is uniquely determined by the condition $n_2 \circ r_2 = p \circ h = p \circ u \circ k_2 = y \circ k_2$, because (7) is a pushout.
- Since (7) and (7) + (8) are pushouts, also (8) is a pushout. Thus q is injective because so is u. Square (8) is also a pullback because all its morphisms are injective.
- Finally, by the so-called "pushout-pullback decomposition property" (see [20]) which holds for category **Graph** (with injective morphism as distinguished class of morphisms **M**), since (6) + (8) is a pushout, (8) is a pullback, and morphisms r_1, u, v, q and p are injective, then (6) is a pushout as well. □

The next easy technical lemma will be helpful in the proof of the main result below.

Lemma 3.9.2 (existence and uniqueness of some constructions)
1. *Given morphisms f, g, f', and g' and isomorphisms ϕ_A, ϕ_B, and ϕ_C, as in Figure 3.27, such that the two resulting squares commute, let $\langle D, h, k\rangle$ be a pushout of $\langle f, g\rangle$. Then*

- *There exists at least one pushout of $\langle f', g'\rangle$, given by $\langle D, h \circ \phi_B^{-1}, k \circ \phi_C^{-1}\rangle$.*
- *For each pushout $\langle D', h', k'\rangle$ of $\langle f', g'\rangle$, there exists a unique isomorphism $\phi_D : D \to D'$ that makes the diagram commutative.*

2. *Given injective morphisms f and f', morphism h and h', and isomorphisms ϕ_A, ϕ_B, and ϕ_D, as in Figure 3.27, such that the two resulting squares commute, let $\langle C, g, k\rangle$ be a pushout complement of $\langle f, h\rangle$. Then:*

- *There exists at least one pushout complement of $\langle f', h'\rangle$, given by $\langle C, g \circ \phi_A^{-1}, \phi_D \circ k\rangle$.*

- *For each pushout complement $\langle C', g', k'\rangle$ of $\langle f', h'\rangle$, there exists an isomorphism $\phi_C : C \to C'$ that makes the diagram commutative.*

Proof
Point 1 holds in any category, and it is a direct consequence of the uniqueness of colimits up to a unique mediating isomorphism [18]. Point 2 follows instead from specific properties of the category **Graph**, where the pushout complement happens to be unique up to isomorphism if the top arrow is injective (see Proposition 3.3.4). □

Finally, we can present the proof of Lemma 3.5.11, stating that analysis, synthesis and shift preserve equivalence $\equiv_1$ on derivations. Since most of the work was in the presentations of the constructive proofs of Lemmas 3.4.8 and 3.4.9, some parts of the proof are just outlined.

Proof of Lemma 3.5.11 (analysis, synthesis and shift preserve $\equiv_1$)
Let $\rho \equiv_1 \rho'$ via $\langle \Pi_i\rangle_{i\leq n}$, and let ϕ_{G_0} and ϕ_{G_n} be the corresponding isomorphisms between their starting and ending graphs (see Figure 3.14).

1. **(Analysis)** *If $\rho_1 \in ANAL_j^i(\rho)$, then there exists at least one derivation $\rho_2 \in ANAL^i_{\Pi_i(j)}(\rho')$. Moreover, for each $\rho_2 \in ANAL^i_{\Pi_i(j)}(\rho')$, it holds that $\rho_2 \equiv_1 \rho_1$ with the same isomorphisms ϕ_{G_0} and ϕ_{G_n} relating their starting and ending graphs.*

For simplicity, we assume that ρ and ρ' are parallel *direct* derivations (thus $\rho \equiv_1 \rho'$ via Π): the generalization to equivalent derivations of arbitrary length is straightforward. The existence of a derivation $\rho_2 \in ANAL_{\Pi(j)}(\rho')$ is ensured by Lemma 3.4.8.

If $\rho = (G \overset{p_1+\ldots+p_k}{\Longrightarrow} H)$ and $\rho' = (G' \overset{p'_1+\ldots+p'_k}{\Longrightarrow} H')$, exploiting the same technique used above at the beginning of the proof of Lemma 3.4.8, the two direct derivations can be transformed into corresponding derivations $\eta = (G \overset{p_j+q_2}{\Longrightarrow} H)$ and $\eta' = (G' \overset{p'_{\Pi(j)}+q'_2}{\Longrightarrow} H')$, where $q_2 = p_1 + \ldots + p_{j-1} + p_{j+1} + \ldots + p_k$, and $q'_2 = p'_1 + \ldots + p'_{\Pi(j)-1} + p'_{\Pi(j)+1} + \cdots + p'_k$.

Suppose now that $p_j : (L_1 \overset{l_1}{\leftarrow} K_1 \overset{r_1}{\rightarrow} R_1)$, $p'_{\Pi(j)} : (L'_1 \overset{l'_1}{\leftarrow} K'_1 \overset{r'_1}{\rightarrow} R'_1)$, $q_2 : (L_2 \overset{l_2}{\leftarrow} K_2 \overset{r_2}{\rightarrow} R_2)$, and $q'_2 : (L'_2 \overset{l'_2}{\leftarrow} K'_2 \overset{r'_2}{\rightarrow} R'_2)$.[f] By construction it is clear that η and η' are 1-equivalent because so are ρ and ρ'. Let Π_1 be the unique permutation

[f] Note that since $p_j = p'_{\Pi(j)}$, we have $X_1 = X'_1$ for $X \in \{L, K, R, l, r\}$. They have different names just for ease of reference.

on $\{1\}$ and Π_2 be the bijective mapping between the productions of q_2 and q_2' induced by Π: Π_1 and Π_2 will be used below.

Without loss of generality, let us assume that derivations η and η' have the same shape of the right double-pushout of Figure 3.23, more precisely, that the direct derivation $(1)+(2)$ of Figure 3.23 is η, that η' is an identical diagram $(1')+(2')$ where all the names of graphs and morphisms have an apex, and that η and η' are related by a family of isomorphisms $\{\phi_X : X \to X' \mid X \text{ is a graph of } \eta\}$ making the diagram commutative. In particular, also the squares obtained by considering individually the components of each coproduct commute (for example, $\phi_G \circ m_1 = m_1' \circ \phi_{L_1}$ and $\phi_G \circ m_2 = m_2' \circ \phi_{L_2}$).

Following the construction of the proof of Lemma 3.4.8, diagrams as in Figure 3.24 plus square (9) of Figure 3.25 can be constructed (in a non-deterministic way) for both η and η'. We show below that each pair of collections of diagrams obtained in this way induces a unique family of isomorphisms between corresponding graphs of the diagrams, which includes the family of isomorphisms relating η and η'. These isomorphisms, together with the permutations Π_1 and Π_2 introduced above, prove that the two sequential derivations obtained from η and η' (like that in the right part of Figure 3.25) are indeed 1-equivalent. This construction shows also that the isomorphisms ϕ_G and ϕ_H relating the starting and ending graphs of ρ and ρ' are not changed.

Let us assume that the diagrams of Figure 3.24 and square (9) of Figure 3.25 are obtained from derivation η, and that similar diagrams (but where all the names of graphs and morphisms have an apex) are obtained from η'. All the corresponding graphs of the two families of diagrams are related by the isomorphisms relating η and η', except of X_1, X_2, Y_1, Y_2 and X. Isomorphism $\phi_{X_1} : X_1 \to X_1'$ is uniquely determined by point 1 of Lemma 3.9.2, because (3) and (3′) are pushouts; using the same argument, isomorphism ϕ_{X_2}, ϕ_{Y_1}, ϕ_{Y_2} and ϕ_X are uniquely determined as well, because their source and target graphs are pushout objects. It is straightforward to check that the resulting family of isomorphisms makes everything commutative.

2. **(Synthesis)** *If $\rho_1 \in SYNT^i(\rho)$, then there exists at least one derivation $\rho_2 \in SYNT^i(\rho')$. Moreover, for each $\rho_2 \in SYNT^i(\rho')$, it holds that $\rho_2 \equiv_1 \rho_1$ with the same isomorphisms ϕ_{G_0} and ϕ_{G_n} relating their starting and ending graphs.*

We assume without loss of generality that ρ and ρ' have length 2, thus $\rho \equiv_1 \rho'$ via $\langle \Pi_1, \Pi_2 \rangle$. Since $\rho_1 \in SYNT(\rho)$ by hypothesis, derivation ρ is sequential independent: let it be the derivation of Figure 3.26. To show the existence of a $\rho_2 \in SYNT(\rho')$, by Lemma 3.4.9 it is sufficient to show that that ρ' is sequential independent as well. Let ρ' be as in Figure 3.26, but with all names of graphs and morphisms having an apex. Since $\rho \equiv_1 \rho'$, there is a family of isomorphisms relating the corresponding graphs of ρ and ρ' so that everything commutes: thus morphism f' and t' making ρ' sequential independent exists, and are uniquely determined (because i' and j' are injective) as $f' = \phi_{X_2} \circ f \circ \phi_{L_2}^{-1}$ and $t' = \phi_{Y_1} \circ t \circ \phi_{R_1}^{-1}$.

Since both ρ and ρ' are sequential independent, the diagrams of Figure 3.24 can be constructed (non-deterministically) for both of them, in such a way that the required squares are pushouts or pullbacks.

We show below that such constructions induce a family of isomorphism between the corresponding graphs of the diagrams, which includes the family of isomorphisms relating ρ and ρ'. The resulting isomorphisms can then be used to show that the two parallel direct derivations (similar to that of Figure 3.23) obtained from ρ and ρ' respectively are 1-equivalent. For the mapping between productions, it is obtained in the obvious way by joining the permutations $\langle \Pi_1, \Pi_2 \rangle$ which relate the productions used in ρ and ρ'.

To prove that the diagrams of Figure 3.24 constructed for ρ and ρ' induce isomorphisms between corresponding graphs, it is sufficient to go through the proof of Lemma 3.4.9. In particular, the missing isomorphisms $\phi_D : D \to D'$, $\phi_{X_1} : X_1 \to X_1'$, and $\phi_{Y_2} : Y_2 \to Y_2'$ are determined by Lemma 3.9.2, because the corresponding graphs are either pushout objects or pushout complement objects (and suitable morphisms are injective). The remaining steps of the mentioned proof are needed to show constructively that all the squares induced by the family of isomorphisms commute.

3. **(Shift)** *If $\rho_1 \in SHIFT_j^i(\rho)$, then there exists at least one derivation $\rho_2 \in SHIFT_{\Pi_i(j)}^i(\rho')$. Moreover, for each $\rho_2 \in SHIFT_{\Pi_i(j)}^i(\rho')$, it holds that $\rho_2 \equiv_1 \rho_1$ with the same isomorphisms ϕ_{G_0} and ϕ_{G_n} relating their starting and ending graphs.*

If $j \geq 1$, the proof is immediate by exploiting points 1) and 2), because $SHIFT_j^i = SYNT^{i-1} \circ ANAL_j^i$ by Definition 3.5.4. If instead $j = 0$ (and thus $\Pi_i(j) = 0$ as well), then both ρ and ρ' end with an empty direct derivation, and ρ_1 and ρ_2 are uniquely determined by the construction described in Definition 3.5.4, because relation $SHIFT_0^i$ is deterministic. It is easy to check that if this construction is applicable to ρ, then it is applicable to ρ' as well, producing 1-equivalent derivations. □

References

1. H. Ehrig, M. Pfender, and H.J. Schneider. Graph-grammars: an algebraic approach. In *Proceedings IEEE Conf. on Automata and Switching Theory*, pages 167–180, 1973.

2. H. Ehrig. Introduction to the Algebraic Theory of Graph Grammars. In V. Claus, H. Ehrig, and G. Rozenberg, editors, *Proceedings of the 1st International Workshop on Graph-Grammars and Their Application to Computer Science and Biology*, volume 73 of *Lecture Notes in Computer Science*, pages 1–69. Springer Verlag, 1979.
3. H. Ehrig. Tutorial introduction to the algebraic approach of graph-grammars. In H. Ehrig, M. Nagl, G. Rozenberg, and A. Rosenfeld, editors, *Proceedings of the 3rd International Workshop on Graph-Grammars and Their Application to Computer Science*, volume 291 of *Lecture Notes in Computer Science*, pages 3–14. Springer Verlag, 1987.
4. H. Ehrig, M. Korff, and M. Löwe. Tutorial introduction to the algebraic approach of graph grammars based on double and single pushouts. In H. Ehrig, H.-J. Kreowski, and G. Rozenberg, editors, *Proceedings of the 4th International Workshop on Graph-Grammars and Their Application to Computer Science*, volume 532 of *Lecture Notes in Computer Science*, pages 24–37. Springer Verlag, 1991.
5. M. Löwe. Algebraic approach to single-pushout graph transformation. *Theoret. Comput. Sci.*, 109:181–224, 1993.
6. H.-J. Kreowski. Is parallelism already concurrency? Part 1: Derivations in graph grammars. In H. Ehrig, M. Nagl, G. Rozenberg, and A. Rosenfeld, editors, *Proceedings of the 3rd International Workshop on Graph-Grammars and Their Application to Computer Science*, volume 291 of *Lecture Notes in Computer Science*, pages 343–360. Springer Verlag, 1987.
7. H.-J. Kreowski and A. Wilharm. Is parallelism already concurrency? Part 2: Non-sequential processes in graph grammars. In *Proceedings of the 3rd International Workshop on Graph-Grammars and Their Application to Computer Science*, volume 291 of *Lecture Notes in Computer Science*, pages 361–377. Springer Verlag, 1987.
8. P. Padawitz. Graph grammars and operational semantics. *Theoret. Comput. Sci.*, 19:37–58, 1982.
9. B. Hoffmann and D. Plump. Implementing Term Rewriting by Jungle Evaluation. *Informatique théorique et Applications/Theoretical Informatics and Applications*, 25:445–472, 1991.
10. A. Corradini and F. Rossi. Hyperedge Replacement Jungle Rewriting for Term Rewriting Systems and Logic Programming. *Theoret. Comput. Sci.*, 109:7–48, 1993.
11. P. Böhm, H.-R. Fonio, and A. Habel. Amalgamation of graph transformations: a synchronization mechanism. *Journal of Computer and System Science*, 34:377–408, 1987.

12. H. Ehrig, P. Böhm, U. Hummert, and M. Löwe. Distributed parallelism of graph transformation. In *13th Int. Workshop on Graph Theoretic Concepts in Computer Science*, volume 314 of *Lecture Notes in Computer Science*, pages 1–19. Springer Verlag, 1988.
13. H.J. Schneider. Describing distributed systems by categorial graph grammars. In *15th International Workshop on Graph-theoretic Concepts in Computer Science*, volume 411 of *Lecture Notes in Computer Science*, pages 121–135. Springer Verlag, 1990.
14. M. Korff. Graph-interpreted graph transformations for concurrent object-oriented systems. Submitted for publication, 1995.
15. H. Ehrig and R. Bardohl. Specification techniques using dynamic abstract data types and application to shipping software. In *Proc. of the International Workshop on Advanced Software Technology*, pages 70–85, 1994.
16. M. Löwe. Implementing algebraic specifications by graph transformations. *Journal for Information Processing and Cybernetics (EIK)*, 26(11/12):615–641, 1990.
17. A. Habel. *Hyperedge Replacement: Grammars and Languages*, volume 643 of *Lecture Notes in Computer Science*. Springer Verlag, 1992.
18. S. Mac Lane. *Categories for the working mathematician*. Springer Verlag, 1971.
19. J. Adamek, H. Herrlich, and G. Strecker. *Abstract and Concrete Categories*. Wiley Interscience, 1990.
20. H. Ehrig, H.-J. Kreowski, and G. Taentzer. Canonical derivations for high-level replacement systems. In H.-J. Schneider and H. Ehrig, editors, *Proceedings of the Dagstuhl Seminar 9301 on Graph Transformations in Computer Science*, volume 776 of *Lecture Notes in Computer Science*, pages 153–169. Springer Verlag, 1994.
21. H. Ehrig, A. Habel, H.-J. Kreowski, and F. Parisi-Presicce. Parallelism and Concurrency in High-Level Replacement Systems. *Mathematical Structures in Computer Science*, 1:361–404, 1991.
22. W. Reisig. *Petri Nets: An Introduction*. EACTS Monographs on Theoretical Computer Science. Springer Verlag, 1985.
23. A. Corradini, U. Montanari, and F. Rossi. Graph processes. *Fundamenta Informaticae*, 1996. To appear.
24. M. Korff. True Concurrency Semantics for Single Pushout Graph Transformations with Applications to Actor Systems. In *Proceedings International Workshop on Information Systems – Corretness and Reusability, IS-CORE'94*, pages 244–258. Vrije Universiteit Press, 1994. Tech. Report IR-357.

25. M. Korff and L. Ribeiro. Concurrent Derivations as Single Pushout Graph Grammar Processes. In A. Corradini and U. Montanari, editors, *Proceedings SEGRAGRA'95*, volume 2 of *Electronic Notes in Theoretical Computer Science*. Elsevier Sciences, 1995. `http://www.elsevier.nl/locate/entcs/volume2.html`.
26. G. Winskel. An introduction to event structures. In *Linear Time, Branching Time and Partial Order in Logics and Models for Concurrency*, volume 354 of *Lecture Notes in Computer Science*, pages 325–392. Springer Verlag, 1989.
27. G. Schied. On relating rewriting systems and graph grammars to event structures. In H.-J. Schneider and H. Ehrig, editors, *Proceedings of the Dagstuhl Seminar 9301 on Graph Transformations in Computer Science*, volume 776 of *Lecture Notes in Computer Science*, pages 326–340. Springer Verlag, 1994.
28. A. Corradini, H. Ehrig, M. Löwe, U. Montanari, and F. Rossi. An event structure semantics for safe graph grammars. In E.-R. Olderog, editor, *Programming Concepts, Methods and Calculi*, IFIP Transactions A-56, pages 423–444. North-Holland, 1994.
29. A. Corradini, H. Ehrig, M. Löwe, U. Montanari, and F. Rossi. An Event Structure Semantics for Graph Grammars with Parallel Productions. In J. Cuny, H. Ehrig, G. Engels and G. Rozenberg, editors, *Proceedings of the Fifth International Workshop on Graph Grammars and their Application to Computer Science*, volume 1073 of *Lecture Notes in Computer Science*, pages 240–256. Springer Verlag, 1996.
30. R. Heckel, J. Müller, G. Taentzer, and A. Wagner. Attributed graph transformations with controlled application of rules. In G. Valiente and F. Rossello Llompart, editors, *Proc. Colloquium on Graph Transformation and its Application in Computer Science*. Technical Report B - 19, Universitat de les Illes Balears, 1995.
31. A. Corradini and R. Heckel. A Compositional Approach to Structuring and Refinement of Typed Graph Grammars. In A. Corradini and U. Montanari, editors, *Proceedings SEGRAGRA'95*, volume 2 of *Electronic Notes in Theoretical Computer Science*. Elsevier Sciences, 1995. `http://www.elsevier.nl/locate/entcs/volume2.html`.
32. R. Heckel, A. Corradini, H. Ehrig, and M. Löwe. Horizontal and Vertical Structuring of Graph Transformation Systems. *Mathematical Structures in Computer Science*, 1996. To appear.
33. A. Corradini, H. Ehrig, M. Löwe, U. Montanari, and J. Padberg. The Category of Typed Graph Grammars and their Adjunction with Categories of Derivations. In J. Cuny, H. Ehrig, G. Engels and G. Rozenberg,

editors, *Proceedings of the Fifth International Workshop on Graph Grammars and their Application to Computer Science*, volume 1073 of *Lecture Notes in Computer Science*, pages 56–74. Springer Verlag, 1996.

34. M. Korff. Single pushout transformations of equationally defined graph structures with applications to actor systems. In H.-J. Schneider and H. Ehrig, editors, *Proceedings of the Dagstuhl Seminar 9301 on Graph Transformations in Computer Science*, volume 776 of *Lecture Notes in Computer Science*, pages 234–247. Springer Verlag, 1994.
35. R. Heckel and A. Wagner. Ensuring consistency of conditional graph grammars – a constructive approach. In A. Corradini and U. Montanari, editors, *Proceedings SEGRAGRA'95*, volume 2 of *Electronic Notes in Theoretical Computer Science*. Elsevier Sciences, 1995. `http://www.elsevier.nl/locate/entcs/volume2.html`.
36. D. Plump. Hypergraph Rewriting: Critical Pairs and Undecidability of Confluence. In M.R Sleep, M.J. Plasmeijer, and M. C.J.D. van Eekelen, editors, *Term Graph Rewriting*, pages 201–214. Wiley, 1993.
37. D. Plump. On termination of graph rewriting. In M. Nagl, editor, *Proc. 21st Workshop on Graph-Theoretic Concepts in Computer Science (WG '95)*. volume 1017 of *Lecture Notes in Computer Science*, pages 88–100. Springer Verlag, 1995.
38. M. Löwe and J. Müller. Critical pair analysis in single-pushout graph rewriting. In G. Valiente Feruglio and F. Rosello Llompart, editors, *Proc. Colloquium on Graph Transformation and its Application in Computer Science*. Technical Report B-19, Universitat de les Illes Balears, 1995.
39. H. Ehrig and M. Löwe. Categorical principles, techniques and results for high-level replacement systems in computer science. *Applied Categorical Structures*, 1(1):21–50, 1993.
40. M. Löwe. *Extended algebraic graph transformation*. PhD thesis, Technische Universität Berlin, 1990.
41. M. Korff. *Generalized graph structure grammars with applications to object-oriented systems*. PhD thesis, Technical University of Berlin, 1995.
42. J.A. Goguen. A categorical manifesto. *Math. Struc. Comput. Sci.*, 1, 1991.
43. H.-J. Kreowski. *Manipulation von Graphmanipulationen*. PhD thesis, Technische Universität Berlin, 1977.
44. D.B. Benson. The basic algebraic structures in categories of derivations. *Info. and Co.*, 28:1–29, 1975.
45. J. Meseguer and U. Montanari. Petri Nets are Monoids. *Info. and Co.*, 88:105–155, 1990.

46. P. Degano, J. Meseguer, and U. Montanari. Axiomatizing Net Computations and Processes. In *Proceedings 4th Annual Symposium on Logic in Computer Science*, pages 175–185, 1989.
47. G. Ferrari and U. Montanari. Towards the Unification of Models for Concurrency. In *Proceedings CAAP 1990*, volume 431 of *Lecture Notes in Computer Science*, pages 162–176. Springer Verlag, 1990.
48. A. Corradini and U. Montanari. An algebraic semantics for structured transition systems and its application to logic programs. *Theoret. Comput. Sci.*, 103:51–106, 1992.
49. A. Corradini and A. Asperti. A categorical model for logic programs: Indexed monoidal categories. In *Proceedings REX Workshop, Beekbergen, The Netherlands, June 1992*, volume 666 of *Lecture Notes in Computer Science*. Springer Verlag, 1993.
50. A. Corradini, H. Ehrig, M. Löwe, U. Montanari, and F. Rossi. Note on Standard Representation of Graphs and Graph Derivations. In H.-J. Schneider and H. Ehrig, editors, *Proceedings of the Dagstuhl Seminar 9301 on Graph Transformations in Computer Science*, volume 776 of *Lecture Notes in Computer Science*, pages 104–118. Springer Verlag, 1994.
51. A. Corradini, H. Ehrig, M. Löwe, U. Montanari, and F. Rossi. Abstract Graph Derivations in the Double-Pushout Approach. In H.-J. Schneider and H. Ehrig, editors, *Proceedings of the Dagstuhl Seminar 9301 on Graph Transformations in Computer Science*, volume 776 of *Lecture Notes in Computer Science*, pages 86–103. Springer Verlag, 1994.
52. H. Ehrig. Aspects of Concurrency in Graph Grammars. In H. Ehrig, M. Nagl, and G. Rozenberg, editors, *Proceedings of the 2nd International Workshop on Graph-Grammars and Their Application to Computer Science*, volume 153 of *Lecture Notes in Computer Science*, pages 58–81. Springer Verlag, 1983.
53. M. Löwe and J. Dingel. Parallelism in single-pushout graph rewriting. *Lecture Notes in Computer Science 776*, pages 234–247, 1994.
54. E. Ehrig and B. K. Rosen. Parallelism and concurrency of graph manipulations. *Theoretical Computer Science*, 11:247–275, 1980.
55. G. Taentzer. Hierarchically distributed graph transformation. In J. Cuny, H. Ehrig, G. Engels, and G. Rozenberg, editors, *Proceedings of Fifth International Workshop on Graph Grammars and their Applications to Computer Science*, volume 1073 of *Lecture Notes in Computer Science*, pages 304–320. Springer Verlag, 1996.
56. G. Taentzer. *Parallel and Distributed Graph Transformation: Formal Description and Application to Communication-Based Systems*. PhD thesis, Technical University Berlin, FB13, 1996.
57. H. Ehrig and M. Löwe. Parallel and distributed derivations in the single pushout approach. *TCS*, 109:123 – 143, 1993.

Chapter 4

ALGEBRAIC APPROACHES TO GRAPH TRANSFORMATION

PART II: SINGLE PUSHOUT APPROACH AND COMPARISON WITH DOUBLE PUSHOUT APPROACH

H. EHRIG, R. HECKEL, M. KORFF, M. LÖWE, L. RIBEIRO, A. WAGNER
Technische Universität Berlin, Fachbereich 13 Informatik, Franklinstraße 28/29, D-10587 Berlin, Germany

A. CORRADINI
Dipartimento di Informatica, Corso Italia 40, I-56125 Pisa, Italy

The algebraic approaches to graph transformation are based on the concept of gluing of graphs corresponding to pushouts in suitable categories of graphs and graph morphisms. This allows one to give not only an explicit algebraic or set theoretical description of the constructions but also to use concepts and results from category theory in order to build up a rich theory and to give elegant proofs even in complex situations.
In the previous chapter we have presented an overview of the basic notions and problems common to the two algebraic approaches, the double-pushout (DPO) approach and the single-pushout (SPO) approach, and their solutions in the DPO approach. In this chapter we introduce the SPO approach to graph transformation and some of its main results. We study application conditions for graph productions and the transformation of more general structures than graphs in the SPO approach, where similar generalizations have been or could be studied also in the DPO approach. Finally, we present a detailed comparison of the DPO and the SPO approach, especially concerning the solutions to the problems discussed for both approaches in the previous chapter.

Contents

4.1 Introduction

The algebraic approach of graph grammars has been invented at TU Berlin in the early 70'ies by H. Ehrig, M. Pfender and H.J. Schneider in order to generalize Chomsky grammars from strings to graphs [1]. Today this algebraic approach is called double-pushout approach, because it is based on two pushout constructions, in contrast to the single-pushout (SPO) approach where direct derivations are defined by a single pushout in the category of graphs and partial graph morphisms. A general introduction with main application areas and an overview of both algebraic approaches is presented in Chapter 3, in this handbook.

The main aim of this chapter is to provide an introduction to the basic notions of the SPO approach, to present some of its main results and relevant extensions and generalizations, and to compare the notions and results of the SPO approach with the corresponding ones presented in Chapter 3 for the DPO approach. After this introduction, we suggest to read first Section 3.2 where many relevant concepts and results common to both algebraic approaches are introduced informally, in terms of problems.

The following section gives an introduction to the basic notions of the SPO approach, like production, match, and (direct) derivation, and describes the historical roots of the SPO approach. In Section 4.3 we give answers to the problems of Section 3.2, concerning independence and parallelism, embedding of derivations, and amalgamation and distribution of derivations. In several graph grammar approaches it is possible to formulate application conditions for individual productions. But in very few cases the theoretical results can be extended to include these conditions. In Section 4.4 the SPO approach and the results concerning independence and parallelism are extended in order to allow for user-defined application conditions, as recently done in [2]. In Section 4.5 we show how the SPO approach can be applied to more general kinds of graphs and structures like attributed graphs [3], graph structures [4] (including for example hypergraphs), generalized graph structures [5], and even more general structures in high-level replacement systems [6].

In Section 4.6 we compare the main concepts and results of the two approaches, presented in Chapter 3 and this chapter, w.r.t. the problems stated in the overview of both approaches in Section 3.2. Finally in the conclusion we point out that both approaches are suitable for different application areas and discuss some main ideas for further research.

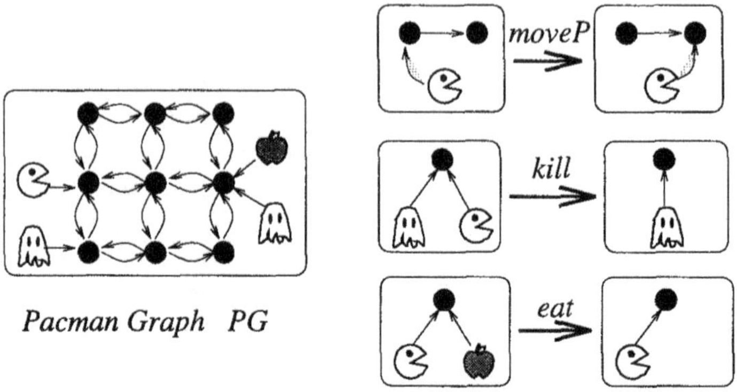

Figure 4.1: Pacman Game

4.2 Graph Transformation Based on the SPO Construction

In this section we introduce the basic notions of the single-pushout approach to graph transformation. In order to be able to follow the formal definitions of this section the reader should be familiar with the notions introduced in Section 3.3 for the double-pushout approach, although the informal motivation in the following can be understood without any preliminary knowledge.

4.2.1 Graph Grammars and Derivations in the SPO Approach

As introductory example we use the model of a (simple version of a) Pacman game.

Example 4.1 (Pacman game)
The game starts with a number of figures (Pacman, ghosts, and apples) placed on a board. Pacman and each of the ghosts may autonomously move from field to field (up, down, left, right). Apples do not move. Pacman wins if he manages to eat all the apples before he is killed by one of the ghosts.
Each of the states of the Pacman game can easily be modeled as a graph (see Figure 4.1). Vertices are used to represent Pacman, ghosts and apples as well as the fields of the board. These different kinds of vertices are distinguished using different graphical layouts (the obvious ones for Pacman, ghosts and apples and the black dots for the fields). Edges pointing from a vertex representing a figure to a field vertex model the current position of the figure on the board. Neighborhood relations between fields are explicitly represented using edges, too.

Each possible activity in the Pacman game causes a change of the current state and hence is modeled by a transformation of the graph representing this state. To describe which activities, i.e., which graph transformations, are possible we use productions. The production *moveP* in Figure 4.1 describes that Pacman is moving one field. All objects are preserved but the edge pointing from the Pacman vertex to the field vertex is deleted and a new one is inserted. The production can be applied to a graph if Pacman is on a field which has a neighbor. As the effect of the application of this production, Pacman moves to a new position while everything else is left unchanged. Such an application of a production to a current graph is called derivation or transformation. To move the ghosts we have a production (*moveG*) which is not drawn but which follows the same scheme. The production *eat* can be applied when Pacman and an apple are on the same field. The result is that the apple and its outgoing edge are deleted. Analogously if Pacman and a ghost are on the same field the production *kill* can be applied and Pacman is deleted. □

Graphs and (total) graph morphisms have been formally introduced in Definition 3.3.1. In contrast to spans of total morphisms in the DPO approach, we use partial morphisms in order to describe, for example, the relationship between the given and derived graph in a derivation. The aim is to model a direct derivation by a single pushout instead of two.

Definition 4.2.1 (partial graph morphism)
Let $G = (G_V, G_E, s^G, t^G, lv^G, le^G)$ be a graph. Recall that G_V, G_E denote the sets of vertices and edges of G, s^G, t^G its source and target mappings, and lv^G, le^G the label assignments for vertices and edges, respectively. A **subgraph** S of G, written $S \subseteq G$ or $S \hookrightarrow G$, is a graph with $S_V \subseteq G_V$, $S_E \subseteq G_E$, $s^S = s^G|_{S_E}$, $t^S = t^G|_{S_E}$, $lv^S = le^G|_{S_E}$, and $le^S = le^G|_{S_E}$. A **(partial) graph morphism** g from G to H is a total graph morphism from some subgraph $dom(g)$ of G to H, and $dom(g)$ is called the **domain** of g.

By defining composition of these morphisms by composition of the components, and identities as pairs of component identities[a], the graphs over a fixed labeling alphabet and the partial morphisms among them form a category denoted by $\boldsymbol{Graph^P}$.
Usually, a production $L \xrightarrow{p} R$ consists of two graphs L and R, called the left- and the right-hand side, respectively, and a partial morphism p between them. The idea is to describe in the left-hand side which objects a graph must contain such that the production can be applied, i.e., all objects (vertices and edges)

[a] Note that a partial morphism can also be considered as a pair of partial functions.

which shall be deleted and the application context which shall be preserved. The right-hand side describes how this part of the graph shall look like after the transformation, i.e., it consists of all the objects which are added together with the application context. The morphism p makes clear which objects in the left-hand side correspond to which ones in the right-hand side. Objects where the morphism is undefined are deleted. All objects which are preserved by the morphism form the application context. If an object in the right-hand side of the production has no preimage under the morphism, it is added. Note that the role of the application context coincides with that of the interface graph K in the DPO productions (see Definition 3.3.2). In the formal definition below the concept of production is enriched with a name, which is used to describe the internal structure of composed productions as, for example, the parallel production (see also the corresponding Definition 3.3.2 for the DPO approach).

Definition 4.2.2 (production, graph grammar)
A **production** $p : (L \stackrel{r}{\rightarrow} R)$ consists of a **production name** p and of an injective partial graph morphism r, called the **production morphism**. The graphs L and R are called the left- and the right-hand side of p, respectively. If no confusion is possible, we will sometimes make reference to a production $p : (L \stackrel{r}{\rightarrow} R)$ simply as p, or also as $L \stackrel{r}{\rightarrow} R$.
A **graph grammar** $\mathcal{G}$ is a pair $\mathcal{G} = \langle (p : r)_{p \in P}, G_0 \rangle$ where $(p : r)_{p \in P}$ is a family of production morphisms indexed by production names, and G_0 is the **start graph** of the grammar.

Since production names are used to identify productions, for example, in direct derivations, they should be unique, i.e., there must not be two different productions with the same name. This is ensured by the above definition of graph grammar. Later on, production names will be used to store the structure of productions which are composed from elementary productions. In such cases the production name is usually a diagram, consisting of the elementary productions and their embeddings in the composed production. In elementary productions, production names are sometimes disregarded. Then we just write $p : L \rightarrow R$ in order to denote the production $p : (L \stackrel{p}{\rightarrow} R)$ where the production morphism p is also used as the name of the production. In this section, all the productions are elementary and we use the notation just introduced.

Example 4.2 (production)
The figure below shows a production r which inserts a loop and deletes two vertices ($\bullet_2$ and $\bullet_3$ resp.). We consider the labeling alphabet to contain only one element $*$, i.e., the graph can be seen as unlabeled. The application context consists of the vertex $\bullet_1$. As indicated by the dashed line, r is defined for vertex

$\bullet_1$ mapping it to the only vertex on the right hand side. r is undefined for the other two vertices. The numbers are used to illustrate the morphism such that we can omit the dashed lines later on if graphs are more complex. Hence graph L can for example be formally denoted by $L = (\{\bullet_1, \bullet_2, \bullet_3\}, \emptyset, s^L, t^L, lv^L, le^L)$, where s^L, t^L and le^L are empty functions. lv^L is defined by $lv^L(\bullet_i) = *$ for $i = 1..3$.

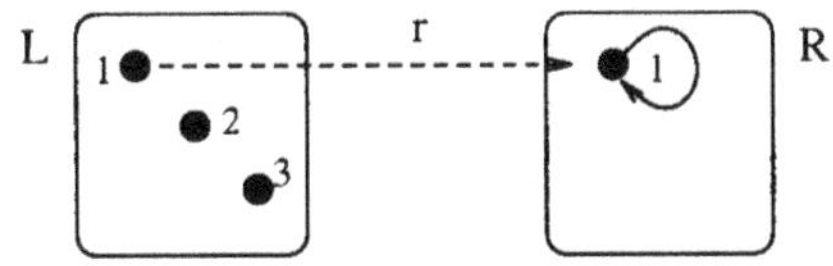

□

Using partial instead of total morphisms, the categorical notion of pushout captures not only the intuitive idea of gluing but that of gluing and deletion. Therefore the construction is much more complicated. In order to make it easier to understand we divided it into three steps. The first two steps are gluings as known from Section 3.3. In the third step the deletion is performed. Intuitively, deletion is not seen as inverse to addition here, but as equalizing morphisms having different domains of definition. Two morphisms with the same source and target objects are reduced to their equal kernel by removing all those items from their range which have different preimages under the morphisms. Formally this is captured by the categorical notion of a co-equalizer.

Definition 4.2.3 (co-equalizer)
Given a category C and two arrows $a : A \to B$ and $b : A \to B$ of C, a tuple $\langle C, c : B \to C\rangle$ is called **co-equalizer** of $\langle a, b\rangle$ if $c \circ a = c \circ b$ and for all objects D and arrows $d : B \to D$, with $d \circ a = d \circ b$, there exists a unique arrow $u : C \to D$ such that $u \circ c = d$.

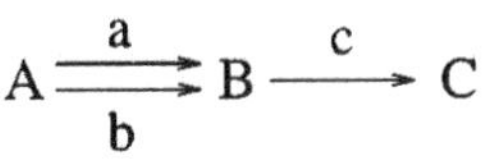

For our pushout construction of partial morphism a very specific co-equalizer is sufficient, which is constructed in the following.

Construction 4.2.4 (specific co-equalizer in $Graph^P$)
Let $a, b : A \to B$ be two (partial) morphisms such that for each $x \in A$, if both a and b are defined on x then $a(x) = b(x)$. Then, the co-equalizer of a and b in $\boldsymbol{Graph^P}$ is given by $\langle C, c\rangle$ where

- *$C \subseteq B$ is the largest subgraph of $[b(A) \cap a(A)] \cup [\overline{a(A)} \cap \overline{b(A)}]$, and*
- *$dom(c) = C$ and c is the identity morphism on its domain.*

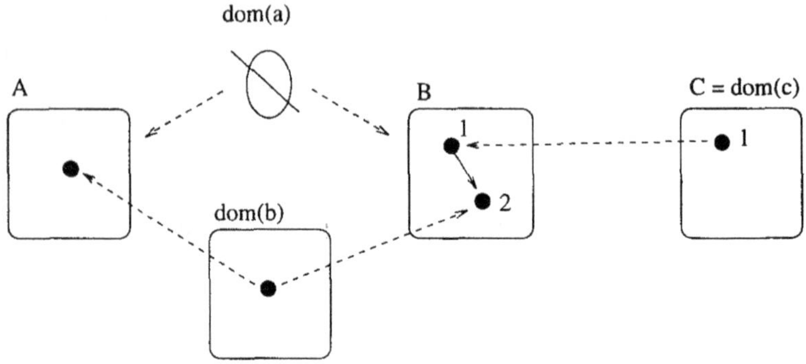

Figure 4.2: Deletion as a co-equalizer

Proof
It is obvious that $c \circ a = c \circ b$. The universal co-equalizer property follows from the fact that for each other morphism $d : B \to D$ with $d \circ a = d \circ b$ we have $dom(d) \subseteq dom(c)$. □

Example 4.3 (deletion as a co-equalizer)
Figure 4.2 shows the co-equalizer of the empty morphism a and a total morphism b in $\boldsymbol{Graph^P}$. According to Construction 4.2.4, C is the maximal subgraph of B such that for all $x \in C$, either $x \in a(A) \cap b(A)$, or $x \notin a(A)$ and $x \notin b(A)$. Thus, vertex 2 is deleted because it has a preimage under b but not under a. In order to obtain a graph, the dangling edge is deleted, too. □

With these preliminaries we are able to construct the pushout of two partial morphisms in $\boldsymbol{Graph^P}$.

Proposition 4.2.5 (pushout in $\boldsymbol{Graph^P}$)
The pushout of two morphisms $b : A \to B$ and $c : A \to C$ in $\boldsymbol{Graph^P}$ always exists and can be computed in three steps, as shown in Figure 4.3:

[gluing 1] *Construct the pushout $\langle C', A \to C', C \to C' \rangle$ of the total morphisms $dom(c) \to C$ and $dom(c) \to A$ in* ***Graph*** *(cf. Example 3.3).*

[gluing 2] *Construct the pushout $\langle D, B \to D, C' \to D \rangle$ of the total morphisms $dom(b) \to A \to C'$ and $dom(b) \to B$ in* ***Graph****.*

[deletion] *Construct the co-equalizer $\langle E, D \to E \rangle$ of the partial morphisms $A \to B \to D$ and $A \to C \to C' \to D$ in $\boldsymbol{Graph^P}$.*

$\langle E, C \to C' \to D \to E, B \to D \to E \rangle$ is the pushout of b and c in $\boldsymbol{Graph^P}$.

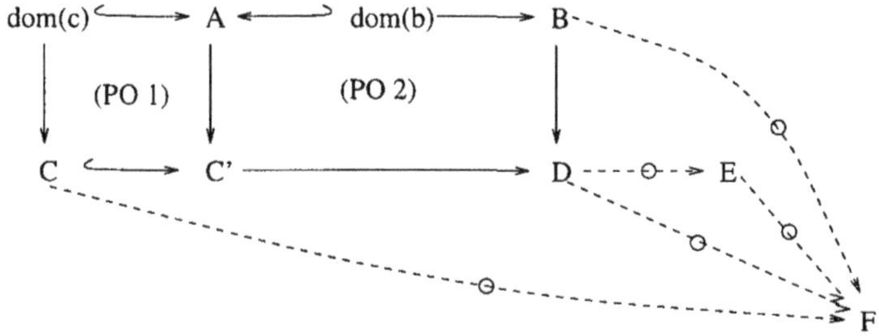

Figure 4.3: Construction of pushout in $\boldsymbol{Graph^P}$

Proof
Commutativity holds by construction. Assume that there is $\langle F, C \to F, B \to F\rangle$ with $A \to B \to F = A \to C \to F$. Prove that $dom(b) \to B \to F = dom(b) \to A \to C' \to F$. Use the fact that pushouts in ***Graph***are pushouts in $\boldsymbol{Graph^P}$ [4] to construct a universal morphism $D \to F$. With the universal co-equalizer property we obtain a unique morphism $E \to F$ with $B \to D \to E \to F = B \to F$ and $C \to C' \to D \to E \to F = C \to F$. □

To apply a production to a graph G, one has to make sure that G contains at least all objects necessary for performing the desired operations and the application context. These are exactly the objects of the production's left-hand side L. Hence we introduce the notion of a match as a total morphism matching the left-hand side of the production with (a part of) G. The match must be total because otherwise not all needed objects have a corresponding image in the graph G. We allow that different items from L are mapped onto the same item in G. This avoids additional productions in some cases. The actual transformation of G is performed by removing the part matched by the production's left-hand side and adding the right-hand side. The result is the derived graph H.

Definition 4.2.6 (match, derivation)
A **match** for $r : L \to R$ in some graph G is a total morphism $m : L \to G$. Given a production r and a match m for r in a graph G, the **direct derivation** from G with r at m, written $G \overset{r,m}{\Longrightarrow} H$, is the pushout of r and m in $\boldsymbol{Graph^P}$, as shown in Figure 4.4. A sequence of direct derivations of the form $\rho = (G_0 \overset{r_1,m_1}{\Longrightarrow} \ldots \overset{r_k,m_k}{\Longrightarrow} G_k)$ constitutes a **derivation** from G_0 to G_k by $r_1, \ldots, r_k$, briefly denoted by $G_0 \Longrightarrow^* G_k$. The **graph language** generated by a graph grammar $\mathcal{G}$ is the set of all graphs G_k such that there is a derivation $G_0 \Longrightarrow^* G_k$ using productions of $\mathcal{G}$.

$$\begin{array}{ccc} L & \xrightarrow{r} & R \\ {\scriptstyle m}\downarrow & (PO) & \downarrow{\scriptstyle m^*} \\ G & \xrightarrow[r^*]{} & H \end{array}$$

Figure 4.4: Direct derivation as pushout in $\boldsymbol{Graph^P}$.

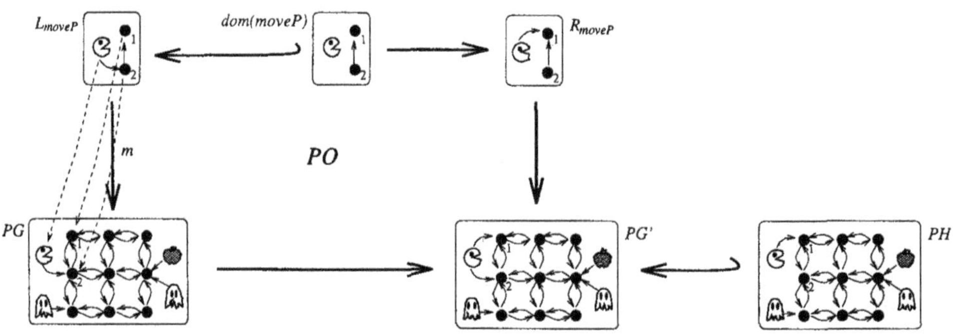

Figure 4.5: Direct derivation (example).

To construct the direct derivation from graph G with r at m we use Proposition 4.2.5. Since match m is a total morphism, step one of the construction is trivial and can be omitted.

Example 4.4 (direct derivation as pushout construction)
The production *moveP* from Example 4.1 can be applied to the graph PG (Figure 4.1) at match m, leading to the direct derivation shown in Figure 4.5. Remember that due to the totality of the match the first step of the construction given in Proposition 4.2.5 is trivial. Hence we start with the second step: An edge between Pacman and the vertex modeling the field he moves to is added by performing the pushout of the total morphisms $dom(moveP) \to L_{moveP} \to PG$ and $dom(moveP) \to R_{moveP}$. Afterwards the co-equalizer of $L_{moveP} \to R_{moveP} \to PG'$ and $L_{moveP} \to PG \to PG'$ leads to the derived graph PH which is the subgraph of PG' where the edge between Pacman and the vertex representing his old position is deleted. □

In the example derivation above deletion is rather intuitive. In contrast Figure 4.6 depicts a situation where a vertex is meant to be preserved as well as deleted on the left. The result of this derivation is the empty graph, i.e., deletion has priority. This property of the approach is caused by the order of

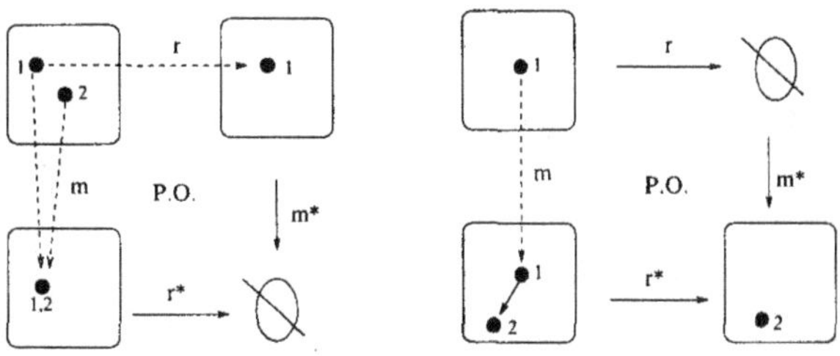

Figure 4.6: Deletion with conflicts.

the construction steps: first gluing of conflicting items and then their deletion. Similarly the problem of the dangling edge on the right of Figure 4.6 is solved: it is deleted, too. This is illustrated in Example 4.3.

In the following we will characterize simple deletion by properties of the match. If a match is *d(elete)-injective* then conflict situations are avoided. If it is *d-complete*, deletion does not cause the implicit deletion of dangling edges.

Definition 4.2.7 (special matches)
Given a match $m : L \to G$ for a production $r : L \to R$. Then m is called **conflict-free** if $m(x) = m(y)$ implies $x, y \in dom(r)$ or $x, y \notin dom(r)$. It is called **d-injective** if $m(x) = m(y)$ implies $x = y$ or $x, y \in dom(r)$. Finally, m is **d-complete** if for each edge $e \in G_E$ with $s^G(e) \in m_V(L_V - dom(r)_V)$ or $t^G(e) \in m_V(L_V - dom(r)_V)$ we have $e \in m_E(L_E - dom(r)_E)$.

Lemma 4.2.8 (pushout properties)
If $(H, r^* : G \to H, m^* : R \to H)$ is the pushout of $r : L \to R$ and $m : G \to H$ in $\boldsymbol{Graph^P}$, then the following properties are fulfilled:

1. *Pushouts preserve surjectivity, i.e. r surjective implies r^* surjective.*
2. *Pushouts preserve injectivity, i.e. r injective implies r^* injective.*
3. *r^* and m^* are jointly surjective.*
4. *m conflict-free implies m^* total.*

Proof
See [7]. ⧉

4.2.2 Historical Roots of the SPO Approach

Single-pushout transformations in a setting of some sort of partial morphisms have been investigated already by Raoult [8] and Kennaway [9]. The following historical roots are taken from [4].

Raoult [8] introduces two conceptually very different approaches. The first one is described in the category of sets and partial mappings. A rule is a partial morphism $r : L \to R$, i.e. a partial map which respects the graph structure on all objects of L it is defined for. A match $m : L \to G$ in some graph G is a total morphism of this type. The result of applying r at m is constructed in two steps. First, the pushout $(H, r^* : G \to H, m^* : R \to H)$ of r and m in the category of sets and partial maps is built. In the second step, a graph structure is established on H such that the pushout mappings r^* and m^* become morphisms. He characterizes the situations in which this graph structure uniquely exists; double-pushout transformations with gluing conditions are special cases of these situations. The second model of graph transformation in [8] uses another kind of partiality for the morphisms: a rule is a total map $r : L \to R$, which is only partially compatible with the graph structure. Let $rewrite(r)$ denote the set of objects which are not homomorphically mapped by r. A match $m : L \to G$ is total which means now $rewrite(m) = \emptyset$. Application of r at m is again defined by two steps. First construct the pushout $(H, r^* : G \to H, m^* : R \to H)$ of r and m in the category of sets and total mappings and then impose a graph structure on H such that the pushout mappings become as compatible as possible, i.e. such that $rewrite(r^*) = m(rewrite(r))$ and $rewrite(m^*) = r(rewrite(m))$. Raoult [8] gives sufficient conditions for the unique existence of this structure. This approach has the major disadvantage that objects cannot be deleted at all.

Kennaway [9] provides a categorical description for the second approach of [8]. Graphs are represented the same way. Morphisms $f : A \to B$ are pairs (f, hom). The first component is a total mapping from A to B. The second component provides a subset of A on which f respects the graph structure. A rule $r : L \to R$ is any morphism in this sense and a match $m : L \to G$ is a total morphism which now means $hom = L$. He shows that under certain conditions the two-step construction of [8] coincides with the pushout construction in the category of graphs and the so-defined morphisms.

Unfortunately, only sufficient conditions for the existence of pushouts are given. Besides that, object deletion remains impossible. The concept in [9] has been further developed in [10]. They introduce "generalized graph rewriting" which uses the same kind of graph morphism. The corresponding transformation concept not only involves a pushout construction but also a coequalizer. Since both constructions are carried out in different categories (of total resp. partial morphisms) theoretical results are difficult to obtain. The SPO approach in [4] as discussed above is closely related to the first approach in [8]. His concept of partial mappings which are compatible with the graph structure on their domain can be generalized to a concept of partial homomorphisms on special categories of algebras such that pushout construction in the categories is always possible. Hence, we get rid of any application conditions. If, however, the necessary and sufficient conditions of [8] are satisfied, the construction of pushout objects coincides with his two-step construction.

Recently, Kennaway [11] independently started to study graph transformation in some categories of partial morphisms of this type. His work is based on the categorical formulation of a partial morphism provided by [12]. While [4] considers concrete algebraic categories, [11] stays in a purely categorical framework. Future research has to show how both approaches can benefit from each other. Van den Broek [13] introduces another kind of single-pushout transformations based on "partial" morphisms. Partiality in this framework is described by total morphisms which map objects "outside their domain" to marked objects in their codomain. In this chapter we follow the SPO approach as introduced in [7,4] and extended in several subsequent papers mentioned below.

4.3 Main Results in the SPO Approach

In this section we present some of the main results of the SPO approach. They are concerned with the conceptual notions of parallelism, context embedding, synchronization and distribution, interpreted into the world of SPO derivations. The main reason for this choice is the fact that a most natural and very basic view on a graph transformation system is to see a graph as a system state and a production as a syntactical description of corresponding state transformations obtained by direct derivations. By further (informal) arguments certain derivations can be classified as to be concurrent, embedded into each other, or (somehow) synchronized or distributed.

Complementary, on the syntactical level, different ways of composing productions can be considered, generating an integrated description, i.e., the composed production. As for derivations, we consider parallel, derived, and synchronized/amalgamated productions. An elementary production essentially provides a description of the direct derivations it may perform. A composed production, additionally, contains all the information about its construction from elementary productions, which is stored in the production name. If, e.g.,

$p : r$ is a parallel production, the production name p contains the elementary productions and their embeddings into the resulting production morphism r. The (complete or partial) correspondence between composed derivations and derivations using (correspondingly) composed productions provides the essential results of this section, which may be seen as answers to the problems stated in Section 3.2. Related results are mentioned briefly.

4.3.1 Parallelism

There are essentially two different ways to model concurrent computations, the interleaving and the truly concurrent model. This subsection provides the basic concepts of these models in the world of SPO graph rewriting, including the solutions to the Local Church-Rosser Problem 3.2.1 and the Parallelism Problem 3.2.2.

Interleaving

In an interleaving model two actions are considered to be concurrent (i.e., potentially in parallel) if they may occur in any order with the same result. Modeling single actions by direct derivations, we introduce two notions of independence, formalizing this concept from two different points of view. The condition of *parallel independence* shall ensure that two *alternative* direct derivations are not mutually exclusive. Safely, we may expect this if these direct derivations act on totally disjoint parts of the given graph. More precisely, a direct derivation $d_1 = (G \overset{p_1,m_1}{\Longrightarrow} H_1)$ does not affect a second $d_2 = (G \overset{p_2,m_2}{\Longrightarrow} H_2)$ if d_1 does not delete elements in G which are accessed by d_2. In other words, the overlapping of the left hand sides of p_1 and p_2 in G must not contain elements which are deleted by p_1. The vertices deleted from G by d_1 are those in $m_1(L_1 - dom(p_1))$. An edge will be deleted from G if it is in $m_1(L_1 - dom(r_1))$ or - which is an additional feature in the SPO approach - one of its incident vertices is deleted. Formally, we obtain the following definition of weakly parallel independent derivations.

Definition 4.3.1 (parallel independence)
Let $d_1 = (G \overset{p_1,m_1}{\Longrightarrow} H_1)$ and $d_2 = (G \overset{p_2,m_2}{\Longrightarrow} H_2)$ be two alternative direct derivations. Then we say that d_2 is **weakly parallel independent** of d_1 iff $m_2(L_2) \cap m_1(L_1 - dom(p_1)) = \emptyset$. We call the derivations d_1 and d_2 **parallel independent** if they are mutually weakly parallel independent.

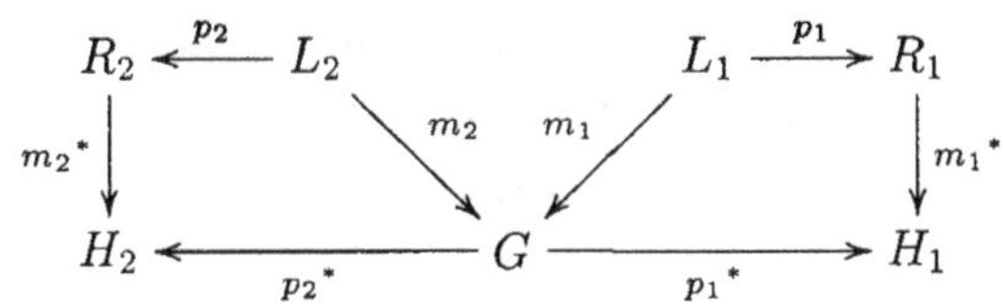

Weak parallel independence can be characterized in more abstract terms on a categorical level.

Characterization 4.3.2 (parallel independence)
Let $d_1 = (G \overset{p_1,m_1}{\Longrightarrow} H_1)$ *and* $d_2 = (G \overset{p_2,m_2}{\Longrightarrow} H_2)$ *be two direct derivations. Then* d_2 *is weakly parallel independent of* d_1 *if and only if* $p_1{}^* \circ m_2 : L_2 \to H_1$ *is a match for* p_2.

Proof
Let there be a match $m_2' = p_1{}^* \circ m_2$ as above. Then $m_2(L_2) \subseteq dom(p_1{}^*)$ by definition of composition. The commutativity of pushouts provides $m_1(L_1 - dom(p_1)) \cap dom(p_1{}^*) = \emptyset$ which is the desired weak parallel independence.

Vice versa, we must show that $m_2' = p_1{}^* \circ m_2$ is total. According to the construction of a derivation, each vertex which is not explicitly deleted by p_1 is preserved, i.e., $v \in G - m_1(L_1 - dom(p_1))$ implies $v \in dom(p_1{}^*)$. This implies that each preimage of v under m_2 has also an image in H_1. Each edge which is not explicitly deleted by p_1 is either (i) preserved or (ii) implicitly deleted by the deletion of an incident vertex i.e., for each edge $e \in G - m_1(L_1 - dom(p_1))$ we either have (i) $e \in dom(p_1{}^*)$ which inherits the arguments for vertices; Otherwise, in case (ii), $s^G(e) \in m_1(L_1 - dom(p_1))$ or $t^G(e) \in m_1(L_1 - dom(p_1))$. By definition of parallel independence this implies $s^G(e) \notin m_2(L_2)$ or $t^G(e) \notin m_2(L_2)$ which, by definition of graph morphisms excludes $e \in m_2(L_2)$. In other words $e \in m_2(L_2)$ implies $e \in dom(p_1{}^*)$. □

The condition of ***sequential independence*** shall ensure that two ***consecutive*** direct derivations are not causally dependent. A direct derivation is weakly sequential independent of a preceding one if it could already have been performed before that. Analogous to the parallel case above, weak sequential independence requires that the overlapping of the right hand side of the first production and the left hand side of the next must not contain elements which were generated by the first. The idea of the stronger notion of sequential independence is that additionally the second will not delete anything which was needed by the first.

Definition 4.3.3 (sequential independence)
Let $d_1 = (G \overset{p_1,m_1}{\Longrightarrow} H_1)$ and $d_2' = (H_1 \overset{p_2,m_2'}{\Longrightarrow} X)$, be two consecutive direct derivations. Then we say that d_2' is **weakly sequentially independent** of d_1

if $m_2'(L_2) \cap m_1{}^*(R_1 - p_1(L_1)) = \emptyset$. If additionally $m_2'(L_2 - dom(p_2)) \cap m_1{}^*(R_1) = \emptyset$, we say that d_2' is **sequentially independent** of d_1, and the derivation $(G \overset{p_1,m_1}{\Longrightarrow} H_1 \overset{p_2,m_2'}{\Longrightarrow} X)$ is called **sequentially independent**.

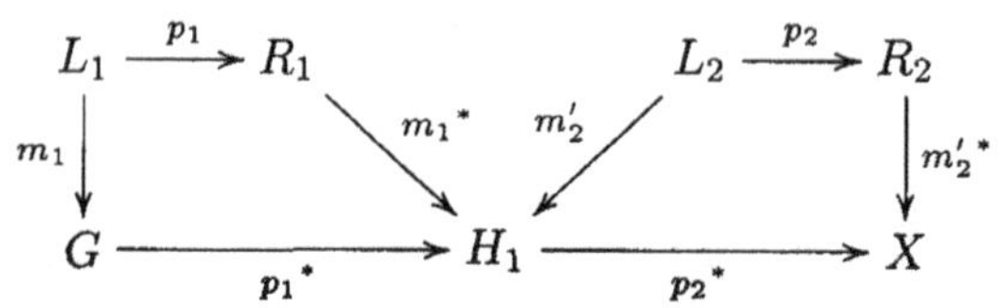

Again we will provide a categorical characterization. The formulation of this statement is analogous to the case of weak parallel independence. The proof has therefore been omitted

Characterization 4.3.4 (sequential independence)
Assume two direct derivations $d_1 = (G \overset{p_1,m_1}{\Longrightarrow} H_1)$ *and* $d_2' = (H_1 \overset{p_2,m_2'}{\Longrightarrow} X)$ *to be given. Then* d_2' *is weakly sequentially independent of* d_1 *iff there is a match* $m_2 : L_2 \to G$ *for* p_2 *such that* $m_2' = p_1{}^* \circ m_2$. *The derivation* d_2' *is sequentially independent of* d_1 *iff* d_2' *is weakly sequentially independent of* d_1 *and* d_1 *is weakly parallel independent of the correspondingly existing derivation* $d_2 = (G \overset{p_2,m_2}{\Longrightarrow} H_2)$. □

By definition, the weakly parallel independence of two direct derivations implies the existence of consecutive direct derivations. The definition of weakly sequential independent derivations contains a symmetric implication. The argumentation could be summarized in the following way: weak parallel independence allows to delay a derivation — complementary, weak sequential independence allows to anticipate a derivation. Formally this is captured by the following lemma.

Lemma 4.3.5 (weak independence)
Given a direct derivation $d_1 = (G \overset{p_1,m_1}{\Longrightarrow} H_1)$, *the following statements are equivalent:*

1. *There is a direct derivation* $d_2 = (G \overset{p_2,m_2}{\Longrightarrow} H_2)$ *which is weakly parallel independent of* d_1.

2. *There is a direct derivation* $d_2' = (H_1 \overset{p_2,m_2'}{\Longrightarrow} X)$ *which is weakly sequentially independent of* d_1.

$$
\begin{array}{ccccc}
 & & L_1 & \xrightarrow{p_1} & R_1 \\
 & & \downarrow m_1 & (1) & \downarrow m_1^* \\
L_2 & \xrightarrow{m_2} & G & \xrightarrow{p_1^*} & H_1 \\
\downarrow p_2 & (2) & \downarrow p_2^* & (3) & \downarrow m_1^{**} \\
R_2 & \xrightarrow{m_2^*} & H_2 & \xrightarrow{m_1^{**}} & X
\end{array}
$$

Figure 4.7: Local Church Rosser

Up to isomorphism, a bijective correspondence between 1. and 2. above is given by $m_2' = p_1{}^* \circ m_2$ *and* $m_2 = (p_2{}^*)^{-1} \circ m_2'$.

Proof

This is a direct consequence of Characterizations 4.3.2 and 4.3.4 together with the fact that, by Lemma 4.2.8, injectivity of productions implies injectivity of $p_1{}^*$ and thus $m_2 = (p_1{}^*)^{-1} \circ p_1{}^* \circ m_2$ as well as $m_2' = p_1{}^* \circ (p_1{}^*)^{-1} \circ m_2'$. □

According to our interpretation, the conditions of parallel and sequential independence formalize the concepts of concurrent computation steps. In a more abstract sense, two such steps are concurrent if they may be performed in any order with the same result. The following Local Church-Rosser Theorem shows the correctness of the characterization of concurrent direct derivations by their parallel and sequential independence. It provides the solution to the Local Church-Rosser Problem 3.2.1.

Theorem 4.3.6 (local Church-Rosser)

Let $d_1 = (G \overset{p_1,m_1}{\Longrightarrow} H_1)$ *and* $d_2 = (G \overset{p_2,m_2}{\Longrightarrow} H_2)$ *be two direct derivations. Then the following statements are equivalent:*

1. *The direct derivations* d_1 *and* d_2 *are parallel independent.*

2. *There is a graph* X *and direct derivations* $H_1 \overset{p_2,m_2'}{\Longrightarrow} X$ *and* $H_2 \overset{p_1,m_1'}{\Longrightarrow} X$ *such that* $G \overset{p_1,m_1}{\Longrightarrow} H_1 \overset{p_2,m_2'}{\Longrightarrow} X$ *and* $G \overset{p_2,m_2}{\Longrightarrow} H_2 \overset{p_1,m_1'}{\Longrightarrow} X$ *are sequentially independent derivations.*

Up to isomorphism, a bijective correspondence between 1. and 2. above is given by $m_2' = p_1{}^* \circ m_2$ *and* $m_1' = p_2{}^* \circ m_1$.

Proof

Consider the diagram in Figure 4.7. Subdiagrams (1) and (2) depict the derivations d_1 and d_2, respectively. Subdiagram (3) represents the pushout of $p_1{}^*$ and

$p_2{}^*$. The composition of pushout diagrams (2) and (3) yields a pushout diagram (2)+(3) which is a derivation $H_1 \overset{p_2,m_2'}{\Longrightarrow} X$ provided that $m_2' = p_1{}^* \circ m_2$ is a match, i.e., total. But this is ensured since d_1 and d_2 have been required to be parallel independent. Analogously we obtain a derivation $H_2 \overset{p_1,m_1'}{\Longrightarrow} X$ by composing pushout diagrams (1) and (3). The stated sequential independence and the bijective correspondence follow from Lemma 4.3.5. □

Explicit Parallelism

In contrast to the interleaving model above, a *truly* parallel computation step has to abstract from any possible interleaving order, i.e., it must not generate any intermediate state. Therefore, a parallel direct derivation is a *simultaneous* application of productions, which are combined into a single ***parallel production***. Constructing this production as the disjoint union of the given elementary productions reflects the view that the overall effect of a parallel production application can be described by a 'most independent' combination of both component descriptions. The formal definition below uses the fact that the disjoint union of graphs (obtained from the disjoint union of carrier sets) can categorically be characterized by a coproduct. The construction of the parallel production as a coproduct of elementary productions is recorded in the production name $p_1 + p_2$, which is given by the coproduct diagram below. This notation is well-defined, i.e., the diagram below is uniquely determined by the term $p_1 + p_2$, if we fix a certain coproduct construction (like, for example, the disjoint union).

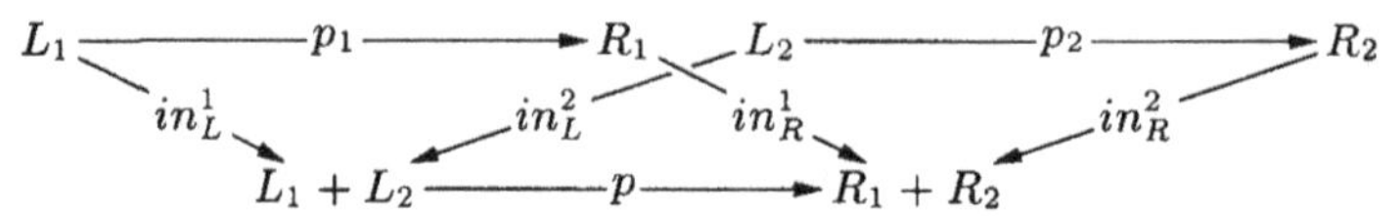

Definition 4.3.7 (parallel productions and derivations)
Given two productions $p_1 : L_1 \rightarrow R_1$ and $p_2 : L_2 \rightarrow R_2$, the **parallel production** $p_1 + p_2 : (L_1 + L_2 \overset{p}{\longrightarrow} R_1 + R_2)$ is composed of the production name $p_1 + p_2$, i.e., the diagram above, and the associated partial morphism p. The graphs $L_1 + L_2$ and $R_1 + R_2$ (together with the corresponding injections in_L^1, in_L^2, in_R^1, and in_R^2) are the coproducts of L_1, L_2 and R_1, R_2 respectively. The partial morphism $L_1 + L_2 \overset{p}{\longrightarrow} R_1 + R_2$ is induced uniquely by the universal property of the coproduct $L_1 + L_2$ such that $p \circ in_L^1 = in_R^1 \circ p_1$ and $p \circ in_L^2 = in_R^2 \circ p_2$. The application of a parallel production $p_1 + p_2$ at a match m constitutes a **direct parallel derivation**, denoted by $G \overset{p_1+p_2,m}{\Longrightarrow} X$.

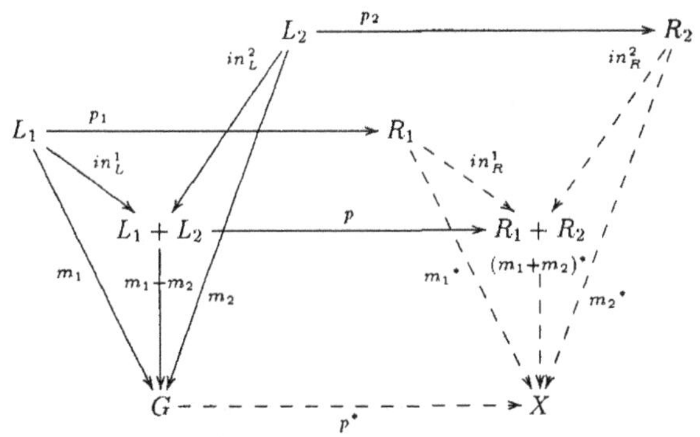

Figure 4.8: Weak Parallelism

By referring to morphisms $m_1 = m \circ in_L^1$ and $m_2 = m \circ in_L^2$ we also write $G \overset{p_1+p_2,m_1+m_2}{\Longrightarrow} X$.

Direct parallel derivations provide us with an explicit notion of parallelism, which shall now be related to the interleaving model that has been formulated before. The Parallelism Problem 3.2.2 asked for conditions allowing the sequentialization of a parallel derivation, and the parallelization of a sequential one. The following Weak Parallelism Theorem answers these questions in the SPO approach.

Theorem 4.3.8 (weak parallelism)
Given productions $p_1 : L_1 \to R_1$ and $p_2 : L_2 \to R_2$, the following statements are equivalent:

1. *There is a direct parallel derivation $G \overset{p_1+p_2,m_1+m_2}{\Longrightarrow} X$ such that $G \overset{p_2,m_2}{\Longrightarrow} H_2$ is weakly parallel independent of $G \overset{p_1,m_1}{\Longrightarrow} H_1$.*

2. *There is a weakly sequential independent derivation $G \overset{p_1,m_1}{\Longrightarrow} H_1 \overset{p_2,m_2'}{\Longrightarrow} X$.*

Up to isomorphism, a bijective correspondence between 1. and 2. above is given by $m_2' = {p_1}^* \circ m_2$.

Proof
Constructing the parallel derivation means first to construct the colimits (coproducts) for in_L^1 and in_L^2 as well as for in_R^1 and in_R^2 then the colimit (pushout) of p and $m_1 + m_2$ in a second step. In other words it means to construct the colimit of the diagram given by p_1, m_1 and p_2, m_2 as depicted in the Figure 4.8 above. Consider now Figure 4.7 of the Local Church Rosser Theorem.

Also this shows a colimit constructed from the diagram given by p_1, m_1 and p_2, m_2. Due to the commutativity of colimit construction (see [14]) ensuring that all colimits can iteratively be obtained from composing partial colimits, we conclude that both of these constructions coincide. [b] So, the result of the parallel derivation can equivalently be obtained by first constructing the colimits (pushout) of p_1, m_1 and p_2, m_2 and second the colimit (pushout) for the resulting morphisms $p_1{}^*$ and $p_2{}^*$. But this means first to construct the direct derivations $d_1 = (G \overset{p_1,m_1}{\Longrightarrow} H_1)$ and $d_2 = (G \overset{p_2,m_2}{\Longrightarrow} H_2)$ represented by sub-diagrams (1) and (2) in Figure 4.7 respectively. Their preassumed weak parallel independence leads then to a derivation $d_2' = (H \overset{p_2,m_2'}{\Longrightarrow} X)$ represented by subdiagram (2)+(3). By Characterization 4.3.4 we finally observe that d_2' is weakly sequentially independent of d_1 as required.

Vice versa, given an arbitrary sequentially independent derivation, we observe that all the arguments above can uniquely be reversed. The bijective correspondence between 1. and 2. is due to Lemma 4.3.5. □

Before looking at the counter-example below, the typical situation of Theorem 4.3.8 shall be illustrated by the Pacman-game.

Example 4.5 (killing is weakly parallel independent of eating)
In Figure 4.9 a situation is shown where Pacman, a ghost and an apple are on the same field. We observe that two production applications are possible: Pacman may eat an apple and the ghost may kill Pacman. Weak parallel independence allows the ghost to be merciful. Pacman may eat his last apple; nevertheless, the ghost will in both situations G and H_1 be sure about his victim. In other words, killing is weakly independent of eating. Contrastingly, eating is not at all weakly independent of killing because killing implies that eating becomes impossible. The Weak Parallelism Theorem 4.3.8 now ensures that the procedure of eating the last apple and being killed can be shortcut by correspondingly applying the parallel production. □

The following example proves that there are parallel derivations which cannot be sequentialized.

Example 4.6 (non-sequentializable parallel derivations)
Consider a production $\emptyset : L \to R$ where both L and R contain exactly one vertex $\bullet$.
Let $L \overset{\emptyset+\emptyset, id+id}{\Longrightarrow} X$ be a parallel derivation with $\emptyset + \emptyset$. Note that the two component matches $m_1 = m_2 = id : L \to L$ are parallel dependent since the

[b] i.e. both colimits may at most differ up to isomorphisms, but each obtained in one way can also be obtained in the other way.

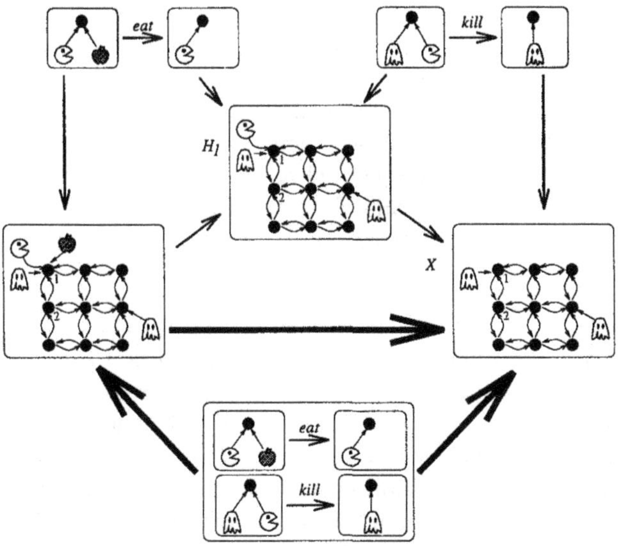

Figure 4.9: Killing and eating can be parallelized

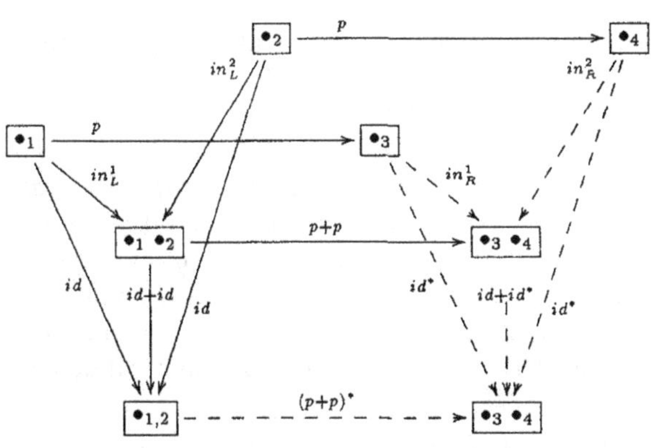

Figure 4.10: A Parallel Derivation which cannot be sequentialized

vertex in L is meant to be deleted. However, the derived graph X contains two vertices. Clearly this effect cannot be obtained by applying $\emptyset$ in two sequential steps. Let us compare this formally obtained result with our intuition: We observe that $\emptyset$ deletes a vertex and re-generates it afterwards. Correspondingly, the parallel production deletes two vertices and re-generates two. Hence we may expect that a match by which the two vertices in $L+L$ are identified leads to a co-match $(m_1 + m_2)^* : R+R \to X$ which identifies the two vertices. However, this does not happen in the parallel derivation which is due to the fact that, formally, the vertices in L and R are completely unrelated (i.e., there is no formal notion of 're'-generation). Hence, the two applications of $\emptyset$ generate two different vertices. □

Additional Remarks: The Weak Parallelism Theorem allows to define an operation on derivations, which shifts the application of a certain production one step towards the beginning of the derivation. Iterated applications of this so-called *shift operation* lead to a derivation in which each production is applied as early as possible. This maximally parallel derivation is introduced in [15] as the *canonical derivation.* Another fundamental theoretical problem addressed in [15] are abstract derivations. It was shown that for each derivation there is a unique abstract canonical derivation. See also Chapter 3 for a detailed discussion of this topic.

In the SPO approach, the results of this section have essentially been formulated in [7,4]. Originally, however, these problems have been investigated in the DPO approach, see Section 3.4 and 3.5. The differences and similarities of the corresponding results are discussed to some depth in Section 4.6.

In [16] a new notion of a concurrent derivation captures the idea of a concurrent history. A complementary notion of a morphism describes a concurrent subhistory relation; it is based on causal rather than sequential dependencies between activities. This leads to a category of abstract concurrent derivations taken as the concurrency semantics of a SPO graph grammar. In addition an interleaving semantics is proposed, given by a subcategory of abstract concurrent derivations, partially ordered by a sequential subcomputation relation. The concurrency semantics is characterized as a configuration domain of a prime event structure. The explicit consideration of infinite derivations leads to a notion of a fair derivation and a corresponding fair concurrency semantics.

4.3.2 Embedding of Derivations and Derived Productions

In this section we answer the question, under which conditions a derivation can be embedded into a larger context (cf. Problem 3.2.3). Therefore, we first have to formalize the notion of "embedding". Below, an embedding is defined as a family of compatible injections between the graphs of the original derivation and the graphs of the embedded one.

Definition 4.3.9 (embedding)
Given derivations $\rho = (G_0 \overset{p_1,m_0}{\Longrightarrow} \cdots \overset{p_k,m_{k-1}}{\Longrightarrow} G_k)$ and $\delta = (X_0 \overset{p_1,n_0}{\Longrightarrow} \cdots \overset{p_k,n_{k-1}}{\Longrightarrow} X_k)$, an **embedding** of ρ into δ is a family of total injective morphisms $e = (G_i \overset{e_i}{\longrightarrow} X_i)_{i\in\{1,\dots,k\}}$ such that $e_i \circ m_i = n_i$ for all $i \in \{1,\dots,k\}$, see also the diagram below. The embedding of ρ into δ is denoted by $\rho \overset{e}{\longrightarrow} \delta$, and the first injection $e_0 : G_0 \to X_0$ is called the **embedding morphism** of e.

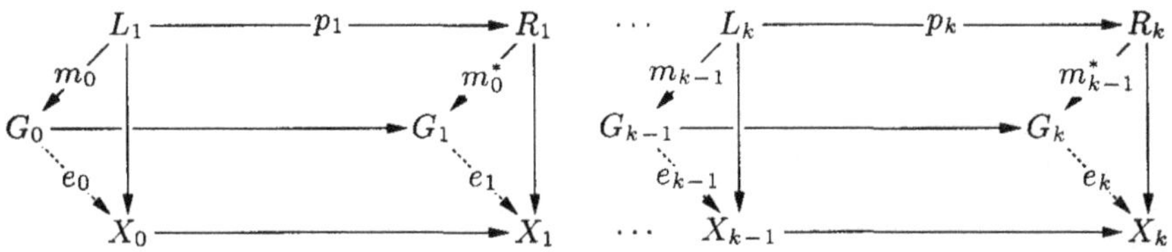

Given a derivation ρ, a possible embedding e of ρ into a derivation δ is completely determined by the the embedding morphism e_0. The existence of the embedding $e : \rho \to \delta$ can therefore be characterized in terms of the derivation ρ and the embedding morphism e_0. In the sequel, a *derived production* will be constructed simulating the effect of the derivation ρ, such that the applicability of this derived production at the match e_0 is equivalent to the embedding e of ρ into δ induced by e_0. First we consider the basic case of a direct derivation leading to the notion of directly derived production:

Definition 4.3.10 (directly derived production)
The **directly derived production** $\langle d\rangle : G \overset{p^*}{\longrightarrow} H$ of a direct derivation $d = (G \overset{p,m}{\Longrightarrow} H)$ is composed of the production name $\langle d\rangle$ and the partial morphism p^*, i.e., the co-production of the direct derivation. A direct derivation $X \overset{\langle d\rangle,e}{\Longrightarrow} Z$ using $\langle d\rangle : p^*$ is called **directly derived derivation.**

The following equivalence, relating directly derived derivations with embeddings of the original derivation, is also known as the vertical composition property of derivation diagrams.

Proposition 4.3.11 (directly derived production)
Given the directly derived production $\langle d\rangle : p^*$ of a derivation $d = (G \overset{p,m}{\Longrightarrow} H)$ as in Figure 4.11 on the left, the following statements are equivalent:

1. *There is a directly derived derivation $d' = (X \overset{\langle d\rangle,e}{\Longrightarrow} Z)$.*

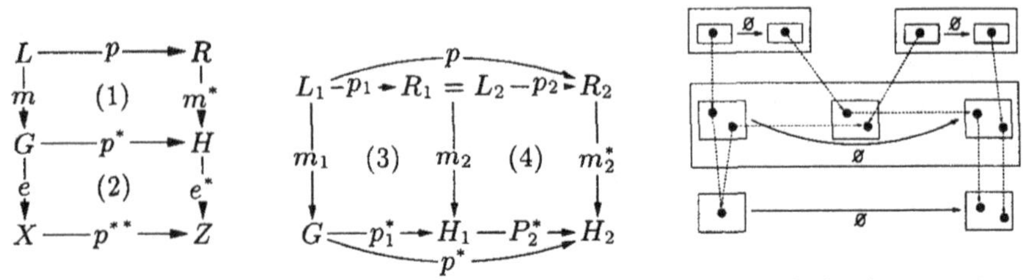

Figure 4.11: Horizontal and sequential composition of direct derivations, and non-sequentializable derived derivation.

2. *There is a direct derivation* $d'' = (X \overset{p,n}{\Longrightarrow} Z)$ *and an embedding* $\langle e, e^* \rangle : d' \to d''$.

Up to isomorphism, a bijective correspondence between 1. and 2. is given by $n = e \circ m$.

Proof

The directly derived derivation in 1. is represented by the pushout diagrams (1) and (2) on the left of Figure 4.11. Due to well-known pushout composition properties, diagram (1+2) is a pushout, too, which represents the direct derivation in 2. Vice versa, we can reconstruct the pushouts (1) and (2) from (1+2) by the pushout (1) of p and m and the pushout (2) of p^* and e. Uniqueness of pushouts up to isomorphism ensures the bijective correspondence between 1. and 2. ∎

Now we want to extend this equivalence to derivations $\rho = (G_0 \overset{p_1,m_0}{\Longrightarrow} \cdots \overset{p_k,m_{k-1}}{\Longrightarrow} G_k)$ with $k > 1$. To this aim we introduce the notion of "sequential composition" of two productions, in order to obtain a single derived production $\langle \rho \rangle$ from the composition of the sequence of its directly derived productions $\langle d_1 \rangle; \langle d_2 \rangle; \ldots; \langle d_k \rangle$. We speak of *sequential composition* because the application of such a composed production has the same effect of a sequential derivation based on the original productions.

Definition 4.3.12 (sequential composition)
Given two productions $p_1 : L_1 \to R_1$ and $p_2 : L_2 \to R_2$ with $R_1 = L_2$, the **sequentially composed production** $p_1; p_2 : L_1 \overset{p}{\to} R_2$ consists of the production name $p_1; p_2$ and the associated partial morphism $p = p_2 \circ p_1$.

Derivations $G \overset{p_1,m_1}{\Longrightarrow} H_1 \overset{p_2,m_2}{\Longrightarrow} H_2$, where the match m_2 of the second direct derivation is the co-match of the first, may now be realized in one step using the sequentially composed production $p_1; p_2$. Vice versa, each direct derivation $G \overset{p_1;p_2,m_1}{\Longrightarrow} H_2$ via $p_1; p_2$ can be decomposed into a derivation $G \overset{p_1,m_1}{\Longrightarrow} H_1 \overset{p_2,m_1^*}{\Longrightarrow}$

H_2, provided that the co-match m_1^* of the first direct derivation is total. In the Proposition below, this is ensured by the assumption that m_1 is conflict-free w.r.t. p_1.

Proposition 4.3.13 (sequential composition)
Given two productions p_1 and p_2 and their sequential composition $p_1;p_2 : p$ as above, the following statements are equivalent:

1. *There is a direct derivation $G \stackrel{p_1;p_2,m_1}{\Longrightarrow} H_2$ where m_1 is conflict-free w.r.t. p_1.*
2. *There is a derivation $G \stackrel{p_1,m_1}{\Longrightarrow} H_1 \stackrel{p_2,m_1^*}{\Longrightarrow} H_2$ such that m_1^* is the co-match of m_1 w.r.t. p_1.*

Up to isomorphism, a bijective correspondence between 1. and 2. above is given by $p = p_2 \circ p_1$.

Proof
By properties of pushouts in $\boldsymbol{Graph^P}$ (see Lemma 4.2.8), the co-match m^* is total if and only if m is conflict-free w.r.t. p_1. Then Proposition 4.3.13 follows from pushout composition and decomposition properties, similar to Proposition 4.3.11. □

Combining directly derived productions by sequential composition we now define derived productions for derivations of length greater than one:

Definition 4.3.14 (derived production)
Let $\rho = (G_0 \stackrel{p_1,m_0}{\Longrightarrow} \cdots \stackrel{p_k,m_{k-1}}{\Longrightarrow} G_k)$ be a derivation consisting of direct derivations $d_i = (G_{i-1} \stackrel{p_i,m_{i-1}}{\Longrightarrow} G_i)$ and $\langle d_i \rangle : p_i^*$ the corresponding directly derived productions. Then, the **derived production** $\langle \rho \rangle : G_0 \stackrel{p^*}{\longrightarrow} G_k$ of ρ is given by the production name $\langle \rho \rangle = \langle d_1 \rangle; \ldots ; \langle d_k \rangle$ and the associated partial morphism $p^* = p_k^* \circ \ldots \circ p_1^*$. A direct derivation $K \stackrel{\langle \rho \rangle, e}{\Longrightarrow} Y$ using $\langle \rho \rangle : p^*$ is called a **derived derivation.**

Each derivation sequence may be shortcut by a corresponding derived derivation. Vice versa, each derived derivation based on a d-injective match corresponds to an embedding of the original derivation. The following theorem provides the solution to the Derived Production Problem 3.2.4.

Theorem 4.3.15 (derived production)
Let $\rho = (G_0 \stackrel{p_1,m_0}{\Longrightarrow} \cdots \stackrel{p_k,m_{k-1}}{\Longrightarrow} G_k)$ be a derivation, $\langle \rho \rangle : p^*$ the corresponding derived production, and $G_0 \stackrel{e_0}{\longrightarrow} X_0$ a d-injective match for $\langle \rho \rangle$. Then the following statements are equivalent:

1. *There is a derived derivation* $X_0 \overset{\langle\rho\rangle,e_0}{\Longrightarrow} X_k$.

2. *There is a derivation* $\delta = (X_0 \overset{p_1,n_0}{\Longrightarrow} \dots \overset{p_k,n_{k-1}}{\Longrightarrow} X_k)$ *and an embedding* $\rho \xrightarrow{e} \delta$, *where* e_0 *is the embedding morphism of* e.

Up to isomorphism, a bijective correspondence between 1. and 2. above is given by $p^* = p_k^* \circ \dots \circ p_1^*$ *and* $e_i \circ m_i = n_i$ *for* $i \in \{0, \dots, k-1\}$.

Proof
By induction over the length k of ρ: Let $k = 1$. Then $\langle\rho\rangle$ is a directly derived production, and Theorem 4.3.15 follows from Proposition 4.3.11. Assume that 1. and 2. are equivalent for derivations ρ of length $k = l$. Let $\rho' = (G_0 \overset{p_0,m_0}{\Longrightarrow} \dots \overset{p_l,m_{l-1}}{\Longrightarrow} G_l \overset{p_{l+1},m_l}{\Longrightarrow} G_{l+1})$ be a derivation of length $k = l + 1$, ρ the prefix of ρ' consisting of the first l direct derivations, and $d_{l+1} = (G_l \overset{p_{l+1},m_l}{\Longrightarrow} G_{l+1})$ the last direct derivation of ρ'. Then, there is a derivation δ' and an embedding $\rho' \xrightarrow{e'} \delta'$ iff there are embeddings $\rho \xrightarrow{e} \delta$ and $\langle e_l, e_{l+1}\rangle : d_{l+1} \to d'_{l+1}$ where e_l is also the last injection of e (cf. Definition 4.3.9). We show that the embeddings e and $\langle e_l, e_{l+1}\rangle$ exist iff there are corresponding derived derivations using the derived productions $\langle\rho\rangle : G_0 \xrightarrow{p^*} G_l$ of the derivation ρ and $\langle d_{l+1}\rangle : p$ of the direct derivation d_{l+1}. Then, Theorem 4.3.15 follows from Proposition 4.3.13 using the fact that, since e_0 is injective, it is in particular conflict-free w.r.t. $\langle\rho\rangle$.

By applying the assumption, there is a derived derivation $X_0 \overset{\langle\rho\rangle,e_0}{\Longrightarrow} X_l$ as in 1. iff there is an embedding $\rho \xrightarrow{e} \delta$ into a derivation $\delta = (X_0 \overset{p_1,n_0}{\Longrightarrow} \dots \overset{p_l,n_{l-1}}{\Longrightarrow} X_l)$ as in 2. Using Proposition 4.3.11, the same equivalence holds between directly derived derivations $X_l \overset{\langle d_{l+1}\rangle,e_l}{\Longrightarrow} X_{l+1}$ and direct derivations $X_l \overset{p_{l+1},n_l}{\Longrightarrow} X_{l+1}$ which are embeddings of d_{l+1}. This concludes the proof of Theorem 4.3.15. □

The following example is taken from [7]. It illustrates that a derived derivation based on a non-d-injective match may not be sequentializable.

Example 4.7 (non-sequentializable derived derivation)
Consider the right diagram of Figure 4.11 on page 270, where the very same production is used twice for generating the derived production. The original production deletes one vertex and generates a new one.

The derived production specifies the deletion of two vertices and the generation of two others. But in fact by mapping the vertices to be deleted onto a single one leads to the generation of two vertices out of one. This is due to the fact that there is no direct connection between deleted and generated items, which in this situation means that, indeed, two instead of one vertex is newly added.
□

Let us come back now to Problem 3.2.3 which asked for an embedding of a derivation into a larger context. This question is now answered using the equivalence above, that is, a derivation ρ may be embedded via an embedding morphism e_0 if and only if the corresponding derived production $\langle\rho\rangle : p^*$ is applicable at e_0. Since in the SPO approach there is no further condition for the application of a production but the existence of a match, this implies that there is an embedding $\rho \stackrel{e}{\longrightarrow} \delta$ of the derivation ρ via each embedding morphism e_0.

Theorem 4.3.16 (embedding)
Let $\rho = (G_0 \stackrel{p_1,m_0}{\Longrightarrow} \cdots \stackrel{p_k,m_{k-1}}{\Longrightarrow} G_k)$ *be a derivation and* $G_0 \stackrel{e_0}{\longrightarrow} X_0$ *an embedding morphism. Then there is a derivation* $\delta = (X_0 \stackrel{p_1,n_0}{\Longrightarrow} \cdots \stackrel{p_k,n_{k-1}}{\Longrightarrow} X_k)$ *and an embedding* $\rho \stackrel{e}{\longrightarrow} \delta$.

Proof

Let $\langle\rho\rangle : p^*$ be the derived production of ρ and $X_0 \stackrel{\langle\rho\rangle,e_0}{\Longrightarrow} X_k$ the derived derivation at the embedding morphism $G_0 \stackrel{e_0}{\longrightarrow} X_0$. Since e_0 is injective, it is in particular d-injective. By Theorem 4.3.15 this implies the existence of the derivation δ and the embedding $\rho \stackrel{e}{\longrightarrow} \delta$. □

Additional Remarks: A derived production represents the overall effect of the derivation it is constructed from, and fixes the interaction of the productions applied in the derivation. This representation, however, can still be reduced by cutting of the unnecessary context, i.e., all those elements of the starting and ending graphs of the derivation which are never touched by any of the productions. Such *minimal derived productions* (minimal w.r.t. the embedding relation) have been introduced in the SPO approach in [4] for two-step derivations, generalized in [17] to derivations of length $k \geq 1$.

4.3.3 Amalgamation and Distribution

In this section we will investigate the simultaneous application of graph productions, which shall not work concurrently (as in a parallel derivation) but cooperatively, synchronizing their applications w.r.t. commonly accessed elements in the graph (cf. Section 3.2). The synchronization of productions w.r.t. certain deleted and/or generated elements is expressed by a common subproduction in the definition below.

Definition 4.3.17 (subproduction, synchronized productions)
Let $p_i : L_i \to R_i$ be three productions for $i = 0, 1, 2$. We say that p_0 is a **subproduction** of p_1 if there is an embedding $in^1 = \langle in^1_L, in^1_R\rangle : p_0 \to p_1$, i.e., a pair of total graph morphisms $in^1_L : L_0 \to L_1$ and $in^1_R : R_0 \to R_1$ such

that $in^1_R \circ p_0 = p_1 \circ in^1_L$. The productions p_1 and p_2 are **synchronized** w.r.t. p_0, shortly denoted by $p_1 \stackrel{in^1}{\leftarrow} p_0 \stackrel{in^2}{\rightarrow} p_2$, if p_0 is a subproduction of both p_1 and p_2 with embeddings in^1 and in^2, respectively.

$$\begin{array}{ccc} L_0 & \xrightarrow{p_0} & R_0 \\ {\scriptstyle in^1_L}\downarrow & & \downarrow{\scriptstyle in^1_R} \\ L_1 & \xrightarrow{p_1} & R_1 \end{array}$$

If synchronized productions $p_1 \stackrel{in^1}{\leftarrow} p_0 \stackrel{in^2}{\rightarrow} p_2$ shall be applied simultaneously to one global graph, this may be modeled by the application of an amalgamated production which is constructed as the gluing of p_1 and p_2 along p_0. Formally such a gluing is described by a pushout construction. Since the amalgamated production is a composed production, its production name has to record this construction. So similar as for the parallel production, the name $p_1 \oplus_{p_0} p_2$ of an amalgamated production is a whole diagram comprising the given synchronized productions $p_1 \stackrel{in^1}{\leftarrow} p_0 \stackrel{in^2}{\rightarrow} p_2$ and their embeddings into the production morphism of the amalgamated production.

Definition 4.3.18 (amalgamated productions and derivations)
Let $p_1 \stackrel{in^1}{\leftarrow} p_0 \stackrel{in^2}{\rightarrow} p_2$ be synchronized productions with $p_i : L_i \to R_i$ for $i \in \{0, 1, 2\}$. Then the **amalgamated production** $p_1 \oplus_{p_0} p_2 : L \xrightarrow{p} R$, consists of the production name $p_1 \oplus_{p_0} p_2$ and the associated partial morphism p. The production name $p_1 \oplus_{p_0} p_2$ is constructed on the left of Figure 4.12, where $\langle in^{2\,*}_L, in^{1\,*}_L \rangle$ and $\langle in^{2\,*}_R, in^{1\,*}_R \rangle$ are pushouts of $\langle in^1_L, in^2_L \rangle$ and $\langle in^1_R, in^2_R \rangle$, respectively, and $L \xrightarrow{p} R$ is obtained as the universal morphism satisfying $p \circ in^{1\,*}_L = in^{1\,*}_R \circ p_1$ and $p \circ in^{2\,*}_L = in^{2\,*}_R \circ p_2$. A direct derivation $G \stackrel{p_1 \oplus_{p_0} p_2, m}{\Longrightarrow} X$ using the amalgamated production $p_1 \oplus_{p_0} p_2$ is called **amalgamated derivation**. By referring to morphisms $m_1 = m \circ in^{2\,*}_L$ and $m_2 = m \circ in^{1\,*}_L$ we also write $G \stackrel{p_1 \oplus_{p_0} p_2, m_1 \oplus m_2}{\Longrightarrow} X$.

A distributed graph models a state which is splitted into substates related by common interface states. Gluing the local states along their interfaces yields the global state again. Below these ideas are formalized for two local states related by one interface.

Definition 4.3.19 (distributed graph)
A **distributed graph** $DG = (G_1 \stackrel{g_1}{\leftarrow} G_0 \stackrel{g_2}{\rightarrow} G_2)$ consists of two *local graphs* G_1 and G_2, an **interface graph** G_0 and graph morphisms g_1 and g_2 embedding the interface graph into the two local graphs. The **global graph** $G = \oplus\, DG = G_1 \oplus_{G_0} G_2$ of DG is defined as the pushout object of g_1 and g_2

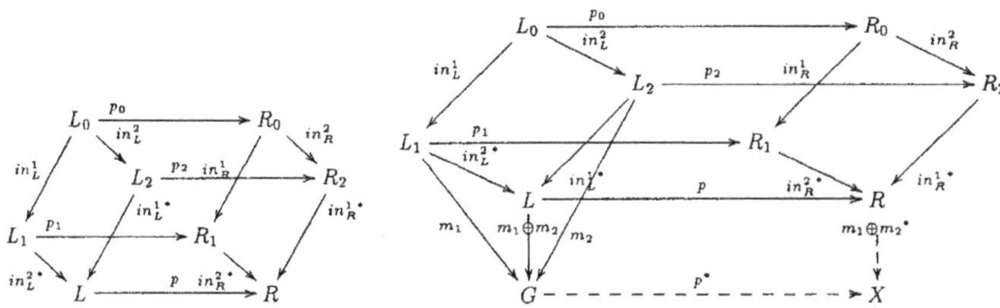

Figure 4.12: Amalgamated Production Amalgamated Derivation

in the diagram below. The distributed graph DG is a **total splitting** of G if the graph morphisms g_1 and g_2 are total. In general DG is called a **partial splitting**.

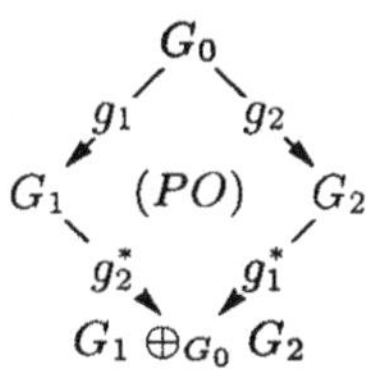

A truly partial splitting models an inconsistent distributed state where, for example, there are dangling references between the local substates. In such situations, there is no general agreement if a certain item belongs to the corresponding global state or not. Constructing the global graph as the pushout object of the splitting, this question is decided in favor of deletion, that is, conflicting items are removed from the global state.
Distributed graphs can be transformed by synchronized productions, which specify a simultaneous update of all local graphs.

Definition 4.3.20 (distributed derivation)
Let $DG = (G_1 \overset{g_1}{\longleftarrow} G_0 \overset{g_2}{\longrightarrow} G_2)$ be a distributed graph and $L_i \overset{m_i}{\longrightarrow} G_i$ for $i \in \{0,1,2\}$ be matches for synchronized productions $p_1 \overset{in^1}{\leftarrow} p_0 \overset{in^2}{\rightarrow} p_2$ as in Figure 4.13 on the left. The matches $(m_i)_{i\in\{0,1,2\}}$ form a **distributed match** for $p_1 \overset{in^1}{\leftarrow} p_0 \overset{in^2}{\rightarrow} p_2$ into DG if $g_k \circ m_0 = m_k \circ in_L^k$ for $k \in \{1,2\}$. In this case, the **distributed derivation** $d_1 ||_{d_0} d_2 : DG \Longrightarrow DH$ with $DH = (H_1 \overset{h_1}{\longleftarrow} H_0 \overset{h_2}{\longrightarrow} H_2)$ is constructed by the **local direct derivations** $d_1 = (G_1 \overset{p_1,m_1}{\Longrightarrow} H_1)$ and $d_2 = (G_2 \overset{p_2,m_2}{\Longrightarrow} H_2)$ and the **interface derivation** $d_0 = (G_0 \overset{p_0,m_0}{\Longrightarrow} H_0)$. The partial morphisms h_1 and h_2 are induced by the universal property of the pushout $\langle p_0^*, m_0^* \rangle$ of $\langle p_0, m_0 \rangle$ such that the left and bottom squares of the right

diagram of Figure 4.13 commute. If g_1 and g_2 as well as h_1 and h_2 are total, we speak of a **synchronous distributed derivation** $d_1||_{d_0}d_2$.

A distributed graph becomes inconsistent (i.e., a partial splitting) if we delete an item in a local graph which has a preimage in the interface that is not deleted. This situation can be characterized in set-theoretical terms:

Characterization 4.3.21 (synchronous distributed derivation)
Let $d_i = (G_i \overset{p_i,m_i}{\Longrightarrow} H_1)$ for $i \in \{0,1,2\}$ be direct derivations at conflict-free matches. A distributed derivation $d_1||_{d_0}d_2 : DG \Longrightarrow DH$ as in Figure 4.13 is synchronous if and only if DG is a total splitting, and for $k \in \{1,2\}$ we have that $y \in G_0$ with $g_k(y) \in m_k(L_k - dom(p_k))$ implies $y \in m_0(L_0 - dom(p_0))$.

Proof sketch
The structures $m_k(L_k - dom(p_k))$ for $k = 1,2$ and $m_0(L_0 - dom(p_0))$ consist of those elements of G_k and G_0 which are explicitly deleted by the direct derivations d_k and d_0, respectively. The distributed derivation $d_1||_{d_0}d_2$ is synchronous if the morphisms h_k in the right diagram of Figure 4.13 are total. We sketch the "if" part of the proof of Characterization 4.3.21. For a complete proof the reader is referred to [18].

Let $x \in H_0$. Since p_0^* and m_0^* are pushout morphisms, they are jointly surjective by Lemma 4.2.8, i.e., x has either a preimage $y \in G_0$ or $z \in R_0$. In the first case there is $g_k(y) \in G_k$ since g_k is total. Since y is preserved by p_0^*, it is not in $m_0(L_0 - dom(p_0))$. Hence $g_k(y) \notin m_k(L_k - dom(p_k))$, which implies by some further arguments that $g_k(y) \in dom(p_k^*)$, i.e., $p_k^* \circ g_k(y)$ is defined. By commutativity of the bottom square this implies that also $h_k(p_0^*(y)) = h_k(x)$ is defined. If x has a preimage $z \in R_0$, $m_k^*(in_R^k(z))$ is defined since m_k^* is total by conflict-freeness of m_k (see Lemma 4.2.8). By commutativity of the left square this implies that also $h_k(m_0^*(z)) = h_k(x)$ is defined. □

Finally, we investigate the relationship between distributed and amalgamated derivations. Amalgamating a distributed derivation means to construct a global observation of the local actions. This is possible only if there is a consistent global view at least of the given distributed state DG. Hence DG is assumed to be a total splitting in the distribution theorem below. On the other hand, distributing an amalgamated derivation means to split a global action into local ones. Therefore, the matches of the elementary productions have to be compatible with the splitting of the global graph. The Distribution Theorem below provides the solution to Problem 3.2.5.

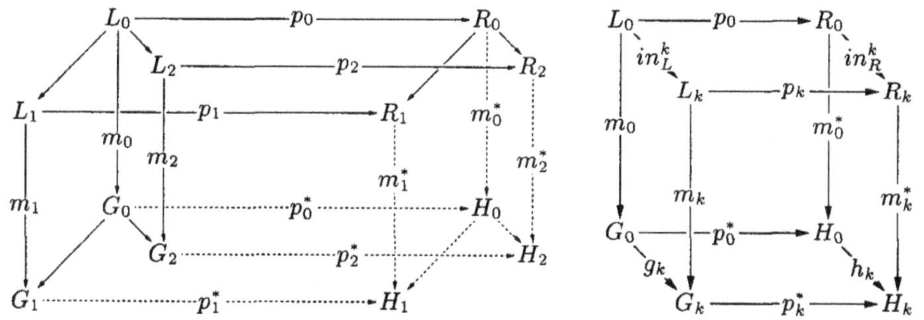

Figure 4.13: A distributed derivation $d_1||_{d_0}d_2 : DG \to DH$ with $d_0 = (G_0 \overset{p_0,m_0}{\Longrightarrow} H_0)$ and $d_k = (G_k \overset{p_k,m_k}{\Longrightarrow} H_k)$ for $k \in \{1,2\}$.

Theorem 4.3.22 (distribution)
Let $DG = (G_1 \overset{g_1}{\longleftarrow} G_0 \overset{g_2}{\longrightarrow} G_2)$ be a total splitting of $G = \oplus DG$, $p_1 \overset{in^1}{\leftarrow} p_0 \overset{in^2}{\rightarrow} p_2$ synchronized productions and $p_1 \oplus_{p_0} p_2 : L \overset{p}{\longrightarrow} R$ their amalgamated production. Then the following statements are equivalent:

1. *There are an amalgamated derivation $G \overset{p_1 \oplus_{p_0} p_2, m}{\Longrightarrow} H$, and a distributed match $(m_i)_{i\in\{0,1,2\}}$ for $p_1 \overset{in^1}{\leftarrow} p_0 \overset{in^2}{\rightarrow} p_2$ into DG, which is compatible with m, i.e., $g_2^* \circ m_1 = m \circ {in_L^2}^*$ and $g_1^* \circ m_2 = m \circ {in_L^1}^*$.*
2. *There is a distributed derivation $d_1||_{d_0}d_2 : DG \Longrightarrow DH$ with $d_i = (G_i \overset{p_i,m_i}{\Longrightarrow} H_i)$ for $i \in \{0,1,2\}$.*

Proof sketch
The proof of this theorem in [19] uses the 4-cube lemma presented in [20], which is valid in every category and can be derived as a special case of the commutativity of colimits. □

Additional Remarks: Amalgamated derivations are introduced in the SPO approach in [4], while corresponding concepts in the DPO approach have been developed in [21]. The Amalgamation Theorem in [4] is concerned with the sequentialization of an amalgamated derivation. It is based on the notion of a remainder (production) which can be roughly be considered as that part of a given production which is not covered by its subproduction. A suitably given amalgamated derivation can then be simulated by a derivation consisting of an application of the common subproduction, followed by a parallel application of the remainders [4]. The concept of amalgamated productions and derivations was generalized to cases including more than two elementary productions. Corresponding developments in [18] were motivated by the idea to specify derivations in which a variable number of mutually interacting productions must be synchronized.

Distributed graphs and derivations in the SPO approach are introduced in [19], where also the Distribution Theorem is formulated. The comparison of several kinds of global and (synchronous and asynchronous) distributed derivations led to a hierarchy theorem for distributed derivations. Splitting a state into two substates with one interface is, of course, a very basic kind of a distribution. In [18] this is generalized to arbitrary many local states, pairwise related by interfaces. Even more general topologies are considered in [22,23], in the DPO approach.

4.4 Application Conditions in the SPO Approach

Using the rule-based formalism introduced so far we may easily and intuitively describe, *how* given graphs shall be transformed into derived graphs. For specifying *when* these transformations should occur, however, we are restricted to positive application conditions concerning the existence of certain nodes and edges, which can be specified within the left-hand side of the productions. In this section we introduce the possibility to specify also negative application conditions for each particular production, and extend the results of Section 4.3.1 concerning independence and parallelism to productions and derivations with application conditions.

4.4.1 Negative Application Conditions

We use the model of the Pacman game to motivate the use of negative application conditions.

Example 4.8
Recall the production *moveP* from Figure 4.1. As effect of this production Pacman moves to a new field not taking into account whether this field is dangerous for him or not. If Pacman moves to a field where a ghost is waiting, he may be killed in the next derivation step. An intelligent player of the Pacman game would like to apply the *moveP* production only if there *is no ghost* on the field Pacman is moving to. Hence we have a *negative application condition* for our production. The production *ImoveP* which models an intelligent moving of Pacman is depicted as the left upper production of Figure 4.14. The forbidden context, i.e., the ghost with the edge pointing to the field Pacman wants to move to, is enclosed by a dotted line and crossed out, in order to denote that these elements must not exist if the production shall be applied. Analogously

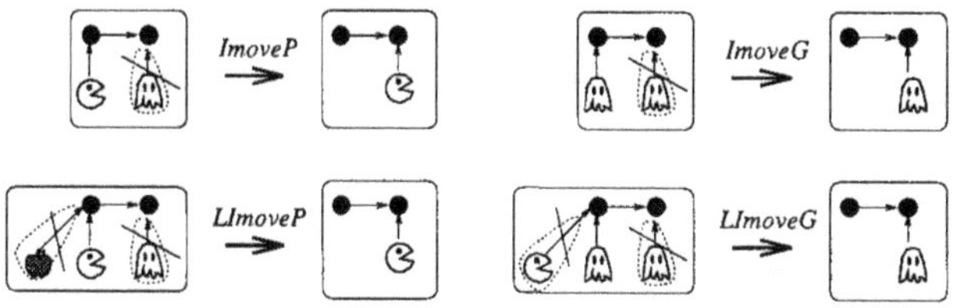

Figure 4.14: Intelligent moving of Pacman and ghosts.

one could think of an intelligent behavior of the ghosts. If they want to kill Pacman they have to occupy as many fields as possible. Hence a ghost should only be moved to a field on which there is not already another ghost (see production *ImoveG* in the right upper part of Figure 4.14).

By now we have shown that negative application conditions can be used to include rather simple strategies of the game in our model. Graph transformations are by default non-deterministic. But for our model it is desirable to restrict this non-determinism, i.e., to have priorities on the productions. Moving has a lower priority than eating an apple for Pacman or than killing Pacman for a ghost. These priorities can also be coded into negative application conditions for the move productions, such that they are only applicable if the *eat* resp. *kill* production is not applicable to Pacman resp. the same ghost (see lower part of Figure 4.14). Note that intelligent moving with lower priority has a negative application condition consisting of two forbidden contexts, called constraints in the following, one taking care of the field to move to and one making sure that moving is of lower priority. These application conditions with more than one constraint are satisfied if each of the constraints is satisfied. □

The general idea is to have a left-hand side not only consisting of one graph but of several ones, connected by morphisms $L \xrightarrow{l} \hat{L}$, called constraints, with the original left-hand L. For each constraint, $\hat{L} - l(L)$ represents the forbidden structure, i.e., the dotted bordered part of the example productions. A match satisfies a constraint if it cannot be extended to the forbidden graph $\hat{L}$, i.e., if the additional elements are not present in the context of the match. If a constraint is non-injective and surjective, i.e., the forbidden structure $\hat{L}-l(L)$ is empty but items of L are identified by l, it can only be satisfied by a match not identifying these items. Thus, negative constraints can also express injectivity requirements for the match. A match satisfies a negative application condition if it satisfies all constraints the application condition consists of.

Definition 4.4.1 (application conditions)

1. A **negative application condition**, or application condition for short, over a graph L is a finite set A of total morphisms $L \xrightarrow{l} \hat{L}$, called **constraints**.

2. A total graph morphism $L \xrightarrow{m} G$ **satisfies** a constraint $L \xrightarrow{l} \hat{L}$, written $m \models l$, if there is no total morphism $\hat{L} \xrightarrow{n} G$ such that $n \circ l = m$. m **satisfies** an application condition A over L, written $m \models A$, if it satisfies all constraints $l \in A$.

3. An application condition A is said to be **consistent** if there is a graph G and a total morphism $L \xrightarrow{m} G$ s.t. m satisfies A.

4. A **production with application condition** $\hat{p} : (L \xrightarrow{p} R, A(p))$, or **conditional production** for short, is composed of a production name $\hat{p}$ and a pair consisting of a partial graph morphism p and an application condition $A(p)$ over L. It is **applicable** to a graph G at $L \xrightarrow{m} G$ if m satisfies $A(p)$. In this case, the direct derivation $G \overset{p,m}{\Longrightarrow} H$ is called **direct conditional derivation** $G \overset{\hat{p},m}{\Longrightarrow} H$.

Example 4.9

In Figure 4.15 we show the formal representation of the production $LImoveP : (moveP, \{lp, int\})$ of Figure 4.14 consisting of the unconditional production *moveP* of Figure 4.1 and two constraints. Morphisms are denoted by numbers at corresponding nodes. The constraint *int* states that there must not be a ghost at the field Pacman wants to move to, and *lp* ensures that there is no apple at Pacmans current position. Accordingly the match m for *moveP* satisfies lp since we cannot extend m to $\hat{L}_{lp}$, while m does not satisfy *int*. Hence m does not satisfy $\{lp, int\}$. If, however, Pacman moves to the left, i.e. vertex $\bullet_2$ of L is mapped to vertex $\bullet_6$ in G, *int* is satisfied as well and the production can be applied. □

It is possible to define application conditions which can not be satisfied by any match, i.e., the corresponding conditional production is never applicable. This is due to the fact that contradictions may appear between the positive requirements of the productions left-hand side and the negative constraints. A trivial example for such a contradiction is a constraint which is an isomorphism. More generally, a contradiction occurs if the forbidden items of $\hat{L} - l(L)$ can be mapped onto the required ones in L, i.e., a total morphism $\hat{L} \xrightarrow{n} L$ exists, extending the identity on L. Hence, a negative application condition A is consistent in the sense of Definition 4.4.1.3 (i.e., free of these contradictions) iff the match $L \xrightarrow{id} L$ satisfies A.

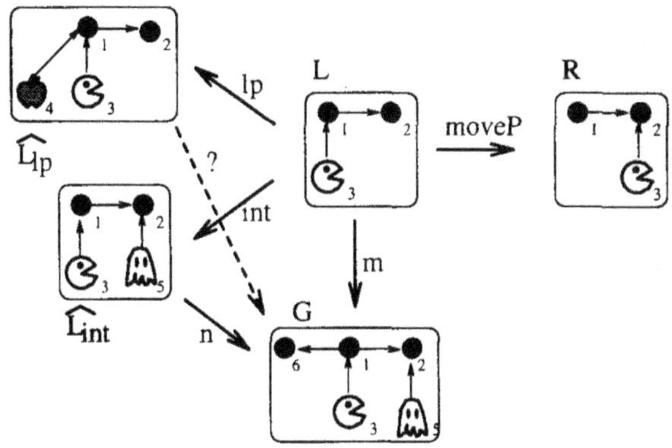

Figure 4.15: Formal representation of the conditional production *LImoveP*

Lemma 4.4.2 (consistency of application conditions)
An application condition A over L is consistent if and only if it is satisfied by id_L.

Proof
If A is satisfied by id_L, A is consistent by Definition 4.4.1. Vice versa, let id_L not satisfy A. Then there are a constraint $L \xrightarrow{l} \hat{L} \in A$ and a total morphism $\hat{L} \xrightarrow{n} L$ s.t. $n \circ l = id_L$. Since for any given match $L \xrightarrow{m} G$ we have that $m = m \circ id_L$, this implies that $(m \circ n) \circ l = m$, i.e., m does not satisfy the constraint $l \in A$, and hence not A. □

Even more powerful application condition, which can express, for example, cardinality restrictions on the number of incoming edges of a given node, are obtained if the the forbidden morphisms $\hat{L} \xrightarrow{n} G$ of Definition 4.4.1.2 are required to be injective. These *application conditions with injective satisfaction*, as they are called in [2], may, for example, express the gluing condition of the DPO approach (see Proposition 3.3.4) consisting of the dangling condition and the identification condition: For a given production $L \xrightarrow{p} R$ we let

- the **identification condition** of p be the set $IC(p)$ of all total surjective morphisms l, except isomorphisms, such that for all l we have $l(x) = l(y)$ for some $x, y \in L$ with $x \neq y$ and $x \notin dom(p)$, and
- the **dangling condition** of p be the set $DC(p)$ of all total morphisms $L \xrightarrow{l} \hat{L}$ such that l is surjective up to an edge e (and possibly a node) with $s(e)$ or $t(e)$ in $l(L - dom(p))$.

Now a match m satisfies the gluing condition if and only if it injectively satisfies the application conditions $IC(p)$ and $DC(p)$, i.e., iff there is no total *injective* morphism $\hat{L} \xrightarrow{n} G$ for any constraint $L \xrightarrow{l} \hat{L} \in IC(p) \cup DC(p)$ satisfying $n \circ l = m$. Using this equivalence, DPO derivations can be characterized by conditional SPO derivations.

4.4.2 Independence and Parallelism of Conditional Derivations

Following the line of [2] we extend the results of Section 4.3.1 on independence and parallelism to conditional derivations. Thereby, we provide solutions to the Local Church-Rosser Problem 3.2.1 and the Parallelism Problem 3.2.2 for the SPO approach with application conditions.

Interleaving

According to Section 4.3, two derivations d_1 and d_2 are considered to be parallel independent if they may occur in any order with the same result. For unconditional derivations this is mainly a problem of deletion, i.e., none of the two derivations should delete something that is needed for the match of its alternative. Taking into account negative application conditions specifying forbidden application contexts, we have to ensure that, in addition, no such forbidden context is established.

Definition 4.4.3 (parallel independence)
Let $d_1 = (G \overset{\hat{p}_1,m_1}{\Longrightarrow} H_1)$ and $d_2 = (G \overset{\hat{p}_2,m_2}{\Longrightarrow} H_2)$ be two direct conditional derivations using $\hat{p}_1 : (p_1, A(p_1))$ and $\hat{p}_2 : (p_2, A(p_2))$, respectively. Then we say that d_2 is **weakly parallel independent** of d_1 if $m_2' = r_1{}^* \circ m_2 : L_2 \to H_1$ is a match for p_2 that satisfies the application condition $A(p_2)$ of $\hat{p}_2$ (see left diagram of Figure 4.17). Direct conditional derivations d_1 and d_2 are **parallel independent** if they are mutually weakly parallel independent.

Example 4.10
Let (1) and (2) in the left side of Figure 4.16 be applications of the productions *moveP* and *moveG* of Figure 4.1, respectively. Then these unconditional direct derivations are parallel independent. If we apply the conditional production *ImoveP* of Figure 4.14 instead of *moveP*, preventing Pacman from approaching the ghost, (1) is not weakly parallel independent of (2). □

A derivation d_2' is sequentially independent of its predecessor d_1 if d_2' may occur also alternatively to, or even before d_1. In Section 4.3 this is shown to be the case if the application context of d_2' has already been present before the occurrence of d_1. In the conditional case we additionally require that no forbidden context of d_2' has been destroyed by d_1.

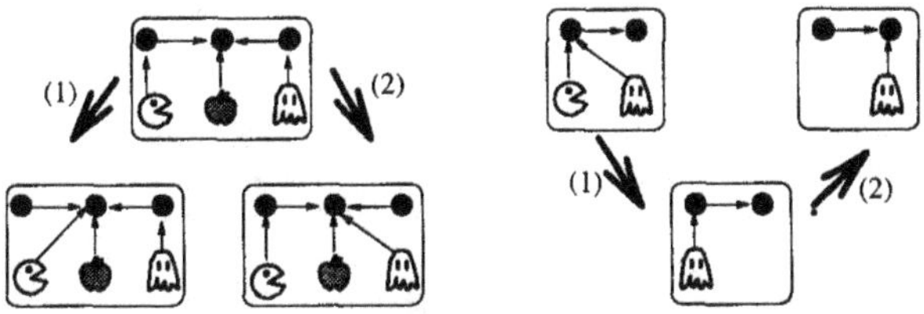

Figure 4.16: Parallel and sequential independence (example).

Definition 4.4.4 (sequential independence)
Let $d_1 = (G \overset{\hat{p}_1, m_1}{\Longrightarrow} H_1)$ and $d'_2 = (H_1 \overset{\hat{p}_2, m'_2}{\Longrightarrow} X)$ be two direct conditional derivations using $\hat{p}_1 : (p_1, A(p_1))$ and $\hat{p}_2 : (p_2, A(p_2))$, respectively. Then we say that d'_2 is **weakly sequentially independent** of d_1 if $m_2 = ({r_1}^*)^{-1} \circ m'_2 : L_2 \to G$ is a match for p_2 that satisfies the application condition $A(p_2)$ of $\hat{p}_2$ (see right diagram in Figure 4.17). Let $d_2 = (G \overset{\hat{p}_2, m_2}{\Longrightarrow} H_2)$ be the corresponding direct derivation. In case that, additionally, d_1 is weakly parallel independent of d_2 the derivation sequence $(G1 \overset{\hat{p}_1, m_1}{\Longrightarrow} H_1 \overset{\hat{p}_2, m'_2}{\Longrightarrow} X)$ is called **sequentially independent**.

Example 4.11
In the right side of Figure 4.16 a sequential independent derivation sequence using *kill* and *moveG* of Figure 4.1 is shown. If, however, (2) results from an application of *LImoveG* of Figure 4.14, (2) is not independent of (1) because the ghost has to kill Pacman before leaving his field. ⊟

The following theorem is a straightforward generalization of the Local Church-Rosser Theorem of Section 4.3 to conditional derivations. It provides the solution to the Local Church-Rosser Problem 3.2.1 for SPO derivations with application conditions.

Theorem 4.4.5 (conditional local Church-Rosser)
Let $d_1 = (G \overset{\hat{p}_1, m_1}{\Longrightarrow} H_1)$ and $d_2 = (G \overset{\hat{p}_2, m_2}{\Longrightarrow} H_2)$ be two direct conditional derivations. Then the following statements are equivalent:

1. *d_1 and d_2 are parallel independent.*

2. *There are direct conditional derivations $H_1 \overset{\hat{p}_2, m'_2}{\Longrightarrow} X$ and $H_2 \overset{\hat{p}_1, m'_1}{\Longrightarrow} X$ such that $G \overset{\hat{p}_1, m_1}{\Longrightarrow} H_1 \overset{\hat{p}_2, m'_2}{\Longrightarrow} X$ and $G \overset{\hat{p}_2, m_2}{\Longrightarrow} H_2 \overset{\hat{p}_1, m'_1}{\Longrightarrow} X$ are sequentially independent conditional derivations.*

Up to isomorphism, a bijective correspondence between 1. and 2. above is given by $m'_2 = {r_1}^* \circ m_2$ and $m'_1 = {r_2}^* \circ m_1$.

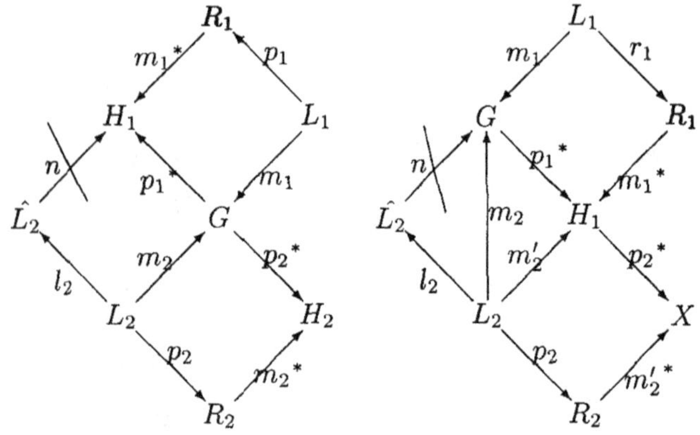

Figure 4.17: Independence of conditional derivations.

Proof
Directly from Definitions 4.4.3 and 4.4.4 and the Local Church-Rosser Theorem 4.3.6. □

Explicit Parallelism

Now we consider the construction of a parallel conditional production from two conditional elementary productions. Starting with the parallel production $p_1 + p_2$ defined in Definition 4.3.7 we have to find a suitable application condition for $p_1 + p_2$. For unconditional single-pushout derivations, the applicability of the parallel production is (trivially) equivalent to the applicability of the elementary productions at their corresponding matches. Generalizing this we have to construct $A(p_1 + p_2)$ as conjunction of $A(p_1)$ and $A(p_2)$.

If two application conditions are defined over the same left-hand side, their conjunction is simply given by their union. In our case however, $A(p_1)$ and $A(p_2)$ are application conditions over L_1 and L_2, respectively. Thus the problem remains, how to extend the application conditions over L_1 and L_2 to the larger graph $L_1 + L_2$. Below the extension of a constraint l along a total morphism m is defined by the pushout of l and m.

Definition 4.4.6 (extension)
If $L \xrightarrow{m} G$ is a total morphism and $L \xrightarrow{l} \hat{L}$ a constraint, the *extension* $m^{\#}(l)$ *of* l *along* m is given by the pushout diagram (1) in the left-hand side of Figure 4.18. The **extension of an application condition** A over L is defined by $m^{\#}(A) = \{m^{\#}(l) | l \in A\}$.

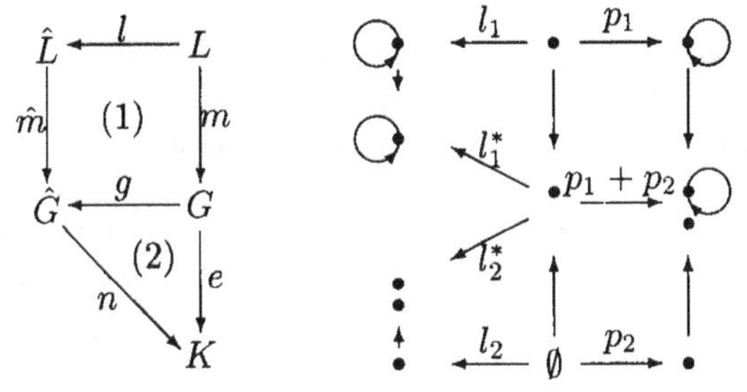

Figure 4.18: Extension of constraints and construction of a parallel conditional production

Proposition 4.4.7 (extension)
Let $L \xrightarrow{m} G$ be a total morphism and A an application condition over L. Then, for all matches $G \xrightarrow{e} K$, $e \models m^{\#}(A)$ iff $e \circ m \models A$.

Proof
We show that for each $l \in A$ we have $e \models m^{\#}(l)$ iff $e \circ m \models l$. Assume n s.t. (2) in Diagram 4.18 commutes. Then $n' = n \circ \hat{m}$ and $n' \circ l = m \circ e$ by commutativity of (1) and (2). Vice versa, let $\hat{L} \xrightarrow{n'} K$ be given with $n' \circ l = e \circ m$. Then n with $n \circ g = e$ exists by the universal property of (1). □

Now we define the parallel conditional production $\hat{p}_1 + \hat{p}_2$ as the parallel production of the underlying productions p_1 and p_2 of $\hat{p}_1$ and $\hat{p}_2$, equipped with the extensions of the application conditions $A(p_1)$ and $A(p_2)$ of the component productions along in_L^1 and in_L^2, respectively.

Definition 4.4.8 (parallel conditional production)
Let $\hat{p}_1 : (L_1 \xrightarrow{r_1} R_1, A(p_1))$ and $\hat{p}_2 : (L_2 \xrightarrow{r_2} R_2, A(p_2))$ be conditional productions and $p_1 + p_2 : L \xrightarrow{p} R$ the parallel production of p_1 and p_2 according to Definition 4.3.7. Then the **parallel conditional production** $\hat{p}_1 + \hat{p}_2 : (p, A(p))$ is composed of the production name $\hat{p}_1 + \hat{p}_2$ and the pair $(p, A(p))$, where $A(p) = {in_L^1}^{\#}(A(p_1)) \cup {in_L^2}^{\#}(A(p_2))$. A conditional derivation $G \overset{\hat{p}_1+\hat{p}_2, m}{\Longrightarrow} X$ using $\hat{p}_1 + \hat{p}_2$ is called **parallel conditional derivation.**

Example 4.12
The construction above does not guarantee, that the application condition $A(p_1 + p_2)$ of $\hat{p}_1 + \hat{p}_2$ is consistent, even if both $A(p_1)$ and $A(p_2)$ are. In the right side of Figure 4.18 the production $\hat{p}_1 : (p_1, \{l_1\})$ adds a loop to a node if there isn't already one. The production $\hat{p}_2 : (p_2, \{l_2\})$ inserts a node in an empty graph. The application condition of the parallel production $\hat{p}_1 + \hat{p}_2 : (p_1 + p_2, \{l_1^*, l_2^*\})$ is not consistent (compare Lemma 4.4.2) because we can

reverse l_2^* by identifying the two nodes. On the other hand, there is no graph to which we can apply both $\hat{p}_1$ and $\hat{p}_2$, i.e., their application domains are disjoint. That this is no mere coincidence is shown by the following proposition. □

Proposition 4.4.9 (applicability of parallel production)
Let $\hat{p}_1, \hat{p}_2$ and $\hat{p}_1 + \hat{p}_2$ be given as above together with matches $L_1 \xrightarrow{m_1} G$ and $L_2 \xrightarrow{m_2} G$ for p_1 and p_2 into G and let $L_1 + L_2 \xrightarrow{m_1+m_2} G$ be the parallel match for $p_1 + p_2$. Then $m_1 + m_2 \models A(p_1 + p_2)$ if and only if $m_1 \models A(p_1)$ and $m_2 \models A(p_2)$.

Proof
$m_k = m_1 + m_2 \circ in_L^k$ for $k = 1, 2$ by universal property of $L_1 + L_2$. Then Proposition 4.4.9 is a direct consequence of Proposition 4.4.7. □

Since we can check consistency of application conditions by Lemma 4.4.2 this provides us with a method to decide whether a parallel production makes sense or not. Furthermore, we may now extend the Weak Parallelism Theorem 4.3.8 to conditional derivations, which solves the Parallelism Problem 3.2.2 for conditional derivations.

Theorem 4.4.10 (conditional weak parallelism)
Given conditional productions $\hat{p}_1$ and $\hat{p}_2$, the following two statements are equivalent:

1. *There is a direct conditional parallel derivation $G \overset{\hat{p}_1+\hat{p}_2, m_1+m_2}{\Longrightarrow} X$, s.t. $G \overset{\hat{p}_2, m_2}{\Longrightarrow} H_2$ is weakly parallel independent of $G \overset{\hat{p}_1, m_1}{\Longrightarrow} H_1$.*

2. *There is a conditional derivation $G \overset{\hat{p}_1, m_1}{\Longrightarrow} H_1 \overset{\hat{p}_2, m_2'}{\Longrightarrow} X$, where $H_1 \overset{\hat{p}_2, m_2'}{\Longrightarrow} X$ is weakly sequentially independent of $G \overset{\hat{p}_1, m_1}{\Longrightarrow} H_1$.*

Up to isomorphism, a unique correspondence between 1. and 2. above is given by $m_2' = r_1{}^* \circ m_2$.

Proof
Directly from Proposition 4.4.9, Definition 4.4.3 and 4.4.4 and the Parallelism Theorem 4.3.8. □

Example 4.13
In the left side of Figure 4.19 the parallel production of *ImoveP* and *ImoveG* is shown, modeling a simultaneous move of Pacman and one of the ghosts. Its graphical representation is given on the right. Applying this production to the graph in the upper left of Figure 4.16 we have an example of a parallel conditional derivation that cannot be sequentialized, because the alternative

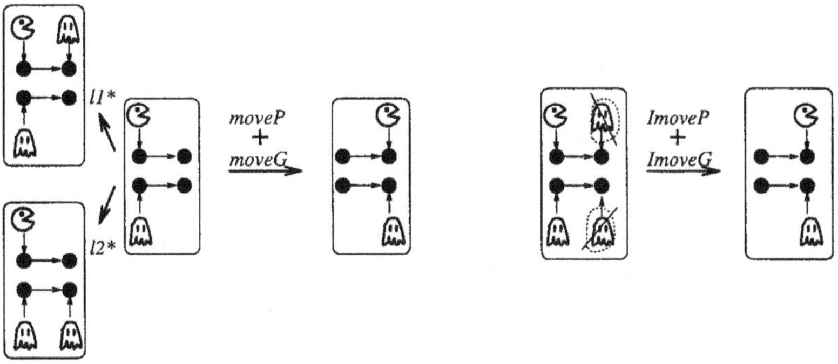

Figure 4.19: Parallel production of *ImoveP* and *ImoveG*.

derivations using *ImoveG* and *ImoveP* (denoted by (1) and (2) in the left side of Figure 4.16) are not parallel independent (cf. Example 4.10). □

Additional Remarks: The results of this section have been obtained in [2]. In the case of application conditions with injective satisfaction (cf. Section 4.4), similar results are possible. Moreover, most of the remaining results of Section 4.3 have already been extended to such application conditions in [24,25]. In addition to the left-sided negative application conditions of Definition 4.4.1, also right-sided application conditions and so-called propositional application conditions, i.e., propositional expressions over constraints, are considered in [25]. This is even further generalized in [26] and [27] by conditional application conditions.

Another interesting line of research is the generative power of graph grammars with (different kinds of) conditional productions. In [2] it has been shown that context-free graph grammars with positive and/or negative application conditions are strictly more powerful than unconditional ones. Similar investigations can be found in [26] for graph grammars without nonterminal labels.

4.5 Transformation of More General Structures in the SPO Approach

Until now we presented SPO transformations and corresponding results based on labeled graphs. In this section we will show that the SPO approach is not restricted to this kind of graphs, but it is also applicable to more sophisticated structures. We will basically consider two kinds of extensions: a more powerful labeling concept – leading to the concept of attributed graphs in Section 4.5.1; and more complex graphical structures – leading to the concept of graph structures in Section 4.5.2. By a generalization of graph structures we will then be able to cope with these two different extensions within the same framework. This framework, introduced in [5,16], opens new possibilities for defining graphs. (For example we can use instead of sets and labeling *functions*, graphs and labeling graph *morphisms* – this idea was worked out in [28] for the definition of class-based graph grammars.) Finally in Section 4.5.3 we review *high-level replacement systems*: a general axiomatic framework for transformation systems and their properties.

4.5.1 Attributed Graphs

The existence of labels allows us to distinguish vertices or edges within a graph according to their associated label. Thus, labels provide a very basic type concept for graphs. Usually in a software system the types of elements do not change during the execution of the system, and therefore the treatment of labels presented in the last sections (where label morphisms were identities) is very reasonable. Nevertheless, often a situation occurs in which the labels (types) of the elements in a graph are not relevant for the application of a production. In this case labels should take the role of parameters (generic types). But actually, the strict typing through labels presented so far requires a number of productions (one for each possible combination of labels) to describe this situation. For practical reasons this is not adequate, a higher-level kind of typing concept is needed. From now on we will refer to this kind of high-level types as attributes. The basic idea of attributes in graph transformations is to allow the handling of labels abstractly. This is realized by the presence of corresponding variables and a concept of assignment of these variables in an actual situation. In particular, the concept of attributes includes the presence of operations on these sets (of attributes).

Attributes are used in all graph grammar proposals for software engineering since they integrate structural (i.e. graphical) aspects of a system with data-type aspects (i.e. calculation of values). This leads to compact descriptions in which e.g. well-known arithmetic operations need not artificially be coded into graphical structures. Similar concepts of combining structural and algebraic aspects can be found in the theory of attributed string grammars [29] in algebraic high-level Petri-nets [30,31,32], which are a combination of Petri-nets and algebraic specifications, and in the specification language LOTOS [33], which integrates CCS-like specifications for the structural part with algebraic specifications for the data type component.

In the SPO approach, attributes have been integrated in [3] (see [34] for a corresponding extension of the DPO approach). This integration preserves the fundamental derivation concept of the algebraic approach i.e., both the manipulation of the graphical structure and the calculation of the new attributes are combined within a (single) pushout construction. In [3] attributes were specified using algebraic specifications in the sense of [35]. Algebraic specifications provide a well-established formalism for treating data-types, variables, evaluations and substitutions. Moreover not only a set of types is available as attributes, but we can make use of the operations of the specification in order to indicate abstractly relationships between types. The proposal for the integration of attributes into graph transformation reviewed here is a simplification of the approach in [3]. It has already been used in [36].

Before introducing formally the concept of an attributed graph, we need to introduce some basic notions of universal algebra. A *signature* $Sig = (S, OP)$ consists of a set of sorts and a family of sets $OP = (OP_{w,s})_{w \in S^*, s \in S}$ of operation symbols. For $op \in OP_{w,s}$, we also write $op : w \to s$. A *Sig-algebra* A is an S-indexed family $(A_s)_{s \in S}$ of carrier sets together with an OP-indexed family of mappings $(op^A)_{op \in OP}$ such that $op^A : A_{s_1} \times \ldots \times A_{s_n} \to A_s$ if $op \in OP_{s_1 \ldots s_n, s}$. If $w = s_1 \ldots s_n \in S^*$, we sometimes write A_w for $A_{s_1} \times \ldots \times A_{s_n}$. A *Sig-homomorphism* $f : A \to B$ between two Sig-algebras A and B is a sort-indexed family of total mappings $f = (f_s : A_s \to B_s)_{s \in S}$ such that $op^B(f(x)) = f(op^A(x))$ for all $x \in A_s$. The category $\boldsymbol{Alg(Sig)}$ has as objects all Sig-algebras and as morphisms all total homomorphisms between them. It is well-know that $\boldsymbol{Alg(Sig)}$ has all colimits. $\mathcal{U} : \boldsymbol{Alg(Sig)} \to \boldsymbol{Set^P}$ is a functor assigning to each Sig-algebra A the disjoint union of its carrier sets A_s, and to each homomorphism f the disjoint union the total functions f_s, for all $s \in S$.

Attributes are labels of graphical objects taken from an *attribute algebra*. Hence, an attributed graph consists of a (labeled) graph and an attribute algebra, together with some *attribute functions* connecting the graphical and the algebraic part.

Definition 4.5.1 (attributed graph)
Given a label alphabet L and a signature $Sig = (S, OP)$. Then $AG = (AG_V, AG_E, s^{AG}, t^{AG}, lv^{AG}, le^{AG}, AG_A, av^{AG}, ae^{AG})$ is a *Sig-attributed graph*, where

i) $AG_G = (AG_V, AG_E, s^{AG}, t^{AG}, lv^{AG}, le^{AG})$ is an L-labeled graph with labeling functions lv^{AG}, le^{AG} for vertices and edges (cf. Definition 3.3.1),

ii) AG_A is a Sig-algebra,

iii) $av^{AG} : AG_V \to \mathcal{U}(AG_A)$ and $ae^{AG} : AG_E \to \mathcal{U}(AG_A)$ are the vertex and edge attributing functions

A (*Sig-attributed graph) morphism* between two *Sig*-attributed graphs $AGi = (AGi_V, AGi_E, s^{AGi}, t^{AGi}, lv^{AGi}, le^{AGi}, AGi_A, av^{AGi}, ae^{AGi})$ for $i = 1, 2$ is a tuple $f = (f_G, f_A)$ where $f_G = (f_V, f_E)$ is a partial graph morphism, and f_A is a total algebra homomorphism such that $\forall v \in dom(f_V) \cdot \mathcal{U}(f_A)(av^{AG1}(v)) = av^{AG2}(f_V(v))$ and $\forall e \in dom(f_E) \cdot \mathcal{U}(f_A)\ (ae^{AG1}(e)) = ae^{AG2}(f_E(e))$; f is total (injective) if f_G and f_A are total (injective).

Proposition 4.5.2 (category *AGraph*^{P})
Sig-attributed graphs and Sig-attributed graph morphisms form a category, called ***AGraph***^{P}.

Proof
The proof is based on the fact that the composition of morphisms is well-defined. □

For the definition of direct derivations of attributed graphs as single-pushout constructions, it is essential that ***AGraph***^{P} has pushouts. The construction of pushouts in ***AGraph***^{P} can be seen as a special case of [3]: pushouts are constructed componentwise in ***Graph***^{P} and ***Alg(Sig)***. Actually, the category ***Agraph***^{P} has not only pushouts but all colimits (due to the fact that coproducts in ***Agraph***^{P} can be constructed componentwise in ***Set***). As most of the results presented in Section 4.3 are based on colimit constructions, they should carry over to the ***Agraph***^{P} setting. For a proof of the following theorem we refer to [3].

Theorem 4.5.3
The category ***AGraph***^{P} *has pushouts.* □

Productions, grammars, matches, and (direct) derivations using attributed graphs are defined analogously to the Definitions 4.2.2 and 4.2.6 in Section 4.2, by replacing the category ***Graph***^{P} of graphs and partial graph morphisms by some category ***AGraph***^{P} of attributed graphs.

Below we extend the Pacman example (Example 4.1 of Section 4.2) using attributes for counting the apples Pacman has eaten.

Example 4.14 (attributed graph transformation)
For the graphical part AG_G we use the same labeled graphs as in Example 4.1. The signature of natural numbers

Signature Nat: **sorts** nat
opns $0 \rightarrow nat$
$succ : nat \rightarrow nat$
$+ : nat \times nat \rightarrow nat$

is used as the signature for the attribute algebras AG_A.

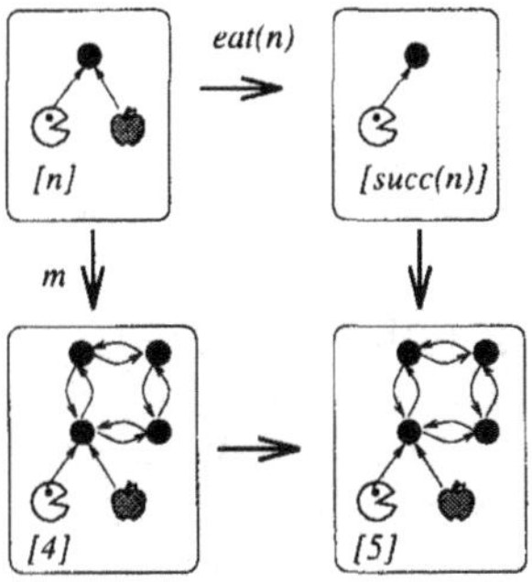

Figure 4.20: Counting Pacmans apples.

Attributes in productions are usually taken from "syntactical" (term) algebras. Figure 4.20 shows the attributed graph production $eat(n)$, where the left- and right-hand side graphs are attributed over the term algebra $T_{Nat}(X)$ with variables in $X = \{n\}$. The graphical part is similar to production *eat* in Figure 4.1 on page 290.[c] On the left-hand side, Pacman is attributed with the variable n representing the number of apples Pacman has eaten before the application of the production. On the right-hand side, n is replaced by $succ(n)$, i.e., the number of apples increases by one.

For the graphs to be rewritten, we fix some "semantical algebra", the algebra $I\!N$ of natural numbers. As a transformation of labeled graphs does not change the label set, this algebra is preserved by an attributed graph transformation as well. Figure 4.20 shows an application of the production $eat(n)$ to a graph representing a state where Pacman has already eaten four apples, and is about to eat the fifth one. The graphical part is matched as usual. Then we have to find an assignment to the variable n which is compatible with the matching of the graphical part, i.e., n is mapped to 4. The derived graph is constructed componentwisely by taking the pushout in $\boldsymbol{Graph^P}$ for the graphical part, and by extending the assignment $n \mapsto 4$ to $succ(n)$, leading to $succ(n) \mapsto 5$. In this way, attributes in the derived graphs are calculated from attributes in the given graphs. □

[c]Since the compatibility conditions for attributed graph morphisms (cf. Definition 4.5.1) do not allow to change attributes, carrier loops have to be introduced at attributed vertices, which are deleted and re-generated each time an attribute is modified. In order to simplify the presentation, these carrier loops are often omitted in the drawings.

4.5.2 Graph Structures and Generalized Graph Structures

In order to be more flexible in system modeling, often hypergraphs are used instead of graphs [37,38,39], see also Chapter 2 in this handbook.

Definition 4.5.4 (hypergraph)
A *hypergraph* $G = (V, E, s, t)$ consists of a set of vertices V, a set of edges E, and two total mappings $s, t : E \to V^*$ from the set of edges into the free monoid over the set of vertices which provide each hyperedge $e \in E$ with a sequences of n source and m target vertices $s(e) = v_1 \dots v_n$ and $t(e) = v_1 \dots v_m$, respectively.

Since in the algebraic approaches graphs are considered as algebras w.r.t. a certain signature, also hypergraphs are defined in this way. The signature for hypergraphs is given below, where hyperedges are sorted according to their number of source vertices n and target vertices m.

Signature $HSig$: **sorts** $V, (E_{n,m})_{n,m \in IN}$
opns $(s_1, \dots, s_n, t_1, \dots, t_m : E_{n,m} \to V)_{n,m \in IN}$

Example 4.15 (hypergraph)
In the following picture a concrete hypergraph of this signature is shown, having 3 non-empty carrier sets, namely $G_V = \{\bullet_1, \bullet_2, \bullet_3\}$, $G_{E1,1} = \{e_{1,1}\}$, and $G_{E1,2} = \{e_{1,2}\}$.

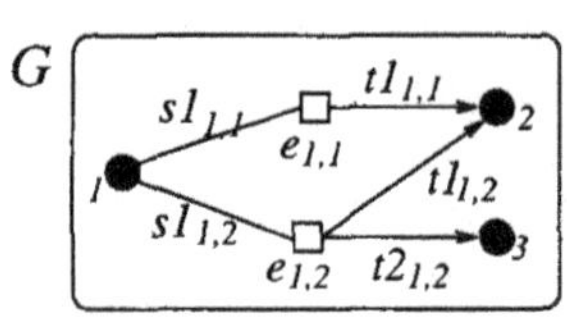

Generalizing partial graph morphisms, a partial homomorphism $f : G \to H$ between two algebras can be defined as total homomorphism $f! : dom(f) \to H$ from some subalgebra $dom(f) \subseteq G$. For each signature Sig this leads to a category of algebras and partial morphisms $\boldsymbol{Alg(Sig)^P}$. In [4] it was shown that this category is cocomplete if and only if all operations in Sig are unary, which motivated the following definition.

Definition 4.5.5 (graph structure signature, graph structure)
A **graph structure signature** GS is a signature which contains unary operator symbols only. A **graph structure** is a GS-algebra.

Most of the results presented in Section 4.3 have originally been elaborated for graph structures [7,4], which do not only include graphs and hypergraphs, but also hierarchical graphs and higher order graphs (having edges between edges).

An even larger framework is obtained by the concept of generalized graph structures [16]. In order to explain this let us reconsider Definition 4.5.4, where a hypergraph is given by two carrier sets G_V, G_E and two operations $s^G, t^G : G_E \to G_V^*$. Such an algebra can not be defined directly as a graph structure since the mappings s^G, t^G are not between the carriers but from the set of edges G_E to the free monoid G_V^* over the set of vertices G_V. This has led to the infinite graph structure signature for hypergraphs above. Using generalized graph structures instead we may realize directly the hypergraphs of Definition 4.5.4: A hypergraph is an algebra $G = (G_V, G_E, s^G, t^G : F_E(G_E) \to F_V(G_V))$, where G_V and G_E are sets of vertices and hyperedges, and s^G, t^G are operations between sets derived from the carrier sets by application of the functors F_E and F_V, associated with the sort symbols E and V, respectively. The functor F_E is the identity functor, i.e., $F_E(G_E) = G_E$, and the functor F_V maps every set of vertices to its free monoid, i.e., $F_V(G_V) = G_V^*$.

On morphisms (partial functions) these functors are defined as follows: F_E is the identity functor, i.e., $F_E(f_E) = f_E$. $F_V(f_V) = f_V^*$ shall be defined pointwisely: for each sequence of vertices $l = v_1 v_2 \dots v_n \in G_V^*$ $f_V^*(l) = f_V(v_1) f_V(v_2) \dots f_V(v_n) \in H_V^*$ if $f_V(v_i)$ is defined for all $i = 1 \dots n$, and otherwise it is undefined.

This allows to define GGS-morphisms. Figure 4.21 depicts a partial morphism $f = (f_E, f_V) : G \to H$ between generalized graph structures $G = (G_V, G_E, s^G, t^G)$ and $H = (H_V, H_E, s^H, t^H)$. It consists of a pair of mappings $f_E : G_E \to H_E$ and $f_V : G_V \to H_V$ between edges and vertices of G and H, respectively, such that the top diagram commutes for each operation symbol, i.e. $f_V^* \circ s^G \circ f_E? = s^H \circ f_E!$ and $f_V^* \circ t^G \circ f_E? = t^H \circ f_E!$, where $f_E? : dom(f_E) \to G_E$ and $f_E! : dom(f_E) \to H_E$ form the span-representation of f_E.

Thus generalized graph structures may be described as comma categories w.r.t. certain signatures where the compatibility requirement of morphisms has been relaxed in order to allow partial morphisms ('weak homomorphisms'). Analogously to comma-categories, colimits in GGS-categories can be constructed componentwisely followed — in the GGS-case — by a free construction ('totalization').

The concept of GGS provides a constructive approach for generating internally structured categories of graph-like structures and partial morphisms from simpler ones. Among the large number of examples there are labeled and unlabeled

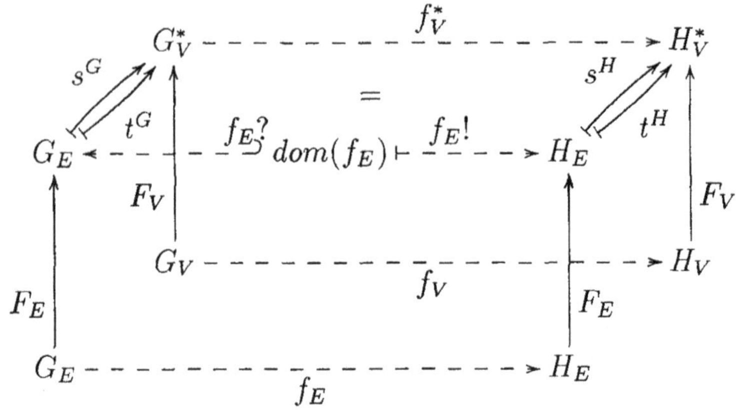

Figure 4.21: Definition of a partial GGS morphism $f : G \to H$

graphs, hypergraphs, attributed graphs [d], attributed and labeled hypergraphs, hierarchical graphs, graph-interpreted (or typed) graphs, graph grammars, Petri Nets, and Algebraic High-Level (AHL) Nets. Properties of these categories, like cocompleteness, can be inferred from properties of the components and the way in which they are composed.

4.5.3 High-Level Replacement Systems

The formal investigation of graph grammars based on different kinds of graphs has often led to similar results based on similar proofs. This fact gave raise to the question whether it is possible to 'reuse' the proofs from one kind of graph grammar to obtain analogous results in another kind. The definition of *High-Level Replacement Systems* [6], shortly HLR systems, was the answer to this question. They provide an abstract framework in which not graphs, but objects of an arbitrary (instance) category are transformed (generalizing the ideas of graph grammars to arbitrary categorical grammars). Obviously, not every instance category gives raise to the same results; they depend very much on the properties of these categories. Hence, due to its categorical nature, the focus of HLR systems is not on the structure of objects (as in the GGS approach) but on the properties of its instance categories. Minimal conditions are extracted which allow for definitions and theorems concerning, for example, parallelism, embedding, or amalgamation, and in this way such theorems can be

[d] Recall that attributed graphs cannot be seen as graph structures, due to the fact that arbitrary signatures, like of booleans and natural numbers, may have non-unary operations.

proven for the biggest class of instance categories. Originally HLR systems were defined following the DPO approach [40]. Here we present the basic definitions of HLR systems of type SPO, developed in [6]. In the following, HLR systems are assumed to be of type SPO.

The basic assumption of HLR systems is that the structures we want to transform belong to some category $\boldsymbol{Cat}$. The theory of HLR systems defines the basic concepts of production, derivation and grammar in a generic (categorical) way: a production is a morphism in $\boldsymbol{Cat}$, a match belongs to a distinguished class of morphism, and direct derivations are given by pushouts. For example, if $\boldsymbol{Cat}=\boldsymbol{Graph^P}$, we obtain the SPO framework as introduced in Section 4.2.

Definition 4.5.6 (HLR system)
Let $\boldsymbol{Cat}$ be a category and $\boldsymbol{Match}$ be a subcategory of $\boldsymbol{Cat}$ such that $\boldsymbol{Match}$ and $\boldsymbol{Cat}$ coincide on objects.

1. A **production** $r : L \rightarrow R$ is a $\boldsymbol{Cat}$ morphism.

2. An object G can be **directly derived** to an object H using a production $r : L \rightarrow R$ if there is a pushout (1) in $\boldsymbol{Cat}$ such that m is a **match**, i.e., it belongs to $\boldsymbol{Match}$.

$$\begin{array}{ccc} L & \xrightarrow{r} & R \\ {\scriptstyle m}\downarrow & (1) & \downarrow{\scriptstyle m^*} \\ G & \xrightarrow[r^*]{} & H \end{array}$$

3. A **derivation** is a sequence of direct derivations.

4. An **HLR system** $HLRS = (I, P, T)$ over $\boldsymbol{Cat}$ is given by a start object I, a set of productions P, and a class of terminal objects $T \subseteq \boldsymbol{Cat}$.

The HLR framework covers several concrete rewrite formalisms, which are obtained by choosing a concrete category $\boldsymbol{Cat}$ of structures, satisfying some basic conditions. The following conditions ensure that the well-known parallelism results are valid in the instance category $\boldsymbol{Cat}$.

Definition 4.5.7 (SPO conditions and categories)
The following conditions (1)–(4) are called *SPO conditions for parallelism of HLR systems*. A category $\boldsymbol{Cat}$ together with a subcategory $\boldsymbol{Match}$ as given in Definition 4.5.6 is called *SPO category* if these conditions are satisfied.

1. Existence of pushouts with ***Match*** morphisms, i.e. ***Cat*** has pushouts if at least one morphism is in ***Match***.

2. ***Match*** has finite coproducts, which are preserved by the inclusion functor $\subseteq$: ***Match*** $\to$ ***Cat***.

3. Prefix closure of coproducts.[e]

4. Prefix closure of ***Match***, i.e. if $f \circ g \in$ ***Match*** then $g \in$ ***Match***.

Examples for SPO categories are $\boldsymbol{Graph^P}$, $\boldsymbol{Alg(GS)^P}$ (with GS being a graph structure signature), $\boldsymbol{SPEC^P}$ (category of algebraic specifications and strict partial morphisms) [6]. The interpretation of productions and derivations depends on the instance category. For example, transformations of algebraic specifications may be interpreted as interconnection of modules in the context of modular system design [41]. In [6] it is shown that slightly different versions of the Local Church Rosser Theorem 4.3.6 and the Parallelism Theorem 4.3.8 hold in HLR-systems over SPO categories, and it is an interesting topic for further research to generalize also the other results of Section 4.3 to the HLR-framework.

4.6 Comparison of DPO and SPO Approach

Section 3.2 provided an informal introduction in the algebraic theory of graph rewriting, where many relevant concepts and results have been introduced in terms of problems. In this section the solutions to these problems – given throughout Chapter 3 and this chapter for the DPO and SPO approaches, respectively – are summarized and compared with each other. Working out the similarities and the differences of the corresponding results will allow to understand the relationships between the two algebraic approaches.

Since in Section 3.2 the concepts are introduced independently of a particular approach, this provides us with an abstract terminology that can be considered as a common "signature" for (algebraic) graph rewriting approaches. This signature has sort symbols for graphs, productions, and derivations, etc., and operation or predicate symbols for (direct) derivation, parallel and sequential composition, etc., and the various relations that may hold between productions or derivations like the subproduction or the embedding relation. Most of

[e] A morphism $p : L \to P$ is called a *prefix* of $r : L \to R$ if there is $x : P \to R$ such that $x \circ p = r$. This type of properties which make reference to prefixes of productions and requires some closure properties for all prefixes is typical for the theory of HLRS. It is the device to control the interaction of "partiality" and "totality" of morphisms in ***Cat*** resp. ***Match*** (see [6] for more details).

the problems of Section 3.2 ask for conditions, i.e., relations on derivations, ensuring that some construction is defined for these derivations. The parallelization condition of Problem 3.2.2, for example, which is a unary relation on two-step sequential derivations, ensures that the synthesis construction is defined, leading to an equivalent direct parallel derivation.

In Chapter 3 and this chapter, the signature of Section 3.2 is interpreted in the DPO and SPO approach, respectively, by providing a definition for its sorts, operation and predicate symbols in terms of the productions, derivations and constructions, etc., of the approach. Hence, the DPO and the SPO approach can be considered as models of the same signature. The interpretation of the parallelization condition of Problem 3.2.2.2 in the DPO approach, for example, is given by Definition 3.4.2 of sequential independence, and the Synthesis Lemma 3.4.9 shows that this interpretation is indeed a solution to Problem 3.2.2.2. In a similar way, many of the definitions of Chapter 3 and this chapter can be seen as interpretations of the signature of Section 3.2, and the corresponding results show that they satisfy the statement of the problems.

This view allows for two different ways of comparing the two algebraic approaches. The first one is w.r.t. to their properties: A property of an approach is a statement using the abstract terminology of Section 3.2 that is valid in this particular approach. The basic property of the SPO approach, for example, is the completeness of its direct derivations, that is, given a match for a production there is always a corresponding direct derivation. DPO direct derivations are not complete because of the gluing condition. On the other hand, in the DPO approach we have that direct derivations are invertible, which is not true in the SPO approach. Summarizing in this way the valid statements leads to a nice abstract characterization of the two algebraic approaches in terms of their properties.

It is well-known that models of the same signature can be compared by homomorphisms. In our case the signature of Section 3.2 defines a notion of "homomorphism" between graph rewriting approaches being interpretations of this signature. Such a homomorphism is given by mappings between the carriers, i.e., productions, matches, derivations, etc., which are compatible with the interpretation of the operation and predicate symbols, i.e., with the parallel or sequential composition of productions, for example. Thereby, it embeds one approach in the other approach, which is necessary in order to compare directly the concepts of the two approaches. Below we show how the DPO approach may be embedded in the SPO approach. This applies not only to the basic concepts, like productions and direct derivations, but also to the constructions and results of the theory, like the parallel production and the parallelism theorem.

Beside the direct comparison of corresponding concepts, such an embedding can be used in several other ways. On the one hand, we may transfer theoretical results, concerning for example analysis techniques for graph grammars, between the two approaches. On the other hand, it shows that we may simulate the constructions of the DPO approach within the SPO approach, which may be quite useful if we aim at a common implementation for the algebraic approaches. Finally, if we restrict the SPO approach by suitable application conditions (as introduced, for example, in Section 4.4), we may use the idea of a homomorphism in order to show the equivalence of the DPO approach and the correspondingly restricted SPO approach.

The section is organized like Section 3.2. In each of the following sections we summarize the solutions to the problems of Section 3.2, state the corresponding properties of the approaches, and discuss the embedding of the DPO in the SPO approach w.r.t. to the concepts of this section.

4.6.1 Graphs, Productions, and Derivations

According to Section 3.2.1 each approach to graph rewriting is distinguished by three characteristics: its notion of a graph, the conditions under which a production may be applied, and the way the result of such an application is constructed. These characteristics define what is called a direct derivation. The DPO and SPO approach use the same kind of graphs (see Definition 3.3.1). The productions of the two approaches are essentially the same except of differences in the representation: A DPO production $L \stackrel{l}{\longleftarrow} K \stackrel{r}{\longrightarrow} R$ is a span of total injective graph morphism (compare Definition 3.3.2). A SPO production is instead a partial graph morphism $L \stackrel{p}{\longrightarrow} R$, i.e., a total graph morphism $dom(p) \stackrel{p!}{\longrightarrow} R$ from some subgraph $dom(p)$ of L to R (cf. Definitions 4.2.2 and 4.2.1). Both concepts of production have been extended by a name, which is used for identifying the production and for storing its internal structure (in case of a composed production).

Definition 4.6.1 (translation of SPO and DPO productions)
Let $p : (L \stackrel{s}{\longrightarrow} R)$ be a SPO production. Then $\mathcal{D}(p) : (L \stackrel{l}{\longleftarrow} dom(r) \stackrel{r}{\longrightarrow} R)$ denotes its **translation to a DPO production**, where l is the inclusion of $dom(s)$ in L and r is the domain restriction of s to $dom(s)$. Conversely, let $p : (L \stackrel{l}{\longleftarrow} K \stackrel{r}{\longrightarrow} R)$ be a DPO production. Then, $\mathcal{S}(p) : (L \stackrel{s}{\longrightarrow} R)$ denotes its **translation to a SPO production**, where $dom(s) = l(K)$ and $s = r \circ l^{-1}$. The partial morphism s is well-defined since l is supposed to be injective in Definition 3.3.2.

Hence, SPO productions can be considered as DPO productions where the interface graph K is represented in a unique way as a subgraph of the left-hand side graph L. The mapping $\mathcal{S}$ of DPO productions p to SPO productions $\mathcal{S}(p)$ will be used throughout this section to translate the concepts and results of the DPO approach to the corresponding ones of the SPO approach. The most basic concept of each graph rewriting approach is that of a direct derivation. In fact, the mapping $\mathcal{S}$ translates each direct DPO derivation to a direct SPO derivation where the match is d-injective and d-complete (compare Definition 4.2.7).

Proposition 4.6.2 (translation of DPO derivations)
Let $p : (L \xleftarrow{l} K \xrightarrow{r} R)$ be a DPO production, $\mathcal{S}(p) : L \rightarrow R$ its translation to a SPO production, and $L \xrightarrow{m} G$ a match for p into a graph G. Then m is also a match for $\mathcal{S}(p)$ and there is a direct DPO derivation $d = (G \overset{p,m}{\Longrightarrow} H)$ if and only if there is a direct SPO derivation $\mathcal{S}(d) = (G \overset{\mathcal{S}(p),m}{\Longrightarrow} H)$ such that m is a d-injective and d-complete match for $\mathcal{S}(p)$. In this way, the mapping $\mathcal{S}$ is extended from productions to derivations.

Proof
See [4]. □

Not every direct derivation in the SPO approach can be obtained from a DPO derivation using the translation $\mathcal{S}$. In contrast to DPO derivations a direct SPO derivation $G \overset{p,m}{\Longrightarrow} H$ always exists if there is a match m for a production p. This first and most important property distinguishing the DPO and the SPO approach is called *completeness of direct derivations.* Indeed, most of the differences between the DPO and the SPO approach are caused by the fact that SPO derivations are complete while DPO derivations are not.

Property 4.6.3 (completeness of SPO derivations)
Direct derivations in the SPO approach are **complete**, *i.e., for each production $p : L \rightarrow R$ and each match $L \xrightarrow{m} G$ for p into a graph G there is a direct derivation $G \overset{p,m}{\Longrightarrow} H$.* □

Because of the gluing condition (see Proposition 3.3.4), DPO derivations are not complete. In fact, being complete essentially means to be free of any application conditions, that is, SPO derivations with application conditions as introduced in Section 4.4 are incomplete as well. The gluing condition, however, causes another interesting property of the DPO approach, the *invertibility of direct derivations.*

Property 4.6.4 (invertibility of DPO derivations)
Direct derivations in the DPO approach are **invertible**, *i.e., for each direct derivation* $G \overset{p,m}{\Longrightarrow} H$ *using production* $p : (L \overset{l}{\longleftarrow} K \overset{r}{\longrightarrow} R)$ *there is an inverse derivation* $H \overset{p^{-1},m^*}{\Longrightarrow} G$ *using the inverse production* $p^{-1} : (R \overset{r}{\longleftarrow} K \overset{l}{\longrightarrow} L)$, *where* $R \overset{m^*}{\longrightarrow} H$ *is the co-match of* $G \overset{p,m}{\Longrightarrow} H$.

Proof
See [42]. □

Constructing the inverse of a direct derivation can be seen as a kind of "undo". SPO direct derivations are not invertible in general since they may model implicit effects, like the deletion of dangling edges, which are not invertible. Because of these "side effects", SPO derivations are more difficult to understand and to control than DPO derivations and may be considered, in certain situations, as "unsafe". If direct derivations are invertible, there are no implicit effects. Hence, DPO derivations show a "safe" behavior that is easier to determine beforehand. In view of the overall complexity of the formalism, this is important if people from other areas shall use graph rewriting techniques for modeling their problems.

Most graph grammar approaches are not dedicated to a particular application area but may be considered as "general purpose" formalisms. For many applications, however, a tailored approach is more adequate which is as powerful as necessary and as simple as possible, in order to allow for a natural modeling of the problems. Then, standard application conditions like in the DPO approach are needed to restrict the expressiveness of the approach if necessary. Exceptions of these conditions, however, should be possible as well, if they are explicitly specified by the user. The graph grammar based specification language PROGRES [43] (see also Chapter 7 in this handbook) provides an example of this concept, where e.g. injective matches are standard, but identification of vertices may be explicitly allowed.

It depends on the choice of the standard application conditions and of the possible exceptions, which approach is the most adequate one for a particular application. A very general solution is provided by user-defined application conditions, as introduced in Section 4.4. In particular in the SPO setting they complement in a nice way the generality of the pure approach.

4.6.2 Independence and Parallelism

Interleaving

The Local Church-Rosser Problem 3.2.1 asked for two conditions formalizing the concept of concurrent direct derivations from two different points of view:

1. Two alternative direct derivations $H_1 \overset{p_1,m_1}{\Longleftarrow} G \overset{p_2,m_2}{\Longrightarrow} H_2$ are concurrent if they are not in conflict (parallel independence).

2. Two consecutive direct derivations $G \overset{p_1,m_1}{\Longrightarrow} H_1 \overset{p_2,m_2'}{\Longrightarrow} X$ are concurrent if they are not causally dependent, i.e., if there are also direct derivations $G \overset{p_2,m_2}{\Longrightarrow} H_2 \overset{p_1,m_1'}{\Longrightarrow} X$ (sequential independence).

For the DPO approach these conditions are given the Definitions 3.4.1 and 3.4.2, and the Local Church-Rosser Theorem 3.4.3 ensures that they indeed solve Problem 3.2.1. The corresponding solution for the SPO approach is provided by the Definitions 4.3.1 and 4.3.3 together with Theorem 4.3.6.

Both solutions formalize the same intuitive idea. Due to the different representation of productions in the DPO and SPO approach, however, the corresponding conditions are stated in a different way. The following proposition shows that they are equivalent via the correspondence of productions and derivations established in Definition 4.6.1 and Proposition 4.6.2.

Proposition 4.6.5 (equivalence of DPO and SPO independence)
Let $p_1 : (L_1 \overset{l_1}{\longleftarrow} K_1 \overset{r_1}{\longrightarrow} R_1)$ and $p_2 : (L_2 \overset{l_2}{\longleftarrow} K_2 \overset{r_2}{\longrightarrow} R_2)$ be two DPO productions, and $\mathcal{S}(p_1) : L_1 \to R_1$ and $\mathcal{S}(p_2) : L_2 \to R_2$ their translation to SPO productions. Two alternative direct DPO derivations $H_1 \overset{p_1,m_1}{\Longleftarrow} G \overset{p_2,m_2}{\Longrightarrow} H_2$ are parallel independent in the sense of Definition 3.4.1 if and only if the corresponding SPO derivations $H_1 \overset{\mathcal{S}(p_1),m_1}{\Longleftarrow} G \overset{\mathcal{S}(p_2),m_2}{\Longrightarrow} H_2$ are parallel independent in the sense of Definition 4.3.1.

Two consecutive direct DPO derivations $G \overset{p_1,m_1}{\Longrightarrow} H_1 \overset{p_2,m_2'}{\Longrightarrow} X$ are sequential independent in the sense of Definition 3.4.2 if and only if the corresponding SPO derivations $G \overset{\mathcal{S}(p_1),m_1}{\Longrightarrow} H_1 \overset{\mathcal{S}(p_2),m_2'}{\Longrightarrow} X$ are sequential independent in the sense of Definition 4.3.3.

Proof
According to Definition 3.4.1 the direct DPO derivations $H_1 \overset{p_1,m_1}{\Longleftarrow} G \overset{p_2,m_2}{\Longrightarrow} H_2$ are parallel independent if $m_1(L_1) \cap m_2(L_2) \subseteq m_1(l_1(K_1)) \cap m_2(l_2(K_2))$. Using the translation $\mathcal{S}$ to SPO productions (see Definition 4.6.1) this is the case iff $m_1(L_1) \cap m_2(L_2) \subseteq m_1(dom(\mathcal{S}(p_1)) \cap m_2(dom(\mathcal{S}(p_2)))$ holds for the

corresponding SPO derivations. The two direct SPO derivations $H_1 \overset{\mathcal{S}(p_1),m_1}{\Longleftarrow} G \overset{\mathcal{S}(p_2),m_2}{\Longrightarrow} H_2$ are parallel independent if $m_1(L_1) \cap m_2(L_2 - dom(\mathcal{S}(p_2))) = \emptyset$ and $m_2(L_2) \cap m_1(L_1 - dom(\mathcal{S}(p_1))) = \emptyset$, which is equivalent to $m_1(L_1) \cap m_2(L_2) \subseteq dom(\mathcal{S}(p_2)))$ and $m_2(L_2) \cap m_1(L_1(\subseteq dom(\mathcal{S}(p_1)))$, and hence to $m_1(L_1) \cap m_2(L_2) \subseteq m_1(dom(\mathcal{S}(p_1)) \cap m_2(dom(\mathcal{S}(p_2)))$.
In a similar way one shows the equivalence of the notions of sequential independence in the DPO and SPO approach. □

In Section 3.2.2 independent derivations have been interpreted as to be concurrent in the interleaving model of concurrency. In this view, Proposition 4.6.5 states that both approaches allow for the same amount of "interleaving parallelism".

Recently, both algebraic approaches came up with a weaker, asymmetric notion of independence, ensuring that two alternative direct derivation can at least be sequentialized into one order. In the DPO approach this concept has been introduced in [44] under the name "serializability", while in the SPO approach we speak of "weak parallel independence" in Definition 4.3.1. In turn, the weak notion of sequential independence, introduced in Definition 4.3.3, ensures that the second direct derivation does not depend on the first one, that is, they may also occur in parallel.

Explicit Parallelism

In order to represent parallel computations in a more explicit way, Section 3.2.2 assumed a parallel composition "+" of productions leading to the notions of parallel production and derivation. In the DPO and SPO approaches this operator is interpreted as the coproduct (disjoint union) of productions, see Definition 3.4.4 [f] and Definition 4.3.7, respectively. In both approaches, a direct parallel derivation is a direct derivation using the parallel production. With respect to these definitions, the Parallelism Problem 3.2.2 asked for the relationship between parallel and sequential derivations. The basic requirement is, of course, that each two direct derivations which are concurrent in the interleaving model can be put in parallel using the explicit notion of parallelism. This property is called completeness of parallel derivations below. On the other hand, if explicit parallelism allows for exactly the same amount of concurrency as the interleaving model, we say that it is safe (w.r.t. interleaving parallelism). In this case, the effect of a parallel derivation can always be obtained by a sequential derivation, too. With the following DPO parallelism properties we

[f] In fact, Definition 3.4.4 introduces the parallel composition of an arbitrary, finite number of productions.

summarize the DPO solution to the Parallelism Problem 3.2.2 provided by the DPO Parallelism Theorem 3.4.6.

Property 4.6.6 (parallelism properties of DPO)
Parallel derivations in the DPO approach satisfy the following completeness and safety properties:

Completeness: *Each sequentially independent DPO derivation* $G \stackrel{p_1,m_1}{\Longrightarrow} H_1 \stackrel{p_2,m_2'}{\Longrightarrow} X$ *satisfies the* parallelization condition, *i.e., there is an equivalent direct parallel DPO derivation* $G \stackrel{p_1+p_2,m}{\Longrightarrow} X$.

Safety: *Each direct parallel DPO derivation* $G \stackrel{p_1+p_2,m}{\Longrightarrow} X$ *satisfies the* sequentialization condition, *i.e., there are equivalent, sequentially independent DPO derivations* $G \stackrel{p_1,m_1}{\Longrightarrow} H_1 \stackrel{p_2,m_2'}{\Longrightarrow} X$ *and* $G \stackrel{p_2,m_2}{\Longrightarrow} H_2 \stackrel{p_1,m_1'}{\Longrightarrow} X$. □

Due to the completeness of SPO direct derivations, also its parallel direct derivations are complete, i.e., a parallel production $L_1 + L_2 \stackrel{p_1+p_2}{\longrightarrow} R_1 + R_2$ is applicable to any match $L_1 + L_2 \stackrel{m}{\longrightarrow} G$. Therefore, each two alternative direct derivations $H_1 \stackrel{p_1,m_1}{\Longleftarrow} G \stackrel{p_2,m_2}{\Longrightarrow} H_2$ can be composed to a parallel direct derivation $G \stackrel{p_1+p_2,m_1+m_2}{\Longrightarrow} X$, regardless of their independence. In particular, this allows for the parallel composition of weakly (parallel or sequential) independent direct derivations using the same notion of parallel production. Hence, SPO parallel derivations are also complete w.r.t. weakly independent sequential derivations, which is shown by the implication from 1. to 2. in the Weak Parallelism Theorem 4.3.8. Contrastingly, in the DPO approach, a more complex, so called synchronized (parallel) composition had to be developed in [44] in order to put two serializable (i.e., weakly parallel independent) direct derivations in parallel.

On the other hand, because of its generality, SPO parallelism is no longer safe w.r.t. to sequential derivations, neither in the symmetric nor asymmetric version. We can, however, distinguish those direct parallel derivations which may be sequentialized at least in one order by considering the corresponding pair of alternative derivations, which is shown by the implication from 2. to 1. in the Weak Parallelism Theorem 4.3.8.

Finally let us investigate the relationship between DPO and SPO parallelism using the translation $\mathcal{S}$ introduced in Section 4.6.1. The first important observation is, that this translation is compatible with parallel composition, that is, given DPO productions p_1 and p_2, we have that $\mathcal{S}(p_1) + \mathcal{S}(p_2) = \mathcal{S}(p_1 + p_2)$.[g]

[g] In fact, this is true only up to isomorphism since there are infinitely many isomorphic parallel productions for the same two elementary productions. This problem can be solved

This gives us the possibility to compare in a more direct way the relationships between parallel and sequential derivations in the two approaches. The following proposition states that the mapping $\mathcal{S}$ preserves the sequentialization and the parallelization conditions and is compatible with the analysis and synthesis construction.

Proposition 4.6.7 (compatibility of $\mathcal{S}$ with analysis and synthesis) *Let $d = (G \overset{p_1+p_2,m}{\Longrightarrow} X)$ be a direct parallel DPO derivation, and $\rho_1 = (G \overset{p_1,m_1}{\Longrightarrow} H_1 \overset{p_2,m_2'}{\Longrightarrow} X)$ and $\rho_2 = (G \overset{p_2,m_2}{\Longrightarrow} H_2 \overset{p_1,m_1'}{\Longrightarrow} X)$ its sequentializations. Then $\mathcal{S}(d) = (G \overset{\mathcal{S}(p_1)+\mathcal{S}(p_2),m}{\Longrightarrow} X)$ may be sequentialized to $\mathcal{S}(\rho_1) = (G \overset{\mathcal{S}(p_1),m_1}{\Longrightarrow} H_1 \overset{\mathcal{S}(p_2),m_2'}{\Longrightarrow} X)$ and $\mathcal{S}(\rho_2) = (G \overset{\mathcal{S}(p_2),m_2}{\Longrightarrow} H_2 \overset{\mathcal{S}(p_1),m_1'}{\Longrightarrow} X)$.*
Moreover, let $\rho = (G \overset{p_1,m_1}{\Longrightarrow} H_1 \overset{p_2,m_2'}{\Longrightarrow} X)$ be a sequentially independent DPO derivation and $d = (G \overset{p_1+p_2,m}{\Longrightarrow} X)$ the direct parallel DPO derivation obtained by the synthesis construction. Then $\mathcal{S}(\rho) = (G \overset{\mathcal{S}(p_1),m_1}{\Longrightarrow} H_1 \overset{\mathcal{S}(p_2),m_2'}{\Longrightarrow} X)$ is sequentially independent, and the corresponding direct parallel SPO derivation is $\mathcal{S}(d) = (G \overset{\mathcal{S}(p_1+p_2),m}{\Longrightarrow} X)$.

Proof

Let d be the direct parallel DPO derivation above. Then there are parallel independent direct derivations $H_1 \overset{p_1,m_1}{\Longleftarrow} G \overset{p_2,m_2}{\Longrightarrow} H_2$. According to the analysis construction in the DPO approach (cf. Lemma 3.4.8), the match m_2' of the second direct derivation in ρ_1 is defined by $m_2' = r_1^* \circ k_1$ in the diagram below, where $p_1^* : (G \overset{l_1^*}{\longleftarrow} D_1 \overset{r_1^*}{\longrightarrow} H_1)$ is the co-production of $G \overset{p_1,m_1}{\Longrightarrow} H_1$ and k_1 exists by the above parallel independence s.t. subdiagram (1) commutes (cf. Definition 3.4.1).

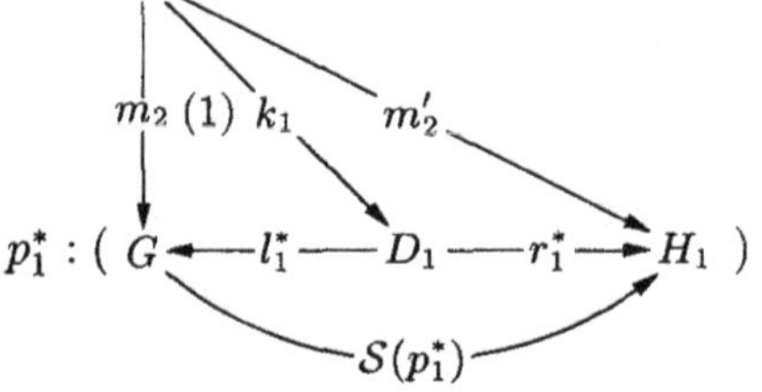

Now let $\mathcal{S}(p_1^*)$ be the translation of p_1^* to a SPO production (cf. Definition 4.6.1). Then, $m_2' = r_1^* \circ k_1 = r_1^* \circ (l_1^*)^{-1} \circ m_2 = \mathcal{S}(p_1^*) \circ m_2$ since $k_1 = (l_1^*)^{-1} \circ m_2$ by commutativity of (1) and $\mathcal{S}(p_1^*) = r_1^* \circ (l_1^*)^{-1}$ by Definition 4.6.1. But this

by assuming a fixed coproduct construction (like, for example, the disjoint union), which induces a coproduct functor as shown in Appendix A.1 of 3.

is exactly the definition of the match of the second direct derivation in the SPO Weak Parallelism Theorem 4.3.8, i.e., $G \stackrel{S(p_1),m_1}{\Longrightarrow} H_1 \stackrel{S(p_2),m_2'}{\Longrightarrow} X$ is indeed a sequentialization of $G \stackrel{S(p_1)+S(p_2),m}{\Longrightarrow} X$.
The second part of Proposition 4.6.7 can be shown in a similar way by exchanging m_2 and m_2'. ∎

4.6.3 Embedding of Derivations and Derived Productions

Section 3.2.3 introduced two problems, the Embedding Problem 3.2.3 and the Derived Production Problem 3.2.4, and it has been anticipated that the solution to the first problem is based on the solution to the second. The Embedding Problem asked for a condition under which an embedding of a graph G_0 via a morphism $e_0 : G_0 \to X_0$ may be extended to a derivation $\rho = (G_0 \stackrel{p_1,m_1}{\Longrightarrow} \cdots \stackrel{p_k,m_k}{\Longrightarrow} G_k)$, leading to an embedded derivation $\delta = (X_0 \stackrel{p_1,n_1}{\Longrightarrow} \cdots \stackrel{p_k,n_k}{\Longrightarrow} X_k)$. The idea was, to represent the informations relevant for the embedding of ρ by a *derived production* $\langle\rho\rangle : G_0 \rightsquigarrow G_n$ such that there is an embedding ρ via e_0 if and only if the derived production $\langle\rho\rangle$ is applicable at this match. The corresponding constructions are given in Section 3.6.1 for the DPO approach and in Section 4.3.2 for the SPO approach. In both cases, the derived production $\langle\rho\rangle$ is obtained as the sequential composition $\langle d_1\rangle; \ldots ; \langle d_k\rangle$ of the directly derived productions $\langle d_i\rangle : G_{i-1} \rightsquigarrow G_i$ of the given derivation ρ. Then, horizontal and vertical composition properties of direct derivations ensure that the derived production is applicable at a given injective match if and only if the original derivation may be embedded along this morphism. This is formulated as the *derived production property* below, see Theorem 3.6.3 and Theorem 4.3.15 for corresponding statements in the DPO and SPO approach, respectively.

Property 4.6.8 (derived production)
DPO and SPO approach satisfy the derived production property, that is, for each derivation $\rho = (G_0 \stackrel{p_1}{\Longrightarrow} \cdots \stackrel{p_n}{\Longrightarrow} G_n)$ and each injection $G_0 \stackrel{e_0}{\longrightarrow} X_0$ there is a derived derivation $X_0 \stackrel{\langle\rho\rangle}{\Longrightarrow} X_n$ using the derived production $\langle\rho\rangle : G_0 \rightsquigarrow G_n$ if and only if there is an an embedding $e : \rho \to \delta$ of ρ into a derivation $\delta = (X_0 \stackrel{p_1}{\Longrightarrow} \cdots \stackrel{p_n}{\Longrightarrow} X_n)$ using the original sequence of productions $p_1, \ldots, p_n$. □

The embedding problem is now reduced to the question if the derived production is applicable at the embedding morphism e_0. In the DPO approach approach this means that a derivation ρ may be embedded via e_0 if this match satisfies the gluing condition of the derived production (which specializes to the dangling condition since e_0 is supposed to be injective). Since in the SPO approach there are no further conditions for the applicability of a production but the existence of a match, the embedding condition is satisfied for each derivation $G_0 \overset{p_1,m_1}{\Longrightarrow} \cdots \overset{p_k,m_k}{\Longrightarrow} G_k$ and each embedding morphism $G_0 \overset{e_0}{\longrightarrow} X_0$. Therefore, we say that the embedding of SPO derivations is *complete.*

Property 4.6.9 (completeness of SPO embedding)
The embedding of derivations in the SPO approach is complete, i.e., for each derivation $\rho = (G_0 \overset{p_1,m_1}{\Longrightarrow} \cdots \overset{p_k,m_k}{\Longrightarrow} G_k)$ *and each embedding morphism* $e_0 : G_0 \to X_0$ *there is a derivation* $\delta = (X_0 \overset{p_1,n_1}{\Longrightarrow} \cdots \overset{p_k,n_k}{\Longrightarrow} X_k)$ *and an embedding* $e : \rho \to \delta$. □

As anticipated above, the DPO approach is not complete w.r.t. the embedding of derivations.
Finally it is worth stressing that the translation $\mathcal{S}$ of productions and derivations from the DPO approach to the SPO approach is also compatible with the concepts of this section. For DPO productions p_1 and p_2, for example, we have that $\mathcal{S}(p_1; p_2) = \mathcal{S}(p_1); \mathcal{S}(p_2)$ provided that the left-hand side is defined. As a consequence, we may show that $\mathcal{S}$ is compatible with the construction of derived productions, i.e., given a DPO derivation ρ we have $\mathcal{S}(\langle\rho\rangle) = \langle\mathcal{S}(\rho)\rangle$.

4.6.4 Amalgamation and Distribution

The concepts of amalgamated and distributed derivations are informally introduced in Section 3.2.4. Their formalization in the DPO approach is given in Section 3.6.2, while the SPO variants are introduced in Section 4.3.3. The main idea is more or less the same for both approaches: Synchronization of productions p_1 and p_2 is modeled by a common subproduction p_0, i.e., a production that is related to the elementary productions by compatible embeddings. The application of two synchronized productions (to a global state) is modeled by applying their amalgamated production, that is, the gluing of the elementary productions along their common subproduction.

A distributed graph $DG = (G_1 \overset{g_1}{\longleftarrow} G_0 \overset{g_2}{\longrightarrow} G_2)$ models a state that is splitted into two local substates, related by a common interface state. Gluing the local graphs G_1 and G_2 along their interface G_0 yields the global graph $\oplus DG = G_1 \oplus_{G_0} G_2$ of the system. The local states are consistent (with each other) if they form a total splitting of the global graph, i.e., if the interface embeddings g_1 and g_2 are total. The transformation of a distributed graph $DG = (G_1 \overset{g_1}{\longleftarrow} G_0 \overset{g_2}{\longrightarrow} G_2)$ is done by local direct derivations $d_i = (G_i \overset{p_i,m_i}{\Longrightarrow} H_i)$ for $i \in \{0,1,2\}$ where $p_1 \overset{in^1}{\leftarrow} p_0 \overset{in^2}{\rightarrow} p_2$ are synchronized productions and $m_i : L_i \to G_i$ are compatible matches for p_i into the two local graphs and the interface graph.

A distributed derivation is synchronous if the given and the derived distributed graphs are total splittings.

Distributed derivations in the DPO approach are synchronous by definition. The distributed gluing condition, which is used as a standard application condition for distributed derivations, ensures that no element in a local graph is deleted as long as it is referenced from the interface graph. This ensures that the derived distributed graph is again a total splitting. Hence, in the DPO approach, distributed derivations show the same safe behavior like "normal" direct derivations. The prize we have to pay is a global application condition, which may not be easy to check in a real distributed system.

A distributed derivation in the SPO approach always exists if there are compatible matches for synchronized productions in a distributed graph. Hence, also in the SPO approach, the basic property of direct derivations (i.e., their completeness) is resembled by distributed derivations. In contrast to the DPO approach, a distributed SPO derivation needs no global application conditions. However, it may cause global effects: Deleting a vertex in a local graph which is referenced from other local components leads to partial interface embeddings. Constructing the corresponding global graph all dangling references are deleted. It depends very much on the problem to be solved, whether global conditions or global effects (i.e., DPO or SPO distributed derivations) are more appropriate. Relying on the distributed gluing condition certainly leads to a more abstract specification, which assumes that dangling references are avoided by some standard mechanism on a lower level. However, if the references itself are subject of the specification, as e.g. for garbage collection in distributed systems, we need to model explicitly the inconsistencies caused by the deletion of referenced items.

These observations are also reflected in the solutions to the Distribution Problem 3.2.5. In the DPO approach, each distributed derivation satisfies the amalgamation condition, which means that each distributed computation can be observed from a global point of view. Vice versa, an amalgamated derivation can be distributed if its match can be splitted according to the splitting of the given graph and if this splitting satisfies the distributed gluing condition, see Theorem 3.6.5. In the SPO approach, a distributed derivation may be amalgamated if at least the given distributed graph represents a consistent state. The distribution of an amalgamated derivation, on the other hand, requires only the existence of compatible matches. We do not state the corresponding properties here but refer to Theorem 3.6.5 for the DPO approach and Theorem 4.3.22 for the SPO approach.

4.7 Conclusion

In Chapter 3 and this chapter we have presented the two algebraic approaches and compared them with each other. The double-pushout (DPO) approach was historically the first and has a built-in application condition, called the gluing condition, which is important in several application areas in order to prevent undesirable possibilities for the application of productions. The single-pushout (SPO) approach allows to apply productions without any application conditions, because the problem of dangling edges, for example, is solved by deletion of these edges. This is adequate in some application areas and problematic in other ones. In general it seems to be important to allow user-defined application conditions for productions in order to prevent application of productions in undesirable cases. In this chapter it is shown how to extend the SPO approach to handle such user-defined application conditions, and in a similar way the DPO approach could be extended.

In fact, the DPO approach could be considered as a special case of the SPO approach, because the SPO approach with gluing condition is equivalent to the DPO approach. However, as shown in the comparison of both approaches, they provide different solutions to the general problems stated in Section 3.2 of Chapter 3. Moreover, it does not seem adequate to specialize all the results in the SPO approach to the case with gluing condition, because the explicit proofs in the DPO approach are much simpler in several cases. For this reason DPO and SPO should be considered as two different graph transformation approaches within the context of this handbook.

Finally, let us point out that most of the concepts and results presented in Chapter 3 and this chapter are concerned with a single graph transformation system. This might be called the "theory of graph transformation systems in the small" in contrast to structuring and refinement concepts combining and relating different graph transformation systems which may form a "theory of graph transformation systems in the large". The "theory in the large" is especially important if graph transformation concepts are used for the specification of concurrent, object oriented and/or distributed systems. In fact, both algebraic approaches seem to be most suitable to handle such "problems in the large", and first attempts and results have been presented within the last couple of years (see [45,46,47,48,49]). Moreover, we believe that graph transformation in the large will be a main topic of research and development in the future.

Acknowledgment The research results presented in this paper have been obtained within the ESPRIT Working Groups COMPUGRAPH (1989-1992) and COMPUGRAPH 2 (1992-1995). We are most grateful to Raoult and Kennaway for initial motivation of the single pushout approach and to the members of the working group for their contributions and stimulating discussions.

References

1. H. Ehrig, M. Pfender, and H. J. Schneider. Graph grammars: an algebraic approach. In *14th Annual IEEE Symposium on Switching and Automata Theory*, pages 167–180, 1973.
2. A. Habel, R. Heckel, and G. Taentzer. Graph grammars with negative application conditions. Accepted for special issue of Fundamenta Informaticae, 1996.
3. M. Löwe, M. Korff, and A. Wagner. An algebraic framework for the transformation of attributed graphs. In M.R. Sleep, M.J. Plasmeijer, and M.C. van Eekelen, editors, *Term Graph Rewriting: Theory and Practice*, chapter 14, pages 185–199. John Wiley & Sons Ltd, 1993.
4. M. Löwe. Algebraic approach to single-pushout graph transformation. *TCS*, 109:181–224, 1993.
5. M. Korff. Single pushout transformations of generalized graph structures. Technical Report RP 220, Federal University of Rio Grande do Sul, Porto Alegre, Brazil, 1993.
6. H. Ehrig and M. Löwe. Categorical principles, techniques and results for high-level replacement systems in computer science. *Applied Categorical Structures*, 1(1):21–50, 1993.
7. M. Löwe. *Extended Algebraic Graph Transformations*. PhD thesis, Technical University of Berlin, 1990. short version in TCS (109):181 - 224.
8. J. C. Raoult. On graph rewriting. *Theoretical Computer Science*, 32:1–24, 1984.
9. R. Kennaway. On "On graph rewriting". *Theoretical Computer Science*, 52:37–58, 1987.
10. J. Glauert, R. Kennaway, and R. Sleep. A categorical construction for generalised graph rewriting. Technical report, School of Information Systems, University of East Anglia, Norwich NR4 7TJ, U.K., 1989.
11. R. Kennaway. Graph rewriting in some categories of partial maps. In Ehrig et al. [50], pages 475–489. Lecture Notes in Computer Science 532.
12. E. Robinson and G. Rosolino. Categories of partial maps. *Information and Computation*, 79:95 - 130, 1988.

13. P.M. van den Broek. Algebraic graph rewriting using a single pushout. In *Int. Joint Conf. on Theory and Practice of Software Development (TAPSOFT'91), LNCS 493*, pages 90–102. Springer Verlag, 1991.
14. H. Herrlich and G. Strecker. *Category Theory.* Allyn and Bacon, Rockleigh, New Jersey, 1973.
15. M. Löwe and J. Dingel. Parallelism in single-pushout graph rewriting. *Lecture Notes in Computer Science 776*, pages 234–247, 1994.
16. M. Korff. *Generalized graph structure grammars with applications to concurrent object-oriented systems.* PhD thesis, Technical University of Berlin, 1995.
17. M. Korff. Minimality of derived rules in single pushout graph rewriting. Technical Report 94/10, Technical University of Berlin, 1994.
18. G. Taentzer. Towards synchronous and asynchronous graph transformations. Accepted for special issue of Fundamenta Informaticae, 1996.
19. H. Ehrig and M. Löwe. Parallel and distributed derivations in the single pushout approach. *Theoretical Computer Science*, 109:123 – 143, 1993. Also in Tech. Rep. 91/01, Technical University of Berlin.
20. H. Ehrig and B. K. Rosen. Parallelism and concurrency of graph manipulations. *Theoretical Computer Science*, 11:247–275, 1980.
21. P. Böhm, H.-R. Fonio, and A. Habel. Amalgamation of graph transformations: a synchronization mechanism. *Journal of Computer and System Science*, 34:377–408, 1987.
22. G. Taentzer. Hierarchically distributed graph transformation. In *5th Int. Workshop on Graph Grammars and their Application to Computer Science, Williamsburg '94, LNCS* , 1996. Accepted.
23. G. Taentzer. *Parallel and Distributed Graph Transformation: Formal Description and Application to Communication-Based Systems.* PhD thesis, Technical University of Berlin, Dep. of Comp. Sci., 1996.
24. R. Heckel. Embedding of conditional graph transformations. In G. Valiente Feruglio and F. Rosello Llompart, editors, *Proc. Colloquium on Graph Transformation and its Application in Computer Science.* Technical Report B-19, Universitat de les Illes Balears, 1995.
25. R. Heckel. Algebraic graph transformations with application conditions. Master's thesis, TU-Berlin, 1995.
26. A. Wagner. On the expressive power of algebraic graph grammars with application conditions. In *Int. Joint Conf. on Theory and Practice of Software Development (TAPSOFT'95), LNCS 915.* Springer Verlag, 1995.
27. R. Heckel and A. Wagner. Ensuring consistency of conditional graph grammars – a constructive approach. *Proc. of SEGRAGRA'95 "Graph Rewriting and Computation", Electronic Notes of TCS*, 2, 1995. http://www.elsevier.nl/locate/entcs .

28. M. Korff. Graph-interpreted graph transformations for concurrent object-oriented systems. Extended abstract for the 5th International Workshop on Graph Grammars and their Application to Computer Science, 1994.
29. K. Räihä. Bibliography of attribute grammars. *SIGPLAN Notices*, 15(3):35–44, 1980.
30. C. Dimitrovici, U. Hummert, and L. Petrucci. Composition and net properties of algebraic high-level nets. In *Advances of Petri Nets*, volume 483 of *Lecture Notes in Computer Science*. Springer Verlag Berlin, 1991.
31. W. Reisig. Petri nets and algebraic specifications. *Theoretical Computer Science*, 80:1–34, 1991.
32. H. Ehrig, J. Padberg, and L. Ribeiro. Algebraic high-level nets: Petri nets revisited. In *Recent Trends in Data Type Specification*, pages 188–206, Caldes de Malavella, Spain, 1994. Springer Verlag. Lecture Notes in Computer Science 785.
33. ISO. Information processing systems – Open Systems Interconnection – LOTOS – A formal description technique based on the temporal ordering of observational behaviour. International Standard ISO 8807, ISO, 1989.
34. G. Schied. *Über Graphgrammatiken, eine Spezifikationsmethode für Programmiersprachen und verteilte Regelsysteme*. Arbeitsberichte des Institus für mathematische Maschinen und Datenverarbeitung (Informatik), University of Erlangen, 1992.
35. H. Ehrig and B. Mahr. *Fundamentals of Algebraic Specification 1: Equations and Initial Semantics*, volume 6 of *EATCS Monographs on Theoretical Computer Science*. Springer, Berlin, 1985.
36. M. Korff. True concurrency semantics for single pushout graph transformations with applications to actor systems. In *Working papers of the International Workshop on Information Systems – Corretness and Reusability IS-CORE'94*, pages 244–258, 1994. Tech. Report IR-357, Free University, Amsterdam.
37. A. Habel and H.-J. Kreowski. May we introduce to you: Hyperedge replacement. In *3rd Int. Workshop on Graph Grammars and their Application to Computer Science, LNCS 291*, Berlin, 1987. Springer Verlag.
38. A. Habel. *Hyperedge Replacement: Grammars and Languages*. PhD thesis, University of Bremen, 1989.
39. A. Habel. *Hyperedge replacement: Grammars and Languages*, volume 643 of *LNCS*. Springer Verlag, Berlin, 1992.
40. H. Ehrig, A. Habel, H.-J. Kreowski, and F. Parisi-Presicce. From graph grammars to High Level Replacement Systems. In Ehrig et al. [50], pages 269–291. Lecture Notes in Computer Science 532.

41. F. Parisi-Presicce. Modular system design applying graph grammar techniques. In *ICALP'89*. Springer Lecture Notes in Computer Science, 1989.
42. H. Ehrig. Introduction to the algebraic theory of graph grammars. In V. Claus, H. Ehrig, and G. Rozenberg, editors, *1st Graph Grammar Workshop, Lecture Notes in Computer Science 73*, pages 1–69. Springer Verlag, 1979.
43. A. Schürr. Progress: A vhl-language based on graph grammars. In *LNCS532*. Springer, 1991.
44. A. Corradini and F. Rossi. Synchronized composition of graph grammar productions. In *5th Int. Workshop on Graph Grammars and their Application to Computer Science, Williamsburg '94, LNCS* , 1996. Accepted.
45. H.-J. Kreowski and S. Kuske. On the interleaving semantics of transformation units - a step into GRACE. In *5th Int. Workshop on Graph Grammars and their Application to Computer Science, Williamsburg '94, LNCS* , 1996. Accepted.
46. H. Ehrig and G. Engels. Pragmatic and semantic aspects of a module concept for graph transformation systems. In *5th Int. Workshop on Graph Grammars and their Application to Computer Science, Williamsburg '94, LNCS* , 1996. Accepted.
47. G. Taentzer and A. Schürr. DIEGO, another step towards a module concept for graph transformation systems. *Proc. of SEGRAGRA'95 "Graph Rewriting and Computation", Electronic Notes of TCS*, 2, 1995. http://www.elsevier.nl/locate/entcs .
48. F. Parisi-Presicce. Transformation of graph grammars. In *5th Int. Workshop on Graph Grammars and their Application to Computer Science, Williamsburg '94, LNCS* , 1996. Accepted.
49. R. Heckel, A. Corradini, H. Ehrig, and M. Löwe. Horizontal and vertical structuring of typed graph transformation systems. Accepted for special issue of MSCS, 1996.
50. H. Ehrig, H.-J. Kreowski, and G. Rozenberg, editors. *4th International Workshop on Graph Grammars and Their Application to Computer Science.* Springer Verlag, 1991. Lecture Notes in Computer Science 532.

Chapter 5

THE EXPRESSION OF GRAPH PROPERTIES AND GRAPH TRANSFORMATIONS IN MONADIC SECOND-ORDER LOGIC

B. COURCELLE
LABRI (URA CNRS 1304), Bordeaux I University
351, Cours de la Libération, 33405 Talence - France
e-mail: courcell@labri.u-bordeaux.fr

By considering graphs as logical structures, one can express formally their properties by logical formulas. We review the use of monadic second-order logic for expressing graph properties, and also, graph transformations. We review the intimate relationships of monadic second-order logic and context-free graph grammars. We also discuss the definition of classes of graphs by forbidden configurations.

Contents

5.1 Introduction

By considering graphs as logical structures, one can express formally their properties by logical formulas. One can thus describe classes of graphs by formulas of appropriate logical languages expressing characteristic properties. There are two main motivations for doing this: the first one, originating from the work by Fagin [37], consists in giving logical characterizations of complexity classes; the second one consists in using logical formulas as finite devices, comparable to grammars or automata, to specify classes of graphs and to establish properties of such classes from their logical descriptions.

We shall only consider here the second of these motivations. The ideal language is in this respect monadic second-order logic, as we shall demonstrate. It is crucial for establishing "easily" results like this one:

the set of planar graphs belonging to a HR set of graphs[a] is HR,

or this one:

the set of Hamiltonian graphs belonging to a HR set of graphs is HR,

by essentially the same proof, using the fact that planarity and Hamiltonicity can both be described by MS (Monadic Second-order) formulas. These two results do not concern logic, but their proofs use logic as a tool. The deep reason why MS logic is so crucial is that it replaces for graphs (and for the development of the theory of context-free graph grammars) the notion of a finite automaton which is very important in the theory of formal languages. It "replaces" because no convenient notion of finite automaton is known for graphs. The notion of a transformation from words or trees to words or trees is also essential in language theory (Berstel [4], Raoult [52]). These transformations are usually defined in terms of finite automata, that produce an output while traversing the given word or tree. Since we have no notion of finite graph automaton, we cannot define graph transformations in terms of automata. However, we can define such transformations in terms of MS-formulas. We call them *definable transductions.* "Definable" refers to logic and "transduction" to the way transformations of words and trees are usually named.

MS logic is thus an essential notion in the extension of formal language theory to graphs, hypergraphs and related structures. Another important notion is that of a graph (or hypergraph) operation. By using it, one can define context-free sets of graphs as components of least solutions of systems of equations

[a] A HR set of graphs is a set of finite graphs generated by a Hyperedge Replacement graph grammar

(without using any graph rewriting rule) and recognizable sets of graphs (without using any notion of graph automaton). Context-free and recognizable sets can thus be defined and investigated in the general framework of Universal Algebra. They instanciate immediately to graphs and hypergraphs of all kinds as soon as appropriate operations on these structures are defined. Furthermore the notion of recognizability establishes the link between Logic and Universal Algebra because every MS-definable set of graphs is recognizable: this result extends the result by Büchi saying that every MS-definable set of finite words is a regular language.

This chapter is organized as follows. Section 5.2 reviews relational structures, first-order logic, second-order logic and monadic second-order logic. Some basic lemmas helping in the construction of logical formulas are proved. Section 5.3 introduces several possible representations of graphs, hypergraphs and partial orders by relational structures, and discusses how the choice of a representation affects the expressive power of the three considered logical languages. Section 5.4 discusses the expressibility in MS logic of the finiteness of a set and of the parity of its cardinality when it is finite. Section 5.5 introduces definable transductions, reviews their basic properties and their application to the comparison of the different relational structures representing partial orders, graphs and hypergraphs reviewed in Section 5.3. Section 5.6 defines the hyperedge replacement (HR) hypergraph grammars and the vertex replacement (VR) graph grammars, in terms of systems of recursive set equations, by means of appropriate operations on hypergraphs and graphs respectively. Logical characterizations of the HR sets of hypergraphs and of the VR sets of graphs in terms of definable transductions are also given. These characterizations yield the stability of the corresponding classes under the relevant definable transductions. Section 5.7 introduces the recognizability of sets of graphs and hypergraphs. This notion is based on finite congruences relative to the operations on graphs and hypergraphs introduced in Section 5.6. In general, the intersection of an equational and a recognizable set is an equational set. (An instance of this result is the fact that the intersection of a context-free language and a regular one is context-free). The major result of this section says that a monadic second-order definable set of graphs or hypergraphs is recognizable. This result yields in particular the two results mentioned at the beginning of the introduction concerning HR sets of graphs, planarity and Hamiltonicity. Section 5.8 deals with the logical aspect of the definition of sets of graphs by forbidden minors. Kuratowski's Theorem stating that a graph is planar iff it does not contain any of the two graphs K_5 and $K_{3,3}$ as a minor is a well-known example of such a definition. By using deep results by Robertson and Seymour, we relate definitions by forbidden minors with definitions by MS formulas and/or by HR grammars.

5.2 Relational structures and logical languages

In order to express graph properties by logical formulas, we shall represent graphs by relational structures, i.e., by logical structures with relations only (without functions). We shall review first-order logic, second-order logic and monadic second-order logic, which is an extension of the former and a fragment of the latter. We shall review some basic tools that will help in the construction of formulas in forthcoming sections.

5.2.1 Structures

Let $\mathcal{R}$ be a finite set of relation symbols. Each symbol $R \in \mathcal{R}$ has an associated positive integer called its arity, denoted by $\rho(R)$. An $\mathcal{R}$-structure is a tuple $S =< D_S, (R_S)_{R\in\mathcal{R}} >$ such that D_S is a possibly empty set called the *domain of* S and each R_S is a $\rho(R)$-ary relation on D_S, i.e., a subset of $D_S^{\rho(R)}$. We shall say that "$R(d_1, \cdots, d_n)$ holds in S" iff $(d_1, \cdots, d_n) \in R_S$, where, of course, $d_1, \cdots, d_n \in D_S$. We shall denote by $STR(\mathcal{R})$ the class of $\mathcal{R}$-structures. Structures may have infinite domains.

We give two examples of the use of structures. A word u in A^* is usually defined as a sequence of letters from A, equivalently as a mapping $\{1, \cdots, n\} \to A$ for some $n \in \mathcal{N}$ (with $n = 0$ for the empty word). In order to represents words by relational structures we let $\mathcal{R}_{w,A} := \{suc, lab_a, \cdots, lab_d\}$ where $A = \{a, \cdots, d\}$ (A can have more than 4 letters), suc is binary and $lab_a, \cdots, lab_d$ are unary. For every word $u \in A^*$, we let $\| u \| \in STR(\mathcal{R}_{w,A})$ be the structure S such that:

D_S	$= \emptyset$	if u is the empty word ε;
		otherwise $D_S = \{1, \cdots, n\}$ if u has length n;
suc_S	$=$	$\{(1,2), (2,3), \cdots, (n-1, n)\}$;
$i \in lab_{yS}$		iff y is the i-th letter of u.

In order to represent graphs by relational structures, we let $\mathcal{R}_s = \{edg\}$ where edg is binary. With a directed graph G, we associate the $\mathcal{R}_s$-structure $\mid G \mid_1 = < V_G, edg_G >$ where V_G is the set of vertices of G (and the domain of $\mid G \mid_1$) and $(x, y) \in edg_G$ iff there is in G an edge from x to y. We do the same if G is undirected and we let $(x, y) \in edg_G$ iff there is in G an edge linking x and y: it follows that edg_G is in this case symmetric. The structure $\mid G \mid_1$ does not contain information concerning the multiplicity of edges. It is thus appropriate to represent simple graphs only because two simple graphs G and G' are isomorphic iff the structures $\mid G \mid_1$ and $\mid G' \mid_1$ are isomorphic. (In Section 5.3, we shall define another representation, denoted by $\mid G \mid_2$, which will be appropriate for graphs with multiple edges). For directed graphs, we shall sometimes use *suc* instead of *edg*. We call y a *successor* of x if $(x, y) \in suc_G$.

Relational structures form the basis of the theory of relational databases (Abiteboul et al. [1]). More precisely, a finite $\mathcal{R}$-structure S can be considered as a state of a relational database describing relations between the objects of D_S. The relations R_S for $R \in \mathcal{R}$ are the various relations of the database. A finite structure S is represented by means of some coding of the objects of D_S and for each $R \in \mathcal{R}$ a list of $\rho(R)$-tuples of "codes" of the elements of D_S. The study of query languages for relational databases has also been a motivation for finite model theory. (See the survey by Kannellakis [46] and the book by Abiteboul et al. [1]).

5.2.2 First-order logic

We let $\mathcal{X}$ be a countable alphabet of lowercase letters called *individual variables*. Let $\mathcal{R}$ be a finite set of relation symbols. The atomic formulas are $x = y, R(x_1, \cdots, x_n)$ for $x, y, x_1, \cdots, x_n \in \mathcal{X}$, $R \in \mathcal{R}$, $n = \rho(R)$. The first-order formulas are formed from atomic formulas with the propositional connectives $\wedge, \vee, \neg, \Rightarrow, \Leftrightarrow$, and quantifications $\exists x, \forall x$ (for $x \in \mathcal{X}$).

We shall denote by $FO(\mathcal{R}, \mathcal{Y})$ where $\mathcal{Y} \subseteq \mathcal{X}$ the set of first-order formulas over $\mathcal{R}$ with free variables in $\mathcal{Y}$. In order to specify the free variables that may occur in a formula φ we shall write it $\varphi(x_1, \cdots, x_n)$ if $\varphi \in FO(\mathcal{R}, \{x_1, \cdots, x_n\})$. (Some variables in $\{x_1, \cdots, x_n\}$ may have no free occurrence in φ). If $\varphi(x_1, \cdots, x_n)$ is a first-order formula over $\mathcal{R}$, if $S \in STR(\mathcal{R})$, and if $d_1, \cdots, d_n \in D_S$, we write $S \models \varphi(d_1, \cdots, d_n)$ or $(S, d_1, \cdots, d_n) \models \varphi$ to mean that φ is true in S if x_i is given the value d_i for $i = 1, \cdots, n$. If φ has no free variables, i.e., if it is *closed*, then it describes a property of S and not of tuples of elements of S. Here is an example. The formula $\varphi(x)$ over $\mathcal{R}_s$ defined as:

$$\forall y_1, y_2, y_3[edg(x, y_1) \wedge edg(x, y_2) \wedge edg(x, y_3) \Rightarrow y_1 = y_2 \vee y_1 = y_3 \vee y_2 = y_3]$$

expresses that the vertex x of the represented graph has out-degree at most 2. The closed formula $\forall x.\varphi(x)$ expresses thus that the considered graph has outdegree at most 2.

We now give an example concerning words. We let $A = \{a, b, c\}$. The formula θ below is constructed in such a way that, for every word $u \in A^*$:

$$\| u \| \models \theta \text{ iff } u \in ab^*c.$$

Here is θ:

$$\exists x[lab_a(x) \wedge \forall y(\neg suc(y, x))] \wedge \forall x[lab_a(x) \Rightarrow \exists y(suc(x, y) \wedge (lab_b(y) \vee lab_c(y)))]$$

$\wedge\forall x[lab_b(x) \Rightarrow \exists y(suc(x,y) \wedge (lab_b(y) \vee lab_c(y)))] \wedge \forall x[lab_c(x) \Rightarrow \forall y(\neg suc(x,y))]$.

Let us remark that θ has models that are not representations of words. So, θ characterizes ab^*c as a subset of A^*; it does not characterize the structures representing the words of ab^*c among those of $STR(\mathcal{R}_{w,A})$. The languages characterized in this way by a first-order formula form a subclass of the class of regular languages, called the class of locally threshold testable languages (see Thomas [61] or the survey by Pin [50]).

5.2.3 *Second-order logic*

We now let $\mathcal{X}$ contain individual variables as in Subsection 5.2.2 and also *relation variables*, denoted by uppercase letters, $X, Y, X_1, \cdots, X_n$. Each relation variable has an arity which is a positive integer ($\rho(X)$ is the arity of X). We let $\mathcal{R}$ be a finite set of relation symbols and we now define the second-order formulas over $\mathcal{R}$. The atomic formulas are: $x = y, R(x_1, \cdots, x_n), X(x_1, \cdots, x_n)$, where $x, y, x_1, \cdots, x_n, X \in \mathcal{X}$, $R \in \mathcal{R}$ and $n = \rho(R) = \rho(X)$. The formulas are constructed from the atomic formulas with the propositional connectives (as in Subsection 5.2.2) and the quantifications $\exists x, \forall x, \exists X, \forall X$ over individual and relation variables. (We do not give a formal syntax; see the examples below). We shall denote by $SO(\mathcal{R}, \mathcal{Y})$ the set of second-order formulas over $\mathcal{R}$ with free variables in $\mathcal{Y}$. The notation $\varphi(x, y, z, X_1, \cdots, X_n)$ indicates that the free variables of φ belong to $\{x, y, z, X_1, \cdots, X_n\}$.

Consider a formula $\varphi(x_1, \cdots, x_m, X_1, \cdots, X_n)$. If $S \in STR(\mathcal{R})$ if $d_1, \cdots, d_m \in D_S$, if $E_1, \cdots, E_n$ are relations on D_S of respective arities $\rho(X_1), \cdots, \rho(X_n)$ then the notation $S \models \varphi(d_1, \cdots, d_m, E_1, \cdots, E_n)$ means that φ is true in S for the values $d_1, \cdots, d_m$ of $x_1, \cdots, x_m$ and $E_1, \cdots, E_n$ of $X_1, \cdots, X_n$ respectively. Let $\mathcal{Y} = \{x_1, \cdots, x_m, X_1, \cdots, X_n\}$. A *$\mathcal{Y}$-assignment* in S is a mapping γ with domain $\mathcal{Y}$ such that $\gamma(x_i) \in D_S$ for $i = 1, \cdots, m$ and $\gamma(X_j)$ is a $\rho(X_j)$-ary relation on D_S for each $j = 1, \cdots, n$. We shall also use the notation $(S, \gamma) \models \varphi$ instead of $S \models \varphi(d_1, \cdots, d_m, E_1, \cdots, E_n)$ where $\gamma(x_i) = d_i$ and $\gamma(X_j) = E_j$.

We now give some examples. Let $\beta(X)$ be the formula in $SO(\emptyset, \{X\})$:

$$\forall x, y, z[X(x, y) \wedge X(x, z) \Longrightarrow y = z] \wedge \forall x, y, z[X(x, z) \wedge X(y, z) \Longrightarrow x = y],$$

where X is binary. It expresses that X is a functional relation, and that the corresponding function is injective on its domain. (This domain may be strictly included in the domain of the considered structure). Hence the following formula $\alpha(X)$ of $SO(\mathcal{R}_s,\{X\})$ expresses that X is an automorphism of a simple graph G represented by the structure $|G|_1$ in $STR(\mathcal{R}_s)$. Here is $\alpha(X)$:

$$\beta(X) \wedge \forall x \exists y X(x,y) \wedge \forall x \exists y X(y,x)$$

$$\wedge \forall x,y,x',y'[edg(x,y) \wedge ((X(x,x') \wedge X(y,y')) \vee (X(x',x) \wedge X(y',y))) \Longrightarrow edg(x',y')].$$

Hence, G has a nontrivial automorphism iff $|G|_1$ satisfies the closed formula

$$\exists X[\alpha(X) \wedge \exists x,y(\neg x = y \wedge X(x,y))].$$

Consider now the formula $\gamma(Y_1,Y_2) \in SO(\emptyset,\{Y_1,Y_2\})$:

$$\exists X[\beta(X) \wedge \forall x(Y_1(x) \Longleftrightarrow \exists y X(x,y)) \wedge \forall x(Y_2(x) \Longleftrightarrow \exists y X(y,x))].$$

It expresses that there exists a bijection (namely the binary relation defined by X) between the sets Y_1, Y_2 (handled as unary relations). One can thus characterize by the following second-order formula the nonregular language $\{a^n b^n / n \geq 1\}$:

$$\exists x[lab_a(x) \wedge \forall y(\neg suc(y,x))] \wedge \forall x[lab_a(x) \Longrightarrow \exists y(suc(x,y) \wedge (lab_a(y) \vee lab_b(y)))]$$

$$\wedge \forall x,y[lab_b(x) \wedge suc(x,y) \Longrightarrow lab_b(y)] \wedge$$

$$\exists Y_1,Y_2[\gamma(Y_1,Y_2) \wedge \forall x[lab_a(x) \Longleftrightarrow Y_1(x)] \wedge \forall x[lab_b(x) \Longleftrightarrow Y_2(x)]].$$

The main logical language to be used in this paper is monadic second-order logic which lies inbetween first-order and second-order logic.

5.2.4 Monadic second-order logic

Let $\mathcal{R}$ be a finite set of relation symbols. Let $\mathcal{X}$ contain individual variables and relation variables of arity one. Since a relation with one argument is nothing but a set, we shall call these variables *set variables*. A monadic second-order formula over $\mathcal{R}$ is a second-order formula written with $\mathcal{R}$ and $\mathcal{X}$: the quantified and free variables are individual or sets variables; there is no restriction on the arity of the symbols in $\mathcal{R}$. In order to get more readable formulas, we shall write $x \in X$ instead of $X(x)$ where X is a set variable. We denote by $MS(\mathcal{R},\mathcal{Y})$ the set of monadic second-order formulas (MS formulas for short) over $\mathcal{R}$, with free variables in $\mathcal{Y}$.

We give some examples. The MS formula $\delta \in MS(\{suc\},\emptyset)$ below expresses that a word in A^* has an odd length. Since this property does not depend on the letters, the formula δ does not use the relations $lab_x, x \in A$. Here is δ:

$$\exists X,Y[\delta'(X,Y) \wedge \exists x(x \in X \wedge \forall y(\neg suc(x,y)))]$$

where $\delta'(X,Y)$ is

$$\exists x(x \in X \wedge \forall y(\neg suc(y,x))) \wedge \forall x,y(x \in X \wedge suc(x,y) \Longrightarrow y \in Y) \wedge$$

$$\forall x,y(x \in Y \wedge suc(x,y) \Longrightarrow y \in X).$$

For every word $w \in A^*$, there is a unique pair X, Y with $X, Y \subseteq D_{\|w\|}$ satisfying δ': the elements of X are the odd rank positions in w and the elements of Y are the even rank ones. The formula δ expresses that the last position is odd, i.e., that the considered word has odd length. A theorem by Büchi and Elgot [8], [30] (see Thomas [60], Thm 3.2) says that the languages defined by MS formulas are exactly the regular languages.

We now consider an example concerning graphs. The following formula $\tau_3 \in MS(\{\mathcal{R}_s\})$ expresses that the considered graph is 3-vertex colorable (with every two adjacent vertices of different colors; one may have loops). Here is τ_3:

$$\exists X_1, X_2, X_3[\forall x(x \in X_1 \vee x \in X_2 \vee x \in X_3) \wedge \forall x(x \in X_1 \Longrightarrow \neg x \in X_2 \wedge \neg x \in X_3)$$

$$\wedge \forall x(x \in X_2 \Longrightarrow \neg x \in X_1 \wedge \neg x \in X_3) \wedge \forall x(x \in X_3 \Longrightarrow \neg x \in X_1 \wedge \neg x \in X_2)$$

$$\wedge \forall x,y(edg(x,y) \wedge \neg x = y$$

$$\Longrightarrow \neg(x \in X_1 \wedge y \in X_1) \wedge \neg(x \in X_2 \wedge y \in X_2) \wedge \neg(x \in X_3 \wedge y \in X_3))]$$

A triple of sets X_1, X_2, X_3 satisfying this condition is a partition of the set of vertices. Considering i such that $x \in X_i$ as the color of x, the formula expresses that adjacent vertices have different colors. For each positive integer k, one can write a similar formula τ_k expressing that the considered graph is k-vertex colorable.

Let L be any of the languages FO, MS and SO. We write $L(\mathcal{R})$ for $L(\mathcal{R}, \emptyset)$. A property P of structures in $STR(\mathcal{R})$ is $L-expressible$ iff there exists a formula φ in $L(\mathcal{R})$ such that, for every structure $S \in STR(\mathcal{R}), P(S)$ holds iff $S \models \varphi$. A class of structures $\mathcal{C}$ is $L-definable$ iff it is the class of structures satisfying an $L-definable$ *property*. These definitions can be relativized to specific classes of structures. Let $\mathcal{T} \subseteq STR(\mathcal{R})$. Then, a property P *of structures in* $\mathcal{T}$ is $L-expressible$ iff there exists a formula φ in $L(\mathcal{R})$ such that, for every S in $\mathcal{T}$, $P(S)$ holds iff $S \models \varphi$. One obtains similarly the notion of a subclass $\mathcal{C}$ of $\mathcal{T}$ that is $L-definable$ *with respect to* $\mathcal{T}$. A property P of tuples $(d_1, \cdots, d_n, E_1, \cdots, E_m)$ in the structures S of a class $\mathcal{T}$ is $L-expressible$ iff there exists a formula $\varphi \in L(\mathcal{R}, \{x_1, \cdots, x_n, X_1, \cdots, X_m\})$ such that for

all $S \in \mathcal{T}$, all $(d_1, \cdots, d_n, E_1, \cdots, E_m)$ of appropriate type, $P(S, d_1, \cdots, E_m)$ holds iff $S \models \varphi(d_1, ..., E_m)$.
One obtains thus a hierarchy of graph properties linked with the hierarchy of languages: $FO \subset MS \subset SO$. In Sections 5.3 and 5.4, we shall introduce intermediate languages between MS and SO and discuss the strictness of the corresponding hierarchy of graph properties, relativized to various classes of graphs or related objects like words, partial orders and hypergraphs. The languages defined by monadic second-order formulas are the regular languages by a theorem of [8] (see also the survey by Pin [50]).

5.2.5 *Decidability questions*

Let $\mathcal{R}$ be a finite set of relation symbols. Let $\mathcal{L}$ be a class of closed formulas expressing properties of the structures in $STR(\mathcal{R})$. Let $\mathcal{C} \subseteq STR(\mathcal{R})$. The $\mathcal{L} - theory\ of\ \mathcal{C}$ is the set:

$$Th_{\mathcal{L}}(\mathcal{C}) := \{\varphi \in \mathcal{L} / S \models \varphi \ for\ every\ S \in \mathcal{C}\}.$$

We say that the $\mathcal{L}$-theory of $\mathcal{C}$ is *decidable* if this set is recursive. The *$\mathcal{L}$-satisfiability problem for $\mathcal{C}$* is the problem of deciding whether a given formula belongs to the set

$$Sat_{\mathcal{L}}(\mathcal{C}) := \{\varphi \in \mathcal{L} / \ S \models \varphi \ for\ some\ S \in \mathcal{C}\}.$$

If $\mathcal{L}$ is closed under negation, the $\mathcal{L}$-theory of $\mathcal{C}$ is decidable iff its $\mathcal{L}$-satisfiability problem is decidable. We shall consider some conditions on a class of graphs $\mathcal{C}$ ensuring that its *monadic* theory (i.e., the MS-theory) is decidable.

If $\mathcal{C} = \{S\}$ where S is a finite (and effectively given) structure then the property $S \models \varphi$ is decidable for every formula φ of the languages $\mathcal{L}$ we have considered, or we will consider. The $\mathcal{L}$-theory of $\mathcal{C}$ is thus (trivially) decidable. An interesting question is thus the complexity of this problem. For an example, the property $S \models \varphi$ is polynomial in *size*(S) if φ is a fixed first-order formula and S is the input; it is NP if φ is existential second-order (Fagin [37]), where again φ is fixed and S is the input. Conversely, every NP property can be described in this way. More generally, a property belongs to the polynomial hierarchy iff it can be described by a second-order formula (see Stockmeyer [58], Immerman [44] or Abiteboul et al. [1]).

If $\mathcal{C} = \{S\}$ where S is infinite, then the $\mathcal{L}$-theory of $\mathcal{C}$ may be undecidable. However the study of the theories of infinite structures and especially of their monadic (second-order) theories is a very rich topic that we cannot even touch here. We refer the reader to survey papers like those of Gurevich [42], Courcelle [10], Thomas [60].

Proposition 5.2.1 (Trakhtenbrot [62])
The first-order theory of the class of finite graphs is undecidable. □

So are a fortiori its monadic and its second-order theories. The following result is a basic tool for obtaining undecidability results. It concerns square grids. A *grid* is a directed graph isomorphic to $G_{n,m}$ for some $n, m \in \mathcal{N}_+$ where $V_{G_{n,m}} = [n] \times [m]$ and $E_{G_{n,m}} = \{((i,j),(i',j'))/1 \leq i \leq i' \leq n, 1 \leq j \leq j' \leq m$ and, either $i' = i+1$ and $j' = j$, or $i' = i$ and $j' = j+1\}$. ($\mathcal{N}_+$ denotes the set of positive integers and for $n \in \mathcal{N}_+$; $[n]$ denotes $\{1, 2, \cdots, n\}$). A *square grid* is a graph of the above form where $m = n$.

Proposition 5.2.2
The monadic (second-order) satisfiability problem of every class of graphs $\mathcal{C}$ containing graphs isomorphic to $G_{n,n}$ for infinitely many integers n is undecidable.

Proof sketch
One first constructs an MS formula γ that defines the square grids among the finite simple loop-free directed graphs (see Subsection 5.2.9). Consider now a Turing machine M with total alphabet A (input letters, states, end markers) and an initial configuration w_M. Let $A = \{a_1, \cdots, a_m\}$. If $(X_1, \cdots, X_m)$ is a partition of the set of vertices of a square grid $G_{n,n}$ then one can consider that this partition defines a sequence of n words of length n in A^*. (Each line of the grid represents a word where a line of $G_{n,n}$ is a subgraph with set of vertices $\{i\} \times [n]$ for some i; the first line represents the configuration w_M). One can construct an MS-formula $\varphi_M(X_1, \cdots, X_m)$ such that, for every $X_1, \cdots, X_m \subseteq V_{G_{n,n}}$

$$G_{n,n} \models \varphi_M(X_1, \cdots, X_m)$$

iff $(X_1, \cdots, X_m)$ is a partition of $V_{G_{n,n}}$ which defines the sequence of configurations of a terminating computation of M. (Configurations of length smaller than n can be extended to the right by a special symbol). It follows that if $\mathcal{C}$ contains infinitely many square grids, then M terminates iff some graph G in $\mathcal{C}$ satisfies the formula

$$\gamma \wedge \exists X_1, \cdots, X_m.\varphi_M \tag{5.1}$$

Hence the halting problem of Turing machine reduces to the monadic satisfiability problem of any class $\mathcal{C}$ containing infinitely many square grids. This latter problem is thus undecidable. □

Remark:
In this construction, one may assume that the machine M is deterministic and one can write φ_M in such a way that the square grid on which its (unique) computation is encoded is minimal. It follows that there exists at most one graph satisfying formula (3.1). Hence, even if we know that a MS-formula φ has only finitely many finite models (up to isomorphism), we cannot construct this set by an algorithm taking φ as input. □

5.2.6 Some tools for constructing formulas

Two formulas $\varphi, \varphi' \in SO(\mathcal{R}, \{x_1, \cdots, x_n, X_1, \cdots, X_m\})$ are *equivalent* if for every $S \in STR(\mathcal{R})$, for all $d_1, \cdots, d_n \in D_S$, for all relations $E_1, \cdots, E_m$ on D_S of respective arities $\rho(X_1), \cdots, \rho(X_m)$ we have:

$$S \models \varphi(d_1, \cdots, d_n, E_1, \cdots, E_m) \text{ iff } S \models \varphi'(d_1, \cdots, d_n, E_1, \cdots, E_m).$$

Clearly, two equivalent formulas express the same properties of the relevant tuples $(d_1, \cdots, d_n, E_1, \cdots, E_m)$ in every structure $S \in STR(\mathcal{R})$.

In some cases equivalence is relativized to a subclass $\mathcal{S}$ of $STR(\mathcal{R})$: equivalent formulas express the same properties of the relevant tuples in the structures of $\mathcal{S}$, not in all structures.

Lemma 5.2.3
Let $\varphi \in SO(\mathcal{R}, \mathcal{Y})$ and $\mathcal{Z}$ be a finite set of variables. One can transform φ into an equivalent formula φ' in $SO(\mathcal{R}, \mathcal{Y})$ in which no variable of $\mathcal{Z}$ is bound. If φ is MS, then φ' is MS; if φ is FO, then φ' is FO.

Proof
One simply renames the bound variables belonging to $\mathcal{Z}$. Of course, one renames a variable into one of same type (individual or relational) and arity (in case of a relation variable). This is possible because our "universal" set of variables contains countably many variables of each type and arity. □

For readibility, one usually writes a formula by choosing bound variables that are not free in the formula. This is always possible without loss of generality.

We now recall the definition of first-order substitutions in formulas. Let φ be a second-order formula; let $x_1, \cdots, x_n$ be pairwise distinct variables; let $y_1, \cdots, y_n$ be variables, not necessarily pairwise distinct. (In order to avoid double subscripts, we do not index the variable in $\mathcal{X}$; hence $x_1, \cdots, x_n$ are metavariables denoting variables of $\mathcal{X}$; they are not *the* variables $x_1, \cdots, x_n$;

in the generic such list, we may have two variables equal). We denote by $\varphi[y_1/x_1, \cdots, y_n/x_n]$ the formula obtained as follows:

- using Lemma 5.2.3, one takes φ' equivalent to φ with no bound variable in $\{y_1, \cdots, y_n\}$;
- then one substitutes in φ the variable y_i for each occurrence of $x_i, i = 1, \cdots, n$.

Finally, if φ has been previously described as a formula $\varphi(x_1, \cdots, x_n, X_1, \cdots, X_m)$ (which indicates that its free variables are among $x_1, \cdots, x_n, X_1, \cdots, X_m$) then, $\varphi(y_1, \cdots, y_n, X_1, \cdots, X_m)$ will be another notation for $\varphi[y_1/x_1, \cdots, y_n/x_n]$. With this notation, we have the following lemma, giving the semantics of substitution.

Lemma 5.2.4
Let $\varphi \in SO(\mathcal{R}, \{x_1, \cdots, x_n, x_{n+1}, \cdots, x_p\})$ with $x_1, \cdots, x_p$ pairwise distinct. Let $y_1, \cdots, y_n$ be variables. Let $\{z_1, \cdots, z_q\}$ be an enumeration of the set of variables $\{y_1, \cdots, y_n, x_{n+1}, \cdots, x_p\}$, so that $\varphi[y_1/x_1, \cdots, y_n/x_n] \in SO(\mathcal{R}, \{z_1, \cdots, z_q\})$. For every $S \in STR(\mathcal{R})$, for every $d_1, \cdots, d_q \in D_S$ we have

$$S \models \varphi[y_1/x_1, \cdots, y_n/x_n](d_1, \cdots, d_q) \text{iff} S \models \varphi(d'_1, \cdots, d'_p)$$

where $d'_i = d_j$ iff $1 \leq i \leq n$ and $z_j = y_i$ or $n+1 \leq i \leq p$ and $z_j = x_i$.

Proof
The sequence $d'_1, \cdots, d'_p$ is well-defined because $\{z_1, \cdots, z_q\}$ is an enumeration of the set $\{y_1, \cdots, y_n, x_{n+1}, \cdots, x_p\}$. The verification is easy. □

We now define second-order substitutions, by which relation symbols can be replaced by formulas, intended to define the corresponding relations. Let $\varphi \in SO(\mathcal{R}, \{x_1, \cdots, x_m, X_1, \cdots, X_p\})$. For each $i = 1, \cdots, p$, we let $\psi_i \in SO(\mathcal{R}, \{x_1, \cdots, x_m, y_1, \cdots, y_{\rho(X_i)}\}$ where the listed variables are pairwise distinct. We let $\varphi[\psi_1/X_1, \cdots, \psi_p/X_p]$ be the formula in $SO(\mathcal{R}, \{x_1, \cdots, x_m\})$ constructed as follows:

- by using Lemma 5.2.3 one first constructs φ' equivalent to φ where no variable of $\{x_1, \cdots, x_m, X_1, \cdots, X_p\}$ is bound;
- then one replaces every atomic formula $X_i(u_1, \cdots, u_{\rho(X_i)})$ of φ' (where the variables $u_1, \cdots, u_{\rho(X_i)}$ need not be pairwise distinct) by the formula $\psi_i[u_1/y_1, \cdots, u_{\rho(X_i)}/y_{\rho(X_i)}]$.

With this notation, we have the following lemma.

Lemma 5.2.5
Let $S \in STR(\mathcal{R})$, let $d_1, \cdots, d_m \in D_S$. For each $i = 1, \cdots, p$ let T_i be the $\rho(X_i)$-ary relation on D_S defined by $(a_1, \cdots, a_{\rho(X_i)}) \in T_i$ iff $S \models \psi_i(d_1, \cdots, d_m, a_1, \cdots, a_{\rho(X_i)})$. Then

$$S \models \varphi[\psi_1/X_1, \cdots, \psi_p/X_p](d_1, \cdots, d_m) \text{ iff } S \models \varphi(d_1, \cdots, d_m, T_1, \cdots, T_p).$$

Proof
Straightforward verification from the definition, by using an induction on the structure of φ. □

Let $S \in STR(\mathcal{R})$ and E be a subset of D_S. We denote by $S[E]$ the *restriction of* S to E, i.e., the structure $S' \in STR(\mathcal{R})$ with domain E and such that $R_{S'} = R_S \cap E^{\rho(R)}$ for every R. (Since our structures are relational the restriction of a structure to a subset of its domain is always well-defined: this would not be the case if we had to ensure that some functions have a well-defined image.) If S represents a graph G, i.e., $S = \mid G \mid_1$, then $S[E]$ represents the induced subgraph of S with set of vertices E which will be denoted by $G[E]$.

Lemma 5.2.6
Let $\varphi \in SO(\mathcal{R}, \{x_1, \cdots, x_n\})$; one can construct a formula $\varphi' \in SO(\mathcal{R}, \{x_1, \cdots, x_n, X\})$ such that the following holds for every $S \in STR(\mathcal{R})$. For every $E \subseteq D_S$, for every $d_1, \cdots, d_n \in E$:

$$S \models \varphi'(d_1, \cdots, d_n, E) \text{ iff } S[E] \models \varphi(d_1, \cdots, d_n)$$

If φ is MS or FO then φ' is MS or FO respectively.

Proof
We can assume (by Lemma 5.2.3) that X does not occur in φ. We let φ' be associated with φ by the following inductive definition, where φ may have free variables of all types (relation variables as well as individual variables).

$(\forall Y.\psi)' = \forall Y.\psi'$

$(\exists Y.\psi)' = \exists Y.\psi'$

$(\forall x.\psi)' = \forall x.[x \in X \Longrightarrow \psi']$

$(\exists x.\psi)' = \exists x[x \in X \wedge \psi']$

$(\psi_1 op \psi_2)' = \psi_1' op \psi_2'$ where $op \in \{\wedge, \vee, \Longrightarrow, \Longleftrightarrow\}$

$(\neg\psi)' = \neg\psi'$

$\psi' = \psi$ for every atomic formula ψ. □

The formula φ' is called the *relativization of* φ *to* X and will be denoted by $\varphi\lceil X$.

5.2.7 Transitive closure and path properties

We denote by T^+ the transitive closure of a binary relation T.

Lemma 5.2.7
Let S be an $\{R\}$-structure where R is binary. The transitive closure of R_S is defined in S by the following formula φ_0 of $MS(\{R\}, \{x_1, x_2\})$:

$$\forall X[\{\forall y, z(y \in X \wedge R(y, z) \Longrightarrow z \in X) \wedge \forall y(R(x_1, y) \Longrightarrow y \in X)\} \Longrightarrow x_2 \in X]$$

Proof
Let $d_1, d_2 \in D_S$ such that $S \models \varphi_0(d_1, d_2)$. The set $D = \{d/R_S^+(d_1, d)\}$ satisfies the property $\forall y, z(y \in X \wedge R(y, z) \Longrightarrow z \in X) \wedge \forall y(R(x_1, y) \Longrightarrow y \in X)$. Hence $d_2 \in D$, i.e. $R_S^+(d_1, d_2)$. Conversely, if $R_S^+(d_1, d_2)$ then d_2 must belong to every set D such that $R_S(d_1, d)$ for all $d \in D$ and that is *closed under* R_S, i.e., is such that $d \in D$ and $R_S(d, d')$ implies $d' \in D$. This shows that $S \models \varphi_0(d_1, d_2)$. □

We now discuss some applications of this lemma to path[b] properties in graphs. We shall write formulas relative to $\mathcal{R}_s$.

Lemma 5.2.8
One can write MS formulas expressing in $|G|_1$ the following properties of vertices x, y and sets of vertices X, where G is an arbitrary directed graph, represented by the structure $|G|_1$ in $STR(\mathcal{R}_s)$.

P1: $x = y$ or there is a nonempty path from x to y,

P2: G is strongly connected,

P3: G is connected,

P4: $x \neq y$ and there is a path from x to y, all vertices of which belong to the set X,

P5: X is a connected component of G,

P6: G has no circuit,

P7: G is a directed tree,

P8: G has no circuit, $x \neq y$ and X is the set of vertices of a path from x to y,

P9: G is planar.

---

[b] A nonempty path in G is here a sequence of vertices $(x_1, x_2, \cdots, x_n)$ with $n \geq 2$ such that $(x_i, x_{i+1}) \in edg_G$ for all i and $x_i = x_j \Rightarrow i = j$ or $\{i, j\} = \{1, n\}$; if $x_1 = x_n$, this path is a *circuit*. We consider (x) as an empty path for every vertex x.

Proof
We shall denote by $\varphi_i, i = 1, \cdots, 9$ the formula expressing property Pi, in a structure $S \in STR(\mathcal{R}_s)$.

(1) φ_1 is the formula $x = y \vee \varphi_0(x, y)$ where φ_0 is constructed in Lemma 5.2.7 with $R = edg$.

(2) φ_2 is the formula $\forall x, y[\varphi_1(x, y)]$.

(3) φ_3 is the formula $\varphi_2[\theta/edg]$ where $\theta(x_1, x_2)$ is $edg(x_1, x_2) \vee edg(x_2, x_1)$.

(4) φ_4 is the formula $\neg x = y \wedge x \in X \wedge y \in X \wedge \varphi_1 \lceil X$ hence $\varphi_4(x, y, X)$ says that x, y are vertices of $G[X]$ and that there is a path in $G[X]$ from x to y; this is equivalent to the property of the statement. (We recall that we denote by $G[X]$ the induced subgraph of G with set of vertices X.)

(5) $\varphi_5(X)$ is the formula $\varphi_3 \lceil X \wedge \forall Y[X \subseteq Y \wedge \varphi_3 \lceil Y \Longrightarrow Y \subseteq X]$; (note that $\varphi_3 \lceil X$ expresses that $G[X]$ is connected); the subformula $X \subseteq Y$ stands for: $\forall u[u \in X \Longrightarrow u \in Y]$.

(6) φ_6 is the formula $\forall x, y[edg(x, y) \Longrightarrow \neg\varphi_1(y, x)]$.

(7) φ_7 is the formula $\varphi_6 \wedge \exists x \forall y[\varphi_1(x, y)] \wedge \forall x, y, z[edg(x, z) \wedge edg(y, z) \Longrightarrow x = y]$ expressing that G has no circuit, that every vertex is reachable from some vertex (the root) by a directed path and every vertex is of indegree at most 1: we consider directed trees with edges directed *from* the root, towards the leaves.

(8) The construction of φ_8 is more complex. We let $\varphi_8'(X)$ express that X is linearly ordered by the relation $edg^*_{G[X]}$ where G is assumed to have no circuit, i.e. $\varphi_8'(X)$ is

$$\forall x, y[x \in X \wedge y \in X \Longrightarrow x = y \vee \varphi_4(x, y, X) \vee \varphi_4(y, x, X)].$$

Let $\varphi_8''(x, y, X)$ be the formula defined as

$$\begin{aligned}&\varphi_6 \wedge \varphi_8'(X) \wedge x \in X \wedge y \in X \wedge \neg x = y \wedge \\ &\quad \forall z \in X[(\varphi_4(x, z, X) \vee x = z) \wedge (\varphi_4(z, y, X) \vee z = y)].\end{aligned}$$

It says that the considered graph has no circuit, that X is linearly ordered by $edg^*_{G[X]}$ with first element x and last element y with $y \neq x$. If X is finite, this is enough to express that X is the set of vertices of a path from x to y. But this is not true if X is infinite. (Take for example G with set of vertices $\mathcal{N}$ and set of edges $\{(i,j)/i \neq 0, j \in \{0, i+1\}\}$, take $X = \mathcal{N}$, $x_1 = 1$ and $x_2 = 0$.) We let $\theta(x_1, x_2, X)$ be the following formula:

$$\varphi_6 \wedge \varphi_8'(X) \wedge \varphi_4(x_1, x_2, X) \wedge \forall w[w \in X \wedge \varphi_4(x_1, w, X) \\ \Longrightarrow x_2 = w \vee \varphi_4(x_2, w, X)]$$

expressing that $x_1, x_2 \in X$ and x_2 is the successor of x_1 in X with respect to the linear order $edg^*_{G[X]}$. Let E be the binary relation on X defined by θ. Then X is the set of vertices of a path from x to y iff $(x,y) \in E^+$. Hence the following formula $\varphi_8(x,y,X)$ expresses the desired property:

$$\varphi_8'' \wedge \varphi_0[\theta/R](x,y,X)$$

where φ_0 is from Lemma 5.2.7.

(9) This will be proved in Section 5.8 by using Kuratowski's theorem. □

The condition that G be acyclic is important in $P8$: we shall prove in the next proposition that one cannot express by an MS formula that, in a finite directed graph, a given set X is the set of vertices of a path from x to y.

We now review some properties that are provably not MS-expressible. A directed graph is Hamiltonian if it has at least 2 vertices and a circuit passing through all vertices.

Proposition 5.2.9
The following properties are not MS-expressible:

1. *two sets X and Y have equal cardinality,*
2. *a directed graph is Hamiltonian,*
3. *in a directed graph a set X of vertices is the set of vertices of a path from x to y,*
4. *a graph has a nontrivial automorphism.*

Proof
We shall use the following result of Büchi and Elgot (see Thomas [60], Thm 3.2): if $L \subseteq \{a,b\}^*$ is the set of words u such that $\| u \| \models \varphi$ where φ is a closed MS-formula, then L is a regular language.

(1) Assume that we have a formula $\psi \in MS(\emptyset, \{X, Y\})$ such that for every set S, for every $V, W \subseteq S$:

$$S \models \psi(V, W) \text{ if and only if } Card(V) = Card(W).$$

Then the MS-formula $\psi[lab_a(x_1)/X, lab_b(x_1)/Y]$ of $MS(\mathcal{R}_{w,\{a,b\}})$ would characterize (as a subset of $\{a, b\}^*$) the language L of words having as many $a's$ as $b's$. This language is not regular, so we get a contradiction.
(2) With every word $w \in \{a, b\}^+$ represented by the structure $\| w \| = < \{1, \cdots, n\}, suc_{\|w\|}, lab_{a\|w\|}, lab_{b\|w\|} >$ we associate the graph K_w with set of vertices $\{1, \cdots, n\}$ and an edge from i to j iff $i \in lab_{a\|w\|}$ and $j \in lab_{b\|w\|}$ or vice-versa. Hence K_w is a complete bipartite directed graph. It is Hamiltonian iff w belongs to the language L already used in (1).

Let us now assume the existence of a formula η in $MS(\mathcal{R}_s, \emptyset)$ that would characterize the Hamiltonian graphs among finite graphs. Let $\mu(x_1, x_2)$ be the FO formula defined as:

$$(lab_a(x_1) \wedge lab_b(x_2)) \vee (lab_b(x_1) \wedge lab_a(x_2))$$

This formula defines in $\| w \|$ the edges of K_w (note that $D_{\|w\|} = V_{K_w}$). It follows that for every word $w \in \{a, b\}^*$ of length $n \geq 2$ we have

$$w \in L \text{ iff } K_w \text{ is Hamiltonian iff } \| w \| \models \eta[\mu/edg],$$

and the set of words in L of length at least 2 is regular, a contradiction.

(3) Assume we have a formula $\varphi \in MS(\mathcal{R}_s, \{x, y, X\})$ expressing that in a directed graph, X is the set of vertices of a directed path from x to y. Then the MS formula:

$$\exists X[\forall u(u \in X) \wedge \exists x, y(\varphi(x, y, X) \wedge \neg x = y \wedge edg(y, x))]$$

expresses that the considered graph has at least two vertices and is Hamiltonian. This is not possible by (2) hence no such formula φ does exist.

(4) The proof is very much like that of (2). With every word of the form $a^n bca^m$ with $n \geq 2, m \geq 2$ one associates the graph $H_{n,m}$ with set of vertices $\{-n, -n+1, \cdots, -1, 0, 1, 2, \cdots, m\} \cup \{0'\}$ and undirected edges between 0 and $0'$ and between i and $i+1$ for every $i, -n \leq i < m$. This graph has a nontrivial automorphism iff $n = m$ iff the given word belongs to the non-regular language $\{a^n bca^n / n \geq 2\}$. Hence, no MS formula can express that a graph has a nontrivial automorphism. We omit details. □

Remark:
The structure representing $H_{n,m}$ can be obtained from the structure $\| a^n bca^m \|$ by a definable transduction using quantifier-free formulas (to be introduced in Section 5.5 below). □

5.2.8 *Monadic second-order logic without individual variables*

This technical subsection may be skipped on first reading. We define a syntactical variant of MS logic where all variables denote sets. The corresponding formulas are less readable than those used up to now, but the proofs concerning them and the properties they define will be easier, because the syntax is limited.

We let $\mathcal{R}$ be as before; we shall only use set variables. We define $MS'(\mathcal{R}, \{X_1, \cdots, X_n\})$ by taking atomic formulas of the forms:

$X \subseteq Y$ where X, Y are variables,

$R(Y_1, \cdots, Y_n)$ where $R \in \mathcal{R}$, $n = \rho(R)$ and $Y_1, \cdots, Y_n$ are variables.

The formulas are constructed with the usual propositional connectives and set-quantifications $\exists X$ and $\forall X$. We denote by $MS'(\mathcal{R}, \{X_1, \cdots, X_n\})$ the set of such formulas with free variables among $X_1, \cdots, X_n$.

Let $S \in STR(\mathcal{R})$. The meaning of $X \subseteq Y$ is set inclusion. If $D_1, \cdots, D_n$ are sets denoted by $Y_1, \cdots, Y_n$ then $R(Y_1, \cdots, Y_n)$ is true in S iff $(d_1, \cdots, d_n) \in R_S$ for some $d_1 \in D_1, \cdots, d_n \in D_n$. The validity of formulas in $MS'(\mathcal{R}, \{X_1, \cdots, X_n\})$ in a structure S and for an assignment $\gamma : \{X_1, \cdots, X_n\} \to S$ follows immediately.

Lemma 5.2.10
For every formula $\varphi \in MS'(\mathcal{R}, \{X_1, \cdots, X_n\})$ one can construct an equivalent formula φ' in $MS(\mathcal{R}, \{X_1, \cdots, X_n\})$.

Proof
One replaces $X \subseteq Y$ by $\forall x[x \in X \Longrightarrow x \in Y]$ and $R(Y_1, \cdots, Y_n)$ by $\exists y_1, \cdots, y_n[y_1 \in Y_1 \wedge \cdots \wedge y_n \in Y_n \wedge R(y_1, \cdots, y_n)]$. We omit details. □

Lemma 5.2.11
For every formula $\varphi \in MS(\mathcal{R}, \{x_1, \cdots, x_n, Y_1, \cdots, Y_m\})$ one can construct a formula $\varphi' \in MS'(\mathcal{R}, \{X_1, \cdots, X_n, Y_1, \cdots, Y_m\})$ such that for every $S \in STR(\mathcal{R})$ for every assignment $\gamma : \{x_1, \cdots, x_n, Y_1, \cdots, Y_m\} \longrightarrow S$ then $(S, \gamma) \models \varphi$ iff $(S, \gamma') \models \varphi'$ where $\gamma'(X_i) = \{\gamma(x_i)\}$ and $\gamma'(Y_i) = \gamma(Y_i)$.

Proof
We first let $X = \emptyset$ be the formula $\forall Y[X \subseteq Y]$ in $MS'(\emptyset, \{X\})$ which characterizes the empty set. We let also $Sing(X) \in MS'(\emptyset, \{X\})$ be the formula $\forall Y[Y \subseteq X \Longrightarrow X \subseteq Y \vee Y = \emptyset]$ which characterizes the singleton sets.

We now translate φ into φ'. For each individual variable x, we let X denote a new set variable. ("New" means distinct of any variable in φ). We obtain φ' from φ by the following inductive construction:

$$(x = y)' = X \subseteq Y \wedge Y \subseteq X,$$

$$(R(y_1, \cdots, y_n))' = R(Y_1, \cdots, Y_n),$$

$$(x \in Y)' = X \subseteq Y,$$

$$(\varphi_1 op \varphi_2)' = \varphi_1' op \varphi_2'$$

for every binary connective *op* equal to $\vee, \wedge, \Longrightarrow$ or $\Longleftrightarrow$,

$$(\neg\varphi_1)' = \neg\varphi_1',$$

$$(\forall X.\varphi_1)' = \forall X.\varphi_1',$$

$$(\exists X.\varphi_1)' = \exists X.\varphi_1',$$

$$(\forall x.\varphi_1)' = \forall X[Sing(X) \Longrightarrow \varphi_1'],$$

$$(\exists x.\varphi_1)' = \exists X[Sing(X) \wedge \varphi_1'].$$

We omit details. □

5.2.9 *A worked example: the definition of square grids in MS logic*

The *rectangular grid* $G_{m,n}$ is the graph with set of vertices $V = \{0, 1, \cdots, m-1\} \times \{0, 1, 2, \cdots, n-1\}$ and set of edges $E = \{((i,j),(i',j'))/(i,j),(i',j') \in V$ and either $i' = i$ and $j' = j+1$ or $i' = i+1$ and $j' = j\}$. Its *north-*, *west-*, *south*, and *east-borders* are the sets of vertices:

$$X_n = \{0, \cdots, m-1\} \times \{n-1\}$$

$$X_w = \{0\} \times \{0, \cdots, n-1\}$$

$$X_s = \{0, \cdots, m-1\} \times \{0\}$$

$$X_e = \{m-1\} \times \{0, \cdots, n-1\}.$$

Its *well-coloring* is the 4-tuple of sets of vertices Y_0, Y_1, Y_2, Y_3 such that:
$(i, j) \in Y_k$ iff $k = mod(i) + 2(mod(j))$
where $mod(i)$ is the remaining (in $\{0, 1\}$) of the integer division of i by 2.

We claim that for every positive integer m there exists an MS formula θ with free variables $Y_0, Y_1, Y_2, Y_3, X_n, X_w, X_s, X_e$ that characterizes (among the finite directed simple graphs) those isomorphic to $G_{2m,2m}$ where, in addition, Y_0, Y_1, Y_2, Y_3 form a well-coloring and X_n, X_w, X_s, X_e are the four borders.

We let θ express the following conditions concerning a simple graph G given with sets of vertices $Y_0, Y_1, Y_2, Y_3, X_n, X_e, X_s, X_w$:

(1) G has no circuit;

(2) Y_0, Y_1, Y_2, Y_3 form a partition of V_G; assuming this we shall call i-vertex, i-successor of y, i-predecessor of y a vertex, or a successor of y, or a predecessor of y that belongs to Y_i;

(3) every vertex has at most one i-successor and at most one i-predecessor for each i;

(4) a 0- or a 3-vertex has no 0- or 3-successors; a 1- or a 2-vertex has no 1- or 2-successor;

(5) $G[X_n]$ is a path consisting alternatively of 2- and 3-vertices; its origin is a 2-vertex and its end is a 3-vertex; the origin is the unique element of $X_n \cap X_w$ and the end is the unique element of $X_n \cap X_e$;

(5') $G[X_e]$ is a path consisting alternatively of 1- and 3- vertices; its origin is a 1-vertex which is the unique element of $X_s \cap X_e$; its end is a 3-vertex which is the unique element of $X_n \cap X_e$;

(5") and (5"') state similar conditions on the sets X_s and X_w;

(6) each 2-vertex in X_n has a 3-successor and no 0-successor; each 3-vertex in $X_n - X_e$ has a 2-successor and no 1-successor;

(6') (6") (6"') state similar properties of X_e, X_s, X_w;

(7) each vertex in $V_G - (X_n \cup X_e \cup X_s \cup X_w)$ has two predecessors and two successors;

(8) there exists a path from the vertex in $X_s \cap X_w$ to the one in $X_n \cap X_e$ the vertices of which have the colors $0, 1, 3, 2, 0, 1, 3, 2, \cdots 0, 1, 3$ in this order.

It is not hard to see that conditions (1) to (7) characterize the well-colored grids of the form $G_{2n,2m}$ for $n \geq 1, m \geq 1$. Condition (8) implies furthermore that $m = n$. It is not hard to modify this construction in order to characterize the grids of the form G_{2n+1}, G_{2n+1} for $n \geq 1$.

In view of the proof of Proposition 5.2.2, a line of G given with $Y_0, Y_1, \cdots, X_w$ as above is a set $L \subseteq V_G$ such that

(1) $G[L]$ is a path from a vertex in X_w to a vertex in X_e

(2) either $L \subseteq Y_0 \cup Y_1$ or $L \subseteq Y_2 \cup Y_3$.

5.3 Representations of partial orders, graphs and hypergraphs by relational structures

In this section we discuss more in detail the *various possibilities* of representing a partial order, a graph or a hypergraph by a relational structure. The choice of the representation is important for the possibility of expressing a given property by some formula of the three logical languages we have introduced.

5.3.1 Partial orders

The simplest way to represent a partial order $\leq_D$ on a set D is by the structure $< D, \leq_D >$. If D is finite, one can also represent $< D, \leq_D >$ by a graph $< D, S >$ where S is binary, such that $\leq_D$ is the reflexive and transitive closure of S. There is even a unique minimal such relation (minimal for inclusion) that we shall denote by suc_D (where suc means successor) and the corresponding graph $< D, suc_D >$ is the classical Hasse-diagram of the partial order $< D, \leq_D >$. A property of partial orders representable by a formula φ with respect to the representation $< D, \leq_D >$ will be representable by another (perhaps more complex) formula ψ with respect to the representation $< D, suc_D >$.

Proposition 5.3.1
(1) A property of a finite partial orders is MS with respect to the representation $< D, \leq_D >$ iff it is MS with respect to the representation $< D, suc_D >$.
(2) Furthermore, it is FO with respect to the representation $< D, \leq_D >$ if it is FO with respect to the representation $< D, suc_D >$. □

If a property is FO with respect to the representation $< D, \leq_D >$ we can only conclude (by (1)) that it is MS with respect to the representation $< D, suc_D >$. We have actually the proper hierarchy:

$$FO_{suc} \subset FO_{\leq} \subset MS_{suc} = MS_{\leq},$$

where the subscripts indicate the considered representation. The languages defined by first-order formulas in terms of $\leq$ (as opposed to in terms of the successor relation) are the star-free languages. This class is a proper subclass of the class of regular languages (definable by MS_{Suc} or $MS_{\leq}$ formulas) and contains properly the class of locally threshold testable languages (definable by FO_{suc} formulas) (see Pin [50]). >From these strict inclusions concerning classes of languages follow the corresponding strict inclusions for partial orders ($FO_{suc} \subset FO_{\leq} \subset MS_{suc}$).

Proof

We let suc be a binary relation symbol. Since $\leq_D$ is the reflexive and transitive closure of suc_D, it follows that $\leq_D$ is defined in $< D, suc_D >$ by the MS formula $\mu(y_1, y_2)$:

$$y_1 = y_2 \vee \varphi_0[suc(x_1, x_2)/R]$$

where φ_0 is from Lemma 5.2.7. Hence if a property is expressed by an MS or a FO formula φ with respect to $< D, \leq_D>$ it can be expressed by the MS formula $\varphi[\mu(y_1, y_2)/ \leq]$.

Now suc_D is defined in $< D, \leq_D>$ by the FO formula: $\mu'(y_1, y_2)$

$$\neg y_1 = y_2 \wedge y_1 \leq y_2 \wedge \forall z[y_1 \leq z \wedge z \leq y_2 \Longrightarrow y_1 = z \vee z = y_2].$$

Hence a property expressed in $< D, suc_D >$ by an MS or FO formula φ' can be expressed by the formula $\varphi'[\mu'/suc]$ which is MS or FO respectively. □

5.3.2 *Edge set quantifications*

As already noted the representation of a graph by a relational structure that we have used up to now is not convenient to express properties of graphs that depend on the multiplicity of edges. We shall define another representation, where the edges are elements of the domain, which is more natural for expressing logically the properties of multigraphs and which, furthermore, makes it possible to express more properties of simple graphs by MS formulas.

We let $\mathcal{R}_m = \{inc\}$ where inc is ternary (inc stands for "incidence"). For every graph G, directed or not, simple or not, we denote by V_G its set of vertices and by E_G its set of edges. The incidence relation between vertices and edges is represented by the ternary relation inc_G such that: $(x, y, z) \in inc_G$ iff x is an edge and either this edge is directed and it links y to z or it is undirected and it links y and z. If x is an undirected edge, we have: $(x, y, z) \in inc_G$ iff $(x, z, y) \in inc_G$.

We let $D_G := V_G \cup E_G$ (we always assume that $V_G \cap E_G = \emptyset$) and we let $\mid G \mid_2$ be the structure $< D_G, inc_G > \in STR(\mathcal{R}_m)$. If G has several edges linking y to z then, they are represented by distinct elements of D_G.

In a structure $S = < D_S, inc_S >$ representing a graph, the edges are the elements x of D_S that satisfy the formula $\exists y, z[inc(x, y, z)]$. The structures in $STR(\mathcal{R}_m)$ representing directed graphs are exactly those which satisfy the following conditions:

(1) $\forall x, y, z[inc(x, y, z) \Longrightarrow \neg\exists u, v(inc(y, u, v) \vee inc(z, u, v))]$, and

(2) $\forall x, y, z, y', z'[inc(x, y, z) \wedge inc(x, y', z') \Longrightarrow y = y' \wedge z = z']$.

In any such structure, if we let $E = \{x \in D_S \mid S \models \exists y, z.inc(x, y, z)\}$ and $V = D_S - E$ then $inc_S \subseteq E \times V \times V$ and there exists a unique directed graph G with $V_G = V, E_G = E$ and $\mid G \mid_2 = S$. Similarly, the structures in $STR(\mathcal{R}_m)$ representing undirected graphs are exactly those that satisfy (1) above together with:

(2') $\forall x, y, z, y', z'[inc(x, y, z) \wedge inc(x, y', z')$
$\Longrightarrow (y = y' \wedge z = z') \vee (y = z' \wedge z = y')]$

(3') $\forall x, y, z[inc(x, y, z) \Longrightarrow inc(x, z, y)]$.

Graph properties can be expressed logically, either via the representation of a graph G by $\mid G \mid_2$, or via the initially defined representation $\mid G \mid_1 := < V_G, edg_G >$. The representation $\mid G \mid_1$ only allows quantification on vertices, sets of vertices, relations on vertices (according to the language we consider), whereas the representation $\mid G \mid_2$ also allows quantifications on edges, sets of edges and relations on edges. We shall distinguish the MS_1-definable classes (which are nothing but the MS-definable ones we have considered up to now), from the MS_2-definable ones, which use formulas in $MS(\{inc\}, \emptyset)$ and the representation of a graph G by $\mid G \mid_2$. Similarly, we have FO_1 -, FO_2 -, SO_1- and SO_2-definable classes of graphs.We shall also speak of FO_i - or MS_i - or SO_i-expressible graph properties, where $i = 1$ or 2.

Since a set of edges in a simple graph can be considered as a binary relation on its set of vertices it is quite clear that every MS_2-expressible property is SO_1-expressible (we shall prove that later). However it is not always MS_1-expressible.

Proposition 5.3.2
The following properties of a directed graph G can be expressed by MS_2-formulas.

(1) X is a set of edges (resp. of vertices) forming a path from x to y, where $x \neq y$.

(2) G is Hamiltonian.

Proof
(1) For every subset X of E_G we denote by $G[[X]]$ the subgraph H of G such that V_H is the set of vertices incident to some edge in X and $E_H = X$. The desired condition is thus that $G[[X]]$ is a path from x to y. This can be expressed by a formula ψ_1' in $MS(\mathcal{R}_m, \{x, y, X\})$ constructed with the help of Lemma 5.2.6 and the formula φ_8 of Lemma 5.2.8. The desired formula ψ_1 is thus

$$\neg x = y \wedge \forall z[z \in X \Longrightarrow \exists u, v(inc(z, u, v))] \wedge \psi_1'.$$

If we want to express that Y is the set of vertices of a path from x to y we take $\psi_2(x, y, Y)$ defined as

$$\exists X[\psi_1(x, y, X) \wedge \forall z[z \in Y \Longleftrightarrow \exists u, v(u \in X \wedge (inc(u, z, v) \vee inc(u, v, z)))]].$$

The quantification on sets of edges is thus crucial in ψ_2 since we have proved that no formula in $MS(\mathcal{R}_s, \{x, y, Y\})$ equivalent to ψ_2 does exist (see Proposition 5.2.9).
(2) follows from (1): see the proof of Proposition 5.2.9, assertion (3). □

Proposition 5.3.3
Let C be a class of simple, directed or undirected graphs:

(1) C is FO_1-definable iff C is FO_1-definable

(2) C is SO_2-definable iff C is SO_1-definable

(3) C is MS_2-definable if C is MS_1-definable and the converse does not always hold.

Proof
The relation edg_G is definable from inc_G by the formula $\exists u.inc(u, x_1, x_2)$. It follows that for each $L \in \{\text{FO, SO, MS}\}$, C is L_2-definable if it is L_1-definable.

That the converse does not hold for MS follows from Propositions 5.2.9 and 2.2: the class of simple Hamiltonian directed graphs (with at least 2 vertices) is MS_2-definable but is not MS_1-definable relatively to the class of simple graphs. It holds for FO because a quantification of the form "there exists an edge e..." can be replaced by a quantification of the form "there exist vertices x and y

that form an edge such that...". The proof is similar for SO: a quantification over n-ary edge relations is replaced by a quantification over $2n$-ary vertex relations. We omit details. □

The next theorem reviews some classes of graphs on which MS_1 and MS_2 are equally expressive. (Tree-width will be defined in Section 5.6).

Theorem 5.3.4
Let $\mathcal{C}$ be the class of planar directed simple graphs, or of directed simple graphs of degree at most k or finite directed simple graphs of tree-width at most k (for any fixed k). A property of graphs in $\mathcal{C}$ is MS_2-expressible iff it is MS_1-expressible. The same holds for the corresponding classes of undirected graphs. □ □

We do not reproduce the full proof of this theorem which is quite long (see Courcelle [14]). We only give the basic lemma and some consequences.

Let G be a directed graph. A *semistrong $k-coloring$* of G is a mapping $\gamma : V_G \to \{1, 2, \cdots, k\}$ such that, for every two vertices $v, v' \neq v$:

1. if v, v' are adjacent, then $\gamma(v) \neq \gamma(v')$,
2. if there are edges linking v to w and v' to w, where w is a third vertex, then $\gamma(v) \neq \gamma(v')$.

The following lemma will be proved in Section 5.5. Let $\mathcal{C}_k$ be the class of finite or infinite simple directed loop-free graphs having a semistrong k-coloring.

Lemma 5.3.5
Let $k \in \mathcal{N}$. A property of graphs in $\mathcal{C}_k$ is MS_1-expressible if it is MS_2-expressible. □ □

The class $\mathcal{C}_2$ contains the directed trees. (Edges are directed in such a way that there is a unique path from the root to each vertex). Hence the languages MS_1 and MS_2 are equally powerful for expressing properties of directed trees. Let $d \in \mathcal{N}$ and $k = d^2 + 1$. Let us prove that $\mathcal{C}_k$ contains the simple directed loop-free graphs of degree at most d. Let G be such a graph. Let G' be the graph obtained by adding an edge between any two vertices at distance 2. This graph has degree at most $d + d(d-1) = d^2$. Hence G' has a k-coloring (in the usual sense, where one requires that any two adjacent vertices have different colors). Such a coloring is a semistrong k-coloring of G. Hence we have proved that Theorem 5.3.4 holds for the classes of directed trees and of simple directed loop-free graphs of degree at most any fixed d. The complete proof of Theorem 5.3.4 is based on Lemma 5.3.5 and constructions of appropriate colorings.

5.3.3 Hypergraphs

We define directed, hyperedge labelled hypergraphs. We let A be a finite ranked set: each symbol $a \in A$ has an associated rank, a nonnegative integer denoted by $\tau(a)$. A hypergraph H has a set of vertices V_H and a set of hyperedges E_H; each hyperedge e has a label $lab_H(e)$ in A and a sequence of vertices of length $\tau(lab_H(e))$. The label of e may have rank 0 and in this case, e has no vertex. We always assume that $V_H \cap E_H = \emptyset$. As for graphs we define two representations of hypergraphs by relational structures.

We let $\mathcal{R}_s(A) := \{edg_a / a \in A\}$ and $\mathcal{R}_m(A) := \{inc_a / a \in A\}$ where edg_a is $\tau(a)$-ary and inc_a is $(\tau(a)+1)$-ary. With a hypergraph H we will associate the structures $\mid H \mid_1 := < V_H, (edg_{aH})_{a \in A} > \in STR(\mathcal{R}_s(A))$ and $\mid H \mid_2 := < V_H \cup E_H, (inc_{aH})_{a \in A} > \in STR(\mathcal{R}_m(A))$ where:

(1) $inc_{aH}(x, y_1, \cdots, y_n)$ holds iff $x \in E_H, y_1, \cdots, y_n \in V_H, lab_H(x) = a, n = \tau(a)$ and $(y_1, \cdots, y_n)$ is the sequence of vertices of x, and

(2) $edg_{aH}(y_1, \cdots, y_n)$ holds iff $inc_{aH}(x, y_1, \cdots, y_n)$ holds for some $x \in E_H$.

The structure $\mid H \mid_1$ contains no information on the hyperedges of type 0, and no information either on the number of hyperedges having the same label and the same sequence of vertices. A hypergraph is *simple* if it has no hyperedge of type 0 and if no two hyperedges have the same label and the same sequence of vertices. The structures $\mid H \mid_2$ are appropriate for representing all hypergraphs H whereas the structures $\mid H \mid_1$ are only appropriate for representing simple hypergraphs or for expressing logically properties of hypergraphs that are independent of the multiplicity of hyperedges and of the existence of hyperedges of type 0.

We shall refer by MS_i to MS logic relative to the representation of a hypergraph H by the structure $\mid H \mid_i$ where $i = 1, 2$. The results of Proposition 5.3.3 hold for hypergraphs built over a fixed finite set A as well as for graphs.

Proposition 5.3.6
Let $k \in \mathcal{N}$ and A be a finite ranked alphabet. The same properties of finite simple hypergraphs over A of tree-width at most k are expressible in MS_1 and in MS_2.

Proof
See Courcelle and Engelfriet [23]. □

In certain cases, hypergraphs are equipped with distinguished vertices called sources or ports (see Section 5.6). These vertices will be represented in relational structures by means of additional unary relations. For instance, if a hypergraph H is given with k sets of distinguished vertices then the structure $\mid H \mid_1$ contains k unary relations $P_{1H}, \cdots, P_{kH}$ where $P_1, \cdots, P_k$ are additional unary symbols.

5.4 The expressive powers of monadic-second order languages

In the preceding section, we have seen that the choice of a representing structure for an object like a partial order, a graph or a hypergraph may affect the first-order or the monadic second-order expressibility of properties of these objects. Here, we shall consider the extension of MS logic by cardinality predicates. In some cases this extension is just a syntactic shorthand, and in others we shall obtain a real extension of expressive power.

5.4.1 Cardinality predicates

We first extend MS logic by constructing formulas with the help of the new atomic formulas of the form $Fin(X)$ where X is a set variable. Such a formula is valid iff X denotes a finite set. We shall denote by $MS^f(\mathcal{R},\mathcal{Y})$ the corresponding sets of formulas. We have $MS(\mathcal{R},\mathcal{Y}) \subseteq MS^f(\mathcal{R},\mathcal{Y})$.

This extension is of interest only in the case where we consider possibly infinite structures. In finite structures, $Fin(X)$ is always true, and every formula in $MS^f(\mathcal{R},\mathcal{Y})$ can be simplified into one in $MS(\mathcal{R},\mathcal{Y})$, that is equivalent in finite structures. Otherwise, we obtain a real increase of expressive power because the finiteness of a set is not, in general, MS-expressible. See Corollary 5.4.3 below. We wrote "in general" because in certain structures like binary directed trees, it is, as we shall see.

We now introduce an extension of MS logic called *counting monadic second-order logic* and denoted by CMS. For every integer $p \geq 2$, for every set variable X, we let $Card_p(X)$ be a new atomic formula expressing that the set denoted by X is finite and that its cardinality is a (possibly null) multiple of p. We denote by $CMS(\mathcal{R},\mathcal{Y})$ the extension of $MS^f(\mathcal{R},\mathcal{Y})$ with atomic formulas of the forms $Card_p(X)$ for $p \geq 2$, where X is a set variable. We have thus a hierarchy of languages $MS \subset MS^f \subset CMS$. We shall discuss cases where the corresponding hierarchy of graph properties is also strict.

Lemma 5.4.1
For each $k \in \mathcal{N}$, there is a first-order formula in $FO(\emptyset, \{X\})$ expressing that the set denoted by X has cardinality k.

Proof
We only give the formula for $k = 3$:
$$\exists x_1, x_2, x_3[x_1 \in X \wedge x_2 \in X \wedge x_3 \in X \wedge \neg x_1 = x_2 \wedge \neg x_1 = x_3 \wedge \neg x_2 = x_3 \wedge \forall y(y \in X \Longrightarrow x_1 = y \vee x_2 = y \vee x_3 = y)].$$ □

It follows that one can express in CMS that a set X is finite and has a cardinality of the form $q + \lambda p$ for some $\lambda \in \mathcal{N}$ where $1 \leq q < p$: it suffices to write that for some Y and Z, $X = Y \cup Z$, $Y \cap Z = \emptyset$, $Card(Y) = q$ and $Card(Z)$ is a multiple of p. We now recall a result from Courcelle [11].

Proposition 5.4.2
Every formula $\varphi \in MS(\emptyset, \{X_1, \cdots, X_n\})$ is equivalent to a finite disjunction of conjunctions of conditions of the forms $Card(Y_1 \cap Y_2 \cap \cdots \cap Y_n) = m$ or $Card(Y_1 \cap Y_2 \cap \cdots \cap Y_n) > m$ where $m \in \mathcal{N}$ and, for each $i = 1, \cdots, n, Y_i$ is either X_i or $D - X_i$, where D is the domain of the considered structure. □

By Lemma 5.4.1 the conditions $Card(Y_1 \cap Y_2 \cap \cdots \cap Y_n) = m$ and $Card(Y_1 \cap Y_2 \cap \cdots \cap Y_n) > m$ can be expressed by formulas in $FO(\emptyset, \{Y_1, \cdots, Y_n\})$. This result shows that MS logic is equivalent to FO logic for "pure" finite or infinite sets.

Corollary 5.4.3
CMS is strictly more expressive than MS on finite sets. MS^f is strictly more expressive than MS on infinite sets.

Proof
Let us assume that a formula $\varphi \in MS(\emptyset, \{X\})$ can express, in every finite set D that a subset X of D has even cardinality. By Proposition 5.4.2, φ can be expressed as a disjunction of conditions of the forms:

(1) $Card(X) = m$ and $Card(D - X) = m'$,

(2) $Card(X) = m$ and $Card(D - X) > m'$,

(3) $Card(X) > m$ and $Card(D - X) = m'$,

(4) $Card(X) > m$ and $Card(D - X) > m'$.

We let M be the maximum integer m or m' occurring in these conditions. No condition of the form (3) or (4) can appear because it can be valid for sets X with large enough odd cardinality. The remaining conditions imply that $Card(X) \leq M$. But then φ is not valid for certain sets X with even cardinality larger than M. Contradiction.

If we now assume that φ as above expresses that X is finite we also get a contradiction by the same case analysis. □

This proves that for finite structures, we have the hierarchy (where $\subset$ indicates a proper increase of expressive power and $\equiv$ indicates an equivalent one)

$$MS \equiv MS^f \subset CMS$$

whereas for general (possibly infinite) structures we have

$$MS \subset MS^f \subset CMS.$$

We shall now investigate classes for which these proper inclusions become equivalences.

5.4.2 Linearly ordered structures

We let $\mathcal{L}$ be the class of finite linear orders; we shall represent them by structures of the form $< D, \leq_D >$ as explained in Subsection 5.3.1.

Lemma 5.4.4
For every $p \geq 2$ one can construct a formula $\gamma_p(X) \in MS(\{\leq\}, \{X\})$ expressing in every $D \in \mathcal{L}$ that X has a cardinality equal to 0 modulo p.

Proof
One can easily write a formula γ_p expressing the following:
either X is empty or there exist sets $Y_1, \cdots, Y_p$ forming a partition of X and such that

1. the first element of X is in Y_1,
2. the last element of X is in Y_p,
3. for every two elements x, y of X such that y is the successor of x for the restriction to X of the order $\leq_D$, then, $y \in Y_{i+1}$ if $x \in Y_i$, for some $i = 1, \cdots, p$, and $y \in Y_1$ if $x \in Y_p$.

□

This lemma extends to any class $\mathcal{C}$ of finite structures on which a linear order is definable by MS-formulas. Let us define precisely this notion.

Let $\mathcal{C} \subseteq STR(\mathcal{R})$ for some finite set $\mathcal{R}$ of relation symbols. Let $\theta_0 \in MS(\mathcal{R}, \{X_1, \cdots, X_n\})$ and $\theta_1 \in MS(\mathcal{R}, \{x_1, x_2, X_1, \cdots, X_n\})$. We say that (θ_0, θ_1) *defines a linear order on the structures of* $\mathcal{C}$ if the following conditions hold, for every $S \in \mathcal{C}$:

1. $S \models \exists X_1, \cdots, X_n.\theta_0$

2. $S \models \forall X_1, \cdots, X_n[\theta_0 \Longrightarrow \forall x, y, z\{\theta_1(x,x) \wedge (\theta_1(x,y) \wedge \theta_1(y,x) \Longrightarrow x = y)$
$\wedge(\theta_1(x,y) \wedge \theta_1(y,z) \Longrightarrow \theta_1(x,z)) \wedge (\theta_1(x,y) \vee \theta_1(y,x))\}]$.

In words this means that for every choice of sets $D_1, \cdots, D_n$ satisfying θ_0, the binary relation defined by θ_1 where $X_1, \cdots, X_n$ take respectively the values $D_1, \cdots, D_n$ is a linear order on D_S, and that for every $S \in \mathcal{C}$, there is at least one such tuple from which a linear order is definable by θ_1.

Proposition 5.4.5
Let $\mathcal{C}$ be a class of finite structures on which a linear order is MS-definable. The CMS-expressible properties of the structures in $\mathcal{C}$ are MS-expressible.

Proof
We let $\mathcal{R}$ be such that $\mathcal{C} \subseteq STR(\mathcal{R})$.
Let $(\theta_0(X_1, \cdots, X_n), \theta_1(x, y, X_1, \cdots, X_n))$ be a pair of formulas that defines a linear order on every structure of $\mathcal{C}$. Let $\varphi \in CMS(\mathcal{R}, \emptyset)$ express some property of the structures of $\mathcal{C}$. We can assume that the variables $X_1, \cdots, X_n$ have no occurrence in φ. For each p, we let $\gamma'_p = \gamma_p[\theta_1/ \leq] \in MS(\mathcal{R}, \{X_1, \cdots, X_n\})$ where γ_p is from Lemma 5.4.4. For every $S \in \mathcal{C}$, for every $D_1, \cdots, D_n \subseteq D_S$ satisfying θ_0, for every $D \subseteq D_S$ then

$$S \models \gamma'_p(D, D_1, \cdots, D_n)$$

iff $Card(D)$ is a multiple of p.

We let φ' be obtained from φ by the replacement of every atomic formula $Card_p(Y)$ by $\gamma'_p[Y/X]$. We let then φ" be the formula of $MS(\mathcal{R}, \{X\})$:

$$\exists X_1, \cdots, X_n[\theta_0(X_1, \cdots, X_n) \wedge \varphi'(X, X_1, \cdots, X_n)].$$

If $S \models \varphi"(D)$ then $S \models \theta_0(D_1, \cdots, D_n)$ for some $D_1, \cdots, D_n$ and $S \models \varphi'(D, D_1, \cdots, D_n)$ hence $S \models \varphi(D)$ by the construction of φ'. Conversely, if $S \models \varphi(D)$ then there exists $(D_1, \cdots, D_n)$ satisfying θ_0 hence $S \models \varphi'(D, D_1, \cdots, D_n)$ and $S \models \varphi"(D)$. Hence φ is equivalent to φ" in every structure S in $\mathcal{C}$. □

A *directed tree* is a directed graph, every vertex of which is reachable from some vertex (called the *root*) by one and only one path. (In particular it is connected and has no circuit).

Proposition 5.4.6
Let $d \in \mathcal{N}$. One can define by means of MS formulas a linear order on (possibly infinite) directed trees of degree at most d, and more generally on forests consisting of at most d trees of degree at most d.

Proof
Let T be a directed tree given by the structure $< V_T, suc_T >$. We denote by $\leq_T$ the reflexive and transitive closure of suc_T. A partition $(V_1, V_2, \cdots, V_d)$ of V_T is *good* if no two successors of a vertex belong to a same set V_i. If T has degree at most d (i.e., if each vertex has at most d successors) then V_T has a good partition in d sets. From a good partition $(V_1, \cdots, V_d)$ of V_T, we can define the following linear order:
$x \leq y$ iff either $x \leq_T y$ or there exist z, x', y' such that $z suc_T x' \leq_T x, z suc_T y' \leq_T y, x' \in V_i, y' \in V_j$ and $i < j$.
It is not hard to see that $\leq$ is a linear order on V_T. An FO_1-formula $\theta_0(X_1, \cdots, X_d)$ can express that a given d-tuple $(V_1, V_2, \cdots, V_d)$ of subsets of V_T is good and an MS_1-formula $\theta_1(x, y, X_1, \cdots, X_d)$ can express that $x \leq y$ holds where X_i takes the value V_i (for $i = 1, \cdots, d$) and $(V_1, \cdots, V_d)$ is the good partition from which $\leq$ is defined. (One can write θ_1 with the help of the formula φ_1 of Lemma 5.2.8.)

Hence (θ_0, θ_1) defines a linear order of V_T (and even a topological sorting) for every tree T of degree at most d. The proof for forests is similar except that in the definition of a good partition, we require that the roots of two trees of the forest do no belong to the same set V_i. □

We refer the reader to Courcelle [18] for extensions of this result to other types of graphs than forests.

5.4.3 Finiteness

We know that the finiteness of a set is not MS-expressible in general. However it is in certain structures. Let $< D, \leq_D >$ be a linear order isomorphic to $\mathcal{N}$. A subset X of D is finite iff it has a maximal element, which is FO-expressible. We shall use this observation in order to define classes of graphs in which finiteness is MS-expressible. For these classes of graphs the languages MS and MS^f are thus of equal expressive power.

Proposition 5.4.7
Let $\mathcal{T}$ be the class of directed trees, each vertex of which has finite degree. The finiteness of a set of vertices of a tree in $\mathcal{T}$ is MS-expressible.

Proof
We represent a directed tree T by the structure $T = < V_T, suc_T >$. We denote by $root_T$ the root of T. Every vertex is accessible from the root by a directed path. The $\leq_T$-maximal elements of V_T are the leaves and $root_T$ is the unique $\leq_T$-minimal element.

For each nonempty $X \subseteq V_T$ we let $I(X) := \{y \in V_T \mid y \leq_T x, x \in X\}$ be the *ideal* generated by X. The graph $T[I(X)]$ is a directed tree and its root is $root_T$.

Claim: X is infinite iff the tree $T[I(X)]$ has an infinite branch.

Proof of Claim: Let $(y_0, y_1, y_2, \cdots, y_i, \cdots)$ be an infinite branch in $T[I(X)]$. We construct as follows an infinite sequence in X, proving thus that X is infinite:

1. we let $x_0 \in X$ be an element such that $y_0 \leq_T x_0$;
2. having defined x_j, we define x_{j+1} as follows: we let i be such that y_i is at a larger distance to the root than x_j. We let then x_{j+1} be any element of X such that $y_i \leq_T x_{j+1}$.

The elements $x_0, x_1, \cdots, x_j, \cdots$ of X are at strictly increasing distances to the root. Hence they are pairwise distinct and X is infinite.

Conversely, if X is infinite then $I(X)$ is infinite (since $X \subseteq I(X)$) hence the tree $T[I(X)]$ has an infinite branch by Koenig's lemma. *Proof of Claim* □

We now complete the proof of the proposition. One can construct an MS formula $\theta(X, Y)$ expressing that, in a directed tree T:
Y is linearly ordered by $\leq_T$, and for every $y \in Y$ there is $z \in Y$ such that $y <_T z$ and $z \leq_T x$ for some $x \in X$.

It follows from the claim that a set X is finite iff $\theta(X, Y)$ does not hold for any set Y. □

Corollary 5.4.3 and Propositions 5.4.6 and 5.4.7 yield the following result (we omit details):

Corollary 5.4.8
(1) Let $d \in \mathcal{N}$. The languages MS, MS^f and CMS are equally powerful for expressing properties of trees of degree at most d.
(2) The languages MS and MS^f are equally powerful for expressing properties of trees of finite degree, and the language CMS is strictly more powerful than them. □

5.5 Monadic second-order definable transductions

In this section, we use MS formulas in order to define certain graph transformations, that are as important in the theory of context-free graph grammars as are rational transductions in language theory.

A binary relation $R \subseteq A \times B$ where A and B are sets, typically of words or of trees, can be considered as a multivalued partial mapping associating with certain elements of A one or more elements of B. It is called in this case a *transduction*: $A \to B$. Transductions of words and trees, defined in terms of finite-state automata with output are essential in Formal Language Theory, especially in constructions concerning context-free grammars. (See Berstel [4]). Does there exist an analogous notion for graphs ? Since there is no convenient notion of graph automaton, there is no "machine-based" notion of graph transduction. However, by using a classical technique of logic called "interpretation", by which a structure is defined in another one, one can define transformations of structures, whence of graphs and hypergraphs via their representations by structures. The formulas defining a structure T inside another one S (or rather: inside the union of k disjoint copies of S) will be MS formulas.

5.5.1 *Transductions of relational structures*

We first define *(monadic second-order) definable transductions of relational structures*. Let $\mathcal{R}$ and $\mathcal{Q}$ be two finite ranked sets of relation symbols. Let $\mathcal{W}$ be a finite set of set variables, called here the set of *parameters*. (It is not a loss of generality to assume that all parameters are set variables.) A $(\mathcal{Q},\mathcal{R})$-*definition scheme* is a tuple of formulas of the form:

$$\begin{aligned}\Delta =&(\varphi, \psi_1, \cdots, \psi_k, (\theta_w)_{w \in \mathcal{Q}*k})\\ &\text{where } k > 0,\ \ \mathcal{Q}^*k := \{(q,\vec{j}) \mid q \in \mathcal{Q}, \vec{j} \in [k]^{\rho(q)}\},\\ &\varphi \in MS(\mathcal{R},\mathcal{W}),\\ &\psi_i \in MS(\mathcal{R},\mathcal{W} \cup \{x_1\}) \text{ for } i = 1, \cdots, k,\\ &\theta_w \in MS(\mathcal{R},\mathcal{W} \cup \{x_1,\cdots,x_{\rho(q)}\}), \text{for } w = (q,\vec{j}) \in \mathcal{Q}^*k.\end{aligned}$$

These formulas are intended to define a structure T in $STR(\mathcal{Q})$ from a structure S in $STR(\mathcal{R})$ and will be used in the following way. The formula φ defines the domain of the corresponding transduction; namely, T is defined only if φ holds true in S for some assignment of values to the parameters. Assuming this condition fulfilled, the formulas $\psi_1, \cdots, \psi_k$, define the domain of T as the disjoint union of the sets $D_1, \cdots, D_k$, where D_i is the set of elements in the domain of S that satisfy ψ_i for the considered assignment. Finally, the formulas θ_w for $w = (q, \vec{j}), \vec{j} \in [k]^{\rho(q)}$ define the relation q_T. Here are the formal definitions.

Let $S \in STR(\mathcal{R})$, let γ be a $\mathcal{W}$-assignment in S. A $\mathcal{Q}$-structure T with domain $D_T \subseteq D_S \times [k]$ is *defined in* (S, γ) by Δ if:

(i) $(S, \gamma) \models \varphi$

(ii) $D_T = \{(d, i) \mid d \in D_S, i \in [k], (S, \gamma, d) \models \psi_i\}$

(iii) for each q in $\mathcal{Q}$:
$q_T = \{((d_1, i_1), \cdots, (d_t, i_t)) \in D_T^t \mid (S, \gamma, d_1, \cdots, d_t) \models \theta_{(q,\vec{j})}\}$,
where $\vec{j} = (i_1, \cdots, i_t)$ and $t = \rho(q)$.

(By $(S, \gamma, d_1, \cdots, d_t) \models \theta_{(q,\vec{j})}$, we mean $(S, \gamma') \models \theta_{(q,\vec{j})}$, where γ' is the assignment extending γ, such that $\gamma'(x_i) = d_i$ for all $i = 1, \cdots, t$; a similar convention is used for $(S, \gamma, d) \models \psi_i$.)

Since T is associated in a unique way with S, γ and Δ whenever it is defined, i.e., whenever $(S, \gamma) \models \varphi$, we can use the functional notation $def_\Delta(S, \gamma)$ for T.

The *transduction defined by* Δ is the relation $def_\Delta := \{(S, T) \mid T = def_\Delta(S, \gamma)$ for some $\mathcal{W}$-assignment γ in $S\} \subseteq STR(\mathcal{R}) \times STR(\mathcal{Q})$. A transduction $f \subseteq STR(\mathcal{R}) \times STR(\mathcal{Q})$ is *definable* if it is equal to def_Δ for some $(\mathcal{Q}, \mathcal{R})$-definition scheme Δ. In the case where $\mathcal{W} = \emptyset$, we say that f is *definable without parameters* (note that it is functional). We shall refer to the integer k by saying that def_Δ is k-copying; if $k = 1$ we say that it is *noncopying* and we can write more simply Δ as $(\varphi, \psi, (\theta_q)_{q \in \mathcal{Q}})$. In this case:

$D_T = \{d \in D_S \mid (S, \gamma, d) \models \psi\}$ and for each q in $\mathcal{Q}$

$q_T = \{(d_1, \cdots d_t) \in D_T^t \mid (S, \gamma, d_1, \cdots d_t) \models \theta_q\}$, where $t = \rho(q)$.

Before applying these definitions to general hypergraphs, we give three examples. Our first example is the transduction that associates with a graph the set of its connected components. A graph G is represented by $\mid G \mid_2$ (see Section 5.3).

The definition scheme Δ given below uses a parameter X. It is constructed in such a way that $def_\Delta(\mid G \mid_2, \{x\})$ is the structure $\mid G' \mid_2$ where G' is the connected component of G containing the vertex x. We let $(\varphi, \psi, \theta_{inc})$ be the noncopying definition scheme where:

φ is a formula with free variable X expressing that X consists of a unique vertex,

ψ is a formula with free variables X and x_1 expressing that either x_1 is a vertex linked by a path (where edges can be traversed in either direction) to the vertex in X, or x_1 is an edge, one end of which is linked by such a path to the vertex in X,

θ_{inc} is the formula $inc(x_1, x_2, x_3)$.

It is straightforward to verify that Δ is as desired.

Our second example is the functional transduction that maps a word u in $\{a, b\}^+$ to the word u^3. (We denote by $\{a, b\}^+$ the set of nonempty words written with a and b.) In order to define it as a transduction of structures, we represent a word u in $\{a, b\}^+$ by the structure $\| u \|$ defined in Subsection 5.2.1. We let Δ be the 3-copying definition scheme without parameter $(\varphi, \psi_1, \psi_2, \psi_3, (\theta_{(suc,i,j)})_{i,j=1,2,3}, (\theta_{(lab_a,i)})_{i=1,2,3}, (\theta_{(lab_b,i)})_{i=1,2,3})$ such that:

φ expresses that the input structure indeed represents a word in $\{a, b\}^+$,

ψ_1, ψ_2, ψ_3 are identical to the Boolean constant *true*,

$\theta_{(suc,i,j)}(x_1, x_2)$ is $suc(x_1, x_2)$ if $i = j$,

$\theta_{(suc,i,j)}(x_1, x_2)$ expresses that x_1 is the last position and that x_2 is the first one if $i = 1$ and $j = 2$, or if $i = 2$ and $j = 3$,

$\theta_{(suc,i,j)}(x_1, x_2)$ is the constant *false* if $i \neq j$ and if $(i, j) \notin \{(1, 2), (2, 3)\}$,

$\theta_{(lab_a,i)}(x_1)$ is $lab_a(x_1)$ for $i = 1, 2, 3$,

$\theta_{(lab_b,i)}(x_1)$ is $lab_b(x_1)$ for $i = 1, 2, 3$.

We claim that $def_\Delta(S) = T$ if and only if S is a structure of the form $\| u \|$ and T is the corresponding structure $\| u^3 \|$. To take a simple example, if $S = < \{1, 2, 3, 4\}, suc, lab_a, lab_b >$ representing the word *abba*, then $def_\Delta(S)$ is the structure

$$T = < \{1_1, 2_1, 3_1, 4_1, 1_2, 2_2, 3_2, 4_2, 1_3, 2_3, 3_3, 4_3\}, suc', lab'_a, lab'_b >$$

where we write i_j instead of (i, j), so that i_j is the jth copy if i for $j = 1, 2, 3$ and

$suc'(i_j, k_m)$ holds if and only if k_m is the successor of i_j in the above enumeration of the domain of T,

$lab'_a(i_j)$ holds if $i \in \{1,4\}, j \in \{1,2,3\}$ and

$lab'_b(i_j)$ holds otherwise.

This is an example of a definable transduction from words to words that is not a rational one. (This transduction will also be used as a counterexample in Proposition 5.5.8 below).

Our last example is the product of a finite-state automaton $\mathcal{A}$ by a *fixed* finite-state automaton $\mathcal{B}$. A finite-state automaton is defined as a 5-tuple $\mathcal{A} = < A, Q, M, I, F >$ where A is the input alphabet, (there we take $A = \{a, b\}$), Q is the set of states, M is the transition relation which is here a subset of $Q \times A \times Q$ (because we consider nondeterministic automata without ε-transitions), I is the set of initial states and F is that of final states. The language recognized by $\mathcal{A}$ is denoted by $L(\mathcal{A})$. The automaton $\mathcal{A}$ is represented by the relational structure: $|\ \mathcal{A}| = < Q, trans_a, trans_b, I, F >$ where $trans_a$ and $trans_b$ are binary relations and:

$trans_a(p,q)$ holds if and only if $(p,a,q) \in M$,

$trans_b(p,q)$ holds if and only if $(p,b,q) \in M$.

Let $\mathcal{B} = < A, Q', M', I', F' >$ be a similar automaton, and $\mathcal{A} \times \mathcal{B} = < A, Q \times Q', M'', I \times I', F \times F' >$ be the product automaton intended to define the language $L(\mathcal{A}) \cap L(\mathcal{B})$. We let Q' be $\{1, \cdots, k\}$ (let us recall that $\mathcal{B}$ is fixed). We shall define a definition scheme Δ such that $def_\Delta(|\ \mathcal{A}|) = |\ \mathcal{A} \times \mathcal{B}|$; it will be k-copying since the set of states of $\mathcal{A} \times \mathcal{B}$ is $Q \times \{1, \cdots, k\}$. We let $\Delta = (\varphi, \psi_1, \cdots, \psi_k, (\theta_w)_{w \in R^* k})$, where $R = \{trans_a, trans_b, I, F\}$ and:

φ is the constant *true* (because every structure in $STR(\mathcal{R})$ represents an automaton; this automaton may have inaccessible states and useless transitions),

$\psi_1, \cdots, \psi_k$ are the constant *true*,

$\theta_{(trans_a, i, j)}(x_1, x_2)$ is the formula $trans_a(x_1, x_2)$ if (i, a, j) is a transition of $\mathcal{B}$ and is the constant *false* otherwise,

$\theta_{(trans_b, i, j)}$ is defined similarly,

$\theta_{(I,i)}(x_1)$ is the formula $I(x_1)$ if i is an initial state of $\mathcal{B}$ and is *false* otherwise,

$\theta_{(F,i)}(x_1)$ is defined similarly.

Note that the language defined by an automaton $\mathcal{A}$ is nonempty if and only if there is a path in $\mathcal{A}$ from some initial state to some final state. This latter property is expressible in monadic second-order logic. Hence it follows from Proposition 5.5.5 below that, for a fixed rational language K, the set of structures representing an automaton $\mathcal{A}$ such that $L(\mathcal{A}) \cap K$ is nonempty is definable. This construction is used systematically in Courcelle [17].

The definitions concerning definable transductions of structures apply to hypergraphs via their representation by relational structures as explained above. However, since we have two representations of hypergraphs by logical structures, we must be more precise. We say that a *hypergraph transduction*[c], i.e., a binary relation f on hypergraphs is *(i,j)-definable*, where i and j belong to $\{1,2\}$ if and only if the transduction of structures $\{(\mid G\mid_i, \mid G'\mid_j) \ / \ (G,G') \in f\}$ is definable. We shall also use transductions from trees to hypergraphs. Since a tree t is a graph, it can be represented either by $\mid t \mid_1$ or by $\mid t \mid_2$. However, both structures are equally powerful for expressing monadic second-order properties of trees and definable transductions from trees to trees and from trees to graphs. (This follows from the proof of Theorem 5.3.4). When we specify a transduction involving trees (or words which are special trees) as input or output we shall use the symbol * instead of the integers 1 and 2, in order to recall that the choice of representation is not important in these cases. We shall use *definable* for (*,*)-*definable.*

Here are a few facts concerning definable transductions of structures.

Fact 5.5.1
If f is a definable transduction, there exists an integer k such that $Card(D_T) \leq k\ Card(D_S)$ whenever T belongs to $f(S)$. □

Fact 5.5.2
The domain of a definable transduction is MS-definable. □

Proof
Let Δ be a definition scheme as in the general definition with $\mathcal{W} = \{X_1, \cdots, X_n\}$. Then $Dom(def_\Delta) = \{S \ / \ S \models \exists X_1, \cdots, X_n \varphi\}$. □

The next two propositions list examples of transductions of words, trees and graphs that are definable.

Proposition 5.5.3
The following mappings are definable transductions: (1) word homomorphisms, (2) inverse nonerasing word homomorphisms, (3) gsm mappings, (4) the mirror-image mapping on words, (5) the mapping $\lambda u.[u^n]$ where u is a word and n

[c] Graphs are special hypergraphs. We shall speak of graph transduction if appropriate.

a fixed integer, (6) the mapping that maps a derivation tree relative to a fixed context-free grammar to the generated word, (7) linear root-to-frontier or frontier-to-root tree transductions. □

The proofs are easy to do. Let us recall that a *gsm mapping* is a transduction from words to words defined by a *generalized sequential machine*, i.e., a (possibly nondeterministic) transducer that reads at least one input symbol and outputs and a (possibly empty) word on each move. A special case of (5) has been constructed in detail in our second example before Fact 5.5.1. See Gecseg and Steinby [40] or the survey by Raoult [52] for tree transductions.
Fact 5.5.1 which limits the sizes of the output structures, shows that certain transductions are *not definable*. This is the case of inverse erasing word homomorphisms and of ground tree transducers except in degenerated cases (see Dauchet *et al.* [28] or [52] on ground tree transducers).

Proposition 5.5.4
The transductions that associate with a graph G: (1) its spanning forests, (2) its connected components, (3) its subgraphs satisfying some fixed 2-definable property, (4) its maximal subgraphs satisfying some fixed MS_2-definable property (maximal for subgraph inclusion), (5) the graph consisting of the union of two disjoint copies of G, (6) its minors, are all (2,2)-definable. The mapping associating with a graph its line graph is (2,1)-definable but not (2,2)-definable. □

We recall that the *line graph* of a graph G has E_G as set of vertices and has undirected edges between any two vertices representing edges sharing a vertex (in G).

Proof
Assertions (1)-(6) are easy consequences of the existence of an MS formula expressing that two given vertices are linked by some path. See Section 5.2. Assertion (6) is proved in Courcelle [12]. In the last assertion, the (2,1)-definability is an easy consequence of the definition of a line graph. See [23] for the negative part of this assertion. □

5.5.2 The fundamental property of definable transductions

The following proposition is the basic fact behind the notion of semantic interpretation ([51]). It says that if $T = def_\Delta(S, \mu)$ i.e., if T is defined in (S, μ) by Δ, then the monadic second-order properties of T can be expressed as monadic second-order properties of (S, μ). The usefulness of definable transductions is based on this proposition.

Let $\Delta = (\varphi, \psi_1, \cdots, \psi_k, (\theta_w)_{w \in Q^* k})$ be a $(\mathcal{Q}, \mathcal{R})$-definition scheme, written with a set of parameters $\mathcal{W}$. Let $\mathcal{V}$ be a set of set variables disjoint from $\mathcal{W}$. For every variable X in $\mathcal{V}$, for every $i = 1, \cdots, k$, we let X_i be a new variable. We let $\mathcal{V}' := \{X_i / X \in \mathcal{V}, i = 1, \cdots, k\}$. For every mapping $\eta : \mathcal{V}' \to \mathcal{P}(D)$, we let $\eta^k : \mathcal{V} \to \mathcal{P}(D \times [k])$ be defined by $\eta^k(X) = \eta(X_1) \times \{1\} \cup \cdots \cup \eta(X_k) \times \{k\}$. With these notations we can state:

Proposition 5.5.5
For every formula β in $MS(\mathcal{Q}, \mathcal{V})$ one can construct a formula β' in $MS(\mathcal{R}, \mathcal{V}' \cup \mathcal{W})$ such that, for every S in $STR(\mathcal{R})$, for every assignment μ : $\mathcal{W} \to S$ for every assignment $\eta : \mathcal{V} \to S$, we have:
$def_\Delta(S, \mu)$ is defined (if it is, we denote it by T), η^k is a $\mathcal{V}$-assignment in T, and $(T, \eta^k) \models \beta$

if and only if $(S, \eta \cup \mu) \models \beta'$. □

Note that, even if T is well-defined, the mapping η^k is not necessarily a $\mathcal{V}$-assignment in T, because $\eta^k(X)$ is not necessarily a subset of the domain of T which is a possibly proper subset of $D_S \times [k]$.

Proof sketch
Let us first consider the case where def_Δ is noncopying. In order to transform β into β', one replaces every atomic formula $q(u_1, \cdots, u_n)$ by the formula $\theta_q(u_1, \cdots, u_n)$ which defines it in terms of the relations of S (See Lemma 5.2.5). One also restricts quantifications to the domain of T, that is, one replaces $\exists x[\mu]$ by $\exists x[\psi(x) \wedge \mu]$ and $\exists X[\mu]$ by $\exists X[\forall x\{x \in X \implies \psi(x)\} \wedge \mu]$. (See Lemma 5.2.6). In the case where $k > 1$, one replaces $\exists X[\mu]$ by a formula of the form $\exists X_1, \exists X_2, \cdots, \exists X_k[\mu']$ where μ' is an appropriate transformation of μ based on the fact that $X = X_1 \times \{1\} \cup \cdots \cup X_k \times \{k\}$. The reader will find a complete construction in [13], Proposition 2.5, p. 166. □

From this proposition, we get easily:

Proposition 5.5.6
(1) The inverse image of an MS-definable class of structures under a definable transduction is MS-definable.

(2) The composition of two definable transductions is definable.

Proof
(1) Let $L \subseteq STR(\mathcal{Q})$ be defined by a closed formula β and def_Δ be a transduction as in Proposition 5.5.5. Then $def_\Delta^{-1}(L) \subseteq STR(\mathcal{R})$ is defined by the formula $\exists Y_1, \cdots, Y_n[\beta']$ where $Y_1, \cdots, Y_n$ are the parameters and β' is constructed from β as in Proposition 5.5.5.

(2) Let $\Delta = (\varphi, \psi_1, \cdots, \psi_k, (\theta_w)_{w\in \mathcal{Q}_{*k}})$ be a k-copying definition scheme and $\Delta' = (\varphi', \psi'_1, \cdots, \psi'_{k'}, (\theta'_w)_{w\in \mathcal{P}*k'})$ be a k'-copying definition scheme such that def_Δ is a transduction from $STR(\mathcal{R})$ to $STR(\mathcal{Q})$ and $def_{\Delta'}$ is a transduction from $STR(\mathcal{Q})$ to $STR(\mathcal{P})$. Let f be the transduction $def_{\Delta'} o def_\Delta$ from $STR(\mathcal{R})$ to $STR(\mathcal{P})$: we shall construct a definition scheme Δ" for it. Just to simplify the notation we shall assume that the parameters of Δ are Y and Y' and that those of Δ' are Z and Z'. We shall also assume that the relations of $\mathcal{P}$ are all binary. The general case will be an obvious extension.

In order to describe Δ" we shall denote by S an $\mathcal{R}$-structure, we shall denote by T the $\mathcal{Q}$-structure $def_\Delta(S, Y, Y')$ where Y and Y' are subsets of D_S and we denote by U the $\mathcal{P}$-structure $def_{\Delta'}(T, Z, Z')$ where Z and Z' are subsets of D_T. Hence D_T is a subset of $D_S \times [k]$, and D_U is a subset of $D_S \times [k] \times [k']$ which is canonically isomorphic to a subset of $D_S \times [kk']$. Hence Δ" will be kk'-copying. The parameters Z and Z' represent sets of the respective forms $Z = Z_1 \times \{1\} \cup \cdots \cup Z_k \times \{k\}$ and $Z' = Z'_1 \times \{1\} \cup \cdots \cup Z'_k \times \{k\}$. Hence the definition scheme Δ" will be written in terms of parameters $Y, Y', Z_1, \cdots, Z_k, Z'_1, \cdots, Z'_k$. It will be of the form:

$$(\varphi", (\psi"_{i,j})_{i\in[k], j\in[k']}, (\theta"_{(p,(i,i'),(j,j'))})_{p\in\mathcal{P}, i,i'\in[k], j,j'\in[k']})$$

so that the domain of D_U will be handled as a subset of $D_S \times [k] \times [k']$ and not of $D_S \times [kk']$. The formulas forming Δ" will be obtained from those forming Δ' by the transformation of Proposition 5.5.5. We first consider φ" which should express that $def_\Delta(S, Y, Y')$ is defined, i.e., that $(S, Y, Y') \models \varphi$, and that if $Z = Z_1 \times \{1\} \cup \cdots \cup Z_k \times \{k\}$ and $Z' = Z'_1 \times \{1\} \cup \cdots \cup Z'_k \times \{k\}$, then $def_\Delta(T, Z, Z')$ is defined, i.e., that $(T, Z, Z') \models \varphi$ which is equivalent to:

$$(S, Y, Y', Z_1, \cdots, Z_k, Z'_1, \cdots, Z'_k) \models \tau$$

where τ is obtained from φ' by the transformation of Proposition 5.5.5. Hence φ'' is the conjunction of φ and τ. We omit the constructions of the other formulas because they are quite similar. □

We could define more powerful transductions by which a structure T would be constructed "inside" $S \times S$ instead of "inside" a structure formed of a fixed number of disjoint copies of S (like in [38]). However, with this variant, one could construct a *second-order* formula β' as in Proposition 5.5.5 (with quantifications on binary relations), but not a *monadic* second-order one (at least in general). We wish to avoid non monadic second-order logic because most constructions and decidability results (like those of [11], [12]) break down. In the third example given before Fact 5.5.1, we have shown that the transduction associating the automaton $\mathcal{A} \times \mathcal{B}$ with an automaton $\mathcal{A}$ is definable (via the chosen representation of finite-state automata by relational structures) *for fixed* $\mathcal{B}$.

Here are some other closure properties of the class of definable transductions.

Proposition 5.5.7
The intersection of a definable transduction with a transduction of the form $A \times B$ *where* A *and* B *are MS-definable classes of structures, is a definable transduction.*

Proof
Straightforward. See [13] □

Proposition 5.5.8
(1) The image of an MS-definable class of structures under a definable transduction is not MS-definable in general.

(2) The inverse of a definable transduction is a transduction that is not definable in general.

(3) The intersection of two definable transductions is a transduction that is not definable in general.

Proof
(1) The transduction of words that maps $a^n b$ to $a^n b a^n b a^n b$ for $n \geq 0$ is definable (this follows from Proposition 5.5.7 and the second example given before Fact 5.5.1). The image of the definable language $a^* b$ is a language that is not regular (and even not context-free), hence not definable by the result of Büchi and Elgot ([60], Thm 3.2) saying that a set of words is MS-definable iff it is regular.

(2) The inverse of this transduction is not definable since, if it were, its domain is would be definable (by Fact 5.5.2), hence regular, which is not the case.

(3) The intersection of the definable transductions of words that map $a^n b^m$ to c^n, and $a^n b^m$ to c^m is the one that maps $a^n b^n$ to c^n. It is not definable because its domain is not a an MS-definable (regular) language. □

5.5.3 Comparisons of representations of partial orders, graphs and hypergraphs by relational structures

We shall formulate in terms of definable transductions some transformations between several relational structures representing the same object .

Fact 5.5.9
The transduction of $< D, \leq_D >$ *into* $< D, suc_D >$ *where* $< D, \leq_D >$ *is a partial order is definable, and so is its inverse. The transduction of* $| G |_2$ *into* $| G |_1$ *where* G *is a graph or a hypergraph is definable.* □

It follows in particular from Proposition 5.5.6 that a class of finite partial orders is MS-definable w.r.t. one of the representations $< D, \leq_D >$ or $< D, suc_D >$ iff it is w.r.t. the other (this has been already proved in Proposition 5.3.1). It follows also that a class of graphs or hypergraphs is MS_2-definable if it is MS_1-definable. We shall prove the converse for any subclass of $\mathcal{C}_k$, the class of simple directed graphs having a semi-strong k-coloring, where k is any (fixed) integer (see Subsection 5.3.2).

Lemma 5.5.10
Let $k \in \mathcal{N}$. *One can construct a definable transduction* δ_k *that associates with* $| G |_1$ *a structure* T *isomorphic to* $| G |_2$, *for every graph* G *in* $\mathcal{C}_k$.

Proof
Let G be a simple directed graph with a semi-strong coloring $\gamma : V_G \to \{1, \cdots, k\}$. A mapping $\gamma : V_G \to \{1, \cdots, k\}$ can be specified by the k-tuple $(V_1, \cdots, V_k)$ of subsets of V_G such that $V_i = \gamma^{-1}(i)$ for every $i = 1, \cdots, k$. Thus a formula $\pi \in MS(\mathcal{R}_s, \{X_1, \cdots, X_k\})$ can express that a given tuple $(V_1, \cdots, V_k)$ represents a semi-strong k-coloring. It follows that the class $\mathcal{C}_k$ is MS-definable (the corresponding formula is π' defined as $\exists X_1, \ldots, X_k.\pi$).

Let $G \in \mathcal{C}_k$ and $(V_1, \cdots, V_k)$ represent a semi-strong k-coloring. We let $rep : E_G \to E \subseteq V_G \times \{1, \cdots, k\}$ be the bijection such that: $rep(e) = (v, i)$ iff e links w to v where w is a vertex in V_i. It is one-to-one since $(V_1, \cdots, V_k)$ represents a semi-strong coloring (the origins of two edges with same target have different colors). We shall thus construct T isomorphic to $| G |_2$ with domain $V_G \times \{0\} \cup E \subseteq V_G \times \{0, 1, \cdots, k\}$ as image of $| G |_1$ by a definable transduction. We let Δ be the following definition scheme with parameters $X_1, \cdots, X_k$:

$$\Delta = (\pi, \psi_0, \psi_1, \cdots, \psi_k, (\theta_{(inc,j)})_{j \in [0,k]^3}),$$

where π has already been defined, and:

ψ_0 is: *true*

ψ_i is: $\exists w[w \in X_i \wedge edg(w, x_1)]$, for $i = 1, \cdots, k$

$\theta_{(inc,i,0,0)}$ is: $x_1 = x_3 \wedge edg(x_2, x_3) \wedge x_2 \in X_i$ for $i = 1, \cdots, k$ and

$\theta_{(inc,n,m,p)}$ is: *false* if $n = 0$ or $m \neq 0$ or $p \neq 0$.

Claim: *For every graph G in C_k, for every* k-*tuple* $V_1, \cdots, V_k$ *representing a semi-strong* k-*coloring of* G, *the structure* $def_\Delta(| G |_1, V_1, \cdots, V_k)$ *is isomorphic to* $| G |_2$.

Proof of Claim: Let $T = def_\Delta(| G |_1, V_1, \cdots, V_k)$. Then $D_T = V_G \times \{0\} \cup E$ from the definitions. From the definition of the formulas $\theta_{(inc,j)}$ we have:
$((x_1, n), (x_2, m), (x_3, p)) \in inc_T$ iff
$n \neq 0$ and $m = p = 0$ and $edg(x_2, x_3)$ and $x_2 \in X_n$ and $x_1 = x_3$ iff
$n \neq 0$ and $m = p = 0$ and $rep(e) = (x_1, n)$ where e is the edge of G that links x_2 to x_3. *Proof of Claim* □

It follows that $| G |_2$ is isomorphic to T by the isomorphism mapping $v \in V_G$ onto $(v, 0)$ and $e \in E_G$ onto $rep(e)$. □

The lemma follows immediately.

From Lemma 5.5.10, one obtains immediately Lemma 5.3.5 because any two isomorphic structures satisfy the same formulas.

5.6 Equational sets of graphs and hypergraphs

The easiest way to define context-free sets of graphs is by systems of recursive equations. These equations are written with set union and the set extensions of operations on graphs that generalize the concatenation of words. By these operations one can generate all finite graphs from elementary graphs more or less as one can generate finite nonempty words from the letters, by means of concatenation. This idea also applies to hypergraphs. In this section, we first present systems of equations in a general algebraic setting. Then we review the operations on graphs and hypergraphs that yield algebraic presentations of the VR sets of graphs and of the HR sets of hypergraphs in terms of systems of recursive set equations. The grammatical definitions of these sets, by context-free graph grammars of various types are investigated in detail in other chapters of this book. The relevant operations are presented in a uniform way in terms of operations on structures. This will be helpful for obtaining theorems relating MS logic and context-free graph grammars. We also state logical characterizations of the VR sets of graphs and of the HR sets of hypergraphs from which follow closure properties under definable transductions.

5.6.1 Equational sets

As in many other works, we shall use the term *magma* borrowed from Bourbaki [7] for what is usually called an *abstract algebra* or an *algebra*. The words "algebra" and "algebraic" are used in many different contexts with different meanings. We prefer to avoid them completely and use fresh words. Many-sorted notions are studied in detail by Ehrig and Mahr [31], Wirsing [64] and Wechler [63]. We mainly review the notation. We shall use *infinite* sets of sorts and *infinite* signatures, which is not usual.

F-magmas

Let $\mathcal{S}$ be a set called the set of *sorts*. An $\mathcal{S}$-*signature* is a set F given with two mappings $\alpha : F \to seq(\mathcal{S})$ (the set of finite sequences of elements of $\mathcal{S}$), called the *arity* mapping, and $\sigma : F \to \mathcal{S}$, called the *sort* mapping. The length of $\alpha(f)$ is called the *rank* of f, and is denoted by $\rho(f)$. The *type* of f in F is the pair $(\alpha(f), \sigma(f))$ that we shall rather write $s_1 \times s_2 \times \cdots \times s_n \to s$ where $\alpha(f) = (s_1, \cdots, s_n)$ and $\sigma(f) = s$. (We may have $n = 0$; in this case f is a constant). If $\mathcal{S}$ has only one sort, we say that F is a *ranked alphabet*, and the arity of a symbol is completely defined by its rank.

An F-*magma*, i.e. an F-*algebra* ([31], [64], [63]), is an object $M = < (M_s)_{s\in\mathcal{S}}, (f_M)_{f\in F} >$, where for each s in $\mathcal{S}$, M_s is a nonempty set, called the *domain of sort* s of M, and for each $f \in F$, the object f_M is a total mapping : $M_{\alpha(f)} \to M_{\sigma(f)}$. These mappings are called the *operations* of M. (For a non-empty sequence of sorts $\mu = (s_1, \cdots, s_n)$, we let $M_\mu := M_{s_1} \times M_{s_2} \times \cdots \times M_{s_n}$). We assume that $M_s \cap M_{s'} = \emptyset$ for $s \neq s'$. We let M also denote $\cup\{M_s / s \in \mathcal{S}\}$, and for $d \in M$, we let $\sigma(d)$ denote the unique s such that $d \in M_s$.

If M and M' are two F-magmas, a homomorphism $h : M \to M'$ is a mapping h that maps M_s into M'_s for each sort s, and commutes with the operations of F. We shall call it an F-*homomorphism* if it is useful to specify the signature F. We denote by $T(F)$ the *initial* F-*magma*, and by $T(F)_s$ its domain of sort s. This set can be identified with the set of well-formed ground terms over F of sort s. If M is an F-magma, we denote by h_M the unique homomorphism: $T(F) \to M$. If $t \in T(F)_s$, then the image of t under h_M is an element of M_s, also denoted by t_M. One can consider t as a term *denoting* t_M, and t_M as *the value of* t in M. We say that F *generates* a subset M' of M if M' is the set of values of the terms in $T(F)$. We say that M' is *finitely generated* if it is the set of values of terms in $T(F')$ where F' is a finite subset of F.

An $\mathcal{S}$-*sorted set of variables* is a pair (X, σ) consisting of a set X, and a *sort mapping* $\sigma : X \to \mathcal{S}$. It will be more simply denoted by X, unless the sort mapping must be specified. We shall denote by $T(F, X)$ the set of well-formed terms written with $F \cup X$ and by $T(F, X)_s$ the subset of those of sort s. Hence, $T(F, X) = T(F \cup X)$ and $T(F, X)_s = T(F \cup X)_s$. However, the notations $T(F, X)$ and $T(F, X)_s$ are more precise because they specify the variables among the nullary symbols of $F \cup X$. Let $\mathcal{X}$ be a finite sequence of pairwise distinct variables from X. We shall denote by $T(F, \mathcal{X})_s$, the set of terms of $T(F, X)_s$ having all their variables in the list $\mathcal{X}$. If $t \in T(F, \mathcal{X})_s$, we denote by $t_{M,\mathcal{X}}$ the mapping: $M_{\sigma(\mathcal{X})} \to M_s$ associated with t in the obvious way, by letting a symbol f from F denote f_M, (where $\sigma(\mathcal{X})$ denotes the sequence of sorts of the elements of the sequence $\mathcal{X}$). We call $t_{M,\mathcal{X}}$ a *derived operation of* M. If $\mathcal{X}$ is known from the context, we write t_M instead of $t_{M,\mathcal{X}}$. This is the case in particular if t is defined as a member of $T(F, \{x_1, \cdots, x_k\})_s$: the sequence $\mathcal{X}$ is implicitely $(x_1, \cdots, x_k)$.

Power-set magmas and polynomial systems

Let F be an $\mathcal{S}$-signature. We enlarge it into F_+ by adding, for every sort s in $\mathcal{S}$, a new symbol $+_s$ of type: $s \times s \to s$, and a new constant Ω_s of sort s. With an F-magma M we associate its *power-set magma*:

$$\mathcal{P}(M) := < (\mathcal{P}(M_s))_{s \in \mathcal{S}}, (f_{\mathcal{P}(M)})_{f \in F_+} >$$

where

$$\Omega_{s\mathcal{P}(M)} = \emptyset$$

$$A_1 +_{s\mathcal{P}(M)} A_2 := A_1 \cup A_2 \text{ (for } A_1, A_2 \subseteq M_s),$$

$f_{\mathcal{P}(M)}(A_1, \cdots, A_k) \quad :=^{\mathcal{P}} \quad f_M(A_1, \cdots, A_k)$, i.e.:= $\{f_M(a_1, \cdots, a_k)/a_1 \in A_1, \cdots, a_k \in A_k\}$

for $A_1 \subseteq M_{s_1}, \cdots, A_k \subseteq M_{s_k}$ where $\alpha(f) = (s_1, \cdots, s_k)$. (We denote by ${}^{\mathcal{P}}g$ the set extension of a mapping g.)

Hence $\mathcal{P}(M)$ is an F_+-magma. A *polynomial system* over F is a sequence of equations $S = < u_1 = p_1, ..., u_n = p_n >$, where $U = \{u_1, \cdots, u_n\}$ is an $\mathcal{S}$-sorted set of variables called the set of *unknowns* of S. Each term p_i is a *polynomial*, i.e., a term of the form Ω_s or $t_1 +_s t_2 +_s \cdots +_s t_m$, where the terms t_j are monomials of sort $s = \sigma(u_i)$. A *monomial* is a term in $T(F \cup U)$. The subscript s is usually omitted in $+_s$ and in Ω_s. A mapping $S_{\mathcal{P}(M)}$ of $\mathcal{P}(M_{\sigma(u_1)}) \times \cdots \times \mathcal{P}(M_{\sigma(u_n)})$ into itself is associated with S and M as follows: for $A_1 \subseteq M_{\sigma(u_1)}, \cdots, A_n \subseteq M_{\sigma(u_n)}$, we let

$S_{\mathcal{P}(M)}(A_1, \cdots, A_n) = (p_{1\mathcal{P}(M)}(A_1, \cdots, A_n), \cdots, p_{n\mathcal{P}(M)}(A_1, \cdots, A_n))$.

A *solution* of S in $\mathcal{P}(M)$ is an n-tuple $(A_1, \cdots, A_n)$ such that $A_i \subseteq M_{\sigma(u_i)}$ for each $i = 1, \cdots, n$ and $(A_1, \cdots, A_n) = S_{\mathcal{P}(M)}(A_1, \cdots, A_n)$, i.e.,

$$A_i = p_{i\mathcal{P}(M)}(A_1, \cdots, A_n), \text{ for every } i = 1, \cdots, n. \tag{5.2}$$

A solution of S is also called a fixed-point of $S_{\mathcal{P}(M)}$. Every such system S has a least solution in $\mathcal{P}(M)$ denoted by $(L((S,M),u_1), \cdots, L((S,M),u_n))$. ("Least" is understood with respect to set inclusion). This n-tuple can be concretely described as follows:

$$L((S,M),u_i) = \cup\{A_i^j / j \geq 0\}$$

where $A_i^0 = \emptyset$ for all $i = 1, \cdots, n$ and $(A_1^{j+1}, \cdots, A_n^{j+1}) = S_{\mathcal{P}(M)}(A_1^j, \cdots, A_n^j)$

The components of the least solutions in $\mathcal{P}(M)$ of polynomial systems are the *M-equational* sets. We denote by $Equat(M)$ the family of M-equational sets and by $Equat(M)_s$ the subfamily of those included in M_s.

We review the classical example of context-free grammars. Let A be a finite alphabet, say $A = \{a_1, \cdots, a_n\}$. Let $F_A = A \cup \{., \varepsilon\}$ be the ranked alphabet where $\rho(a_i) = 0$ for all $i, \rho(\varepsilon) = 0, \rho(.) = 2$. We denote by $\mathcal{W}_A$ the F_A-magma $< A^*, ., \varepsilon, a_1, \cdots, a_n >$ where A^* is the set of words over A, . is the concatenation, ε is the empty word, $a_1, \cdots, a_n$ denote themselves as words. It is a monoid (with a binary associative operation having a unit) augmented with constants. The set $Equat(\mathcal{W}_A)$ is the set of *context-free languages* over A. This follows from the theorem of Ginsburg and Rice [41] that characterizes these languages as the components of the least solutions of systems of recursive equations written with the nullary symbols $\varepsilon, a_1, \cdots, a_n$, the concatenation and, of course, set union (denoted here by $+$). Take for example, the context-free grammar $G = \{u \to auuv, u \to avb, v \to avb, v \to ab\}$ with nonterminal symbols u and v and terminal symbols a and b. The corresponding system of equations is:

$$S = \ < u = a.(u.(u.v)) + a.(v.b), v = a.(v.b) + a.b > .$$

We now give an example concerning trees. By a *tree* T we mean a finite connected undirected graph without multiple edges and cycles. The set of trees is denoted by $\mathcal{T}$. A *rooted tree* is a pair $R = (T, r)$ consisting of a tree T and a distinguished node r called the *root*. The set of rooted trees is denoted by $\mathcal{R}$. The set of nodes of a rooted or unrooted tree T is denoted by N_T. Any two isomorphic trees are considered as equal. (The set of trees is actually a quotient set w.r.t. isomorphism but we shall not detail this technical point).

We now define a few operations on trees and rooted trees. The types of these operations will be given in terms of two sorts, t and r, namely, the sort t of trees and the sort r of rooted trees. The first operation is the *root gluing* $//$: $r \times r \to r$. For S and T in $\mathcal{R}$, we let $S//T$ be the rooted tree obtained by fusing the roots of S and T (or rather, of two disjoint isomorphic copies of S

and T). The second operation is the *extension* denoted by $ext : r \to r$. For T in $\mathcal{R}$, we let $ext(T)$ be the rooted tree obtained from T by the addition of a new node that becomes the root of $ext(T)$, linked by a new edge to the root of T. We denote by 1 the rooted tree consisting of a single node (the root). Finally, we let $fg : r \to t$ be the mapping that "forgets" the root of a rooted tree R. Formally, $fg(R) = T$ where $R = (T, r) \in \mathcal{R}$.

Hence, we have an $\{r, t\}$-sorted signature $F := \{//, ext, 1, fg\}$ and a many-sorted F-magma $\mathcal{TREE}$ having $\mathcal{R}$ as domain of sort $r, \mathcal{T}$ as a domain of sort t, and the operations defined above. Hence, $\mathcal{TREE} = < \mathcal{R}, \mathcal{T}, //, ext, 1, fg >$.

The set of trees of odd degree, i.e., such that the degree of every node is odd, is equal to $L((S, \mathcal{TREE}), u)$ where S is the system

$$S = \begin{cases} u = & fg(ext(v)) \\ v = & w//w + v//w//w \\ w = & ext(v) + ext(1) \end{cases}$$

The sorts of u, v and w are t, r and r. It is not hard to see that $L((S, \mathcal{TREE}), u)$ is the set of (finite) trees of odd dregree. As a hint observe that $L((S, \mathcal{TREE}), v)$ is the set of rooted trees different from 1, all nodes of which except the root have odd degree, and that $L((S, \mathcal{TREE}), w)$ is the set of rooted trees such that all nodes have odd degree and the root has degree one.

The following result is due to Mezei and Wright [48].

Proposition 5.6.1
Let M and M' be F-magmas. If $h : M \to M'$ is an F-homomorphism, if S is a polynomial system over F, then $L((S, M'), u) = h(L((S, M), u))$ for every unknown u.

Proof
The homomorphism $h : M \to M'$ extends into an F_+-homomorphism ${}^{\mathcal{P}}h : \mathcal{P}(M) \to \mathcal{P}(M')$ defined by ${}^{\mathcal{P}}h(A) = \{h(a)/a \in A\}$ for $A \subseteq M_s, s \in \mathcal{S}$. It is easy to verify that for every $j \in \mathcal{N}$:

$$S^j_{\mathcal{P}(M')}(\emptyset, \cdots, \emptyset) =^{\mathcal{P}} h(S^j_{\mathcal{P}(M)}(\emptyset, \cdots, \emptyset))$$

where of course ${}^{\mathcal{P}}h(A_1, \cdots, A_n) = ({}^{\mathcal{P}}h(A_1), \cdots, {}^{\mathcal{P}}h(A_n))$ for every $A_1, \cdots, A_n \subseteq M$. (The proof is by induction on j, using the fact that ${}^{\mathcal{P}}h$ is an F_+-homomorphism). The result follows immediately. □

Here is a consequence of Proposition 5.6.1. (We recall that h_M denotes the unique homomorphism $T(F) \to M$).

Corollary 5.6.2
For every polynomial system S and for every unknown u of S we have $L((S,M),u) = h_M(L((S,T(F)),u))$. The emptiness of $L((S,M),u)$ depends only on S and u and is decidable. □

This means that an equational set, defined as a component of the least solution of a polynomial system S, is the image of the corresponding component of the least solution of S in the F-magma of terms $T(F)$, under the canonical homomorphism. Hence the study of the equational subsets of M can, in a sense, be reduced to the study of the equational sets of terms and the homomorphism h_M. In particular, the emptiness of $L((S,T(F)),u)$ can be decided by tree automata techniques (see below); by the first assertion this set is nonempty iff $L((S,M),u)$ is nonempty, where M is *any* F-magma.

Sets of terms can be defined in several ways, by grammars of various types, by automata (usually called "tree automata"), or in terms of congruences on the magmas of terms $T(F)$. Of special importance is the class of recognizable subsets of $T(F)$ characterized in terms of finite-state tree automata, of regular tree-grammars, of congruences having finitely many classes and also of monadic-second order formula. (See Courcelle [22]). We do not recall these characterizations here; we only refer the reader to the book [40] for automata, grammars and congruences and to [60], Theorem 3.2 for the characterization in terms of MS formulas.

In this section, we shall use the following result by Mezei and Wright [48], [40].

Proposition 5.6.3
Let F be a finite signature. A subset of $T(F)$ is equational iff it is recognizable. □

The main part of its proof is the determinization of finite-state frontier-to-root tree-automata.

Corollary 5.6.4
Let M be an F-magma (where F may be infinite). A set L is M-equational iff $L = h_M(K)$ for a recognizable subset K of $T(F')$ where F' is a finite subsignature of F. □

If $t \in K$ and $m = h_M(t)$ then t is a syntactic description of m, m is the value of t in M. If $m \in L$ is given, the parsing problem relative to the system of equations (equivalently the grammar) from which K comes, consists in finding t in K such that $m = h_M(t)$. If such t does not exist, then $m \notin L$. Otherwise the found t is the desired syntactic analysis of m.

5.6.2 *Graphs with ports and VR sets of graphs*

Graphs will be simple and directed, as in Section 5.2. They will have distinguished vertices called *ports* and designated by special labels. We let $\mathcal{C}$ be a countable set called the set of *port labels.* A *graph with ports* is a pair $G = < H, P >$ where H is a graph and $P \subseteq V_H \times C$ for some finite subset C of $\mathcal{C}$. We denote P by $port_G, V_H$ by V_G, E_H by E_G, inc_H by inc_G and edg_H by edg_G. We let also G^o denote H. A vertex v is a *p-port* if $(v,p) \in port_G$. We consider p as a label attached to v and marking it as a p-port. A vertex may have several port labels. We let $\tau(G)$ be the set of labels of ports of G and we call this set the *type* of G.

We denote by $\mathcal{GP}(C)$ the class of graphs with ports G such that $\tau(G) \subseteq C$ where C is a subset of $\mathcal{C}$. We let $\mathcal{GP}$ denote the class of all graphs with ports. We now define some operations on $\mathcal{GP}$. If $G, G' \in \mathcal{GP}$ and G, G' are disjoint (i.e., $V_G \cap V_{G'} = \emptyset$) then $G \oplus G'$ is the *disjoint union* of G and G', and is defined as $< G^o \cup G'^o, \ port_G \cup port_{G'} >$. Clearly $\tau(G \oplus G') = \tau(G) \cup \tau(G')$.

The next operation adds edges to a given graph with ports. If $p, q \in \mathcal{C}$ and $G \in \mathcal{GP}$, we let $add_{p,q}(G) =< H, port_G >$ where H is G^o augmented with the edges (x, y) such that:

x is a p-port, y is a q-port, $x \neq y, (x,y)$ is not already an edge of G^o.

Hence H is simple and the operation $add_{p,q}$ does not add loops to G. The type of $add_{p,q}(G)$ is equal to that of G and $add_{p,q}(G) = G$ if G has no p-port or no q-port. Let us note finally that the number of edges added to G by $add_{p,q}$ depends on the numbers of p-ports and of q-ports and is not uniformly bounded.

A third operation will be useful, in order to modify port labels. Let P be a finite subset of $\mathcal{C} \times \mathcal{C}$. For $G \in \mathcal{GP}$ we let: $mdf_P(G) = < G^o, port_G.P >$. A vertex v is a q-port of $mdf_P(G)$ iff it is a p-port of G for some p such that $(p,q) \in P$. We have $\tau(mdf_P(G)) = \{q \ / \ (p,q) \in P$ for some $p \in \tau(G)\}$. Two special cases of this operation will be useful and deserve special notation: if C is a finite subset of $\mathcal{C}$, we let for $G \in \mathcal{GP}(D)$, $fg_C(G) = mdf_P(G)$ where $P = \{(d,d) \ / \ d \in D - C\}$: this operation "forgets" the p-ports for $p \in C$. It simply "removes" the label p on the p-ports for each $p \in C$; hence $\tau(fg_C(G)) = \tau(G) - C$. If $p_1, p_2, \cdots, p_k, q_1, \cdots, q_k$ are labels in $\mathcal{C}$ with $p_1, \cdots, p_k$ pairwise distinct, we let for $G \in \mathcal{GP}(D)$, $ren_{p_1 \to q_1, p_2 \to q_2, \cdots, p_k \to q_k}(G) = mdf_P(G)$ where $P = \{(d,d) \ / \ d \in D - \{p_1, \cdots, p_k\}\} \cup \{(p_1, q_1), \cdots, (p_k, q_k)\}$. This operation "renames" the port labels p_i into q_i, for all $i = 1, \cdots, k$. Hence $\tau(ren_{p_1 \to q_1, \cdots, p_k \to q_k}(G)) = (\tau(G) - \{p_1, \cdots, p_k\}) \cup \{q_i \ / \ p_i \in \tau(G), 1 \leq i \leq k\}$.

Equational sets of graphs with ports

Having defined operations on $\mathcal{GP}$ we can now write systems of equations intended to define equational subsets of $\mathcal{GP}$. Here is an example. The equation

$$u = p + ren_{q \to p}(add_{p,q}(u \oplus q))$$

where p denotes the graph with a single vertex that is a p-port, and similarly for q, is intended to define (in the sense of Subsection 5.6.1) the set of finite tournaments without circuits, all vertices of which are p-ports. However we are facing a difficulty: the operation $\oplus$ is partial because $G \oplus G'$ is defined if and only if G and G' are disjoint. It follows for instance that $G \oplus G$ is undefined. We can make $\oplus$ into a total operation by letting G' be "replaced in $G \oplus G'$ by an isomorphic copy G'_1 disjoint with G". But now comes another difficulty: we may have $G \oplus (G' \oplus G'_1)$ not equal to $(G \oplus G') \oplus G'_1$ (depending on how the isomorphic copies of the second arguments of the operations $\oplus$ are chosen) which is clearly unsatisfactory. However they are isomorphic and, actually, we are interested in properties of graphs that are invariant under isomorphism. Let us call *concrete* a graph in $\mathcal{GP}$ and *abstract* the isomorphism class of such a graph. We shall denote by GP the class of abstract graphs with ports. We have thus a magma structure on GP because isomorphism is a congruence for the operations $\oplus, mdf_P$ and $add_{p,q}$. Graph grammars and systems of equations define subsets of GP. Logical formulas do the same because any two isomorphic relational structures satisfy the same formulas of any of the logical languages we have considered. However, in order to simplify the presentation we shall omit the distinction between $\mathcal{GP}$ and GP, and we shall do as if $\oplus$ was total on $\mathcal{GP}$.

We let F_{VR} denote the set of all operations $\oplus, mdf_P, fg_C, ren_{p_1 \to q_1, \cdots, p_k \to q_k}, add_{p,q}$ together with the constants p and p^ℓ for $p \in \mathcal{C}$. If $K \subseteq \mathcal{C}$, we denote by $F_{VR}(K)$ the subset of F_{VR} consisting of the above operations such that $P \subseteq K \times K, C \subseteq K, p_1, q_1, \cdots, p_k, q_k, p, q \in K$. The constant p^ℓ denotes a graph consisting of one loop, the unique vertex of which is a p-source. A term in $T(F_{VR})$ will be called a VR-*(graph) expression*. It denotes a graph with ports (actually an element of GP to be precise). We shall denote this graph by $val(t)$ for $t \in T(F_{VR})$.

Proposition 5.6.5
Every finite graph with port G is the value of some VR-expression in $T(F_{VR}(K))$ where $Card(K) \leq Max\{Card(V_G), Card(\tau(G))\}$.

Proof sketch
For each vertex v of G one chooses a port label $p(v)$ (a different one for each vertex). One defines then G by:

$$G = mdf_P(...add_{p(v),p(v')}(...(... \oplus p(v) \oplus ...)))$$

where the sum $... \oplus p(v) \oplus ...$ extends to all $v \in V_G$, the operations $add_{p(v),p(v')}$ are inserted for all edges from v to v' and P is the set $\{(p(v), q)/(v, q) \in port_G\}$. One uses $p(v)^\ell$ instead of $p(v)$ if v has a loop. We omit further details. □

A complexity measure on graphs follows from this definition:

$$c(G) = Min\{Card(K)/G = val(t), t \in T(F_{VR}(K))\}$$

This complexity measure is investigated in Courcelle and Olariu [26]. In particular it is compared with tree-width (defined below in Subsection 5.6.3).

An equational subset of $\mathcal{GP}$ will be called a VR *set of graphs*. Here VR stands for "Vertex-Replacement" and refers to the generation of the same sets by certain context-free graph grammars. These grammars are considered in chapter 1 of this book. (The equivalence between equational systems on $\mathcal{GP}$ and these grammars was first proved in Courcelle et al. [24]).

A logical characterization of the VR sets of graphs

We obtain immediately from Corollary 5.6.4 that:
a set of graphs with ports L is VR iff it is the set of values of the VR-expressions of

a recognizable set $K \subseteq T(F_{VR}(C))$ for some finite set $C \subseteq \mathcal{C}$.

We have already recalled the theorem of Doner, Thatcher and Wright stating that a subset of $T(F)$ (where F is finite) is recognizable iff it is MS-definable (see Thomas [60], thm 11.1). We shall now prove that the mapping $val : T(F_{VR}(C)) \to \mathcal{GP}$ is a definable transduction (see Section 5.5). More precisely:

Proposition 5.6.6
Let C be a finite subset of $\mathcal{C}$. The mapping $val : T(F_{VR}(C)) \to \mathcal{GP}(C)$ is (, 1)-definable.*

Proof
A term $t \in T(F_{VR}(C))$ is considered as a finite directed tree the nodes of which are labelled in $F = F_{VR}(C)$ and represented by the structure $\mid t \mid = < N_t, suc_1, suc_2, (lab_f)_{f \in F} >$ where

N_t is the set of nodes,

$suc_1(x, y) :\Longleftrightarrow y$ is the first (leftmost) successor of x,

$suc_2(x, y) :\Longleftrightarrow y$ is the second successor of x,

$lab_f(x) :\Longleftrightarrow f$ is the label of x.

The nodes have at most two successors since the symbols in F have arity at most 2 ($\oplus$ has arity two and all others have arity 1 or 0). A graph G in $\mathcal{GP}(C)$ is represented by the structure $\mid G \mid_1 \;=\; < V_G, edg_G, (pt_{cG})_{c \in C} >$ where V_G is the set of vertices, edg_G is the set of edges ($edg_G \subseteq V_G \times V_G$), pt_{cG} is the set of c-ports for $c \in C$. We let $\mathcal{R}_s(C) = \{edg\} \cup \{pt_c / c \in C\}$.

Let us now consider $t \in T(F_{VR}(C))$ and $G = val(t) \in \mathcal{GP}(C)$. We shall define V_G as a subset of N_t, edg_G as a binary relation on N_t (such that $edg_G \subseteq V_G \times V_G$), the sets pt_{cG} as subsets of V_G (hence of N_t), all this by MS-formulas. We take V_G equal to the set of leaves of t (i.e., of nodes without successors). If t has no symbol mdf_P or $ren_{p \to q, \cdots}$ then we can define:
$edg_G = \{(x, y) \in N_t / x, y$ are leaves, x has label p, y has label q, there is a node z in N_t labelled by $add_{p,q}$, a path from z to x on which there is no symbol fg_C with $p \in C$ and a path from z to y on which there is no symbol fg_C with $q \in C$.$\}$

Let us take the following example:

$$t = add_{p,r}(add_{p,q}[p \oplus q \oplus fg_{\{p\}}(add_{q,p}(p \oplus q))] \oplus add_{r,q}[q \oplus r]).$$

There are 6 leaves, $x_1, x_2, \cdots, x_6$ corresponding respectively to the constants p, q, p, q, q, r in t. Because of the "$add_{p,q}$" operation there is an edge from x_1 to x_2 but there is no edge from x_3 to x_2 because of the "$fg_{\{p\}}$" operation which "kills" the label p of x_3.

The general construction must handle the fact that the operations mdf_P (and $ren_{p \to q, \cdots}$) can modify certain port labels. Hence if x is a leaf of t with label p, if y is an ancestor of x then x is a vertex of the graph $val(t/y)$ (where we denote by t/y the subterm of t corresponding to the subtree issued from node y) but it may not be a p-port: it may be a q-port with $q \neq p$ or be not a port. In the last example x_3 is not a p-port of $val(t)$. The construction can be completed with the help of the following lemma. We shall omit all further details. □

Lemma 5.6.7
We fix C as in Proposition 5.6.6. For each $p \in C$,one can construct a formula $\varphi_p(x,y)$ such that, for every $t \in T(F_{VR}(C))$ and $x, y \in N_t$:
$| t | \quad \models \varphi_p(x,y)$
iff x is a leaf, y is an ancestor of x in the tree t and x is a p-port of $val(t/y)$.

Proof
Without loss of generality we let $C = \{1, \cdots, n\}$ and $p = 1$. Let us take $t \in T(F_{VR}(C))$; let x be a leaf of t and y an ancestor of x. There exists a unique n tuple of sets $X_1, \cdots, X_n$ satisfying the following conditions:

(1) $X_1 \cup \cdots \cup X_n$ is the set of nodes of t of the path from y to x,

(2) for every $i = 1, \cdots, n : x \in X_i$ iff x has label i (in t),

(3) for every z with $y \leq_T z <_T x$ (see Section 5.3 for notation) for every $i = 1, \cdots, n$, letting z' be the successor of z such that $z' \leq_T x$ we have: $z \in X_i$ iff

(3.1) either z has label $\oplus$ or $add_{j,j'}$ for some j, j' and $z' \in X_i$
(3.2) or z has label mdf_P for some $P \subseteq C \times C$ and $(j,i) \in P$ for some j and $z' \in X_j$.

(Since fg_C and $ren_{...}$ are special cases of mdf_P this last case also applies to them). It is clear that for every z with $y \leq_T z \leq_T x, x$ is an i-port in $val(t/z)$ iff $z \in X_i$.

Conditions (1) (2) (3) above can be expressed by an MS formula $\psi(x, y, X_1, \cdots, X_n)$. The desired formula is thus $\varphi_p(x,y)$ defined as

$$\theta(x,y) \wedge \exists X_1, \cdots, X_n[\psi(x,y,X_1,\cdots,X_n) \wedge y \in X_p]$$

where $\theta(x,y)$ expresses that x is a leaf and y is an ancestor of x. □

We denote by $\mathcal{B}$ the set of finite binary directed trees. Formally (and more precisely) $\mathcal{B} = T(\{f,a\})$ where f is a binary function symbol and a is nullary.

Theorem 5.6.8 ([33], [15])
Let C be a finite subset of $\mathcal{C}$. A subset of $\mathcal{GP}(C)$ is VR iff it is the image of $\mathcal{B}$ under a $(*,1)$-definable transduction.

Proof
"Only if". From Corollary (5.6.4), one obtains that if L is a VR set then $L = val(K)$ where K is a recognizable subset of $T(F_{VR}(C))$. By Doner et

al.'s theorem, K is MS-definable and *val* is $(*, 1)$-definable. It follows that L is the image of $T(F_{VR}(C))$ by a $(*, 1)$-definable transduction (namely the restriction of *val* to K, see Proposition 5.5.7). The classical encoding of terms by binary trees can be adapted and yields a bijection k of K onto a definable subset K' of $\mathcal{B}$ such that k and k^{-1} are both $(*, *)$-definable (See [23], Lemma 2.4 for details). It follows that L is the image of $\mathcal{B}$ under a $(*, 1)$-definable transduction, namely the restriction of $val \circ k^{-1}$ to K'.

"If". The proof is quite complicated and we refer the reader to [33] or [15], Corollary 4.9. □

Corollary 5.6.9
The family of VR sets of graphs is closed under (1,1)-definable transductions.

Proof
Immediate consequence of Theorem 5.6.8 and the fact that the composition of two (1,1)-definable transductions is (1,1)-definable. □

It follows for instance that the set of connected components of the graphs of a VR set is a VR set.

5.6.3 Hypergraphs with sources and HR sets of hypergraphs

This section is fully analogous to the preceding one. We present operations on hypergraphs upon which the (context-free) HR ("Hyperedge Replacement") hypergraph grammars can be defined as systems of equations.

Hypergraphs have been defined in Section 5.3.3. We assume that the ranked set A of hyperedge labels is fixed and we shall not specify it in notation. We shall use operations on hypergraphs needing to distinguish certain vertices by means of labels. We let $\mathcal{C}$ be a fixed countable set of *source labels*, not to be confused with the set of hyperedge labels A.

A *hypergraph with sources* is a pair $H = \langle G, s \rangle$ consisting of a hypergraph G and a total mapping $s : C \to V_G$ called its *source mapping*, where C is a finite subset of $\mathcal{C}$. We say that $s(C) \subseteq V_G$ is *the set of sources of* H and that $s(c)$ is its *c-source* where $c \in C$. We shall also say that the vertex $s(c)$ has *source label* c. A vertex that is not a source is an *internal vertex*. The set C is called *the type of* H and is denoted by $\tau(H)$. The source mapping of H is also denoted by src_H. This mapping is not necessarily injective.

We shall denote by $\mathcal{HS}(C)$ the class of all hypergraphs of type C and by $\mathcal{HS}_d(C)$ the subclass of those having *distinct sources*, i.e., that have an injective source mapping.

We shall use the set $\mathcal{S}$ of finite subsets of $\mathcal{C}$ as a set of sorts. We now define some operations on hypergraphs with sources. These operations will form an $\mathcal{S}$-signature. Note that $\mathcal{S}$ is infinite. The definitions will be given here directly in terms of *abstract hypergraphs* (i.e., of isomorphism classes of hypergraphs).

(1) Parallel composition

If $G \in \mathcal{HS}(C)$ and $G' \in \mathcal{HS}(C')$ we let $H = G//_{C,C'}G'$ be the hypergraph in $\mathcal{HS}(C \cup C')$ obtained as follows: one first constructs a hypergraph H' by taking the union of two disjoint hypergraphs K and K' respectively isomorphic to G and G', and by fusing any two vertices v and v' that are respectively the c-source of K and the c-source of K' for some $c \in C \cap C'$; the hypergraph H is defined as the isomorphism class of H'. Clearly, H does not depend on the choices of K and K'. We shall use the simplified (overloaded) notation $G//G'$ when C and C' are known from the context or are irrelevant. We say that $G//G'$ is the *parallel composition* of G and G'. In the case where G and G' have distinct sources, we have the following characterization:
a hypergraph H of type $\tau(G) \cup \tau(G')$ is isomorphic to $G//G'$ if and only if H has subhypergraphs K and K' respectively isomorphic to G and G' such that: $V_K \cup V_{K'} = V_H, E_K \cup E_{K'} = E_H, E_K \cap E_{K'} = \emptyset$ and $V_K \cap V_{K'}$ is the set of sources of H that have label c where $c \in C \cap C'$.

(2) Source renaming

For every mapping $h : C \to C'$, we let $ren_h : \mathcal{HS}(C') \to \mathcal{HS}(C)$ be the mapping such that $ren_h(< G, s >) =< G, s \circ h >$. In other words, the c-source of $ren_h(G)$ is defined as the $h(c)$-source of G. If a vertex v of G is a c-source with $c \in C' - h(C)$ and is not a c'-source for any c' in $h(C)$, then it is an internal vertex in $ren_h(G)$. We shall say that $ren_h(G)$ is the *source renaming* of G defined by h. Note that when we write ren_h, we assume that h is given together with the sets C and C', so that the type of ren_h (namely $C' \to C$) is defined in a unique way. Nevertheless, we shall use the overloaded notation $ren_\emptyset$ for ren_h when h is the *empty mapping*: $\emptyset \to C$ and C need not be specified (or is known from the context).

(3) Source fusion

For every set B, we denote by $Eq(B)$ the set of equivalence relations on B. Let C be a subset of $\mathcal{C}$. For every $\delta \in Eq(C)$ we let $fuse_\delta$ be the mapping $\mathcal{HS}(C) \to \mathcal{HS}(C)$ such that:

$H = fuse_\delta(G)$ if and only if

$V_H = V_G/\sim$, where $\sim$ is the equivalence relation on V_G generated by the set of pairs $\{(src_G(i), src_G(j)) \mid (i, j) \in \delta\}$,

$E_H = E_G$,

$vert_H(e,i) = [vert_G(e,i)]$, where $[v]$ denotes the equivalence class of v with respect to $\sim$,

$lab_H = lab_G$,

$[v]$ is a c-source of H whenever some $v' \sim v$ is a c-source of G.

Intuitively, H is obtained from G by fusing its c- and c'-sources for all c, c' for which c' is δ-equivalent to c. We shall say that H is obtained from G by a *source fusion.*

For every $p \in \mathcal{C}$, we denote by p the graph with a single vertex which is the p-source. Hence $p \in \mathcal{HS}(\{p\})$. For every $a \in A$, for every $p_1, \cdots, p_n \in \mathcal{C}$ where $n = \tau(a)$, we denote by $a(p_1, \cdots, p_n)$ the hypergraph consisting of a single hyperedge with label a and a sequence of vertices $(x_1, \cdots, x_n)$ such that x_i is the p_i-source for every $i = 1, \cdots, n$. Note that we have $x_i = x_j$ iff $p_i = p_j$. Clearly $a(p_1, \cdots, p_n) \in \mathcal{HS}(\{p_1, \cdots, p_n\})$.

We let F_{HR} be the $\mathcal{S}$-signature consisting of $//_{C,C'}$, $fuse_\delta$, ren_h, p, $a(p_1, \cdots, p_n)$ for all relevant $C, C', \delta, h, p, a, p_1, \cdots, p_n$. For $K \subseteq \mathcal{C}$, we denote by $F_{HR}(K)$ the subsignature consisting of the above symbols with $C, C' \subseteq K, p, p_1, \cdots, p_n \in K$ etc... We obtain thus an F_{HR}-magma $\mathcal{HS}$.The terms in $T(F_{HR})$ are called the *HR(hypergraph)-expressions.* Each of them, say t, denotes a hypergraph $val(t)$ in $\mathcal{HS}$ called its value. (Hypergraph expressions based on different operations were first introduced in [3]).

Proposition 5.6.10
Every finite hypergraph in $\mathcal{HS}(C)$ is the value of some HR-expression in $T(F_{HR})_C$.

Proof
Quite similar to that of Proposition 5.6.3. One takes a source label $p(v)$ for each $v \in V_G$. One defines G by an HR-expression of the form

$$ren_h(\cdots//a(p_1, \cdots, p_n)//\cdots//p(v_1)//\cdots//p(v_m))$$

where $V_G = \{v_1, \cdots, v_m\}$ and the expression contains one factor $a(p_1, \cdots, p_n)$ for each hyperedge. We omit the details. □

The minimum cardinality of $C \subseteq \mathcal{C}$ such that $G = val(t)$ for some $t \in T(F_{HR}(C))$ is connected with an important complexity measure on graphs and hypergraphs called *treewidth*. We refer the reader to [53] and to [20], Theorem 1.1 for more details. We only indicate that the minimum cardinality of C such that $G = val(t), t \in T(F_{HR}(C))$ is $Max\{Card(\tau(G)), twd(G) + 1\}$ where $twd(G)$ denotes the tree-width of G.

A subset of $\mathcal{HS}(C)$ is called an HR *set of hypergraphs* iff it is $\mathcal{HS}$-equational. The terminology HR refers to an equivalent characterization in terms of "Hyperedge Replacement grammars" studied in chapter ?? of this book.

We give here the example of series-parallel graphs. We shall use the source labels 1, 2, 3, and one edge label a of type 2. We define from the operations of $F_{HR}(\{1,2,3\})$ the following operations:
$G//G' = G//_{C,C}G'$ where $C = \{1,2\}$ and $G, G' \in \mathcal{HS}(C)$
$G.G' = ren_h(G//_{\{1,2\},\{2,3\}}ren_{h'}(G'))$
where $h : \{1,2\} \to \{1,3\}$ maps $1 \mapsto 1$ and $2 \mapsto 3$, and $h' : \{2,3\} \to \{1,2\}$ maps $3 \mapsto 2$ and $2 \mapsto 1$. The equation

$$u = u//u + u.u' + a(1,2)$$

defines the class SP of series-parallel graphs. (These graphs are directed and all their edges are labelled by a).

A hypergraph $H \in \mathcal{HS}(C)$ will be represented by the relational structure $| H |_2 = < V_H \cup E_H, (inc_{aH})_{a \in A}, (ps_{cH})_{c \in C} >$ where inc_{aH} has been defined in Section 5.3.3 and: $ps_{cH}(x)$ is true iff x is the c-source. Hence $| H |_2 \in STR(\mathcal{R}_m(A,C))$ where $\mathcal{R}_m(A,C) = \mathcal{R}_m(A) \cup \{ps_c / c \in C\}$ and each ps_c is unary. Clearly H and H' are isomorphic iff $| H |_2$ and $| H' |_2$ are isomorphic.

Proposition 5.6.11
Let K be a finite subset of $\mathcal{C}$ and $C \subseteq K$. The mapping $val : T(F_{HR}(K))_C \to \mathcal{HS}(C)$ is a $(*,2)$-definable transduction.

Proof
We refer the reader to Courcelle [13], Lemma 4.3, p. 177. □

Theorem 5.6.12
Let C be a finite subset of $\mathcal{C}$. A subset of $\mathcal{HS}(C)$ is HR iff it is the image of the set of finite binary trees $\mathcal{B}$ under a $(*,2)$-definable transduction.

Proof
The "only if" direction follows from the preceding proposition like that of Theorem 5.6.8 from Proposition 5.6.6. The "if" direction is quite difficult to prove and we refer the reader to Courcelle and Engelfriet [23]. □

Corollary 5.6.13
The family of HR sets of hypergraphs with sources is closed under $(2,2)$*-definable transductions.*

Proof
Similar to that of Corollary 5.6.9. □

Here are some examples of $(2,2)$-definable transductions that are meaningful in graph theory: the transduction from a graph to its spanning trees, or of a hypergraph to its connected components, or of a graph to its maximal planar subgraphs. (See Proposition 5.5.4 for other examples).

Corollary 5.6.14
The set of line graphs of the graphs of an HR set is a VR set of graphs.

Proof
Immediate consequence of Theorems 5.6.8 and 5.6.12 and the last assertion of Proposition 5.5.4. □

We collect below some properties, the proofs of which are too long to be included in this chapter. The reader will note that the statements do not concern logic but only graph properties. However monadic second-order logic is an essential tool for the proofs. We denote by $und(G)$ the undirected graph obtained from G by forgetting the orientations of edges.

Theorem 5.6.15
Every HR set of simple directed graphs is VR. Conversely, for a VR set L of directed graphs, the following conditions are equivalent:

1. *L is HR*
2. *there exists an integer k such that $Card(E_G) \leq k.Card(V_G)$ for every $G \in L$,*
3. *there exists an integer n, such that the complete bipartite graph $K_{n,n}$ is not a subgraph of $und(G)$ for any $G \in L$.* □

Proof hints: The first assertion is an easy consequence of Theorems 5.6.8 and 5.6.12 because the transduction $\{(\mid G \mid_2, \mid G \mid_1) \ / \ G \text{ is a graph}\}$ is definable (one can also say that the identity on simple directed graphs is $(2,1)$-definable; see Fact 5.5.9).

Implications (1) ⇒ (2) and (1) ⇒ (3) follow easily from general properties of HR sets of graphs (see chapter **??** in this book). The implications (2) ⇒ (1) and (3) ⇒ (1) are difficult: see Courcelle [19]. □

One can extend Theorem 5.6.15 as follows. A set $L \subseteq \mathcal{H}(A)$ of simple hypergraphs (without sources) is said to be VR iff it is the image of $\mathcal{B}$ under a $(*, 1)$-definable transduction. Hence we use here the characterization of Theorem 5.6.8 of VR sets of graphs and we make it into a definition of *VR sets of hypergraphs.* (Equivalent characterizations in terms of systems of equations can be found in Courcelle [15], Thm 4.6 but no equivalent context-free grammar, based on appropriate rewriting rules has yet been defined.)

For every hypergraph H we denote by $K(H)$ the undirected graph obtained by substituting a complete undirected graph K_m for every hyperedge of type m.

Theorem 5.6.16 ([19])
Every HR set of simple hypergraphs is VR. Conversely, for every VR set L of simple hypergraphs, the following conditions are equivalent:

1. *L is HR,*
2. *there exists an integer k such that $Card(E_H) \leq k.Card(V_H)$ for every $H \in L$,*
3. *there exists an integer n such that $K_{n,n}$ is not a subgraph of $K(H)$ for any $H \in L$.*

Again, the only difficult implications are (2) $\Rightarrow$ (1) and (3) $\Rightarrow$ (1). □

5.7 Inductive computations and recognizability

The notion of a recognizable set is due to Mezei and Wright [48]; it extends the notion of a regular language like the notion of an equational set extends that of a context-free one. It was originally defined for one-sort structures, and we adapt it to many-sorted ones, possibly with infinitely many sorts. We begin with a study of recognizability in a general algebraic framework. Then we state the basic result that MS-definable sets of graphs are recognizable. Finally we state a generalized version of Parikh's theorem based on inductive computations, (the fundamental notion behind recognizability) and give some applications to VR and HR grammars.

5.7.1 Inductive sets of predicates and recognizable sets

Let F be an $\mathcal{S}$-signature. An F-magma A is *locally finite* if each domain A_s is finite. Let M be an F-magma and $s \in \mathcal{S}$. A subset B of M_s is *M-recognizable* if there exists a locally finite F-magma A, a homomorphism $h : M \rightarrow A$, and a (finite) subset C of A_s such that $B = h^{-1}(C)$. We denote by $Rec(M)_s$ the family of M-recognizable subsets of M_s.

If F is a finite signature, the recognizable subsets of terms over F, i.e., the $T(F)$-recognizable sets can be characterized by finite-state *tree*-automata (see Gecseg and Steinby [40]). The classical identification of terms with finite ordered ranked trees explains the qualification of "tree"-automaton.

By a *predicate* on a set E, we mean a unary relation, i.e., a mapping $E \to \{true, false\}$. If M is a many-sorted F-magma with set of sorts $\mathcal{S}$, a *family of predicates* on M is an indexed set $\{\hat{p}/p \in P\}$, given with a mapping $\sigma : P \to \mathcal{S}$ such that each $\hat{p}$ is a predicate on $M_{\sigma(p)}$. We call $\sigma(p)$ the *sort* of p. Such a family will also be denoted by P. For $p \in P$, we let $L_p = \{d \in M_{\sigma(p)} \ / \ \hat{p}(d) = true\}$. The family P is *locally finite* if, for each $s \in \mathcal{S}$ the set $\{p \in P \ / \ \sigma(p) = s\}$ is finite. We say that P is *f-inductive* where f is an operation in F, if for every $p \in P$ of sort $s = \sigma(f)$ there exist $m_1, \cdots, m_n$ in $\mathcal{N}$, (where n is the rank of f), an $(m_1 + \cdots + m_n)$-place Boolean expression B, and a sequence of $(m_1 + \cdots + m_n)$ elements of P, $(p_{1,1}, \cdots, p_{1,m_1}, p_{2,1}, \cdots, p_{2,m_2}, \cdots, p_{n,m_n})$, such that, if the type of f is $s_1 \times s_2 \times \cdots \times s_n \to s$ we have:

1. $\sigma(p_{i,j}) = s_i$ for all $j = 1, \cdots, m_i$ and $i = 1, \cdots, n$,
2. for all $d_1 \in M_{s_1}, \cdots, d_n \in M_{s_n}$

$$\hat{p}(f_M(d_1, \cdots, d_n)) = B[\hat{p}_{1,1}(d_1), \cdots, \hat{p}_{1,m_1}(d_1), \hat{p}_{2,m_2}(d_2), \cdots, \hat{p}_{n,m_n}(d_n)].$$

The sequence $(B, p_{1,1}, \cdots, p_{2,1}, \cdots, p_{n,m_n})$ is called a *decomposition of p relative to f*. In words, the existence of such a decomposition means that the validity of p for any object of the form $f_M(d_1, \cdots, d_n)$ can be computed from the truth values of finitely many predicates of P for the objects $d_1, \cdots, d_n$. This computation can be done by a Boolean expression that depends only on p and f. We say that P is *F-inductive* if it is f-inductive for every f in F.

Proposition 5.7.1
Let M be an F-magma. For every $s \in \mathcal{S}$ a subset L of M_s is recognizable if and only if $L = L_{p_0}$ for some predicate p_0 belonging to a locally finite F-inductive family of predicates on M.

Proof
"Only if". Let $L = h^{-1}(C) \subseteq M_s$, for some homomorphism $h : M \to A$ where A is locally finite and $C \subseteq A_s$. We let $P = \cup\{A_t/t \in \mathcal{S}\} \cup \{p_0\}$. Each element a of A_t is of sort t (considered as a member of P). The domains of A are pairwise disjoint and p_0 is of sort s. For $d \in M_t$, and $a \in A_t$, we let:

$\hat{a}(d) \quad = true \quad$ if $h(d) = a$,
$\quad\quad\quad = false \quad$ otherwise,

For $d \in M_s$, we let

$\hat{p}_0(d) \quad = true \quad$ if $h(d) \in C$,
$\quad\quad\quad = false \quad$ otherwise.

It is clear that P is locally finite. It is F-inductive, and, clearly, $L = L_{p_0}$.

"If". Let P be a locally finite F-inductive family of predicates. Let $L = L_{p_0}$ for some $p_0 \in P$. For every $t \in S$ we let P_t be the set of predicates in P of sort t. We let A_t be the set of mappings: $P_t \to \{true, false\}$. We let h be the mapping: $\cup\{M_t \ / \ t \in S\} \to \cup\{A_t \ / \ t \in S\}$ such that, for every $t \in S$ and $m \in M_t, h(m)$ is the mapping: $P_t \to \{true, false\}$ such that $h(m)(p) = \hat{p}(m)$. We want to find operations on the sets A_t such that $A = < (A_t)_{t\in S}, (f_A)_{f\in F} >$ is an F-magma and $h : M \to A$ defined above is a homomorphism. We need to define f_A where f is of type $s_1 \times \cdots \times s_n \to t$ in such a way that for all $(m_1, \cdots, m_m) \in M_{s_1} \times \cdots \times M_{s_n}$ we have:

$$h(f_M(m_1, \cdots, m_n)) = f_A(h(m_1), \cdots, h(m_n)).$$

But it is possible to find such f_A by using the decompositions of the predicates in P_t relative to f. Hence, $L = L_{p_0} = h^{-1}(C)$ where $C = \{\theta \in A_t \ / \ \theta(p_0) = true\}$. It follows that L is M-recognizable since, by construction, A is locally finite. □

Example 5.1

Let L be the set of rooted trees with a number of nodes that is not a multiple of 3. Let p be the corresponding predicate on $\mathcal{R}$. (We use the notation and definitions of Subsection 5.6.1) Let us consider the following predicates: for $i = 0, 1, 2$ we let $q_i(t)$ hold iff the number of nodes of t is of the form $3k + i$ for some k. It is easy to check that $P = \{p, q_0, q_1, q_2\}$ is inductive with respect to the operations ext and $//$ on rooted trees; this check uses in particular the following facts which hold for all rooted trees t and t':

$$q_0(t//t') = (q_0(t) \wedge q_1(t')) \vee (q_1(t) \wedge q_0(t')) \vee (q_2(t) \wedge q_2(t')),$$

$$q_1(t//t') = (q_1(t) \wedge q_1(t')) \vee (q_0(t) \wedge q_2(t')) \vee (q_2(t) \wedge q_0(t')),$$

$$q_2(t//t') = (q_1(t) \wedge q_2(t')) \vee (q_2(t) \wedge q_1(t')) \vee (q_0(t) \wedge q_0(t')),$$

$$q_0(ext(t)) = q_2(t),$$

$$q_1(ext(t)) = q_0(t),$$

$$q_2(ext(t)) = q_1(t).$$

Since p is equivalent to $q_1 \vee q_2$, we have:
$p(ext(t)) = q_0(t) \vee q_1(t)$
and one can easily write a similar definition of $p(t//t')$. Hence $L = L_p$ is recognizable. Note that $\{q_0, q_1\}$ is inductive because q_2 is definable as $\neg q_0 \wedge \neg q_1$. □

Proposition 5.7.2
Let M be generated by F and t be a sort. A subset L of M_t is M-recognizable iff $h_M^{-1}(L)$ is $T(F)$-recognizable.

Proof
We prove the "only if" direction. If $L = h^{-1}(C)$ for some homomorphism $h : M \to A$, where A is locally finite, then $h_M^{-1}(L) = (hoh_M)^{-1}(C)$, and, since hoh_M is a homomorphism: $T(F) \to A$, the set $h_M^{-1}(L)$ is $T(F)$-recognizable. For the "if" direction, see Courcelle [22], Proposition 5.5.6. □

This proposition means that, in order to decide whether an element m of M belongs to L, it suffices to take *any* term t in $T(F)$ denoting m and to decide whether it belongs to the recognizable set $h_M^{-1}(L)$ (for instance by running an automaton on this term). The key point is that the answer is the same for any term t denoting m. This should be contrasted with the characterization of equational sets of Corollary 5.6.4, which says that, if L is equational, then it is of the form $h_M(K)$ for some recognizable set of terms K: in this case, in order to establish that m belongs to L, one must find a specific term t denoting m, and the decision cannot be made from an arbitrary term in $h_M^{-1}(m)$.

Relationships between equational sets and recognizable sets.

We recall that we denote by $Equat(M)_s$ the family of M-equational subsets of M_s.

Theorem 5.7.3
If $K \in Rec(M)_s$ and $L \in Equat(M)_s$ then $L \cap K \in Equat(M)_s$.

Proof
We can assume that $L = L((S, M), u_0)$ for some $u_0 \in U$ where S is a uniform polynomial system over F with set of unknowns U. A polynomial system is *uniform* if its equations are of the form $u = t_1 + t_2 + \cdots + t_m$, where each t_i is of the form $f(u_1, u_2, \cdots, u_k)$ for some $f \in F$, some unknowns $u_1, \cdots, u_k \in U$; the transformation of an arbitrary system into an equivalent one which is uniform is the essence of the transformation of an arbitrary context-free grammar into one in Chomsky normal form. An example will illustrate this step later. See Courcelle [22], Proposition 5.19 for the proof.

Let $F' \subseteq F$ be the finite set of symbols occurring in S, and let $S' \subseteq S$ be the finite set of sorts of these symbols together with those of the unknowns of S. Hence F' is an S'-signature. Let $h : M \to A$ be an F'-homomorphism (with A locally finite), such that $K = h^{-1}(C)$ for some $C \subseteq A_s$.

For every $u \in U$, we let $L_u := L((S, M), u)$. Let W be the new set of unknowns $\{[u, a]/u \in U, a \in A_{\sigma(u)}\}$. It is finite. We shall define a system S', with set of unknowns W, such that:

$$L((S', M), [u, a]) = L_u \cap h^{-1}(a)$$

for every $[u, a] \in W$. Let $u \in U$ and $a \in A_{\sigma(u)}$. Let us assume that the defining equation of u in S is of the form $u = t_1 + \cdots + t_k$. Consider one of the monomials, say t_i. Let us assume that it is of the form $f(u_1, \cdots, u_n)$ for some unknowns $u_1, \cdots, u_n$. For every $a \in A_{\sigma(u)}, a_1 \in A_{\sigma(u_1)}, \cdots, a_n \in A_{\sigma(u_n)}$ such that $f_A(a_1, \cdots, a_n) = a$, we form the monomial $f([u_1, a_1], \cdots, [u_n, a_n])$, and we let $\hat{t_i}$ denote the sum of these monomials. If no such n-tuple $(a_1, \cdots, a_n)$ exists, then $\hat{t_i}$ is defined as Ω. The defining equation of $[u, a]$ in S' is taken as:

$$[u, a] = \hat{t}_1 + \hat{t}_2 + \cdots + \hat{t}_k.$$

It is clear from this construction that the W-indexed family of sets $(L_u \cap h^{-1}(a))_{[u,a]\in W}$ is a solution of S' in $\mathcal{P}(M)$. Hence $L_u \cap h^{-1}(a) \supseteq L_{u,a}$, where $(L_{u,a})_{[u,a]\in W}$ denotes the least solution of S' in $\mathcal{P}(M)$.

In order to establish the opposite inclusion, we define from $L_{u,a}$ the sets $L'_u = \cup\{L_{u,a}/a \in A_{\sigma(u)}\}$ for $u \in U$. Clearly $(L'_u)_{u\in U}$ is a solution of S in $\mathcal{P}(M)$. Hence $L_u \subseteq L'_u$ for all u. For all $a \in A_{\sigma(u)}$, we have:

$$L_u \cap h^{-1}(a) \subseteq L'_u \cap h^{-1}(a) = (\cup\{L_{u,b}/b \in A\}) \cap h^{-1}(a).$$

The latter set is equal to $L_{u,a} \cap h^{-1}(a)$ since $L_{u,a} \subseteq L_u \cap h^{-1}(a)$ and, $h^{-1}(b) \cap h^{-1}(b') = \emptyset$ for all b, b' with $b \neq b'$. Hence $L_u \cap h^{-1}(a) \subseteq L_{u,a}$. By the first part of the proof, we have an equality, and $(L_u \cap h^{-1}(a))_{[u,a]\in W}$ is the least solution of S' in $\mathcal{P}(M)$. We have:

$$\begin{aligned} L \cap K &= L_{u_0} \cap h^{-1}(C) \\ &= \cup\{L((S', M), [u_0, a])/a \in C\}. \end{aligned}$$

Finally, we get

$$L \cap K = L((S'', M), w)$$

where S'' is S' augmented with the equation

$$w = [u_0, a_1] + [u_0, a_2] + \cdots + [u_0, a_r],$$

w is a new unknown and $C = \{a_1, a_2, \cdots, a_r\}$. Hence $L \cap K$ is M-equational. □

This theorem extends the classical result saying that the intersection of a context-free language and a regular one is context-free. The construction is effective if K is effectively given (i.e., if h is computable, if the finite sets A_s are computable, etc...) and L is defined by a given system. Hence, since the emptiness of an equational set (defined by a system of equations) is decidable, we have the following corollary.

Corollary 5.7.4
If K is an effectively given M-recognizable set, and if L is an M-equational set defined by a given system, one can test whether $L \cap K = \emptyset$. □

Example 5.2
We consider the set $\mathcal{R}$ of rooted trees with the operations $//, ext$ and 1 defined in Subsection 5.6.1. We consider the following two predicates on $\mathcal{R}$:

$$p(t) :\Longleftrightarrow \text{all branches of } t \text{ are of even length,}$$

$$q(t) :\Longleftrightarrow \text{all branches of } t \text{ are of odd length.}$$

The following clauses give a precise definition of p and q and show that $\{p, q\}$ is $\{//, ext\}$-inductive. For all $t, t' \in \mathcal{R}$ we have:

$$p(t//t') = p(t) \wedge p(t'),$$

$$q(t//t') = q(t) \wedge q(t'),$$

$$p(ext(t)) = q(t),$$

$$q(ext(t)) = p(t),$$

$$p(1) = false,$$

$$q(1) = true.$$

□

Let us perform on this example the construction of the proof of the "if" part of Proposition 5.7.1. We take $A = \{true, false\}^2$ where in $(x, y) \in A$ the component x is the truth value of p and y is that of q. Actually, since p and q cannot be both *true*, we can restrict ourselves to $A = \{(t, f), (f, t), (f, f)\}$ where t stands for *true* and f for *false*. We let $1_A = (f, t)$. The mappings $//_A$ and ext_A are defined by the following tables:

$//_A$	(t, f)	(f, t)	(f, f)
(t, f)	(t, f)	(f, f)	(f, f)
(f, t)	(f, f)	(f, t)	(f, f)
(f, f)	(f, f)	(f, f)	(f, f)

ext_A	
(t,f)	(f,t)
(f,t)	(t,f)
(f,f)	(f,f)

Take for example the following system S which is not uniform:

$$\begin{cases} u = & 1 + ext(ext(v)) \\ v = & u + ext(v)//ext(v). \end{cases}$$

We transform it into the following uniform system S':

$$\begin{cases} u = & 1 + ext(w) \\ v = & 1 + ext(w) + w//w \\ w = & ext(v) \end{cases}$$

and clearly $L((S',\mathcal{R}),u) = L((S,\mathcal{R}),u)$. In order to construct $L' = L((S',\mathcal{R}),u) \cap L_p$ we introduce the unknowns $[x,y,z]$ with $x \in \{u,v,w\}, y,z \in \{t,f\}, y$ and z not both *true*. We obtain the following equations forming a new system S":

$$\begin{array}{lcl} [u,t,f] & = & ext([w,f,t]) \\ [u,f,t] & = & 1 + ext([w,t,f]) \\ [u,f,f] & = & ext([w,f,f]) \\ [v,t,f] & = & ext([w,f,t]) + [w,t,f]//[w,t,f] \\ [v,f,t] & = & 1 + ext([w,t,f]) + [w,f,t]//[w,f,t] \\ [v,f,f] & = & \dots \\ [w,t,f] & = & ext([v,f,t]) \\ [w,f,t] & = & ext([v,t,f]) \\ [w,f,f] & = & ext([w,f,f]) \end{array}$$

We are interested in $L' = L((S",\mathcal{R}),[u,t,f])$ and it is not hard to see that the unknowns of the form $[x,f,f]$, $x = u,v,w$ are useless for the generation of the elements of L'. So is $[u,f,t]$. We can thus delete the corresponding equations. Letting x_1 replace $[x,t,f]$ and x_2 replace $[x,f,t]$ for $x \in \{u,v,w\}$ we obtain the more readable system T:

$$\begin{cases} u_1 & = & ext(w_2) \\ v_1 & = & ext(w_2) + w_1//w_1 \\ v_2 & = & 1 + ext(w_1) + w_2//w_2 \\ w_1 & = & ext(v_2) \\ w_2 & = & ext(v_1) \end{cases}$$

and $L' = L((T,\mathcal{R}), u_1)$. This system can be reduced into the following one, call it T'.

$$\begin{cases} u_1 &= ext(ext(v_1)) \\ v_1 &= ext(ext(v_1)) + ext(v_2)//ext(v_2) \\ v_2 &= 1 + ext(ext(v_2)) + ext(v_1)//ext(v_1) \end{cases}$$

Finally, we have $L((T', \mathcal{R}), u_1) = L((S, \mathcal{R}), u) \cap L_p$.

5.7.2 *Inductivity of monadic second-order predicates*

In this section, we present the fundamental result saying that, roughly speaking, every monadic second-order definable set of graphs or hypergraphs is recognizable. However, in order to get a sensible statement, we must specify two things:

1. the representation of graphs and hypergraphs we are using (because we have defined two of them, yielding two different notions of monadic second-order definability),
2. the relevant operations on graphs and hypergraphs (we have defined two signatures of operations denoted by F_{VR} and F_{HR}, defining respectively the magma $\mathcal{GP}$ of simple graphs with ports and the magma $\mathcal{HS}$ of hypergraphs with sources; see Subsections 5.6.2 and 5.6.3.

We recall that if C is a set of port labels, we denote by $F_{VR}(C)$ the subsignature of F_{VR} consisting of the operations involving only port labels from C. We denote by $\mathcal{GP}(C)$ the $F_{VR}(C)$-magma with domain $\mathcal{GP}(C)$ and operations of $F_{VR}(C)$. Hypergraphs will be over a fixed finite ranked set A of hyperedge labels that we will not specify in notation. We recall that CMS_i refers to counting monadic second-order logic where a graph or a hypergraph G is represented by the structure $\mid G \mid_i$. (See Sections 5.3 and 5.4). We only consider finite graphs and hypergraphs, but we keep the notations $\mathcal{GP}(C)$ and $\mathcal{HS}(C)$.

Theorem 5.7.5
(1) Let C be a finite set of port labels. Every CMS_1-definable subset of $\mathcal{GP}(C)$ is $\mathcal{GP}(C)$-recognizable.
(2) Let C be a finite set of source labels. Every CMS_2-definable subset of $\mathcal{HS}(C)$ is $\mathcal{HS}$-recognizable. □

We shall sketch the proof of the first assertion. For the second assertion, we shall refer the reader to the proofs given in [11] and [15].

We need some notation. For every $n \in \mathcal{N}$, $q \in \mathcal{N}$-$\{0,1\}$, for every finite subset C of $\mathcal{C}$, for every finite set $\mathcal{W}$ of set variables, $L(n, q, C, \mathcal{W})$ denotes the subset of $CMS(\mathcal{R}_s(C), \mathcal{W})$ consisting of formulas of quantification depth at most

n, possibly written with the atomic formulas $Card_p(X)$ for $2 \leq p \leq q$ (see Subsection 5.4.1).

We shall say that two formulas are ***tautologically equivalent*** if they can be transformed into each other by renamings of bound variables and uses of the laws of Boolean calculus. (For instance φ is tautologically equivalent to $\varphi \wedge \varphi$). Every two tautologically equivalent formulas are equivalent and we shall assume that every formula is transformed into a tautologically equivalent formula of some standard form. With this convention, each set $L(n, q, C, \mathcal{W})$ is finite. Each formula $\varphi \in L(n, q, C, \emptyset)$ defines a predicate $\hat{\varphi}$ on $\mathcal{GP}(C)$ such that for every $G \in \mathcal{GP}(C)$:

$$\hat{\varphi}(G) \text{ is true iff } | G |_1 \models \varphi.$$

We let C be fixed and we denote by $\hat{L}(n, q)$ the corresponding family of predicates. It is finite.

Lemma 5.7.6
For every $n, q \in \mathcal{N}$ with $q \geq 2$, the family of predicates $\hat{L}(n, q)$ is $F_{VR}(C)$-inductive.

Proof sketch
We first consider the inductivity condition for the operation $add_{c,d}$ where $c, d \in C$. Let us recall that $G' = add_{c,d}(G)$ iff G' is obtained from G by the addition of an edge from x to y where x is any c-port and y is any d-port such that $(x, y) \notin edg_G$ and $x \neq y$. For every formula φ in $CMS(\mathcal{R}_s(C), \emptyset)$ one can construct a formula ψ in the same set and of no larger quantification depth such that, for every $G \in \mathcal{GP}(C)$

$$\hat{\varphi}(add_{c,d}(G)) = \hat{\psi}(G) \tag{5.3}$$

i.e.,

$$| add_{c,d}(G) |_1 \models \varphi \quad \textit{iff} \quad | G |_1 \models \psi \tag{5.4}$$

In other words the family of predicates on $\mathcal{GP}(C)$ associated with the formulas in $L(n, q, C, \emptyset)$ is $\{add_{c,d}\}$-inductive. The decomposition of $\hat{\varphi}$ relative to $add_{c,d}$ is simply $(B, \hat{\psi})$ (where B is the trivial Boolean combination such that $B[\psi]$ is ψ).

The proof of (3.4) is actually an immediate consequence of Lemma 5.2.5 because if $G' = add_{c,d}(G)$ we have the following definition of $edg_{G'}$:

$$edg_{G'}(x, y) \Longleftrightarrow edg_G(x, y) \vee (pt_{c_G}(x) \wedge pt_{d_G}(y) \wedge \neg x = y).$$

Hence, one can take ψ equal to $\varphi[\theta(x_1,x_2)/edg]$ where θ is the quantifier-free formula:

$$edg(x_1,x_2) \vee (pt_c(x_1) \wedge pt_d(x_2) \wedge \neg x_1 = x_2).$$

It is important to note that this transformation does not increase the quantification depth because the formula substituted for *edg* has no quantifier. The same holds (with similar argument) for the operations mdf_P. The inductivity of $\hat{L}(n,q)$ with respect to the operations mdf_P can be proved similarly.

Next we consider the case of $\oplus$, the disjoint union operation. The statement analogous to (3.3) is the following one: for every formula φ in $L(n,q,C,\emptyset)$ one can find $m \in \mathcal{N}$ and construct formulas $\psi_1, \psi'_1, \cdots, \psi_m, \psi'_m$ in $L(n,q,C,\emptyset)$ such that, for every G and $G' \in \mathcal{GP}(C)$:

$$\hat{\varphi}(G \oplus G') = \bigvee_{1 \leq i \leq m} \hat{\psi}_i(G) \wedge \hat{\psi'}_i(G'). \tag{5.5}$$

(The validity of (3.5) is actually known from Shelah [57], Section 5.3, given without proof as a easy modification of a previous result by Fefermann and Vaught [38] for first-order formulas). In this case the decomposition of $\hat{\varphi}$ with respect to $\oplus$ is:

$$((x_1 \wedge x_2) \vee (x_3 \wedge x_4) \vee \cdots \vee (x_{2m-1} \wedge x_{2m}), \hat{\psi}_1, \hat{\psi'}_1, \hat{\psi}_2, \hat{\psi'}_2, \cdots, \hat{\psi}_m, \hat{\psi'}_m).$$

The proof is more complicated than for (3.3); it can be done by induction on the structure of φ and the cases where $\varphi = \varphi_1 \wedge \varphi_2, \varphi = \varphi_1 \vee \varphi_2, \varphi = \neg\varphi_1$ are rather straightforward. However, in order to handle inductively the cases where $\varphi = \exists X \varphi_1$, one needs a formulation of (3.5) where φ has free variables. Let $\mathcal{W} = \{X_1, \cdots, X_r\}$. Let $\varphi \in L(n,q,C,\mathcal{W})$. Let $G \in \mathcal{GP}(C)$ and $V_1, \cdots, V_r$ be sets of vertices of G (i.e., subsets of the domain of the structure $\mid G \mid_1$). We let $\hat{\varphi}(G, V_1, \cdots, V_r) = true$ if

$$(\mid G \mid_1, V_1, \cdots, V_r) \models \varphi(X_1, \cdots, X_r)$$

and $\hat{\varphi}(G, V_1, \cdots, V_r) = false$ otherwise. The generalized version of (3.5) is now: for every formula $\varphi \in L(n,q,C,\mathcal{W})$, one can find $m \in \mathcal{N}$ and construct $\psi_1, \psi'_1, \cdots, \psi_m, \psi'_m \in L(n,q,C,\mathcal{W})$ such that, for every $G, G' \in \mathcal{GP}(C)$ such that $V_G \cap V'_G = \emptyset$, for every $V_1, \cdots, V_r \subseteq V_G$, for every $V'_1, \cdots, V'_r \subseteq V'_G$:

$$\hat{\varphi}(G \oplus G', V_1 \cup V'_1, \cdots, V_r \cup V'_r) = \bigvee_{1 \leq i \leq m} \hat{\psi}_i(G, V_1, \cdots, V_r) \wedge \hat{\psi'}_i(G', V'_1, \cdots, V'_r) \tag{5.6}$$

This can now be proved by induction on the structure of φ. □

Proof of Theorem 5.7.5

(1) Let C be finite, let $L \subseteq \mathcal{GP}(C)$ be defined by a formula φ of CMS relatively to the representation of the graphs G in $\mathcal{GP}(C)$ by structures of the form $\mid G \mid_1$. For some large enough n and q the formula φ belongs to $L(n, q, C, \emptyset)$. The predicate $\hat{\varphi}$ belongs to the finite family $\hat{L}(n, q)$ of predicates which is $F_{VR}(C)$-inductive (Lemma 5.7.6). Hence L is $\mathcal{GP}(C)$-recognizable.

(2) The proof is similar. We indicate the main ideas. The finite set A is fixed. The F_{HR}-magma of hypergraphs $\mathcal{HS}$ is many-sorted and its set of sorts is the set $\mathcal{S}$ of finite sets of source labels. For every $n, q, C, \mathcal{W}$ as above, we denote by $L'(n, q, C, \mathcal{W})$ the set of formulas in $CMS(\mathcal{R}_m(A, C), \mathcal{W})$ of quantification depth at most n and using the special atomic formulas $Card_p(X)$ for $p \leq q$. This set is finite. For each C, we consider the finite family of predicates $\{\hat{\varphi} \ / \ \varphi \in L'(n, q, C, \emptyset)\}$ where for $H \in \mathcal{HS}(C)$:

$$\hat{\varphi}(H) \text{ is true iff } \mid H \mid_2 \models \varphi$$

We denote it by $\hat{L}'(n, q, C)$. Hence we obtain a locally finite family of predicates on $\mathcal{HS}$ where $\hat{\varphi} \in \hat{L}'(n, q, C)$ is of sort C. For every $n, q \in \mathcal{N}$ with $q \geq 2$, the family $\cup\{\hat{L}'(n, q, C) \ / \ C \in \mathcal{S}\}$ is F_{HR}-inductive. It follows as above (see [11] or [15] for details) that every CMS_2-definable subset of $\mathcal{HS}(C)$ is $\mathcal{HS}$-recognizable. □

Corollary 5.7.7

(1) The intersection of a VR set of graphs and a CMS_1-definable subset of $\mathcal{GP}(C)$ (where C is finite) is a VR set. The CMS_1-satisfiability problem for a VR set is decidable.

(2) The intersection of an HR subset of $\mathcal{HS}(C)$ (where C is finite) and a CMS_2-definable one is HR. The CMS_2-satisfiability problem for an HR set is decidable.

Proof

(1) Let L be a VR set of graphs. Let D be the finite set consisting of all port labels occuring in the operation of some system of equations defining L. Then, L is $\mathcal{GP}(D)$-equational and $L \subseteq \mathcal{GP}(D)$. Let K be a CMS_1-definable subset of $\mathcal{GP}(C)$ for some finite $C \subseteq \mathcal{C}$. Then L is $\mathcal{GP}(C \cup D)$-equational and K is a CMS_1-definable subset of $\mathcal{GP}(C \cup D)$. Hence $L \cap K$ is $\mathcal{GP}(C \cup D)$-equational by Theorem 5.7.3 hence VR since K is $\mathcal{GP}(C \cup D)$-recognizable. (Theorem 5.7.5).

(2) Let L be an HR-subset of $\mathcal{HS}(C)$ and K be a CMS_2-definable one. Then K is $\mathcal{HS}$-recognizable by Theorem 5.7.5 and $L \cap K$ is $\mathcal{HS}$-equational (Theorem 5.7.3) hence HR. □

Corollary 5.7.8

Let C be a finite set of port or source labels. Let A be a finite ranked alphabet.
(1) For every formula $\varphi \in CMS(\mathcal{R}_s(C), \emptyset)$ the property $| G |_1 \models \varphi$ can be decided in time $O(size(t))$ where $G = val(t)$ and $t \in T(F_{VR}(C))$.
(2) For every formula $\varphi \in CMS(\mathcal{R}_m(A, C), \emptyset)$ the property $| H |_2 \models \varphi$ can be decided in time $O(size(t))$ where $H = val(t)$ and $t \in T(F_{HR}(A))_C$. □

In these two assertions, the input is a term t denoting G or H. Since not all finite graphs (hypergraphs) are the values of terms in $T(F_{VR}(C))(T(F_{HR}(A))_C)$, this corollary gives efficient algorithms only for CMS properties on (finite) graphs (hypergraphs) in special classes. Furthermore, the graphs or hypergraphs must be "parsed", i.e., given by terms.

Proof

Let F be a finite signature. The membership of a term t in a recognizable subset of $T(F)$ is decidable in time $O(size(t))$ because recognizable sets of terms are defined by finite-state deterministic frontier-to-root automata. The two assertions follow then from Proposition 5.7.2 and Theorem 5.7.5. □

5.7.3 Inductively computable functions and a generalization of Parikh's theorem

Let F be an $\mathcal{S}$-signature and M be an F-magma. Let N be a set and let $\mathcal{E}$ be a family of mappings called *evaluations* where each $e \in \mathcal{E}$ maps $M_s \rightarrow N$ for some $s \in S$. We call the sort s the *type* of e. We say that $\mathcal{E}$ is *F-inductive* if for every $e \in \mathcal{E}$ and $f \in F$ there exists a partial function $g_{e,f} : N^m \rightarrow N$ and k sequences $(e_1^1, \cdots, e_{n_1}^1), \cdots, e(_1^k \cdots, e_{n_k}^k)$ where $k = \rho(f)$ such that $m = n_1 + \cdots + n_k$, each function e_i^j has type $\alpha(f)_j$ and we have, for every $d_1, \cdots, d_k \in M$ with $\sigma(d_i) = \sigma(f)_i$, for all $i = 1, \cdots, k$:

$$e(f_M(d_1, \cdots, d_k)) = g_{e,f}(e_1^1(d_1), \cdots, e_{n_1}^1(d_1), e_1^2(d_2), \cdots, e_1^k(d_k), \cdots, e_{n_k}^k(d_k)). \tag{5.7}$$

This means that the value of e at $f_M(d_1, \cdots, d_k)$ can be computed, by means of some fixed functions $g_{e,f}$, from the values at $d_1, \cdots, d_k$ of m mappings of $\mathcal{E}$ (the mappings e_i^j are not necessarily pairwise distinct and some of them may be equal to e). We shall call the tuple

$$(g_{e,f}, (e_1^1, \cdots, e_{n_1}^1), (e_1^2, \cdots), \cdots, (e_1^k, \cdots, e_{n_k}^k)) \tag{5.8}$$

the *decomposition of e relative to f_M*. In Subsection 5.7.1, we have introduced inductive families of predicates: they are just the special case of inductive families of evaluations where N consists of the truth values *true* and *false*. If $\mathcal{E}$ is F-inductive, if $d \in M$ is given as t_M for some $t \in T(F)$ (i.e., this means that d has been "parsed" in terms of the operations of M), then one can evaluate $e(d)$ by the following algorithm using two traversals of t.

Algorithm:

Input: a term t given as a tree, an evaluation e_0 in $\mathcal{E}$:
Output: the value $e_0(t_M)$.
Method:
First traversal (top-down): One associates with every node u of the tree t a set of evaluations $\mathcal{E}(u)$, that will have to be computed at u, i.e., for the argument defined by the subtree of t issued of u. For the root r, we let $\mathcal{E}(r) := \{e_0\}$. For every node u such that $\mathcal{E}(u)$ is already known and that has successors $u_1, \cdots, u_k$, for every e in $\mathcal{E}(u)$, if f is the operation labelling u, and

$$(g, (e_1^1, \cdots, e_{n_1}^1), (e_1^2, \cdots), \cdots, (e_1^k, \cdots, e_{n_k}^k)) \tag{5.9}$$

is the decomposition of e relative to f_M, we add to each set $\mathcal{E}(u_r), i = 1, \cdots, k$, the evaluations $e_1^i, \cdots, e_{n_i}^i$.

Second traversal (bottom-up): Starting from the leaves, one computes at each node u the values of each function $e \in \mathcal{E}(u)$ by using the decomposition of e relative to f_M (see (3.7)). One obtains at the root the desired value of e_0. Denoting by $e(u)$ such a value (i.e., we denote by $e(u)$ the value of e for $(t/u)_M$, we use the formula:

$$e(u) := g(e_1^1(u_1), \cdots, e_{n_1}^1(u_1), e_1^2(u_2), \cdots, e_1^k(u_k), \cdots, e_{n_k}^k(u_k)).$$

based of the decomposition of e relative to f_M of the form (3.9). One obtains at the root the desired value of e_0.

This technique is useful for certain graph algorithms, taking as input graphs or hypergraphs expressed as values of VR or HR expressions. We refer the reader to [11], [12], [2], [25] for the construction of efficient graph algorithms based on this algorithm.

We shall now be interested in understanding the structure of sets of the form $e(L) := \{e(d) \ / \ d \in L\} \subseteq N$ where e is a mapping belonging to an F-inductive family $\mathcal{E}$ and L is M-equational. We shall need the following assumptions:

(H1) $\mathcal{E}$ has finitely many mappings of each type $s \in \mathcal{S}$,

(H2) N is an H-magma for some signature H with set of sorts $\mathcal{S}'$ ($\mathcal{S}'$ is not necessarily equal to $\mathcal{S}$),

(H3) the functions $g_{e,f}$ are derived operations of N; each of them is defined by a term $t_{e,f}$.

A term is *linear* if no variable has more than one occurrence.

Proposition 5.7.9
If conditions (H1) - (H3) hold, if in Equalities (3.7) we have $n_1 \leq 1, \cdots, n_k \leq 1$ *and if the terms* $t_{e,f}$ *are linear, then* $e(L) \in Equat(N)$ *for every* $L \in Equat(M)$.

Proof
Let $L = L((S, M), u_1)$ where S is a uniform polynomial system with unknowns $u_1, \cdots, u_n$. We shall assume that M and N have only one sort: this simplifies the notation and is not a loss of generality. For every $e \in \mathcal{E}$ and $i \in [1, n]$, we let $[e, u_i]$ be a new unknown. We shall build a polynomial system S' such that the component of its least solution corresponding to $[e, u_i]$ is the set $e(L((S, M), u_i))$ for each $e \in \mathcal{E}$ and $i \in [1, n]$. For every equation

$$u_i = \cdots + f(u_{i_1}, \cdots, u_{i_k}) + \cdots$$

of S, every $e \in \mathcal{E}$ we create the equation

$$[e, u_i] = \cdots + t_{e,f}([e^1, u_{i_1}], \cdots, [e^k, u_{i_k}]) + \cdots$$

noindent where $(g, (e^1), \cdots, (e^k))$ is the decomposition of e relative to f_M and $t_{e,f}$ is a linear term defining in N the function g. We let $(A_1^j, \cdots, A_n^j)$ be the j-th iterate (where $j \geq 0$) approximating the least solution of S in $\mathcal{P}(M)$ (see Subsection 5.6.1). Similarly, we denote by $A_{e,i}^j$ for $e \in \mathcal{E}$ and $i \in [1, n]$ the component corresponding to $[e, u_i]$ of the j-th iterate of $S'_{\mathcal{P}(N)}$. □

Claim: *For every* $j \in \mathcal{N}$, *every* $i \in [1, n]$, *every* $e \in \mathcal{E}$, *we have* $e(A_i^j) = A_{e,i}^j$.

Proof
By induction on j. See Courcelle [22], Theorem 6.2 for details. □

As an application, we obtain a form of Parikh's theorem. We let $q \in \mathcal{N}_+$; we let N be the commutative monoid $< \mathcal{N}^q, \mathbf{0}, \oplus, \mathbf{1}_1, \cdots, \mathbf{1}_q >$ where $\mathbf{0} = (0, \cdots, 0), \mathbf{1}_i = (0, 0, \cdots, 1, \cdots, 0)$ (and 1 is at position i), and $\oplus$ is the vector addition: $(a_1, a_2, \cdots, a_q) \oplus (b_1, \cdots, b_q) = (a_1 \oplus b_1, \cdots, a_q \oplus b_q)$. We assume that for every $s \in \mathcal{S}$ there is in $\mathcal{E}$ a unique evaluation mapping $e_s : M_s \to N$, and that the functions $g_{e,f}$ of the decompositions are of the form:

$$g_{e,f}(x_1, \cdots, x_k) = x_1 \oplus \cdots \oplus x_k \oplus b$$

where $b \in \mathcal{N}^q$. Since b can be written as a finite sum of constants $\mathbf{0}, \mathbf{1}_1, \cdots, \mathbf{1}_q$ the operation $g_{e,f}$ is defined by a linear term in $T(\{\oplus, \mathbf{0}, \mathbf{1}_1, \cdots, \mathbf{1}_q\})$. It follows that Equalities (3.7) have the form:

$$e(f_M(d_1,\cdots,d_k)) = e_1(d_1)\oplus\cdots\oplus e_k(d_k)\oplus b \tag{5.10}$$

where e_i is the evaluation in $\mathcal{E}$ of type $\sigma(d_i)$. We shall say that $\mathcal{E}$ is a *Parikh family of evaluations* on M.

We recall a few definitions. A set $A\subseteq\mathcal{N}^q$ is *linear* if it is of the form $A = \{\lambda_1 a_1\oplus\cdots\oplus\lambda_n a_n\oplus b\ /\ \lambda_1,\cdots,\lambda_n\in\mathcal{N}\}$ for some $a_1,\cdots,a_n,b\in\mathcal{N}^q$ and where, for $\lambda\in\mathcal{N}$, $a\in\mathcal{N}^q, \lambda a = \mathbf{0}$ if $\lambda = 0$ and $\lambda a = a\oplus a\oplus\cdots\oplus a$ with λ times a if $\lambda\geq 1$. A subset of $\mathcal{N}^q$ is *semi-linear* if it is a finite union of linear sets. Since a linear set as above can be written $A = a_1^*a_2^*\cdots a_n^*b$ it follows that every linear set, hence, every semi-linear set is rational. Conversely, by using the laws: $(A+B)^* = A^*B^*$ and $(A^*B)^* = \varepsilon + A^*B^*B$ which hold for arbitrary subsets A and B of a commutative monoid, one can transform a rational expression defining $A\in Rat(\mathcal{N}^q)$ into a sum of terms of the form $a_1^*a_2^*\cdots a_n^*b$ for $a_1,\cdots,a_n,b\in\mathcal{N}^q$. Hence, every rational subset of $\mathcal{N}^q$ is semi-linear.

Corollary 5.7.10
If $L\in Equat(M)$ and e belongs to a Parikh family of evaluations then $e(L)\in Equat(\mathcal{N}^q)$ and is semi-linear.

Proof
That $e(L)\in Equat(\mathcal{N}^q)$ follows from Proposition 5.7.9. Pilling has proved in [49] that every equational set of a commutative monoid is rational, hence, we get that $e(L)$ is semi-linear. □

Example 5.3
We consider the magma of series-parallel graphs $< \mathcal{SP}, ., //, e >$ of Subsection 5.6.1. We consider the evaluation $\#: \mathcal{SP}\to\mathcal{N}^2$ such that

$\#(G) =$ (number of edges of G, number of internal vertices of G).

Since $\#(G//G') = \#(G)\oplus\#(G')$, $\#(G.G') = \#(G)\oplus\#(G')\oplus\mathbf{1}_2$, and $\#(e) = \mathbf{1}_1$, from the equation $u = u//u + u.u.u + e$ which defines a proper subset L of $\mathcal{SP}$, we get the following equation that defines $\#(L)\subseteq\mathcal{N}^2$:

$$u = u\oplus u + u\oplus u\oplus u\oplus\mathbf{1}_2\oplus\mathbf{1}_2 + \mathbf{1}_1.$$ □

We now define a more powerful extension of Parikh's theorem which has also applications in graph grammars. We let M be an F-magma, we let N be a G-magma, we let $\mathcal{E}$ be a finite family of mappings: $\mathcal{P}(M)\to\mathcal{P}(N)$ satisfying the following conditions, for all sets $A_1,\cdots,A_i,\cdots\subseteq M$ of appropriate sorts, and for all $f\in F$:

1. $e(A_1 \cup A_2 \cup \cdots) = e(A_1) \cup e(A_2) \cup \cdots$
2. $e(f_{P,M}(A_1, \cdots, A_k)) = \bigcup_{1 \leq i \leq m} t_{i_{\mathcal{P}(N)}}(e_{i,1}(A_1), e_{i,2}(A_2), \cdots, e_{i,k}(A_k))$

where $t_1, \cdots, t_m$ are linear terms in $T(G, \{x_1, \cdots, x_k\})$ and $e_{i,j}, 1 \leq i \leq m, 1 \leq j \leq k$ are elements of $\mathcal{E}$.

Proposition 5.7.11
With these conditions, for every $L \in Equat(M)$, we have $e(L) \in Equat(N)$.

Proof
Essentially the same as the proof of Proposition 5.7.9. □

Example 5.4
As in Example 5.3, we use series-parallel graphs, but directed ones: this means that the basic graph e is a single edge directed from the first source to the second one. Series-parallel graphs have no circuit. We denote their set by $\mathcal{SP}'$. For every graph G in $\mathcal{SP}'$, we let $\pi(G)$ be the set of lengths of directed paths in G that link the first source to the second one. Hence π maps $\mathcal{SP}'$ into $\mathcal{P}(\mathcal{N})$. For $L \subseteq \mathcal{SP}'$ we let $\Pi(L) = \bigcup\{\pi(G) \ / \ G \in L\}$. This mapping satisfies the conditions of Proposition 5.7.11 because it a homomorphism for union and:

$$\begin{aligned} \pi(G//G') &= \pi(G) \cup \pi(G') = t_{1_{\mathcal{P}(\mathcal{N})}}(\pi(G)) \cup t_{2_{\mathcal{P}(\mathcal{N})}}(\pi(G')) \\ \pi(G.G') &= \pi(G) \oplus \pi(G') \\ &= \{n \oplus n'/n \in \pi(G), n' \in \pi(G')\} \\ &= t_{3_{\mathcal{P}(\mathcal{N})}}(\pi(G), \pi(G')) \end{aligned}$$

where $t_1 = x_1, t_2 = x_2$ and $t_3 = x_1 \oplus x_2$. Hence, for every $\{//, .\}$-equational set of series-parallel graphs L, the set $\Pi(L)$ is a semi-linear subset of $\mathcal{N}$. For the equation of Example 5.4 we obtain:

$$u = u \oplus u \oplus u + \mathbf{1}_1$$

Its least solution is the set of odd numbers {1, 3, 5, 7...}. □

This example is actually a special case of a general result of [19] that we recall and that we shall derive from Proposition 5.7.11 and some results of [25].
If G is a hypergraph represented by the structure $| G |_2$, if φ is an MS_2-formula with free set variables $X_1, \cdots, X_k$, we let

$$sat(G, \varphi) = \{(D_1, \cdots, D_k) \ / \ D_i \subseteq V_G \cup E_G, (| G |_2, D_1, \cdots, D_k) \models \varphi\}$$

$$\#sat(G, \varphi) = \{(Card(D_1), \cdots, Card(D_k)) \ / \ (D_1, \cdots, D_k) \in sat(G, \varphi)\} \subseteq \mathcal{N}^k.$$

If L is a set of hypergraphs, we let

$$\#sat(L,\varphi) = \bigcup\{sat(G,\varphi) \ / \ G \in L\}.$$

Proposition 5.7.12
If L is a HR set of hypergraphs and φ is an MS_2-formula with free variables $X_1, \cdots, X_k$, then the subset $\#sat(L,\varphi) \subseteq \mathcal{N}^k$ is semi-linear. A similar statement holds for VR sets of graphs and MS_1-formulas. □

This result is proved in [19] as Corollary 5.4.3. It yields immediately the result of Example 5.4 if one takes the formula $\varphi(X_1)$ saying that "X_1 is the set of edges of a path from the first source to the second one". We now derive the first assertion of Proposition 5.7.12 from Proposition 5.7.11 and some results of [25].

Proof of the first assertion of 5.7.12: Let C, C' be finite sets of source labels; let φ be an MS_2-formula of quantification depth at most h and free variables $X_1, \cdots, X_k$. One can find MS_2-formulas $\psi_1, \cdots, \psi_m, \psi'_1, \cdots, \psi'_m$ of quantification depth at most h and vectors $n_1, \cdots, n_m \in \mathcal{N}^k$ such that, for all hypergraphs G and G' of respective types C and C' we have:

$$\#sat(G//_{C,C'}G',\varphi) = \bigcup_{1\leq j\leq m} \#sat(G,\psi_j) \oplus \#sat(G',\psi'_j) \oplus n_j.$$

Similarly, for every finite set C of source labels, for each unary operation f in F_{HR}, one can find MS_2 formulas $\psi_1, \cdots, \psi_m$ of quantification depth at most h, and vectors $n_1, \cdots, n_m$ in $\mathcal{N}^k$ such that, for every hypergraph G of type C:

$$sat(f(G),\varphi) = \bigcup_{1\leq j\leq m} \#sat(G,\psi_j) \oplus n_j.$$

These facts follow from Lemmas 2.4, 2.5 and 2.6 of [25]. Letting $\mathcal{E}$ be the family of evaluations e of the form $e(G) = \#sat(G,\varphi)$ for G of type C where C is a subset of a fixed finite set of source labels, we obtain that $\mathcal{E}$ satisfies the conditions of Proposition 5.7.11. It follows from this proposition that if L is an HR set of hypergraphs, and φ is an MS_2-formula with free variables $X_1, \cdots, X_k$ then $\#sat(L,\varphi)$ is in $Equat(\mathcal{N}^k)$ hence, by [49], is semi-linear.

The proof of Proposition 5.7.12 is effective. This means that from φ and a system of equations defining L one can construct an expression of $\#sat(L,\varphi)$ as a finite union of linear sets where the linear sets are given by coefficients $a_1, \cdots, a_m, b$ (see the definition). It follows that a number of "boundness" questions can be solved effectively. It can be decided whether $Sup\{x_i \ / \ (x_1, \cdots, x_k) \in sat(L, \varphi)\}$ is finite or not, and if it is, its value can be computed. Much more generally, for every sequence integers $c_1, \cdots, c_k$, one can decide whether $Sup\{x_1c_1 + \cdots + x_kc_k \ / \ (x_1, \cdots, x_k) \in sat(L, \varphi)\}$ is finite or not, and if it is, its value can be computed.

5.7.4 Logical characterizations of recognizability

Theorem 5.7.5 has established that every set of finite graphs or hypergraphs defined by a formula of an appropriate monadic second-order language is recognizable with respect to an appropriate set of operations. It is thus natural to ask for the converse, which holds in several known cases summarized in the following theorem.

Theorem 5.7.13
(1) Let A be a finite alphabet. A language $L \subseteq A^*$ is regular iff it is MS_1-definable.
(2) Let F be a finite ranked alphabet. A set of terms $L \subseteq T(F)$ is $T(F)$-recognizable iff it is MS_1-definable.
(3) A set of finite trees is $\mathcal{TREE}$-recognizable iff it is CMS_1-definable.

Proof
Assertion (1) has been proved by Büchi and Elgot ([8], [30]), assertion (2) by Thatcher, Wright and Doner ([29], [59]); see [60], Thm 11.1), assertion (3) by Courcelle ([11]). ∎

However, no logical characterization of recognizability can exist by the following result.

Proposition 5.7.14
Every set of finite square grids is $\mathcal{HS}$-recognizable and $\mathcal{GP}(C)$-recognizable for every finite set of port labels C. ∎

Square grids have been defined in Section 5.2.9. This result is proved in Courcelle [11] for $\mathcal{HS}$-recognizability. The proof is easily adapted to yield the case of $\mathcal{GP}(C)$-recognizability. It follows that the family of $\mathcal{HS}$-recognizable (or of $\mathcal{GP}(C)$-recognizable) sets of graphs is uncountable. Hence it cannot be characterized by the countably many formulas of any logical language like those we have considered here (including second-order logic).

Courcelle conjectured in [13] that for every k, every recognizable set of graphs of tree-width at most k is CMS_2 definable. This was proved in [13] for $k = 2$ and by Kaller [45] for $k = 3$.

5.8 Forbidden configurations

Many classes of graphs can be characterized in terms of *forbidden configurations*, i.e., of certain graphs that cannot appear as subgraphs. The basic example is that of planar graphs: they can be characterized as the graphs that do not contain K_5 or $K_{3,3}$ as a minor. Such characterizations yield logical descriptions that could not be obtained otherwise: for example, the definition of the planarity of a graph G in terms of an embedding in the plane cannot be expressed logically in a structure like $\mid G \mid_1$ or $\mid G \mid_2$ which describes the graph G (its vertices and edges) but not the plane. One could also try to express the existence of an embedding of G (assumed finite) in a large enough rectangular grid but this is not possible (at least immediately) because this requires to express the existence of vertices and edges forming a grid lying *outside of* G. The logical formalism allows to express properties of a graph G in terms of its vertices and edges, not in terms of objects outside of $\mid G \mid_1$ or $\mid G \mid_2$.

All graphs in this section will be finite. Hence, "graph" will mean "finite graph". We shall review the links between finitary descriptions of sets of graphs in terms of grammars, MS formulas and forbidden minors. We also show how the notion of a minor and the results of Robertson and Seymour help in understanding the structure of the sets of graphs having a decidable MS_2-theory.

5.8.1 *Minors*

Let G and H be undirected graphs. We say that G *is a minor* of H and we write this $G \trianglelefteq H$ iff there exist mappings $f : V_G \to \mathcal{P}(V_H), f' : V_G \to \mathcal{P}(E_H)$ and $g : E_G \to E_H$ satisfying the following conditions:

(1) for every $x \in V_G, (f(x), f'(x))$ is a connected subgraph of H,

(2) for every $x, x' \in V_G$ with $x \neq x'$, we have $f(x) \cap f(x') = \emptyset$,

(3) g is injective and $g(e)$ does not belong to any set $f'(x)$ for any $x \in V_G, e \in E_G$,

(4) if $e \in E_G$ links x and x', then $g(e)$ links a vertex of $f(x)$ and a vertex of $f(x')$.

It is clear that this definition depends only on the isomorphism classes of G and H. Minor inclusion is transitive and reflexive. Because graphs are finite, if $G \trianglelefteq H \trianglelefteq G$ then it is not hard to see that G and H are isomorphic. Hence minor inclusion is a partial order on graphs. (The corresponding strict order is denoted by $\triangleleft$).

Fact 5.8.1
$G \trianglelefteq H$ iff G is obtained from a subgraph H' of H by a sequence of edge contractions. □

Contracting an edge linking x and y results in the fusion of x and y and deletion of the resulting loop. Edge contraction may create loops and multiple edges.

Proof
Consider $G \trianglelefteq H$ with functions f, f', g as in the definition. We let H' be the subgraph of H such that

$$V'_H = \bigcup\{f(x) \ / \ x \in V_G\}, E_{H'} = \bigcup\{f'(x) \ / \ x \in V_G\} \cup \{g(e) \ / \ e \in E_G\}.$$

Then one obtains G (up to isomorphism) by contracting (in any order) the edges in the sets $f'(x), x \in V_G$. The converse is proved similarly. □

Let M be a set of graphs. We define $FORB(M)$ as the set of undirected graphs H such that $G \trianglelefteq H$ for no $G \in M$. For example the set of undirected planar graphs can be characterized as the class $FORB(K_{3,3}, K_5)$. This is a variant of Kuratowski's theorem (see Bollobas [5]). We recall that K_n is the n-clique consisting of n vertices and an edge linking any two distinct vertices, and that $K_{n,m}$ is the complete bipartite graph with $n+m$ vertices. (Formally its vertices are $-n, -(n-1), ..., -1, 1, 2, \cdots, m$ and there is an edge linking any "negative" vertex and any "positive" one). It is clear that every class of graphs of the form $FORB(M)$ is *minor-closed* which means that every minor of every graph in the class is also in the class. In order to get a converse we let, for every class $\mathcal{C}$

$OBST(\mathcal{C}) = \{G \ / \ G \notin \mathcal{C}$, every graph H such that $H \triangleleft G$ belongs to $\mathcal{C}\}$.

$OBST(\mathcal{C})$ is defined as a set of abstract graphs: its elements are by definition pairwise nonisomorphic. This set is called the set of *obstructions of* $\mathcal{C}$.

Fact 5.8.2
For every minor-closed class $\mathcal{C}$ of graphs: $\mathcal{C} = FORB(OBST(\mathcal{C}))$. □

The major result in this field is the following result by Robertson and Seymour [54].

Theorem 5.8.3 (Graph Minor Theorem [54])
Every infinite set of undirected graphs contains two graphs comparable by $\trianglelefteq$. Hence for every class $\mathcal{C}$ of graphs, the set $OBST(\mathcal{C})$ is finite. □

The obstructions of minor-closed classes are known in relatively few cases. We have already mentioned planar graphs. The set of obstructions of forests is the singleton consisting of the graph reduced to a loop. That of graphs of tree-width at most 2 is $\{K_4\}$. That of graphs of tree-width at most 3 consists of 4 graphs one of which is K_5. The obstructions of the classes of graphs of tree-width at most k are not known for $k \geq 4$. Intractable computation methods are known to compute them ([47]). In the following case, one does not even know such a method. Let $\mathcal{C}$ be the class of *almost planar* graphs, i.e. of graphs G such that $G[V_G - \{v\}]$ is planar for some vertex v. (Every planar graph is almost planar and so are K_5 and $K_{3,3}$; K_6 is not almost planar because one must delete at least two vertices in order to obtain from K_6 a planar graph). R. Thomas knows already 60 graphs in $OBST(\mathcal{C})$ one of which is K_6, but does not know whether this list is complete.

We now discuss the logical aspects of minor inclusion.

Proposition 5.8.4
Let H be a finite graph. The property that a graph G contains H as a minor is MS_2-definable. If H is simple and loop-free, this property is MS_1-definable. □

It follows in particular that planarity is MS_1-definable.

Proof
Let H be given with vertices $1, 2, \cdots, k$ and edges $e_1, \cdots, e_m$. Let G be an arbitrary graph. In order to verify that $H \trianglelefteq G$, we have to find appropriate mappings f, f', g. Hence we need only find k subsets of V_G namely $f(1), \cdots, f(k)$, k subsets of E_G namely $f'(1), \cdots, f'(k)$, m edges of G: namely $g(e_1), \cdots, g(e_m)$ and to check the conditions of the definition. This can be done easily by an MS_2 formula.

Assuming H simple and loop-free we now check that one can avoid quantifications on sets of edges. We claim that $H \trianglelefteq G$ iff there exist $X_1, \cdots, X_k \subseteq V_G$ satisfying the following conditions:

(1') the induced subgraph $G[X_i]$ of G with set of vertices X_i is connected for each $i = 1, \cdots, k$.

(2') $X_i \cap X_j = \emptyset$ for $i \neq j$,

(3') for every edge of H linking i and j there is an edge of G linking a vertex of X_i and one of X_j.

If $H \trianglelefteq G$ with corresponding mappings f, f', g then these conditions hold with $X_i = f(i)$. Conversely, if $X_1, \cdots, X_k$ satisfy them, then we take:

$f(i) = X_i$,

$f'(i) = E_{G[X_i]}$, for every $i = 1, \cdots, k$,

$g(e_j)$ to be an edge satisfying condition (3').

These mappings satisfy the conditions of minor inclusion: (1) follows from (1'), (2) follows from (2'), (3) holds from (3') and, because H is simple and loop-free, (4) follows from (3'). The existence of $X_1, \cdots, X_k$ satisfying (1') - (3') is easily expressed by an MS_1-formula. □

We leave as an exercise the verification that the property of a graph G that it contains a fixed graph H as a minor is MS_1 where H may have loops and multiple edges, and G is assumed to be simple. With Theorem 5.8.3 we obtain:

Corollary 5.8.5
Every minor-closed class of graphs is MS_2-definable. Its subclass of loopfree simple graphs is MS_1-definable.

Proof
The first assertion is immediate from the Graph Minor Theorem and Proposition 5.8.4. For the second one, we let:
$OBST'(\mathcal{C}) := \{G \;/\; G$ is simple and loop-free, $G \notin \mathcal{C}$, every simple loop-free graph H with $H \triangleleft G$ belongs to $\mathcal{C}\}$.

If $\mathcal{C}$ is a minor-closed class of simple loop-free graphs, then $\mathcal{C} = FORB'(OBST'(\mathcal{C}))$ where $FORB'(M)$ is the class of simple loop-free graphs in $FORB(M)$. The set $OBST'(\mathcal{C})$ is finite by the Graph Minor Theorem and the result follows by Proposition 5.8.4. □

We shall denote by **Minor**(G) the set of minors of a graph G. Thus **Minor** is a transduction from graphs to graphs. We show it is (2,2)-definable. Let G be a graph. Let $X \subseteq V_G$, let $Y, Z \subseteq E_G$. We say that (X, Y, Z) *defines a minor of* G if the following conditions hold:

(α) if $x, x' \in X, x \neq x'$, then there is no Z-path in G linking them (a *Z-path* is a path all edges of which are in Z);

(β) each end y of an edge in Y is linked to some vertex x of X by some Z-path; (one may have $x = y$);

(γ) $Y \cap Z = \emptyset$.

Notice that by condition (α), the vertex x in condition (β) is uniquely defined. We shall denote it by $\widehat{y}$. The *minor defined by* the triple (X, Y, Z) is the graph H such that $V_H = X, E_H = Y$ and $vert_H(e) = \{\widehat{y}, \widehat{y'}\}$ where y and y' are such that $vert_G(e) = \{y, y'\}$. We shall denote it by $\mathbf{Minor}(X, Y, Z)$. The corresponding mappings f, f', and g are as follows: $f(v)$ is the set of vertices of G linked to v by some Z-path (including v); $f'(v)$ is the set of edges of Z having their ends in $f(v)$; and g is the identity: $V_H \to V_{G'}$. Hence, $\mathbf{Minor}(X, Y, Z)$ is indeed a minor of G. It is not hard to see that every minor H of G is of this form for appropriate sets X, Y, Z. Moreover, $\mathbf{Minor}(X, Y, Z)$ is a strict minor iff $Z \neq \emptyset$ or $Y \neq E_G$ or $X \neq V_G$. We have thus proved the following

Fact 5.8.6
Minor is a (2,2)-definable transduction. □

Proposition 5.8.7
For every MS_2-definable class M of graphs, each of the classes $FORB(M)$ and $OBST(M)$ is MS_2-definable. MS_2-formulas defining them can be effectively constructed from an MS_2-formula defining M.

Proof
Let φ be an MS_2-formula such that $M = \{G \ / \ \mid G \mid_2 \ \models \varphi\}$. Then $H \in FORB(M)$ iff:

$\mid H \mid_2 \ \models \forall X, Y, Z[$"$X, Y, Z$ define a minor of H" $\Longrightarrow$
"**Minor** (X, Y, Z) does not satisfy φ"$]$.

From the proof of Fact 5.8.6 this condition can be expressed by an MS_2-formula. From the definition of $OBST$ we get that for every graph H, $H \in OBST(M)$ iff:

$\mid H \mid_2 \ \models \neg\varphi \wedge \forall X, Y, Z[$"$X, Y, Z$ define a strict minor of H" $\Longrightarrow$
"**Minor**(X, Y, Z) satisfies φ"$]$

Again this latter condition is MS_2-expressible. □

One might hope to be able to construct the sets of obstructions of MS_2-definable classes of graphs. Consider for instance the class of almost planar graphs. It is not hard to prove that it is minor-closed, and to construct an MS_2-formula defining it. One obtains thus an MS_2-formula characterizing its finite set of obstructions. However this is not enough to make it possible to construct (even by an intractable algorithm) this set itself. (See the remark following Proposition 5.2.2.). The best one can do presently is the following

Proposition 5.8.8
Let M be an MS_2-definable class of finite graphs. For every HR set of graphs L one can compute the set $L \cap OBST(M)$.

In particular, taking L to be the set of trees (or of graphs of tree-width at most k), one can obtain effectively the obstructions of M that are trees or that have tree-width at most k. If one knows some upper-bound on the tree-width of the graphs in $OBST(M)$ one can thus (at least in principle) compute $OBST(M)$ explicitely.

Proof
This follows immediately from Propositon 5.8.7, Corollary 5.7.7 and the fact that if a finite set of graphs is HR (given by a known grammar or system of equations), then this set can be explicitely enumerated. (See Habel [43]). □

This result was proved first in a different way by Fellows and Langston [39].

Sets of finite graphs can be specified effectively in various ways: by grammars of several types, by forbidden minors, by monadic second-order formulas. The following theorem summarizes the situation.

Theorem 5.8.9
Let L be a set of loop-free simple graphs containing the simple and loop-free minors of its members. The following properties are equivalent:
(1) L is HR,
(2) L is VR,
(3) L has bounded tree-width,
(4) $OBST(L)$ contains a planar graph.

Assuming these conditions satisfied, from any of the following devices defining L one can construct all others, namely:
(5) an HR grammar (or F_{HR}-system of equations),
(6) a VR grammar (or F_{VR}-system of equations),
(7) an MS_1-formula,
(8) the set $OBST(L)$.

Proof
The set L is MS_1-definable by Corollary 5.8.5 (based on the Graph Minor Theorem). We have the following implications:

(3) $\Longleftrightarrow$ (4) by Robertson and Seymour [53]
(3) $\Longleftrightarrow$ (1) by Corollary 5.7.7 (2) since the set of graphs of tree-width at most k is HR (see [20])
(1) $\Longrightarrow$ (2) by Theorem 5.6.15
(2) $\Longrightarrow$ (1) by a result of [23].

We now consider effective constructions of such devices.
(5) $\Longleftrightarrow$ (6) because the proofs of Theorem 5.6.15 and [23] are effective.
(8) $\Longleftrightarrow$ (7) by Proposition 5.8.4
(7) $\Longleftrightarrow$ (5) because Corollary 5.7.7 is effective and one can find the smallest square grid not in L; from this grid, one gets an upper bound on the tree-width of L by [53].
(5) $\Longleftrightarrow$ (8) by a result of Courcelle and Sénizergues [27]. □

5.8.2 The structure of sets of graphs having decidable monadic theories

By combining techniques from logic (definable graph transductions) and from graph theory (minor inclusion), one obtains the following result of Seese [56].

Theorem 5.8.10
If a set of graphs has a decidable MS_2-theory, then it is a subset of some HR set of graphs.

Proof sketch
The mapping from a graph to the set of its minors is (2,2)-definable. Hence, if a set L of graphs has a decidable MS_2-theory, then so has the set of its minors. But this set cannot contain all square grids by Proposition 5.2.2. Hence L has bounded tree-width (by [53]), hence L is a subset of the set of all graphs of tree-width at most k for some k which is HR. □

This result is a kind of converse of the one (Corollary 5.7.7) saying that the MS_2-theory of a HR set is decidable. We make the following conjecture.

Conjecture 5.8.11
If a set of graphs has a decidable MS_1-theory then it is a subset of some VR set of graphs. □

Special cases of this conjecture are proved in Courcelle [16]. Seese [56] has made a conjecture which is equivalent to Conjecture 5.8.11 by the result of [33] in the generalized form given in [23].

Acknowledgements

Many thanks to Mrs. A. Dupont and to A. Pariès for the Latex typing, and to E. Grandjean and the referees for comments.

References

The reference list below will be updated regularly. The updates can be obtained by email request to:

courcell@labri.u-bordeaux.fr

or downloaded from my home page on the Web:

http://www.labri.u-bordeaux.fr/~courcell/courcell.html

[1] ABITEBOUL S., HULL R., VIANU V., Foundations of Data bases, Addison Wesley, 1994.

[2] ARNBORG S., LAGERGREN J., SEESE D., Problems easy for tree-decomposable graphs, J. of Algorithms **12** (1991) 308-340.

[3] BAUDERON M., COURCELLE B., Graph expressions and graph rewritings, Mathematical Systems Theory **20** (1987) 83-127.

[4] BERSTEL J., Transductions and context-free languages, Teubner Verlag, Stuttgart, 1979.

[5] BOLLOBAS B., Extremal Graph Theory, Academic press, New York, 1978.

[6] BOSSUT F., DAUCHET M., WARIN B., A Kleene theorem for a class of planar acyclic graphs, Information and Computation **117** (1995) 251-265.

[7] BOURBAKI N., Algèbre, Chap. 1, Hermann, Paris, 1970.

[8] BUCHI J., Weak second-order logic and finite automata, Z. Math. Logik Grundlagen Math. **5** (1960) 66-92.

[9] COURCELLE B., An axiomatic definition of context-free rewriting and its application to NLC graph grammars, Theoret. Comput. Sci. **55** (1987) 141-181.

[10] COURCELLE B., Graph rewriting: An algebraic and logic approach, in "Handbook of Theoretical Computer Science, Volume B", J. Van Leeuwen ed., Elsevier, 1990, pp. 193-242.

[11] COURCELLE B., The monadic second-order logic of graphs I: Recognizable sets of finite graphs. Information and Computation **85** (1990) 12-75.

[12] COURCELLE B., The monadic second-order logic of graphs III: Tree-decompositions, minors and complexity issues, Informatique Théorique et Applications **26** (1992) 257-286.

[13] COURCELLE B., The monadic second-order logic of graphs V: On closing the gap between definability and recognizability, Theoret. Comput. Sci. **80** (1991) 153-202.

[14] COURCELLE B., The monadic second-order logic of graphs VI: On several representations of graphs by relational structures, Discrete Applied Mathematics, **54** (1994) 117-149 and erratum in **63** (1995) 199-200.

[15] COURCELLE B., The monadic second-order logic of graphs VII: Graphs as relational structures, Theoret. Comput. Sci. **101** (1992) 3-33.

[16] COURCELLE B., The monadic second-order logic of graphs VIII: Orientations, Annals Pure Applied Logic **72** (1995) 103-143.

[17] COURCELLE B., The monadic second-order logic of graphs IX: Machines and their behaviours, Theoret. Comput. Sci. **151** (1995) 125-162.

[18] COURCELLE B.,The monadic second-order logic of graphs X: Linear orderings, Theoret. Comput. Sci., **160** (1996) 87-143.

[19] COURCELLE B., Structural properties of context-free sets of graphs generated by vertex replacement, Information and Computation, **116** (1995) 275-293.

[20] COURCELLE B., Graph grammars, monadic second-order logic and the theory of graph minors in "Graph Structure Theory", Contemporary Mathematics **147** American Mathematical Society (1993) 565-590.

[21] COURCELLE B., Monadic second-order definable transductions: a survey, Theoret. Comput. Sci., **126** (1994) 53-75.

[22] COURCELLE B., Basic notions of universal algebra for language theory and graph grammars, Theoret. Comput. Sci., **163** (1996) 1-54.

[23] COURCELLE B., ENGELFRIET J., A logical characterization of the sets of hypergraphs defined by hyperedge replacement grammars, Mathematical Systems Theory **28** (1995) 515-522.

[24] COURCELLE B., ENGELFRIET J., ROZENBERG G., Handle-rewriting hypergraph grammars **46** (1993) 218-246.

[25] COURCELLE B., MOSBAH M., Monadic second-order evaluations on tree-decomposable graphs, Theoret. Comput. Sci. **109** (1993) 49-82.

[26] COURCELLE B., OLARIU S., Clique-width, a new complexity measure on graphs, preprint, 1995.

[27] COURCELLE B., SENIZERGUES G., The obstructions of minor-closed sets of graphs defined by a context-free grammars, Proceedings of the international workshop on Graph Grammars, Williamsburg, 1994, L.N.C.S., Vol. 1073, 1996, pp. 351-367.

[28] DAUCHET M., HEUILLARD T., LESCANNE P., TISON S., Decidability of the confluence of finite ground term rewrite systems and of other related term rewrite systems, Information and Computation **88** (1990) 187-201.

[29] DONER J., Tree acceptors and some of their applications, J. Comput. Syst. Sci. **4** (1970) 406-451.

[30] ELGOT C., Decision problems of finite automata design and related arithmetics, Trans. A.M.S. **98** (1961) 21-52.

[31] EHRIG H., MAHR B., Fundamentals of algebraic specifications 1: equations and initial semantics, Springer, 1985.

[32] ENGELFRIET J., Context-free NCE graph grammars, Proc. FCT'89, Lec. Notes Comp.Sci. **380** (1989) 148-161.

[33] ENGELFRIET J., A characterization of context-free NCE graph languages by monadic second-order logic on trees, Lec. Notes Comp.Sci. **532** (1991) 311-327.

[34] ENGELFRIET J., HEYKER L., The string generating power of context-free hypergraph grammars, J. Comp. Syst. Sci. **43** (1991) 328-360.

[35] ENGELFRIET J., HEYKER L., Hypergraph languages of bounded degree, J. Comp. Syst. Sci. **48** (1994) 58-89.

[36] ENGELFRIET J., ROZENBERG G., Graph grammars based on node rewriting: an introduction to NLC graph grammars, Lec. Notes Comp.Sci. **532** (1991) 12-23.

[37] FAGIN R., Generalized first-order spectra and polynomial-time recognizable sets, in "Complexity of Computation", SIAM-AMS Proceedings (1974) 43-73.

[38] FEFERMAN S., VAUGHT R., The first-order properties of products of algebraic systems, Fund. Math. **47** (1959) 57-103.

[39] FELLOWS M., LANGSTON M., An analogue of the Myhill-Nerode theorem and its use in computing finite basis characterizations, 30th symp. FOCS (1989) 520-525.

[40] GECSEG F., STEINBY M., Tree Automata, Akademiai Kiado, Budapest, 1984.

[41] GINSBURG S., RICE , Two families of languages related to ALGOL, J. ACM **9** (1962) 350-371.

[42] GUREVICH Y., Monadic second-order theories, in J. Barwise and S. Feferman eds., "Model theoretic logic", Springer, Berlin, 1985, pp. 479-506.

[43] HABEL A., Hyperedge replacement: grammars and languages, Lect. Notes Comput. Sci. **643**, 1992.

[44] IMMERMAN N., Languages which capture complexity classes, SIAM J. Comput. **16** (1987) 760-778.

[45] KALLER D., Definability equals recognizability of partial 3-trees, preprint, 1995.

[46] KANNELLAKIS P., Elements of relational data base theory, "Handbook of Theoretical Computer Science, Vol. B", J. Van Leeuwen ed., Elsevier 1990, pp. 1073-1156.

[47] LAGERGREN J., ARNBORG S., Finding minimal forbidden minors using a finite congruence, LNCS **510** (1991) 532-543.

[48] MEZEI J., WRIGHT J., Algebraic automata and context-free sets, Information and Control **11** (1967) 3-29.

[49] PILLING D., Commutative regular equations and Parikh's theorem, J. London Math. Soc. **6** (1973) 663-666.

[50] PIN J.-E., Logic, semigroups and automata on words, Annals of Mathematics and Artificial Intelligence, **16** (1996).

[51] RABIN M., A simple method for undecidability proofs and some applications, in "Logic, Methodology and Philosophy of Science II", Y. Bar-Hillel ed., North-Holland, Amsterdam, 1965, pp.58-68.

[52] RAOULT J.-C., A survey of tree transductions, in "Tree automata and languages", M. Nivat and A. Podelski eds., Elsevier, 1992, pp. 311-326.

[53] ROBERTSON N., SEYMOUR P., Graph minors V: Excluding a planar graph, J. Comb. Theory Ser. B **52** (1986) 92-114.

[54] ROBERTSON N., SEYMOUR P., Graph minors XX: Wagner's conjecture, September 1988.

[55] ROZENBERG G., WELZL E., Boundary NLC grammars, Basic definitions, normal forms and complexity, Information and Control **69** (1986) 136-167.

[56] SEESE D., The structure of the models of decidable monadic theories of graphs, Annals Pure Applied Logic **53** (1991) 169-195.

[57] SHELAH S., The monadic theory of order, Annals of Maths **102** (1975) 379-419.

[58] STOCKMEYER L., The polynomial-time hierarchy, TCS **3** (1977) 1-22.

[59] THATCHER J., WRIGHT J., Generalized finite automata theory with an application to a decision problem in second-order logic, Math. Systems Theory **3** (1968) 57-81.

[60] THOMAS W., Automata on infinite objects, "Handbook of Theoretical Computer Science, Volume B" Elsevier 1990, pp. 133-192.

[61] THOMAS W., Classifying regular events in symbolic logic, J. Comput. System Sci. **25** (1982) 360-376.

[62] TRAKHTENBROT B., Impossibility of an algorithm for the decision problem on finite classes, Dokl. Akad. Nauk.SSSR **70** (1950) 569-572.

[63] WECHLER W., Universal Algebra for Computer Scientists, Springer, 1992.

[64] WIRSING M., Algebraic Specifications, "Handbook of Theoretical Computer Science, Volume B" J. Van Leeuwen ed., Elsevier 1990, 675-779.

Chapter 6

2-STRUCTURES - A FRAMEWORK FOR DECOMPOSITION AND TRANSFORMATION OF GRAPHS†

A. EHRENFEUCHT

Department of Computer Science, University of Colorado at Boulder
Boulder, Co 80309, U.S.A.

T. HARJU

Department of Mathematics, University of Turku
FIN-20500 Turku, Finland

G. ROZENBERG

Department of Computer Science, Leiden University
P.O.Box 9512, 2300 RA Leiden, The Netherlands

and

Department of Computer Science, University of Colorado at Boulder
Boulder, Co 80309, U.S.A.

A 2-structure g is a pair (D, R) consisting of a finite set D and an equivalence relation R on the ordered pairs $(x, y) \in D \times D$ with $x \neq y$. These structures can be used to study various combinatorial properties of systems such as graphs, partially ordered sets and communication networks. We present the basic theory of 2-structures, where the main interest is in the clan decomposition of these structures providing a generalization of the modular decomposition (or the substitution decomposition) of graphs and directed graphs.

We consider also local transformations on 2-structures with a labeling of the edges, where the labels come from a group. For these one uses the group operations locally in the nodes to induce transformations of the labels of the edges and thus generating from a single labeled 2-structure a set of labeled 2-structures, which is referred to as a dynamic labeled 2-structure. Since the transformations are based on groups, one obtains an elegant way to define graph transformations. This approach is a generalization of Seidel switching for undirected graphs. Also here the notion of a clan remains central. Now a clan of a dynamic labeled 2-structure is a clan of any of its component labeled 2-structures. As a matter of fact one of the main themes investigated in the theory of dynamic labeled 2-structures is the behaviour of the clan decomposition under the group induced by the local transformations.

† The authors are grateful to BRA Working Groups ASMICS, COMPUGRAPH and Stieltjes Institute for their support.

Contents

6.1 Introduction

This paper is a tutorial on the theory of 2-structures as initiated in Ehrenfeucht and Rozenberg [16]. We have included only the most illustrative proofs from the literature together with new proofs for some of the older results. We assume that the reader is familiar with the basic concepts of graph theory, see Bondy and Murty [1].

A 2-structure $g = (D, R)$ consists of a finite domain D together with an equivalence relation R on the ordered pairs $(x, y) \in D \times D$ with $x \neq y$. Hence a 2-structure can be considered as a complete directed graph with an 'abstract colouring' of the edges.

To make a colouring of the edges concrete one adds a labeling function to a 2-structure obtaining in this way a labeled 2-structure. In the literature the term '2-structure' is used both as a technical term as defined above and as a generic term referring to the theory of labeled and 'unlabeled' 2-structures.

These structures can be used to study various combinatorial properties of systems such as graphs, partially ordered sets and communication networks, where all kinds of (binary) relationships hold between the objects of a system. We refer to Buer and Möhring [3], Ehrenfeucht and Rozenberg [18] and Schmerl and Trotter [40] for some of the intuition and motivation.

In this paper we present the basic theory of 2-structures. We cover both the 'static' part of the theory concerned mainly with the hierarchical representations of 2-structures, and the 'dynamic' part of the theory concerned with the local transformations of 2-structures.

The general results on the static properties of 2-structures are presented in Sections 6.2 through 6.6, where the main interest is in the *clan decomposition* of these structures. A clan X of a 2-structure g is a subset of the domain of g such that an element $y \notin X$ cannot make a distinction between the elements of X by using the equivalence relation R of g, *i.e.*, for all $y \notin X$ and $x_1, x_2 \in X$ we have that $(y, x_1)R(y, x_2)$ and $(x_1, y)R(x_2, y)$.

Decompositions of combinatorial and algebraic structures are special cases of the *divide-and-conquer method*, where a large problem is partitioned into smaller problems, and a method is given how to link the solutions of the subproblems into a solution of the original problem. The *clan decomposition* of 2-structures is an example of such a method, which provides a generalization of the *modular decomposition* (or *substitution decomposition*) of graphs and directed graphs, see Gallai [25], Habib and Maurer [27], Cunningham [4] and Muller and Spinrad [37]. In the clan decomposition a 2-structure g is divided into disjoint substructures $g_1, \ldots, g_k$, and a quotient 2-structure is constructed to indicate the relationships between the substructures g_i. For another kind of a general approach to the clan decomposition theorem we refer to Möhring and Radermacher [36] and Möhring [35].

The results on the local transformations are presented in Sections 6.7 through 6.10. Here one assumes that the labels used by a labeled 2-structure form a group. One uses then the group operations, applied locally in the nodes, to induce transformations of the labels of the edges. The domain, *i.e.*, the set of nodes, does not change in these transformations. In this way a single labeled 2-structure generates a whole set of labeled 2-structures, which is referred to as a dynamic labeled 2-structure. Since the transformations are based on groups, one obtains in this way an elegant way to define graph transformations.

Also here the notion of a clan remains central. Now a clan of a dynamic labeled 2-structure is a clan of any of its component labeled 2-structures. As a matter of fact one of the main themes investigated in the theory of dynamic labeled 2-structures is the behaviour of the clan decomposition under the group induced by the local transformations.

It is interesting to notice that the notion of a clan has been rediscovered quite many times under various names. In particular, a clan corresponds to the domain of the right hand side of a production in node-rewriting graph grammars with hereditary connection relation, see Chapter 1 of the present book. For other appearances and applications of the notion of a clan, see Buer and Möhring [3] or Muller and Spinrad [37].

6.2 2-Structures and Their Clans

6.2.1 Definition of a 2-structure

Let D be a finite nonempty set, and let

$$E_2(D) = \{(x, y) \mid x \in D, y \in D, x \neq y\}$$

be the set of all (*directed*) *edges* between the elements of D. For an edge $e = (x, y)$, we denote by $e^{-1} = (y, x)$ the *reverse edge* of e.

A *2-structure* $g = (D, R)$ consists of a finite nonempty set D, called its *domain*, and an equivalence relation $R \subseteq E_2(D) \times E_2(D)$ on its *edges*. The domain of g is denoted by $\text{dom}(g)$. The elements of the domain D of g are called *nodes*. The equivalence relation of a 2-structure g is sometimes written as R_g.

Equivalently, a 2-structure $g = (D, R)$ can be defined as a triple $G = (D, E, R')$, where the pair (D, E) is a directed graph, *i.e.*, $E \subseteq E_2(D)$, and

R' is an equivalence relation defined on E such that the equivalence relation R of g is obtained as an extension of R' by additionally setting $e_1 R e_2$ for all 'nonedges' $e_1, e_2 \notin E$.

For a 2-structure $g = (D, R)$ and an edge $e \in E_2(D)$, we let $eR = \{e' \mid eRe'\}$ denote the equivalence class of R containing $e \in E_2(D)$, and we call eR an *edge class* of g. Hence $e_1 R e_2$ if and only if $e_1 R = e_2 R$ for the edges e_1 and e_2 of g.

A 2-structure $g = (D, R)$ can be represented as a complete directed graph with nodes D and (directed) edges $E_2(D)$ together with a colouring of the edges corresponding to the edge classes. Here a *colouring* is a mapping $\alpha\colon E_2(D) \to K$ of the edges into a set K of colours such that $e_1 R e_2$ if and only if $\alpha(e_1) = \alpha(e_2)$. Notice, however, that such a colouring α for a 2-structure g is by no means unique, since the choice of the colours is *arbitrary* in the representation. The function α that maps each edge e to its edge class eR is just one possible colouring of g.

Example 6.1
Let $D = \{x_1, x_2, x_3, x_4\}$ be a domain of four nodes, and let $e_{ij} = (x_i, x_j)$ for $i \neq j$. Define a 2-structure $g = (D, R)$ by its edge classes:

$$\begin{aligned} e_{12}R &= \{e_{12}, e_{21}, e_{13}, e_{23}, e_{24}, e_{42}, e_{41}\}, \\ e_{14}R &= \{e_{14}\}, \\ e_{31}R &= \{e_{31}, e_{32}, e_{34}, e_{43}\}. \end{aligned}$$

Clearly, these edge classes are disjoint, and $E_2(D) = e_{12}R \cup e_{14}R \cup e_{31}R$.

Both of the complete coloured directed graphs below represent g. In these representations the edge classes have received the colours a,b,c and b,a,d, respectively. □

We notice that a 2-structure $g = (D, R)$ can also be considered as a *packing* of the directed graphs $g_{eR} = (D, eR)$ with the edge sets eR for $e \in E_2(D)$. For an edge class $a = eR$ we call g_a a *packed component* of g. See, *e.g.*, Yap [43] for more information on the topic of packing (of trees).

Example 6.2
In Figure 6.2 we have unpacked the 2-structure of Example 6.1. Here the edge classes of g are $(x_1, x_2)R$, $(x_3, x_1)R$ and $(x_1, x_4)R$. □

An edge e of g is said to be *symmetric*, if eRe^{-1} (*i.e.*, $eR = e^{-1}R$), and an edge class a is *symmetric*, if all edges $e \in a$ are symmetric, *i.e.*, $e \in a$ if and

Figure 6.1: Two colourings of a 2-structure

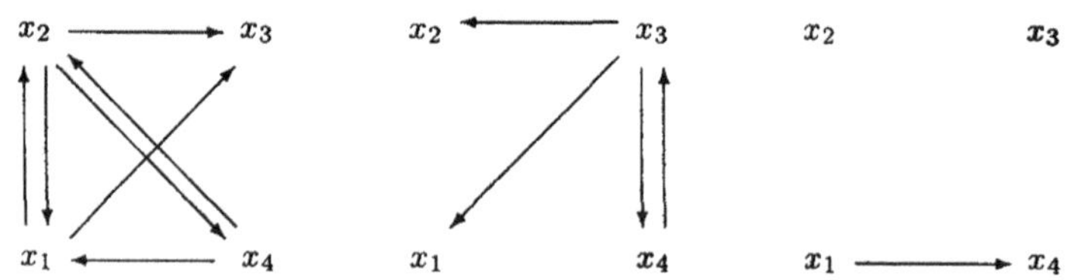

Figure 6.2: The packed components of g

only if $e^{-1} \in a$. If a is a symmetric edge class then the packed component g_a can be thought of as an undirected graph.

In Example 6.1, the edge (x_1, x_2) is symmetric, but the edge class $(x_1, x_2)R$ is not symmetric, because $(x_1, x_2)R = (x_4, x_1)R$ and $(x_1, x_4)R \neq (x_4, x_1)R$. In fact, the 2-structure g from Example 6.1 does not have any symmetric edge classes.

Example 6.3
One can represent various properties of graphs using 2-structures. Let $G = (D, E)$ be an undirected graph. Here E is a subset of the set $S_2(D)$ of 2-element subsets $\{x, y\}$ of D with $x \neq y$.
(1) The graph G can be *represented* as a 2-structure $g(G) = (D, R)$ with two symmetric edge classes by defining $e_1 R e_2$ for all $e_1, e_2 \in E$ and for all $e_1, e_2 \notin E$. Notice, however, that the representing 2-structure $g(G)$ cannot make a distinction between a graph G and the complement graph $(D, S_2(D) \setminus E)$ of G.
(2) We can define a 2-structure $g = (D, R)$ for G by setting $e_1 R e_2$ if and only if the edges $e_1, e_2 \in E_2(D)$ belong to a common cycle of G. (Recall that a cycle in a graph is a sequence $e_1 e_2 \dots e_k$ of edges $e_i = \{x_i, x_{i+1}\}$ for $i = 1, 2, \dots, k$,

where $x_1 = x_{k+1}$ and the nodes $x_1, x_2, \ldots, x_k$ are distinct from each other). Clearly, the so defined relation R is an equivalence relation on $E_2(D)$, and the 2-structure g that represents the cycle structure of G is usually quite different from the representing 2-structure $g(G)$ of G. □

Each 2-structure g with n nodes has $n(n-1)$ edges, and thus *drawing* a larger g is usually inconvenient. However, drawing reversible 2-structures (defined below) is easier.

A 2-structure $g = (D, R)$ is *reversible*, if g satisfies

$$e_1 R e_2 \iff e_1^{-1} R e_2^{-1}$$

for all edges $e_1, e_2 \in E_2(D)$.

Equivalently, g is reversible, if for each edge class a there corresponds a unique edge class a^{-1}, called the *reverse class* of a, such that for each edge e,

$$e \in a \iff e^{-1} \in a^{-1}.$$

Notice that we may have $a = a^{-1}$, in which case the edge class a is symmetric.

We simplify the drawings of reversible 2-structures by
— omitting the reverse edge classes of the chosen edge classes,
— drawing a line segment between x and y, if the edge (x, y) is symmetric, and
— often omitting one symmetric edge class.

Example 6.4
The 2-structure g in Figure 6.3 with $\mathrm{dom}(g) = \{x_1, \ldots, x_8\}$ has two symmetric edge classes c and d. The edges in the class d have not been drawn, *e.g.*, the edges (x_1, x_3) and (x_2, x_7) together with their reverse edges are in the edge class d among quite many other edges (and their reverse edges). The 2-structure g has four nonsymmetric edge classes a, a^{-1}, b and b^{-1}, of which we have drawn the classes a and b only. □

6.2.2 Clans

Let X be a nonempty subset of the domain D of a 2-structure $g = (D, R)$. Then the *substructure* induced by X is defined as the 2-structure

$$\mathrm{sub}_g(X) = (X, R \cap (E_2(X) \times E_2(X))),$$

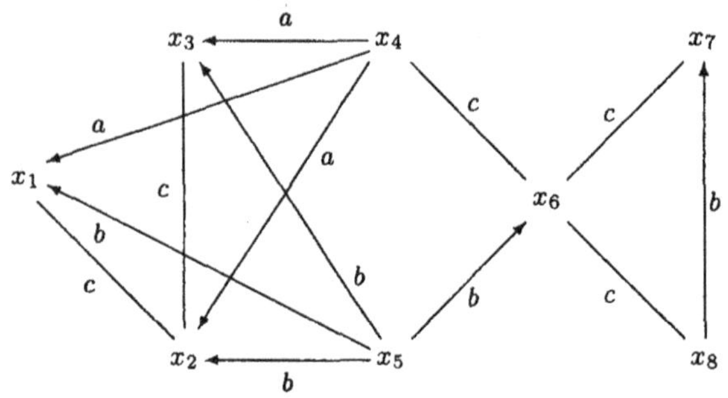

Figure 6.3: A simplified representation of a reversible 2-structure g

which is obtained from g by restricting the relation R to the subset $E_2(X)$ of $E_2(D)$.

The edge classes of $h = \text{sub}_g(X)$ are $eR_h = eR_g \cap E_2(X)$.

The most important technical notion of this paper is that of a clan. We call a subset $X \subseteq \text{dom}(g)$ a *clan*, if every node $y \notin X$ sees the nodes of X in the same way,

$$(y, x_1)R(y, x_2) \quad \text{for all} \quad x_1, x_2 \in X,\ y \notin X,$$

and every two nodes $x_1, x_2 \in X$ see every node $y \notin X$ in the same way,

$$(x_1, y)R(x_2, y) \quad \text{for all} \quad x_1, x_2 \in X,\ y \notin X.$$

The set of all clans of g is denoted by $\mathcal{C}(g)$.

A clan $X \in \mathcal{C}(g)$ is said to be *proper*, if X is a proper subset of the domain of g.

It follows directly from the definition that the empty set $\emptyset$ and the domain $\text{dom}(g)$ of g are clans of g. Also, it is equally clear that the singletons $\{x\}$, $x \in \text{dom}(g)$, are clans. The sets $\emptyset$, $\text{dom}(g)$ and the singletons are the *trivial clans* of g. A 2-structure g that has no other than the trivial clans is called *primitive.*

The substructure $\text{sub}_g(X)$ induced by a nonempty clan $X \in \mathcal{C}(g)$ is called a *factor* of g.

Example 6.5
Consider the 2-structure $g = (D, R)$ of Example 6.4, see Figure 6.3. The two

substructures $\mathrm{sub}_g(X)$ and $\mathrm{sub}_g(Y)$, that have been indicated by rectangles in Figure 6.4, are factors of g. □

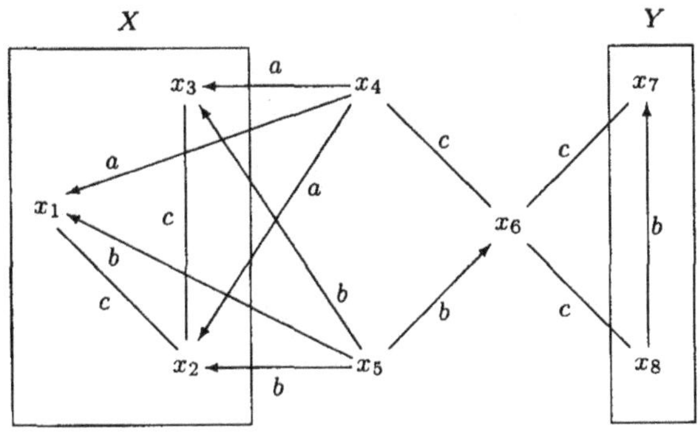

Figure 6.4: Two clans X and Y, and their factors in g

We show now that restricting ourselves to reversible 2-structures does not imply a loss of generality as far as the clans are concerned.

Consider a 2-structure $g = (D, R)$, and define a relation S on $E_2(D)$ by

$$e_1 S e_2 \iff e_1 R e_2 \text{ and } e_1^{-1} R e_2^{-1} \qquad (e_1, e_2 \in E_2(D)).$$

Clearly, S is an equivalence relation. We observe that the new 2-structure $h = (D, S)$ is reversible. Indeed, if $e_1 S e_2$, then by the definition of S, also $e_1^{-1} S e_2^{-1}$. On the other hand, if y, x_1 and $x_2 \in D$ are nodes such that either $(y, x_1)R \neq (y, x_2)R$ or $(x_1, y)R \neq (x_2, y)R$ in g, then by the definition of S, also $(y, x_1)S \neq (y, x_2)S$ in h. The converse holds equally well, and therefore we have the following 'normal form theorem' for 2-structures.

Theorem 6.2.1
For each $g = (D, R)$ there exists a reversible 2-structure $h = (D, S)$ such that for all $X \subseteq D$, $\mathcal{C}(\mathrm{sub}_g(X)) = \mathcal{C}(\mathrm{sub}_h(X))$. In particular, $\mathcal{C}(g) = \mathcal{C}(h)$. □

Reversible 2-structures are simpler to handle than the general 2-structures. As an example, we have the following one-way condition for clans:

Lemma 6.2.2
A subset $X \subseteq \mathrm{dom}(g)$ of a reversible g is a clan if and only if for all $y \notin X$ and $x_1, x_2 \in X$, $(y, x_1)R = (y, x_2)R$. □

We take the advantage of Theorem 6.2.1, and *consider only* reversible *2-structures in the rest of the paper.*

6.2.3 Basic properties of clans

Let $g = (D, R)$ be a (reversible) 2-structure. The following result gives the basic closure properties of clans.

Lemma 6.2.3
Let $X, Y \in \mathcal{C}(g)$. *Then*
(1) $X \cap Y \in \mathcal{C}(g)$;
(2) *if* $X \cap Y \neq \emptyset$, *then* $X \cup Y \in \mathcal{C}(g)$;
(3) *if* $Y \setminus X \neq \emptyset$, *then* $X \setminus Y \in \mathcal{C}(g)$.
Proof
The first two cases of the claim are rather easy to prove, and hence we prove only claim (3). Suppose that $y \in Y \setminus X$. Denote $Z = X \setminus Y$, and let $x \notin Z$, $z_1, z_2 \in Z$. If $x \notin X$, then $(x, z_1)R = (x, z_2)R$, because $z_1, z_2 \in X$ and X is a clan.
If, on the other hand, $x \in X$, then $x \in X \cap Y$. Because Y is a clan and $z_1, z_2 \notin Y$ and $x, y \in Y$, we have $(x, z_1)R = (y, z_1)R$ and $(x, z_2)R = (y, z_2)R$. Moreover, X is a clan with $y \notin X$ and $z_1, z_2 \in X$, and therefore $(y, z_1)R = (y, z_2)R$. Combining these we obtain

$$(x, z_1)R = (y, z_1)R = (y, z_2)R = (x, z_2)R,$$

and therefore $Z = X \setminus Y$ is a clan. □

We say that two subsets X and Y of a domain D *overlap*, if $X \cap Y \neq \emptyset$, $X \setminus Y \neq \emptyset$ and $Y \setminus X \neq \emptyset$, *i.e.*, neither $X \subseteq Y$ nor $Y \subseteq X$ nor $X \cap Y = \emptyset$.
By Lemma 6.2.3, we have

Lemma 6.2.4
If $X, Y \in \mathcal{C}(g)$ *overlap, then* $X \cap Y,\ X \cup Y,\ X \setminus Y \in \mathcal{C}(g)$. □

In fact, the above reasoning can be sharpened to show

Lemma 6.2.5
If the clans X *and* Y *overlap, then the edges in*

$$(X \setminus Y) \times (X \cap Y)\ \cup\ (X \setminus Y) \times (Y \setminus X)\ \cup\ (X \cap Y) \times (Y \setminus X)$$

belong to the same edge class. □

On the other hand, the clans of a 2-structure g need not be closed under complement. Indeed, consider the primitive 2-structure g of four nodes and two symmetric edge classes from Figure 6.5. Since g has only the trivial clans, the complement $\mathrm{dom}(g) \setminus \{x\}$ of the singleton clan $\{x\}$, for $x \in \mathrm{dom}(g)$, is not a clan of g. Also, $\mathcal{C}(g)$ is not necessarily closed under symmetric difference. However, if g has only symmetric edge classes, then $\mathcal{C}(g)$ is also closed under symmetric difference.

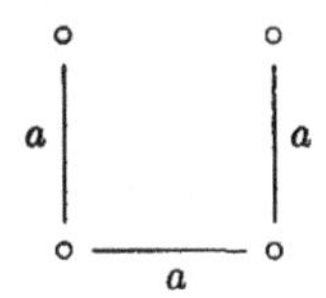

Figure 6.5: A primitive g

The following rather obvious lemma is crucial for the definition of a quotient.

Lemma 6.2.6
If $X, Y \in \mathcal{C}(g)$ and $X \cap Y = \emptyset$, then $(x_1, y_1)R = (x_2, y_2)R$ for all $x_1, x_2 \in X$ and $y_1, y_2 \in Y$. □

By the following result the clans of the factors $\mathrm{sub}_g(X)$ are exactly those clans of g that are contained in X.

Theorem 6.2.7
Let $h = \mathrm{sub}_g(X)$ be a factor of g. Then

$$\mathcal{C}(h) = \{Y \mid Y \subseteq X,\ Y \in \mathcal{C}(g)\}.$$ □

Example 6.6
The 2-structure g from Example 6.4 (see Figure 6.3) has a factor $h = \mathrm{sub}_g(X)$ for $X = \{x_1, x_2, x_3\}$, and $Z = \{x_1, x_3\}$ is a clan of h. Hence Z is also a clan of g. □

6.3 Decompositions of 2-Structures

6.3.1 Prime clans

A nonempty clan $P \in \mathcal{C}(g)$ is a *prime clan*, if it does not overlap with any clan of g. Clearly, the nonempty trivial clans are prime clans.

We denote by $\mathcal{P}(g)$ the set of all prime clans of g. For a prime clan P of g, the factor $\text{sub}_g(P)$ is called a *prime factor* of g.

The next lemma follows immediately from the definition of a prime clan.

Lemma 6.3.1
Let $g = (D, R)$. If P is a prime clan and $X \in \mathcal{C}(g)$ intersects with P, then either $P \subseteq X$ or $X \subseteq P$. □

By Theorem 6.2.7, we may easily deduce that if a prime clan $P \in \mathcal{P}(g)$ is contained in a clan $X \in \mathcal{C}(g)$, then P is a prime clan of the factor $\text{sub}_g(X)$. Also the converse holds for proper prime clans of the factor $\text{sub}_g(X)$ as can been seen using Lemma 6.3.1. Hence we have

Theorem 6.3.2
If $h = \text{sub}_g(X)$ is a factor of g, then

$$\mathcal{P}(h) = \{P \mid P \in \mathcal{P}(g),\ P \subseteq X\} \cup \{X\}.$$ □

6.3.2 Quotients

Let g be a (reversible) 2-structure and let $\mathcal{X} = \{X_1, X_2, \ldots, X_k\}$ be a partition of dom(g) into nonempty clans. We call $\mathcal{X}$ a *factorization* of g, and the substructures $\text{sub}_g(X_i)$ are the *factors* of this factorization.

Each 2-structure g has at least the *trivial factorizations* $\mathcal{X} = \{\text{dom}(g)\}$ and $\mathcal{X} = \{\{x\} \mid x \in \text{dom}(g)\}$.

The *quotient* of a $g = (D, R)$ by a factorization $\mathcal{X}$ is defined to be the 2-structure $g/\mathcal{X}$ with the domain $\mathcal{X}$, and the equivalence relation $R_{\mathcal{X}}$, for which

$$(X_1, Y_1) R_{\mathcal{X}} (X_2, Y_2) \iff (x_1, y_1) R (x_2, y_2) \text{ for some } x_i \in X_i,\ y_i \in Y_i,$$

where each $X_i, Y_i \in \mathcal{X}$. Thus a quotient $g/\mathcal{X}$ is obtained from g by contracting each clan X in the partition $\mathcal{X}$ into a single node, and then inheriting the edge classes from g.

By Lemma 6.2.6, the quotient $g/\mathcal{X}$ is well-defined, *i.e.*, it is independent of the choice of the representatives: if $(x_1, y_1) R (x_2, y_2)$ for *some* $x_1 \in X_1$, $x_2 \in X_2$, and $y_1 \in Y_1$, $y_2 \in Y_2$, then $(x_1, y_1) R (x_2, y_2)$ for *all* $x_1 \in X_1$, $x_2 \in X_2$ and $y_1 \in Y_1$, $y_2 \in Y_2$ (where X_i, Y_i are in $\mathcal{X}$).

Example 6.7
The 2-structure g from Example 6.4 (see Figures 6.3 and 6.4) has a factorization

$$\mathcal{X} = \{X, Y, \{x_4\}, \{x_5\}, \{x_6\}\}.$$

We can redraw g with respect to $\mathcal{X}$ as in Figure 6.6. The quotient $g/\mathcal{X}$ is now obtained by contracting the factors into nodes as in Figure 6.7. □

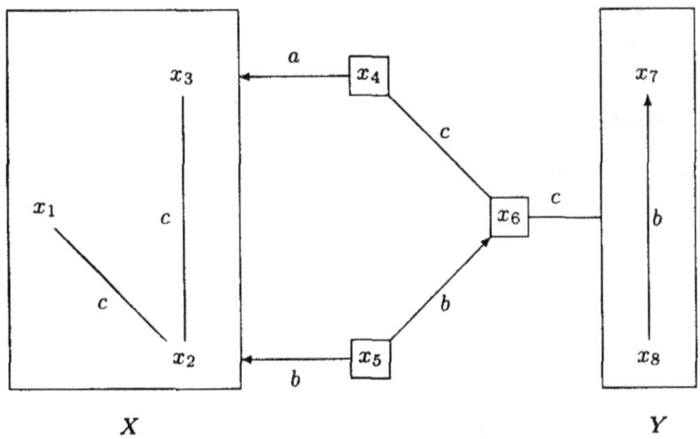

Figure 6.6: Disjoint factors of g

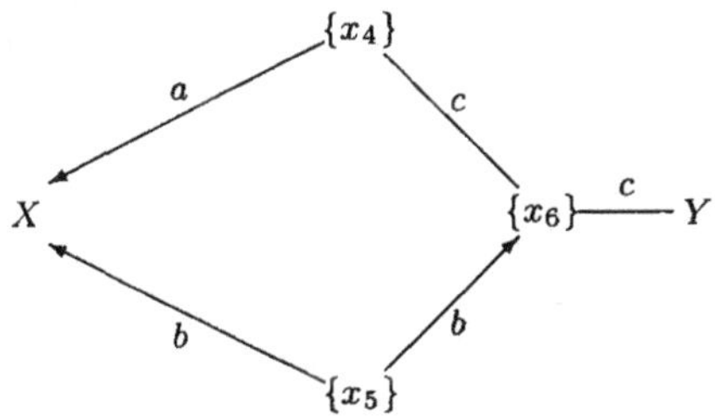

Figure 6.7: $g/\mathcal{X}$

A *decomposition*

$$g/\mathcal{X}(\mathrm{sub}_g(X_1), \ldots, \mathrm{sub}_g(X_k))$$

of a 2-structure g consists of the factors $\mathrm{sub}_g(X_i)$ with respect to a factorization $\mathcal{X} = \{X_1, \ldots, X_k\}$ together with the quotient $g/\mathcal{X}$ that gives the relationships between the factors.

In the general case of 2-structures (that we have been studying now in contrast to the *labeled* 2-structures of Section 6.6) we may lose some information in the formation of a decomposition. Indeed, as the next example shows, two different 2-structures may possess a common decomposition.

Example 6.8
The 2-structures g_1 and g_2 of Figure 6.8 have a common factorization $\mathcal{X} = \{\{x_1, x_2\}, \{x_3, x_4\}\}$. Moreover,

$$\text{sub}_{g_1}(\{x_1, x_2\}) = \text{sub}_{g_2}(\{x_1, x_2\}) \quad \text{and} \quad \text{sub}_{g_1}(\{x_3, x_4\}) = \text{sub}_{g_2}(\{x_3, x_4\}).$$

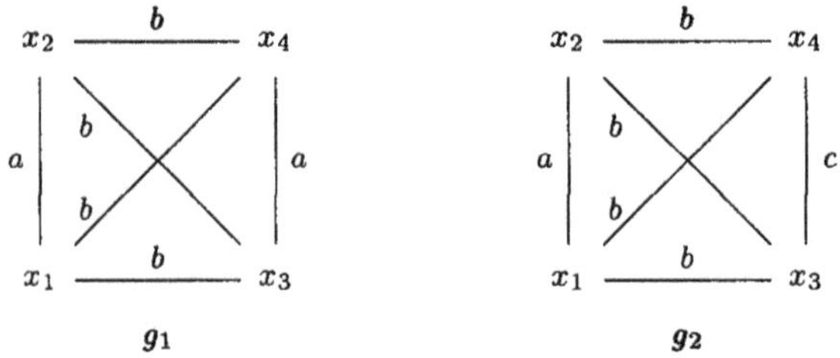

Figure 6.8: g_1 and g_2

Also, the quotients $g_1/\mathcal{X}$ and $g_2/\mathcal{X}$ are equal as shown in Figure 6.9. However, g_1 and g_2 are different 2-structures, since $(x_1, x_2)R_{g_1} = (x_3, x_4)R_{g_1}$, but $(x_1, x_2)R_{g_2} \neq (x_3, x_4)R_{g_2}$. □

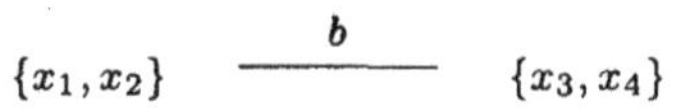

Figure 6.9: A common factorization $\mathcal{X}$

Next we shall study the clans of the quotients $g/\mathcal{X}$ for factorizations $\mathcal{X} \subseteq \mathcal{C}(g)$. For this we adopt the following notations. If $\mathcal{A}$ is a family of sets, then let

$$\cup\mathcal{A} = \bigcup_{A \in \mathcal{A}} A, \quad \text{and} \quad \cap\mathcal{A} = \bigcap_{A \in \mathcal{A}} A.$$

Let $g = (D, R)$ and let $\mathcal{X}$ be a factorization of g. For each subset $Y \subseteq D$, we let

$$K_{\mathcal{X}}(Y) = \{X \mid X \in \mathcal{X}, X \cap Y \neq \emptyset\}$$

be the *frame* of Y in $g/\mathcal{X}$; the situation is illustrated in Figure 6.12.

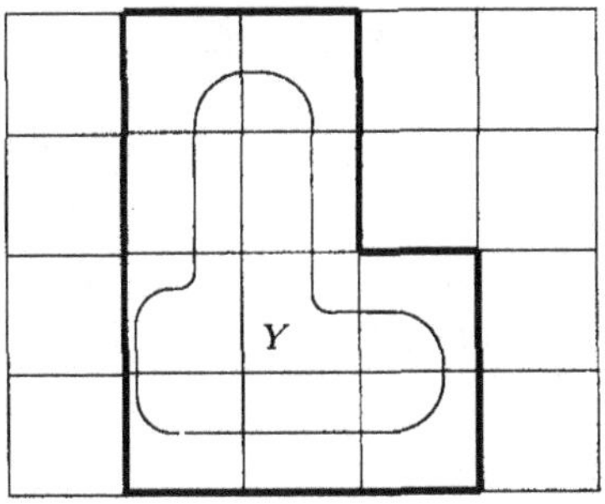

Figure 6.10: The frame of Y in $g/\mathcal{X}$

We begin with an immediate result.

Lemma 6.3.3
Let $g = (D, R)$ and let $\mathcal{X}$ be a factorization of g. If $Y \in \mathcal{C}(g)$, then $K_\mathcal{X}(Y) \in \mathcal{C}(g/\mathcal{X})$ and $\cup K_\mathcal{X}(Y) \in \mathcal{C}(g)$. □

It can happen that for a prime clan $\mathcal{Z}$ of a quotient $g/\mathcal{X}$, the 'inverse image' $\cup\mathcal{Z}$ is not a prime clan of g. Indeed, consider a factorization $\mathcal{X}$ and a clan $X \in \mathcal{X}$. Now, X is a singleton clan of the quotient $g/\mathcal{X}$, and hence a prime clan in this quotient, but $\cup\{X\} \in \mathcal{P}(g)$ only if $X \in \mathcal{P}(g)$. The following lemma states that the singletons are the only examples of such a situation.

Theorem 6.3.4
Let $g = (D, R)$ and let $\mathcal{X}$ be a factorization of g.
(1) *If $\mathcal{Z} \in \mathcal{C}(g/\mathcal{X})$, then $\cup Z \in \mathcal{C}(g)$.*
(2) *If $\mathcal{Z} \in \mathcal{P}(g/\mathcal{X})$ is not a singleton, then $\cup Z \in \mathcal{P}(g)$.*
(3) *If $\mathcal{X} \subseteq \mathcal{P}(g)$, then for each prime clan $\mathcal{Z} \in \mathcal{P}(g/\mathcal{X})$, also $\cup\mathcal{Z} \in \mathcal{P}(g)$.*

Proof
Case (1) follows easily from the definition of clan, and Case (3) is a consequence of (2) and the discussion above. Hence we need to prove only (2).

Suppose $\mathcal{Z}$ is a proper prime clan of the quotient $g/\mathcal{X}$ (*i.e.*, $\mathcal{Z} \neq \mathcal{X}$). By (1), $\cup\mathcal{Z} \in \mathcal{C}(g)$. Assume now that there exists a clan $Y \in \mathcal{C}(g)$ that overlaps with $\cup\mathcal{Z}$. By Lemma 6.3.3, the frame $K_\mathcal{X}(Y)$ is a clan of $g/\mathcal{X}$, and $\cup K_\mathcal{X}(Y) \in \mathcal{C}(g)$.

Since $\mathcal{Z}$ is a prime clan in the quotient, $K_{\mathcal{X}}(Y)$ does not overlap with it. But clearly,

$$K_{\mathcal{X}}(Y) \cap \mathcal{Z} \neq \emptyset \qquad \text{and} \qquad K_{\mathcal{X}}(Y) \setminus \mathcal{Z} \neq \emptyset,$$

since $Y \cap \cup\mathcal{Z} \neq \emptyset$ and $Y \setminus \cup\mathcal{Z} \neq \emptyset$. Therefore $\mathcal{Z} \subset K_{\mathcal{X}}(Y)$, where the inclusion is proper.

By above, for each $z \in \cup\mathcal{Z}$ there exists an $X \in \mathcal{Z}$ such that $X \cap Y \neq \emptyset$ and $z \in X$.

Let $z \in \cup\mathcal{Z} \setminus Y$. Such a z exists, because $\cup\mathcal{Z}$ and Y overlap. Let $X \in \mathcal{Z}$ be such that $X \cap Y \neq \emptyset$ and $z \in X \setminus Y$. The situation is now as in Figure 6.11.

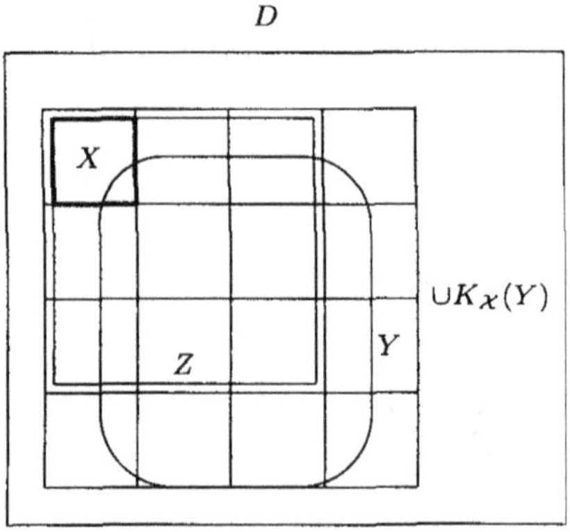

Figure 6.11: X and Y overlap

It follows that X and Y overlap with each other, and, in particular, $Y \setminus X$ is a clan of g by Lemma 6.2.4. By Lemma 6.3.3, $K_{\mathcal{X}}(Y \setminus X) \in \mathcal{C}(g/\mathcal{X})$. However, $K_{\mathcal{X}}(Y \setminus X) = K_{\mathcal{X}}(Y) \setminus \{X\}$, since $X \in \mathcal{X}$. Now, $K_{\mathcal{X}}(Y \setminus X)$ overlaps with $\mathcal{Z}$ unless $\mathcal{Z} = \{X\}$ is a singleton clan of $g/\mathcal{X}$. This proves the claim. □

6.3.3 *Maximal prime clans*

Let $g = (D, R)$. A clan $M \in \mathcal{C}(g)$ is called *maximal*, if it is a proper clan and for all proper clans X, $M \subseteq X$ implies that $X = M$.

A prime clan $P \in \mathcal{P}(g)$ is a *maximal prime clan*, if it is a proper clan, and maximal among the proper prime clans of g: for all proper prime clans Q, $P \subseteq Q$ implies that $Q = P$.

We denote by $\mathcal{P}_{\max}(g)$ the set of all maximal prime clans of g. For a maximal prime clan $P \in \mathcal{P}_{\max}(g)$, the factor $\mathrm{sub}_g(P)$ is called a *maximal prime factor* of g.

We observe that all 2-structures $g = (D, R)$ with $|D| \geq 2$ have maximal prime clans, because every singleton is a prime clan. By definition of maximality, this is not true in the trivial case, where $|D| = 1$. For this reason we set $\mathcal{P}_{\max}(g) = \{D\}$, if $|D| = 1$.

Lemma 6.3.5
Let $g = (D, R)$. Every $x \in D$ belongs to a unique maximal prime clan P_x of g.

Proof
Suppose $|D| \geq 2$. For each $x \in D$ there is a proper prime clan containing x, namely $\{x\}$. Let P_x be then a proper prime clan so that P_x is maximal with respect to the condition $x \in P_x$. Such a P_x exists because the domain D is finite. Now, if $P \neq P_x$ is a proper prime clan with $x \in P$, then $P \cap P_x \neq \emptyset$, and by the maximality of P_x, $P_x \subset P$ cannot hold. Hence since P is a prime clan, $P \subset P_x$, and therefore P_x is the unique maximal prime clan containing x. □

Theorem 6.3.6
Let g be a 2-structure. The maximal prime clans form a partition $\mathcal{P}_{\max}(g)$ of dom(g), and therefore the quotient $g/\mathcal{P}_{\max}(g)$ is well-defined.

Proof
By Lemma 6.3.5, each node x belongs to a unique maximal prime clan P_x, and thus dom(g) is a disjoint union of the maximal prime clans, because for each $x, y \in \text{dom}(g)$ either $P_x = P_y$ or $P_x \cap P_y = \emptyset$. □

Lemma 6.3.7
For each clan $Y \in \mathcal{C}(g)$ either $Y \subseteq P$ for a maximal prime clan $P \in \mathcal{P}_{\max}(g)$ or Y is a union of maximal prime clans.

Proof
Assume Y is not a subset of any maximal prime clan of g. If $Y \cap P \neq \emptyset$ for some $P \in \mathcal{P}_{\max}(g)$, then Lemma 6.3.1 implies that $P \subset Y$. Since every $y \in Y$ belongs to a unique maximal prime clan P_y and $P_y \cap Y \neq \emptyset$, it follows that

$$Y = \bigcup_{y \in Y} P_y,$$

which proves the claim. □

Since each singleton is a prime clan, it follows that every clan $Y \in \mathcal{C}(g)$ is the union of prime clans $P \in \mathcal{P}(g)$ that are contained in Y. By Theorem 6.3.2, the prime clans P of the factor $\text{sub}_g(Y)$ are exactly the prime clans of g that are contained in Y. Therefore we have the following result.

Theorem 6.3.8
Each nonsingleton clan $Y \in \mathcal{C}(g)$ is the union of the prime clans $P \in \mathcal{P}(g)$ that are the maximal prime clans of the factor $\text{sub}_g(Y)$. □

6.3.4 Special 2-structures

Recall that a 2-structure g is primitive if it has only the trivial clans. We generalize the notion of primitivity by defining a 2-structure g to be *special*, if all its prime clans are trivial. Hence if g is special, then $\mathcal{P}_{\max}(g)$ consists of the singletons of the domain.

The next result from [16] is crucial for the clan decomposition theorem.

Theorem 6.3.9
The quotient $g/\mathcal{P}_{\max}(g)$ is special.

Proof
Assume that $g/\mathcal{P}_{\max}(g)$ has a prime clan $\mathcal{X} = \{P_1, \ldots, P_k\}$ for $k \geq 2$, where each P_i is a maximal prime clan of g. By Theorem 6.3.4(2),

$$X = \bigcup_{i=1}^{k} P_i \in \mathcal{P}(g).$$

Hence, because each $P_i \subset X$ $(i = 1, 2, \ldots, k)$ is a maximal prime clan of g, we must have that $X = \text{dom}(g)$. Thus $\mathcal{X}$ is the domain of the quotient $g/\mathcal{X}$, and hence not a proper prime clan. □

Lemma 6.3.10
Let g be a special 2-structure. Then either
(1) the maximal clans M are all singletons, or
(2) all the complements $\text{dom}(g) \setminus M$ of the maximal c[illegible] M are singletons.

Proof
Suppose $M \in \mathcal{C}(g)$ is a maximal clan with at least two nodes. Since in a special g all prime clans are trivial, M is not a prime clan, and hence there exists a clan X that overlaps with M. By Lemma 6.2.4, $M \cup X \in \mathcal{C}(g)$ and $X \setminus M \in \mathcal{C}(g)$.

Now, since M is maximal and $M \subset M \cup X$, it follows that $M \cup X = \text{dom}(g)$, and hence, in particular, $\text{dom}(g) \setminus M = X \setminus M$. By above, the complement of M, $\overline{M} = \text{dom}(g) \setminus M$, is a clan. If $\overline{M}$ is a prime clan, then it must be a singleton because g is special. On the other hand, if $\overline{M}$ overlaps with another clan Y, then Y overlaps also with M by maximality of M, and in this case $M \cup Y \in \mathcal{C}(g)$ would be a proper clan contradicting the maximality of M. We conclude that $\overline{M}$ is a singleton clan.

Finally, if N is any other maximal clan, then the above argument shows that either its complement $\overline{N}$ is a singleton or N itself is a singleton. The latter condition cannot hold, because $|N| = 1$ would imply $N = \overline{M}$ or $N \subseteq M$. The inclusion $N \subseteq M$ and the assumption $N \neq M$ imply $N \subset M$; this contradicts the maximality of N. Hence $N = \overline{M}$ holds. Since X overlaps with M, we get $N \subset X$ contradicting the maximality of N. □

6.3.5 *The clan decomposition theorem*

We shall now give a new proof of the clan decomposition theorem of [16] using maximal clans.

A 2-structure $g = (D, R)$ is said to be *complete*, if it has only one edge class, *i.e.*, for all $e_1, e_2 \in E_2(D)$, $e_1 R e_2$. In particular, the only edge class is symmetric, *i.e.*, eRe^{-1} for all $e \in E_2(D)$. Clearly, in a complete 2-structure g each subset of the domain is a clan. It follows from this that a complete 2-structure is special.

A 2-structure g is said to be *linear*, if g has a nonsymmetric edge class a which linearly orders the domain D of g, *i.e.*, D has an ordering $x_1, x_2, \ldots, x_n$ such that $(x_i, x_j) \in a$ if and only if $i < j$. In this case the nonempty clans of g are exactly the segments $\{x_i, x_{i+1}, \ldots, x_{i+k}\}$ for $1 \leq i \leq i + k \leq n$. Consequently, a linear 2-structure is special.

Recall also that a g is *primitive*, if it has no nontrivial clans. Notice that a 2-structure g with at most two nodes is always primitive and at the same time either complete or linear. For this reason we say that g is *truly primitive*, if it is primitive and it has at least three nodes.

Figure 6.12: A complete and a linear 2-structure on four nodes

As stated in the following theorem (from [16]) there are only three types of special 2-structures.

Theorem 6.3.11
A 2-structure g is special if and only if it is either linear, or complete, or truly primitive.

Proof
It is easy to show that if g is complete, linear or primitive, then its prime clans are trivial.

Suppose then that $g = (D, R)$ is special. We show by induction on the number $|D|$ of nodes in the domain that g is of one of the types as claimed. The claim clearly holds if $|D| \leq 3$ (by simple case analysis).

Assume that $|D| > 3$, and let M be a maximal clan of g. If $|M| = 1$, then by Lemma 6.3.10 it follows that g is primitive, since now the maximal clans are singletons, and so all the proper clans are singletons.

We may thus assume that $|M| > 1$, and hence, again by Lemma 6.3.10, that the complement $\overline{M}$ of M is a singleton. Let $D \setminus M = \{v\}$. There exists a maximal clan N that contains v, and, as above, $|N| > 1$ and the complement $\overline{N}$ is a singleton, say $D \setminus N = \{w\}$, see Figure 6.13.

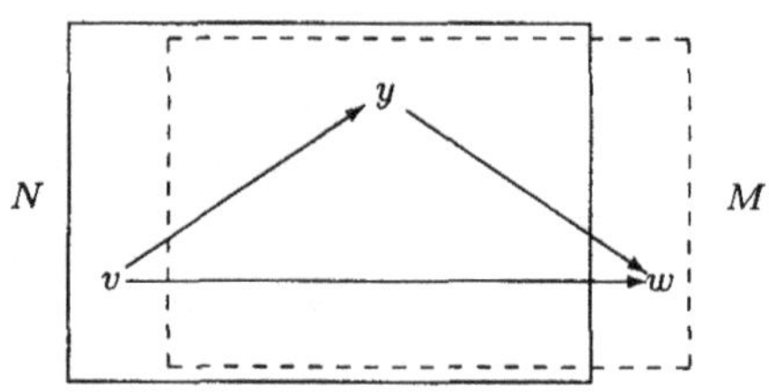

Figure 6.13: Maximal M and N

Consider the maximal factor $\text{sub}_g(M)$. If P is a proper prime clan of $\text{sub}_g(M)$, then P is also a proper prime clan of g by Theorem 6.3.2, and hence P is necessarily a singleton. Therefore $\text{sub}_g(M)$ is special, and by the induction hypothesis $\text{sub}_g(M)$ is complete, linear or primitive.

The factor $\text{sub}_g(M)$ cannot be primitive, since it has the nontrivial clan $N \cap M$ by Theorem 6.2.7 and by the fact that $|D| > 3$.

Assume that $\text{sub}_g(M)$ is linear or complete. Since $M \in \mathcal{C}(g)$ and $v \notin M$,

$$(v, y)R = (v, w)R \quad \text{for all} \quad y \in M.$$

Also, since $N \in \mathcal{C}(g)$ and $w \notin N$,

$$(v, w)R = (y, w)R \quad \text{for all} \quad y \in N.$$

From these it follows that $(v, y)R = (y, w)R$ for all $y \in M \cap N$, and thus g is linear if $\text{sub}_g(M)$ is linear, and g is complete if $\text{sub}_g(M)$ is complete. □

We have thus proved the following *clan decomposition theorem* of [16].

Theorem 6.3.12 (Clan decomposition theorem)
For any 2-structure g, the quotient $g/\mathcal{P}_{\max}(g)$ is either linear, or complete, or truly primitive. □

There are also other proofs of the clan decomposition theorem, see Ehrenfeucht and Rozenberg [16], Engelfriet, Harju, Proskurowski and Rozenberg [23] and Harju and Rozenberg [30]. In a general setting the decomposition theorem is proved also in Möhring and Radermacher [36] and Möhring [35]. The decomposition theorem for submodular functions as presented by Cunningham [5] is closely related to the clan decomposition theorem. In [36] and [30] the proofs are given for infinite domains, for which the theorem has a minor modification due to the fact that maximal prime clans may not exist, if the domain of a 2-structure is infinite.

For a 2-structure g with the maximal prime clans $P_1, \ldots, P_m$, the *clan decomposition* $g/\mathcal{P}_{\max}(g)(\text{sub}_g(P_1), \ldots, \text{sub}_g(P_m))$ of g consists of the maximal prime factors $\text{sub}_g(P_i)$ together with the special quotient $g/\mathcal{P}_{\max}(g)$.

Example 6.9
Consider the 2-structure g of Figure 6.14 with five vertices and five edge classes, a, a^{-1}, b, b^{-1}, c. In Figure 6.14 we have also extracted the maximal prime factors, and the linear quotient $g/\mathcal{P}_{\max}(g)$. □

6.3.6 The shape of a 2-structure

By the clan decomposition theorem, each 2-structure g can be decomposed into its maximal prime factors so that the resulting quotient is special. Also, the maximal prime factors can be decomposed, if they are not special already. This allows us to decompose g completely into its prime factors, not only into its maximal prime factors. Indeed, by Theorem 6.3.2, we have the following immediate result.

Lemma 6.3.13
(1) *Let P be a nonsingleton prime clan of g. The maximal prime clans of the factor $\text{sub}_g(P)$ are exactly the prime clans of g that are properly contained in P, and are maximal in this respect.*

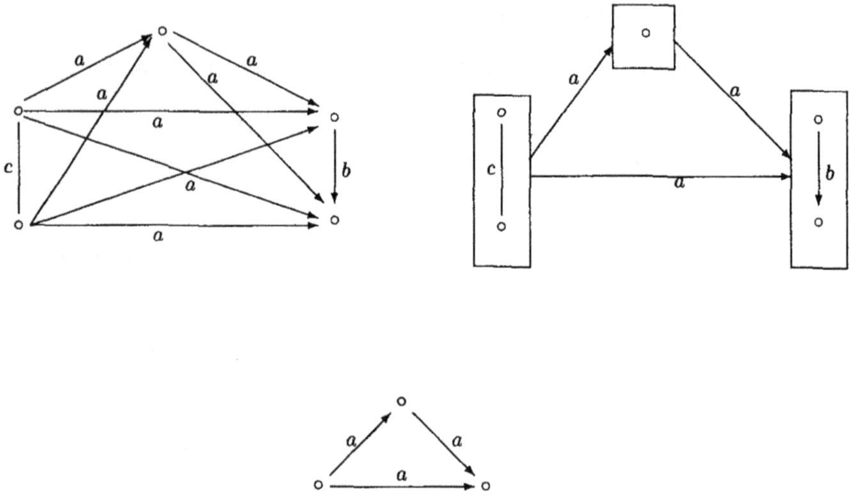

Figure 6.14: A clan decomposition

(2) *Every proper prime clan X of g is a maximal prime clan of some prime factor $sub_g(P)$.* □

The partially ordered set $T(g) = (\mathcal{P}(g), \subseteq)$ of the prime clans forms a rooted tree, where the domain $\text{dom}(g)$ is the root and the singletons are the leaves. We call the tree $T(g)$ the *prime tree* of g.

The *shape* of a 2-structure g is defined as a pair

$$\text{shape}(g) = (T(g), \Psi_g),$$

where $T(g)$ is the prime tree of g and

$$\Psi_g(P) = \text{sub}_g(P)/\mathcal{P}_{\max}(\text{sub}_g(P))$$

is a function, which associates the special quotient $\text{sub}_g(P)/\mathcal{P}_{\max}(\text{sub}_g(P))$ with each prime clan $P \in \mathcal{P}(g)$.

The shape of a 2-structure g can conveniently be represented as a tree-like object by contracting each prime factor $\text{sub}_g(P)$ to the corresponding quotient $\Psi_g(P)$, and then drawing a line down from a node Q of $\Psi_g(P)$ to the corresponding quotient $\Psi_g(Q)$ of Q.

Example 6.10
The shape of the 2-structure g from Figure 6.16 is given in Figure 6.15. We have not identified the inner nodes (circles) of the quotients (rectangles). Indeed, each node $Q \in \Psi_g(P)$ is the prime clan consisting of the singletons x_i that are descendants of this node. Also, we have omitted the colours from this figure, since (in this example) each quotient $\Psi_g(P)$ has only one pair of edges, and hence it has either one symmetric edge class or two edge classes that are reverses of each other. □

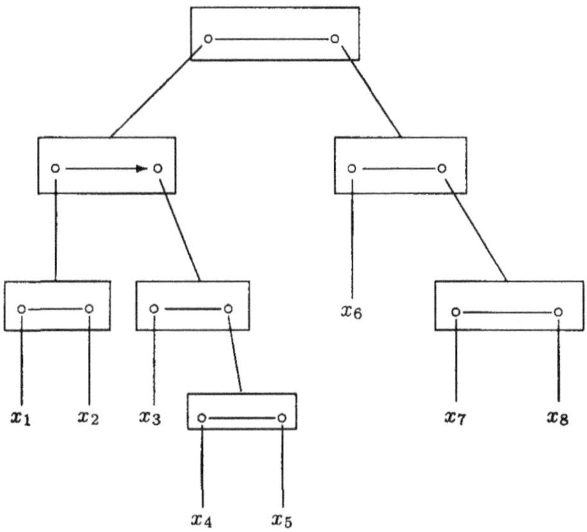

Figure 6.15: The shape of g

The observation of Example 6.8 is valid also for the shapes. We may lose some information on the 2-structure g while forming the shape of g, *i.e.*, two different 2-structures may have the same shape.

We can also draw the shape of a 2-structure g by nesting the special quotients $\Psi_g(P)$: In this representation a quotient (rectangle) $\Psi_g(Q)$ is drawn inside of quotient $\Psi_g(P)$ replacing the node Q of $\Psi_g(P)$.

Example 6.11
In Figure 6.17 there is a nested representation of the shape of the 2-structure g from Figure 6.16. The result is again rather economical compared to the full drawing of g. □

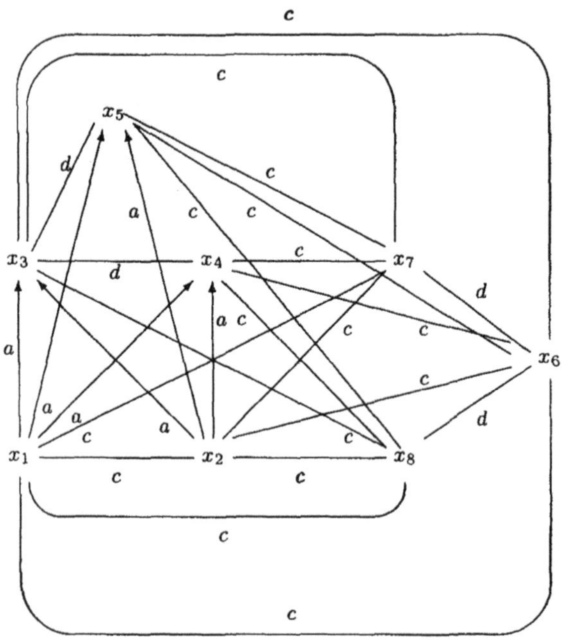

Figure 6.16: A 2-structure g

6.3.7 Constructions of clans and prime clans

The algorithmic complexity of forming the shape of a 2-structure was shown to be of order $O(n^2)$ in Ehrenfeucht, Gabow, McConnell and Sullivan [8]. Here n refers to the number of nodes in the 2-structures.

The algorithm given in [8] is not incremental. In an *incremental algorithm* the shape of a 2-structure g with a domain $\{x_1, x_2, \ldots, x_n\}$ is constructed starting from the shape of a single node substructure $\text{sub}_g(\{x_1\})$ and then for each $i = 1, 2, \ldots, n-1$ constructing the shape of the substructure $\text{sub}_g(\{x_1, x_2, \ldots, x_{i+1}\})$ by updating the shape of the substructure $\text{sub}_g(\{x_1, x_2, \ldots, x_i\})$.

An incremental algorithm of complexity $O(n^2)$ for the construction of the shape of a 2-structure is given in McConnell [34]. When restricted to undi-

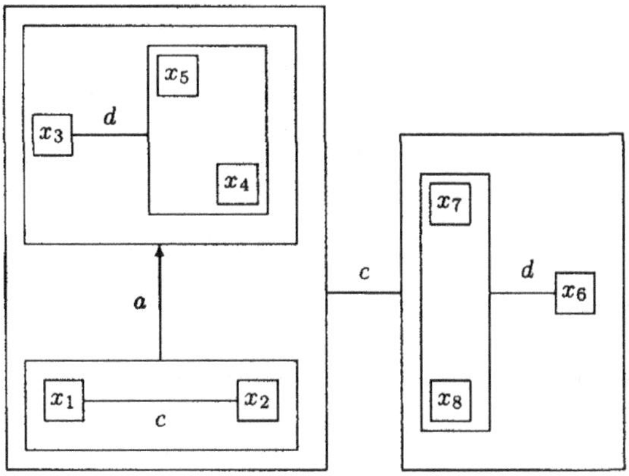

Figure 6.17: The nested representation of the shape of g

rected graphs this algorithm gives another method of constructing the modular decomposition of graphs in time $O(n^2)$, see also Muller and Spinrad [37].

We shall now give some constructive characterizations of clans and prime clans. These results were used in Engelfriet, Harju, Proskurowski and Rozenberg [23] in algorithmic constructions of shapes.

Let $g = (D, R)$ be a 2-structure, and let $X \subseteq D$. We shall denote by $C(X)$ the smallest clan of g that contains the subset X. Such a clan exists since $\mathcal{C}(g)$ is closed under intersection.

For each $x \in D$, define a relation R_x on D as follows,

$$R_x = \{(u, v) \mid (v, u)R \neq (v, x)R\}.$$

In words, if uR_xv, then v 'sees' u and x differently. A subset $X \subseteq D$ is said to be *closed* under R_x, if $u \in X$ and uR_xv imply $v \in X$. Further, let R_x^* be the reflexive and transitive closure of the relation R_x.

Lemma 6.3.14
The following statements are equivalent for each nonempty subset $X \subseteq dom(g)$:

(1) *X is a clan of g,*
(2) *X is closed under R_x for all $x \in X$,*
(3) *X is closed under R_x for some $x \in X$.*

Proof
If $X \in \mathcal{C}(g)$, then clearly X is closed under R_x for all $x \in X$. Hence it suffices to show that (3) implies (1).

Assume that (3) holds, and let $x \in X$ be a node such that X is closed under R_x. Suppose to the contrary of the claim that X is not a clan. Hence there are nodes $u \notin X$ and $y, z \in X$ such that $(u,y)R \neq (u,z)R$. It follows that either $(u,y)R \neq (u,x)R$ or $(u,z)R \neq (u,x)R$. Hence yR_xu or zR_xu, which implies that $u \in X$; a contradiction. □

The following characterization of $C(X)$ was proved in [23].

Theorem 6.3.15
For each subset $X \subseteq dom(g)$ and node $x \in X$,

$$C(X) = \{u \mid yR_x^*u \text{ for some } y \in X\}.$$

Proof
Denote $W(X) = \{u|yR_x^*u \text{ for some } y \in X\}$. First of all, $X \subseteq W(X)$, because R_x^* is reflexive. Moreover, by definition, $W(X)$ is the smallest set containing X and closed under R_x, and hence, $W(X) = C(X)$ by Lemma 6.3.14. □

Let $x, y \in \text{dom}(g)$. We write simply $C(x,y)$ instead of $C(\{x,y\})$. Not every clan is of the form $C(x,y)$, but as the next lemma shows a clan is a prime clan just in case it does not overlap with a clan of this form.

Lemma 6.3.16
Let $X \in \mathcal{C}(g)$ for a 2-structure g. Then $X \in \mathcal{P}(g)$ if and only if X does not overlap with any $C(x,y)$, where $x \in X$ and $y \notin X$.

Proof
In the one direction the claim follows by the definition of a prime clan. On the other hand, if a clan X is not a prime clan, then it overlaps with a clan $Y \in \mathcal{C}(g)$. Let $x \in X \cap Y$ and $y \in Y \setminus X$. Now, $C(x,y) \subseteq Y$, and hence $C(x,y)$ is a clan that overlaps with X. This proves the claim. □

Denote by $P(X)$ the smallest prime clan containing $X \subseteq \text{dom}(g)$. Such a prime clan exists since the prime clans do not overlap, and $\text{dom}(g) \in \mathcal{P}(g)$. Define for each $x \in \text{dom}(g)$,

$$S_x = \{(u,v) \mid uR_xv \text{ or } u \notin C(x,v) \text{ or } x \notin C(u,v)\}.$$

Notice that the requirement $u \notin C(x,v)$ or $x \notin C(u,v)$ is equivalent with $C(x,v) \neq C(u,v)$. Also, denote by S_x^* the reflexive and transitive closure of S_x. Again, we write $P(x,y)$ instead of $P(\{x,y\})$.

In [23] the sets $P(X)$ were characterized as follows.

Theorem 6.3.17
For each subset $X \subseteq dom(g)$ and node $x \in X$,

$$P(X) = \{u \mid yS_x^*u \text{ for some } y \in X\}.$$

Proof
Let $W(X) = \{u \mid yS_x^*u \text{ for some } y \in X\}$. Clearly, $X \subseteq W(X)$. Moreover, by Lemma 6.3.14, $W(X)$ is a clan, since it is closed under the relation R_x.

We prove that $W(X)$ is a prime clan. For this let $u \notin W(X)$. Hence, for all $v \in W(X)$, vS_xu does not hold. By definition of S_x we have that $v \in C(x,u)$ and $x \in C(v,u)$. Hence $C(x,u) = C(v,u)$ for all $v \in W(X)$. Consequently, $W(X) \subseteq C(v,u)$ for all $u \notin W(X)$ and $v \in W(X)$, which proves that $W(X)$ is a prime clan by Lemma 6.3.16. Thus $P(X) \subseteq W(X)$.

Suppose then that $u \notin P(X)$, and let $v \in P(X)$. Clearly, vS_xu does not hold since $P(X)$ is a prime clan, and hence $u \notin W(X)$, which proves that $W(X) \subseteq P(X)$. □

The following theorem (from [23]) shows that every prime clan is of the form $P(x,y)$.

Theorem 6.3.18
Let g be a 2-structure. For each prime clan $P \in \mathcal{P}(g)$ there exist $x, y \in P$ such that $P = P(x,y)$.

Proof
If $P = \{x\}$ is a singleton prime clan, then clearly $P = P(x,x)$. Suppose then that $|P| \geq 2$, and let P_1 and P_2 be two maximal prime clans of $\text{sub}_g(P)$ with $P_1 \neq P_2$. Let $x \in P_1$ and $y \in P_2$. Now, $P = P(x,y)$ as required. □

6.3.8 *Clans and sibas*

We shall now describe how the clan decomposition theorem can be applied to systems which possess clan-like structures in the sense defined below.

Let $\mathcal{S}$ be a family of subsets of a finite set D. We say that $\mathcal{S}$ is a *siba* on D (or a *semi-independent partial boolean algebra*, see [15]), if it satisfies the following conditions:

(1) $\emptyset \in \mathcal{S}$, $D \in \mathcal{S}$ and $\{x\} \in \mathcal{S}$ for all $x \in D$;
(2) if $X, Y \in \mathcal{S}$, then $X \cap Y \in \mathcal{S}$;
(3) if $X, Y \in \mathcal{S}$ and $X \cap Y \neq \emptyset$, then $X \cup Y \in \mathcal{S}$;
(4) if $X, Y \in \mathcal{S}$ and $Y \setminus X \neq \emptyset$, then $X \setminus Y \in \mathcal{S}$.

The sets in (1) are called the *trivial elements* of $\mathcal{S}$.

Let $\mathcal{S}$ be a siba on D. We say that an element $X \in \mathcal{S}$ is a *prime element*, if X does not overlap with any element of $\mathcal{S}$, and that $X \in \mathcal{S}$ is a *maximal* (*prime*) *element*, if it is maximal among the proper (prime) elements of $\mathcal{S}$ with respect to inclusion.

Clearly, the maximal prime elements of $\mathcal{S}$ form a partition of the set D. Further, as in Lemma 6.3.7, we obtain

Lemma 6.3.19
Let $\mathcal{S}$ be a siba on D. For each $Y \in \mathcal{S}$ either $Y \subseteq P$ for a maximal prime element P of $\mathcal{S}$ or Y is a union of maximal prime elements. □

A siba $\mathcal{S}$ is said to be *trivial*, if its elements are trivial, and $\mathcal{S}$ is *special*, if its prime elements are trivial. Further, $\mathcal{S}$ is *complete*, if it is the power set 2^D, and *linear*, if there exists a linear order $\{x_1, x_2, \ldots, x_n\}$ of D such that $\mathcal{S}$ consists of the emptyset and the segments $\{x_i, \ldots, x_j\}$ for $1 \leq i \leq j \leq n$.

We notice that the proof of Lemma 6.3.10 applies as such to sibas, and the proof of Theorem 6.3.11 can be rather easily modified to prove the following result, see also Möhring [35] and Deutz [7].

Lemma 6.3.20
A siba is special if and only if it is either linear, or complete, or trivial. □

Let $\mathcal{S}$ be a siba on D. We denote

$$\mathcal{S}|X = \{X \cap Y \mid Y \in \mathcal{S}\}$$

for each $X \in \mathcal{S}$. It is easy to show that $\mathcal{S}|X$ is a siba on X.

If $\mathcal{X} = \{X_1, X_2, \ldots, X_k\}$ is a partition of D into elements $X_i \in \mathcal{S}$, then we define the *quotient* $\mathcal{S}/\mathcal{X}$ as the family of subsets of $\mathcal{X}$ such that $\mathcal{Y} \in \mathcal{S}/\mathcal{X}$ if and only if $\cup\mathcal{Y} \in \mathcal{S}$. It is again easy to show that the quotient $\mathcal{S}/\mathcal{X}$ is a siba on $\mathcal{X}$.

By Lemma 6.2.3, $\mathcal{C}(g)$ is a siba for each 2-structure g. The following theorem shows that also the converse is true.

Theorem 6.3.21
A family $\mathcal{S}$ of subsets of a finite set D is a siba if and only if there is a 2-structure g such that $\mathcal{S} = \mathcal{C}(g)$.

Proof
Let $\mathcal{S}$ be a siba on D. We show by induction on the cardinality of D that there exists a 2-structure g with $\mathcal{C}(g) = \mathcal{S}$. If $|D| \leq 2$, then the claim is obvious. Suppose thus that $|D| \geq 3$.

Clearly, if $\mathcal{S}$ is special, then the claim holds by Lemma 6.3.20, since there exist linear, complete and primitive 2-structures on the domain D.

Assume now that $\mathcal{S}$ is not special, and let $\mathcal{P}$ be the partition of D into the maximal prime elements of $\mathcal{S}$. Denote $\mathcal{Q} = \mathcal{S}/\mathcal{P}$, for short. By induction hypothesis there are 2-structures $g_P = (P, R_P)$ for each $P \in \mathcal{P}$ such that $\mathcal{C}(g_P) = \mathcal{S}|P$, and a 2-structure $h = (\mathcal{P}, R_h)$ on the domain $\mathcal{P}$ such that $\mathcal{C}(h) = \mathcal{Q}$. We define a 2-structure $g = (D, R)$ on the domain D such that

$$(x_1, y_1)R(x_2, y_2) \iff \begin{cases} (x_1, y_1)R_P(x_2, y_2), & \text{if } (x_i, y_i) \in E_2(P) \text{ for } P \in \mathcal{P}, i = 1, 2, \text{ or} \\ (P_1, Q_1)R_h(P_2, Q_2), & \text{if } x_i \in P_i,\ y_i \in Q_i \text{ for some } (P_i, Q_i) \in E_2(\mathcal{P}), i = 1, 2. \end{cases} \tag{6.1}$$

It is now straightforward to show that $\mathcal{P} = \mathcal{P}_{\max}(g)$, from which one can deduce that $\mathcal{C}(g) = \mathcal{S}$. □

As shown by Möhring and Radermacher [36] the congruence classes of various systems such as set systems, relations and boolean functions form sibas.

Example 6.12
We consider here set systems as in Möhring and Radermacher [36]. A (*normal*) *set system* on a finite domain D is simply a family of subsets $\mathcal{L}$ of D such that $\cup\mathcal{L} = D$. Because of its generality, set systems are applicable in various contexts, see *e.g.* Shapley [39] for game theoretic applications.

Let $\mathcal{S}$ be a set system on A, and let $\mathcal{S}_x$ be a set system on A_x for each $x \in A$, where the domains A_x are nonempty and disjoint. The *substitution composition of* $\{\mathcal{S}_x\}_{x\in A}$ *for* $\mathcal{S}$ is defined as the set system $\mathcal{S}[\mathcal{S}_x; x \in A]$ consisting of the sets $\bigcup_{y\in Y} \tau(y)$, where Y runs through the sets in $\mathcal{S}$ and τ the functions such that $\tau(y) \in \mathcal{S}_y$ for each $y \in A$. The domain of $\mathcal{S}[\mathcal{S}_x; x \in A]$ is $\cup_{x\in A} A_x$.

For a set system $\mathcal{L}$ on D a subset $X \subseteq D$ is called an *autonomous set*, if it is empty or, in the above notations, $\mathcal{L} = \mathcal{S}[\mathcal{S}_x; x \in A]$ and $X = A_x$ for some $x \in A$. A partition of D into nonempty autonomous sets is called a *congruence partition* of $\mathcal{L}$.

It is now a straightforward task to verify that the autonomous sets of a set system $\mathcal{L}$ form a siba. Therefore each set system $\mathcal{L}$ can be represented as a 2-structure such that the autonomous sets of $\mathcal{L}$ are exactly the clans of g. This representation g of $\mathcal{L}$ is easily seen to be faithful in the sense that the congruence partitions of $\mathcal{L}$ are exactly the clan partitions of g, and hence, in a sense, the 'rough' algebraic structure of $\mathcal{L}$ is faithfully represented by the 2-structure g.

The autonomous sets of a set system $\mathcal{L}$ have further closure properties that are not necessarily true for arbitrary 2-structures. In fact, see [36], X is an autonomous set of $\mathcal{L}$ if and only if for all $S_1, S_2 \in \mathcal{L}$ with $S_i \cap X \neq \emptyset$, also $(S_1 \setminus X) \cup (S_2 \cap X) \in \mathcal{L}$. □

6.4 Primitive 2-Structures

6.4.1 Hereditary properties

Of the three special 2-structures linear and complete are determined by (an ordering of) their domains, and therefore they need no further inspection. The only interesting special 2-structures are thus the primitive ones. Indeed, there are arbitrarily complex primitive 2-structures.

By definition, all 2-structures with two nodes are primitive, but these are also either complete or linear. We shall thus consider those primitive 2-structures that are truly primitive.

Figure 6.18: A path of length three, and a cycle of length three

In Figure 6.18 there are two simple truly primitive 2-structures. The first of these is a path of length three and the second is a cyclic triangle.

We shall state now without proofs some interesting properties of the primitive 2-structures. The first of these is from Ehrenfeucht and Rozenberg [17].

Theorem 6.4.1 (Downward hereditarity)
Let $g = (D, R)$ be a truly primitive 2-structure. There are nodes x and y (possibly $x = y$) such that $\mathrm{sub}_g(D \setminus \{x, y\})$ is primitive. □

As an example, from the path of length three (in Figure 6.18) one cannot remove a node without violating primitivity. Hence in this case two nodes have to be removed in order to obtain a primitive substructure.

The proof of Theorem 6.4.1 is based on the following lemma, which gives a simple condition ensuring that a primitive 2-structure $g = (D, R)$ has no primitive substructures of the form $\text{sub}_g(D \setminus \{x\})$ with $x \in D$. We call such a 2-structure *critically primitive.*

Lemma 6.4.2
Let $g = (D, R)$ be a truly primitive 2-structure, and suppose $\text{sub}_g(X)$ is a truly primitive substructure such that $\text{sub}_g(X \cup \{x\})$ is non-primitive for all $x \in D \setminus X$. Then for each $x \in D \setminus X$ the substructure $\text{sub}_g(X \cup \{x\})$ has a unique nontrivial clan Z_x such that either $|Z_x| = 2$ with $x \in Z_x$ or $Z_x = X$. □

The critically primitive 2-structures were characterized in Schmerl and Trotter [40] and Bonizzoni [2]. In particular, the following result was proved in [40].

Theorem 6.4.3
Let $g = (D, R)$ be a critically primitive 2-structure with $|D| = n$. If the substructure $h = \text{sub}_g(X)$ is primitive with $|X| = m \geq 3$, then $n \equiv m \pmod 2$. □

This means that if a critically primitive g has an even (odd) number of nodes, then all its substructures with an odd (even) number of nodes are non-primitive. Moreover,

Corollary 6.4.4
Let g be a critically primitive 2-structure, and $\text{sub}_g(X)$ a primitive substructure of g with $|X| \geq 4$. Then $\text{sub}_g(X)$ is critically primitive. □

Example 6.13
In Figure 6.19 we have a critically primitive 2-structure g. Notice that g has only two edge classes, a class together with its reverse class. In graph theoretical terms, g is a tournament. □

Lemma 6.4.2 can be also used to prove the following hereditary result, see Harju and Rozenberg [30] or Schmerl and Trotter [40]. (Again, in [30] the result is proved for infinite 2-structures, for which Theorem 6.4.1 does not hold).

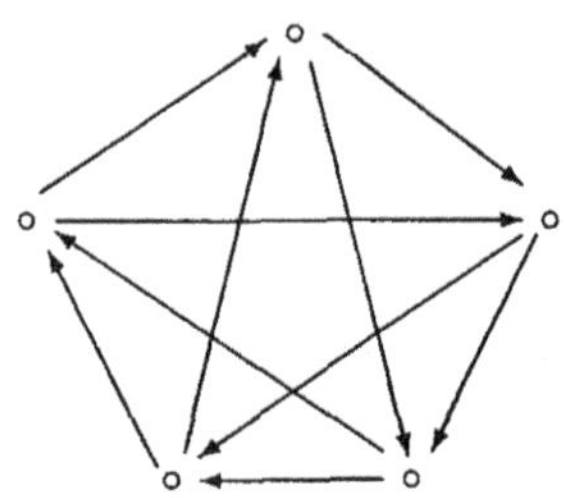

Figure 6.19: Critically primitive 2-structure g

Theorem 6.4.5 (Upward hereditarity)
Let $g = (D, R)$ be primitive, and let $h = \text{sub}_g(X)$ be a proper truly primitive substructure of g. Then there are nodes $x, y \in D \setminus X$ (possibly $x = y$) such that $\text{sub}_g(X \cup \{x, y\})$ is primitive. □

Consider, for a while, a 2-structure g as a directed graph $g = (D, E, R)$ with an equivalence relation R on the edge set $E \subseteq E_2(D)$ as mentioned in the beginning of Section 6.2.

For an edge $e \in E$ of g we let $g - e = (D, E \setminus \{e, e^{-1}\}, R_e)$, where R_e is R restricted to $E \setminus \{e, e^{-1}\}$. (This means that e is classified as 'nonedge' by the equivalence relation R_e).

Theorem 6.4.1 gives a hereditary property of primitivity concerning removals of nodes from a primitive 2-structure. We ask now whether one can remove an edge e from a primitive 2-structure g so that the result $g - e$ is still primitive.

We call a primitive $g = (D, E, R)$ *unstable* (after Sumner [42]), if $g - e$ is non-primitive for all edges $e \in E$. Figure 6.20 gives two examples of unstable 2-structures.

For unstable 2-structures we have the following result from Harju and Rozenberg [29]. A *leaf* of a 2-structure $g = (D, E, R)$ is a node of degree one.

Theorem 6.4.6
Let $g = (D, E, R)$ be a primitive 2-structure. Either

(1) g has an edge e such that $g - e$ is primitive, or

(2) g is unstable and it is either a path of length three or it has a leaf x such that $\text{sub}_g(D \setminus \{x\})$ is primitive. □

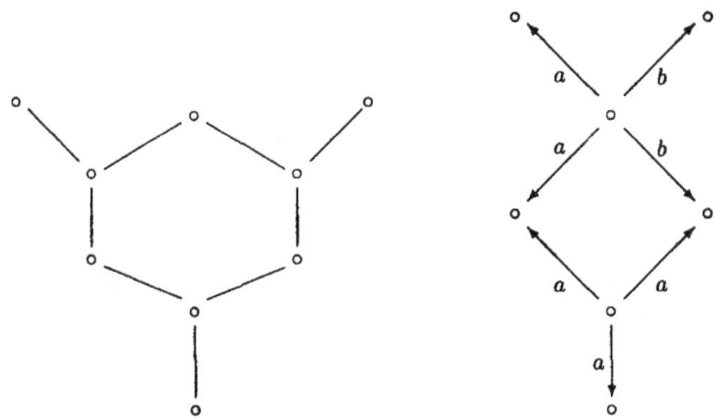

Figure 6.20: Two unstable 2-structures

6.4.2 Uniformly non-primitive 2-structures

We shall now consider those 2-structures that do not contain primitive components. We call a 2-structure g *uniformly non-primitive* if g has no truly primitive substructures.

A number of characterizations of the uniformly non-primitive 2-structures were proved by Engelfriet, Harju, Proskurowski and Rozenberg in [23]. In particular, the graphs and directed graphs that are representable by uniformly non-primitive 2-structures were characterized by forbidden subgraphs in [23]. In the following we shall state only one general characterization result (Theorem 6.4.10) concerning uniformly non-primitive 2-structures.

Theorem 6.4.7
Let g be a uniformly primitive 2-structure. Then the substructures and the quotients of g are uniformly non-primitive. □

By Theorem 6.4.1, if g is a primitive 2-structure with n nodes, then g has a primitive substructure on $n-1$ or $n-2$ nodes. In particular,

Lemma 6.4.8
Every truly primitive 2-structure g has a primitive substructure consisting of three or four nodes. □

Recall that $P(X)$ is the smallest prime clan of a 2-structure $g = (D, R)$ that contains the subset $X \subseteq D$. The following result was proved by Ehrenfeucht and Rozenberg [16].

Theorem 6.4.9
Let $g = (D, R)$ be a 2-structure. If $\text{sub}_g(X)$ is a truly primitive substructure of g, then the quotient $\Psi_g(P(X))$ is primitive. □

We say that a 2-structure g has the *2-block property*, if every substructure h of g with $|\text{dom}(h)| \geq 2$ has a partition into two nonempty clans, *i.e.*, if every substructure h has a nonempty clan $X \in \mathcal{C}(h)$ the complement $\text{dom}(h) \setminus X$ of which is also a nonempty clan of h. Also, we say that g has the *doubleton clan property*, if every substructure h with at least two nodes has a clan of size 2. The following theorem was proved in [23].

Theorem 6.4.10
The following statements are equivalent for a 2-structure g:
(1) *g is uniformly non-primitive,*
(2) *g has no primitive substructures of size 3 and 4,*
(3) *each node of* shape(g) *is linear or complete,*
(4) *g has the 2-block property,*
(5) *g has the doubleton clan property.* □

In fact, the proof in [23] reveals that g is uniformly non-primitive if and only if each substructure $\text{sub}_g(X)$ with $|X| \geq 2$ has a prime clan, the complement of which is a clan.

Example 6.14
Let g be the 2-structure from Figure 6.21. The shape of g is in Figure 6.22. By Theorem 6.4.10, g is uniformly non-primitive. □

6.5 Angular 2-structures and T-structures

6.5.1 Angular 2-structures

By Lemma 6.4.8 a truly primitive 2-structure contains a primitive substructure with 3 or 4 nodes. We call a $g = (D, R)$ *angular*, if $\text{sub}_g(X)$ is non-primitive for every subset X with $|X| = 3$.

Example 6.15
Clearly, if a primitive $g = (D, R)$ with $|D| \geq 4$ has only two edge classes, a and b, both of which are symmetric, then each $\text{sub}_g(X)$ with $|X| = 3$ is

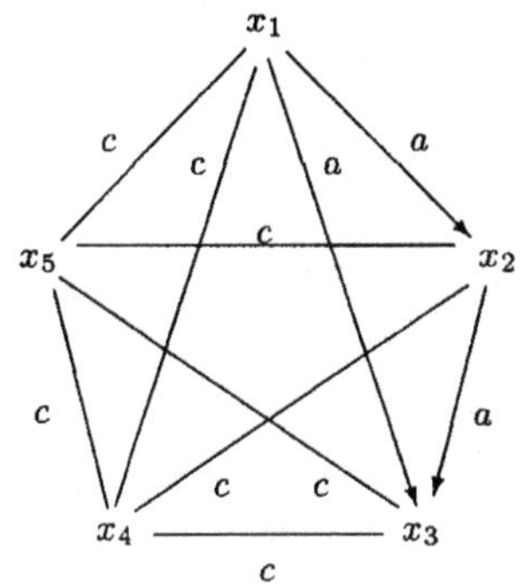

Figure 6.21: A uniformly non-primitive 2-structure g

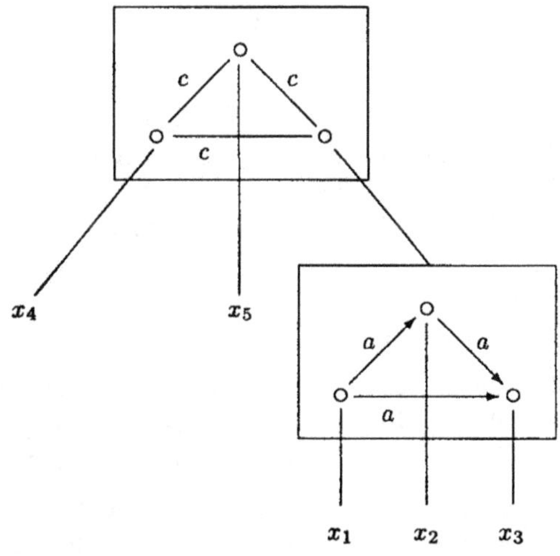

Figure 6.22: The shape of g

non-primitive. Therefore g is angular, and, by Theorem 6.4.1, g has a primitive substructure on four nodes. It is straightforward to verify that then this primitive substructure is isomorphic to the path of length three. □

We define the *feature* of an edge class a of g as the set $F_a = \{a, a^{-1}\}$. Note that a feature $\{a, a^{-1}\}$ is a singleton, if the edge class a is symmetric.
We start with some immediate observations.

Lemma 6.5.1
Let g be an angular 2-structure.
(1) *If g has only one feature, then it is either linear or complete.*
(2) *The substructures and the quotients of g are angular.* □

By Theorem 6.5.3 below primitivity of angular 2-structures is preserved if the edge classes are made symmetric. For a 2-structure $g = (D, R)$ define its *symmetric version* as the 2-structure $g^s = (D, R^s)$, where

$$e_1 R^s e_2 \iff e_1 R e_2 \text{ or } e_1 R e_2^{-1}.$$

Clearly, g^s has only symmetric edge classes,

$$eR_{g^s} = eR_g \cup e^{-1}R_g = e^{-1}R_{g^s}.$$

Example 6.16
Consider the 2-structure g from the left part of Figure 6.23. Now, g is angular and primitive. The edge classes a and b of g are nonsymmetric, and the features $\{a, a^{-1}\}$ and $\{b, b^{-1}\}$ are disjoint. The symmetric version g^s (the 2-structure from the right part of Figure 6.23) with two symmetric edge classes is also angular and primitive. □

For an edge e of a 2-structure $g = (D, R)$, let $g_e = (D, eR_{g^s})$ be the (undirected) graph, which is obtained from g by setting an edge $\{x, y\}$ if and only if (x, y) belongs to the edge class eR_{g^s} of the symmetric version g^s.

We say that a 2-structure g is *connected*, if g_e is connected for all e. Further, a subset $X \subseteq D$ is an *a-connected component* for an edge class a, if X is a connected component of the graph g_e for some $e \in a$.

Lemma 6.5.2
If g is an angular 2-structure, then the a-connected components are clans of g. In particular, if g is primitive, then it is connected.

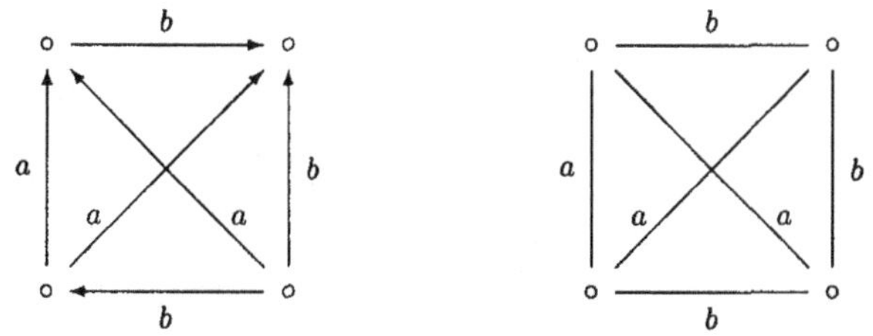

Figure 6.23: Primitive and angular g, and its symmetric version g^s.

Proof
Assume that X is an a-connected component of g for some a. If $X = D$ or $|X| = 1$, then the claim holds. Assume thus that $X \neq D$ and $|X| \geq 2$.

Let $y \notin X$ and $x_1, x_2 \in X$. Since X is an a-connected component, there are nodes $z_1, \ldots, z_k \in X$ such that $z_1 = x_1$, $z_k = x_2$, and $(z_i, z_{i+1})R \in F_a$ for all $i = 1, 2, \ldots, k-1$. Moreover, $(y, z_i)R \notin F_a$ for $i = 1, 2, \ldots, k$, since X is an a-connected component. By angularity, it follows that $(y, z_i)R = (y, z_{i+1})R$ for $i = 1, 2, \ldots, k-1$, and thus also $(y, x_1)R = (y, x_2)R$. We conclude that $X \in \mathcal{C}(g)$. □

The next theorem reduces the primitivity of angular 2-structures to their symmetric versions.

Theorem 6.5.3
Let $g = (D, R)$ be angular. Then g is truly primitive if and only if the symmetric version g^s is truly primitive. □

Finally, we have the following result, for the proof of which we refer again to [19].

Theorem 6.5.4
Let $g = (D, R)$ be a truly primitive angular 2-structure. Then g has exactly two features. □

Theorem 6.5.4 generalizes to all angular 2-structures in the following sense, see Harju and Rozenberg [30].

Theorem 6.5.5
If an angular g is connected, then it has at most two features.

Proof
Let $g = (D, R)$ be angular with k features with $k \geq 2$. If it is primitive, then the claim holds by Theorem 6.5.4. Otherwise, consider the special quotient $g/\mathcal{P}_{\max}(g)$. By Lemma 6.5.1(2) $g/\mathcal{P}_{\max}(g)$ is also angular. Since g is connected, by the formation of a quotient also $g/\mathcal{P}_{\max}(g)$ is connected and therefore it has the same number of features as g. Now, since $k \geq 2$, $g/\mathcal{P}_{\max}(g)$ cannot be linear or complete, and hence it is primitive. Consequently, by Theorem 6.5.4, we have that $k = 2$. □

Theorem 6.5.3 and Theorem 6.5.4 reduce essentially the study of primitive angular 2-structures to that of graphs. One can then use various methods of graph theory to decompose primitive angular 2-structures into smaller 2-structures. We mention here only the *split decomposition* of Cunningham and Edmonds [6] and the decomposition of graphs into P_4-*components* of Jamison and Olariu [32].

6.5.2 T-structures and texts

We say that a 2-structure g is a *partial order*, if it has two features one of which is symmetric and the other is nonsymmetric and such that its edge classes partially order the domain of g. It was shown in [19] that a truly primitive angular 2-structure g is either a graph (*i.e.*, g has two symmetric edge classes), or a partial order, or a T-structure. Here a 2-structure g is a *T-structure*, if g is angular and it has only nonsymmetric edge classes.

The following lemma is obvious.

Lemma 6.5.6
Let g be a T-structure. Then the substructures and quotients of g are T-structures. Further, the special quotient $g/\mathcal{P}_{\max}(g)$ is either linear or truly primitive. □

Assume that g is a truly primitive T-structure. By Lemma 6.4.8 and the fact that g is angular, g has a primitive substructure $h = \mathrm{sub}_g(X)$ with $|X| = 4$. Moreover, by Theorem 6.5.4, g (and h) has exactly two (nonsymmetric) features. One can show that the 2-structure N from Figure 6.24 is the only primitive T-structure with four nodes (up to isomorphism).

Lemma 6.5.7
If g is a truly primitive T-structure, then g contains a substructure N. □

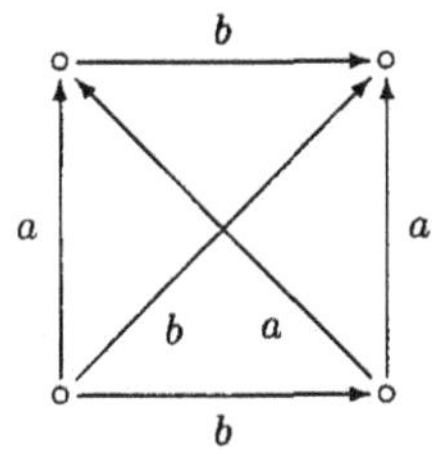

Figure 6.24: The primitive T-structure N

Consider then a 2-structure g without symmetric edge classes. A subset σ of edge classes is a *selection*, if for all features F, $|\sigma \cap F| = 1$. For a selection σ we denote by $\sigma(g)$ the directed graph $(\text{dom}(g), \bigcup\sigma)$.

The next result was proved in Ehrenfeucht and Rozenberg [20].

Lemma 6.5.8
Let g be a 2-structure having only nonsymmetric edge classes. Then g is a T-structure if and only if $\sigma(g)$ is a linear order of dom(g) for all selections σ. □

In general a T-structure may have arbitrarily many features, but by Theorem 6.5.4 a truly primitive T-structure has exactly two features. Therefore the nodes (the special quotients $\Psi_g(P)$ for prime clans P) in the shape of a T-structure g have at most two features. This fact was generalized in [20] to

Theorem 6.5.9
For each T-structure g there exists a T-structure g' with at most two features such that shape(g) = shape(g'). □

Remark:
Notice that in [20] Theorem 6.5.9 is stated using 'similarity' instead of equality of shapes. However, one can easily check that for (unlabeled) 2-structures the two statements are equivalent. Therefore, we shall now consider T-structures with at most two features. For a partial order ρ we let ρ^{-1} denote the dual partial order of ρ, *i.e.*, $\rho^{-1} = \{(x, y) \mid (y, x) \in \rho\}$. □ □
For the proof of the following result see [20].

Theorem 6.5.10
Let g be a 2-structure with at most two features. Then g is a T-structure if

and only if there exist two linear orders ρ_1 and ρ_2 such that the edge classes of g are included in the set $\{\rho_1 \cap \rho_2,\ \rho_1 \cap \rho_2^{-1},\ \rho_1^{-1} \cap \rho_2$ and $\rho_1^{-1} \cap \rho_2^{-1}\}$. □

In Theorem 6.5.10 some of the intersections may be empty.

Notice that in the above each of the intersections of the linear orders ρ_1 and ρ_2 is a partial order of the domain, and moreover $(\rho_1 \cap \rho_2)^{-1} = \rho_1^{-1} \cap \rho_2^{-1}$ and $(\rho_1 \cap \rho_2^{-1})^{-1} = \rho_1^{-1} \cap \rho_2$.

From Theorem 6.5.10 it follows that each T-structure with at most two features is determined by two linear orders of its domain.

Texts were defined in [20] as triples $(\lambda, \rho_1, \rho_2)$, where $\lambda\colon D \to A$ is a function from the domain D into an alphabet A, and ρ_1, ρ_2 are two linear orders of D. Hence a text τ consists of a T-structure $g(\tau)$ (specified by Theorem 6.5.10) together with a labeling function λ.

A text $\tau = (\lambda, \rho_1, \rho_2)$ with $\lambda\colon D \to A$ can be considered as a 'structured word' over A. Indeed, let $D = \{x_1, x_2, \ldots, x_n\}$ and $\rho_1 = (x_{i_1}, x_{i_2}, \ldots, x_{i_n})$, $\rho_2 = (x_{j_1}, x_{j_2}, \ldots, x_{j_n})$. The first linear order gives an ordinary word over the alphabet A,

$$\lambda(\rho_1) = (\lambda(x_{i_1}), \lambda(x_{i_2}), \ldots, \lambda(x_{i_n})),$$

and the linear orders ρ_1 and ρ_2 together assign to the word $\lambda(\rho_1)$ its structure as specified by the shape of the corresponding T-structure $g(\tau)$.

Example 6.17
Let $D = \{1,2,3,4,5\}$, and $A = \{x,y,z\}$. Define a text $\tau = (\lambda, \rho_1, \rho_2)$ as follows:

$$\begin{aligned} \lambda(1) &= x = \lambda(2),\ \lambda(3) = y = \lambda(4),\ \lambda(5) = z, \\ \rho_1 &= (2,3,5,1,4),\ \rho_2 = (3,2,4,1,5). \end{aligned}$$

The word defined by τ is $\lambda(\rho_1) = (x,y,z,x,y)$, and its structure is given by ρ_2. The T-structure $g(\tau)$ has the features $\{a, a^{-1}\}$ and $\{b, b^{-1}\}$, where

$$\begin{aligned} a &= \rho_1 \cap \rho_2 = \{(2,1),(2,4),(2,5),(3,1),(3,4),(3,5)\}, \\ b &= \rho_1 \cap \rho_2^{-1} = \{(1,4),(2,3),(5,1),(5,4)\}, \end{aligned}$$

and hence $g(\tau)$ is as given in Figure 6.25.

The maximal prime clans of $g(\tau)$ are $\{2,3\}$ and $\{1,4,5\}$, and the shape of $g(\tau)$ is given in Figure 6.26. □

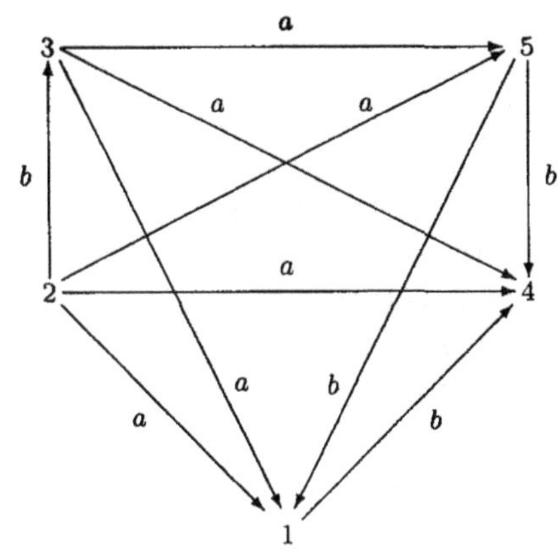

Figure 6.25: The T-structure $g(\tau)$

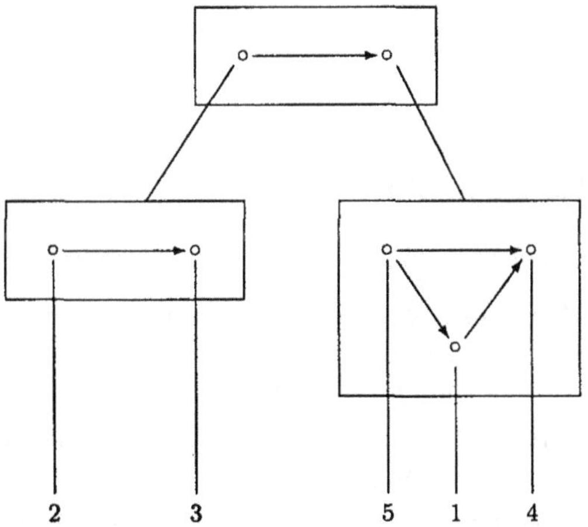

Figure 6.26: The shape of $g(\tau)$

The texts can be classified with respect to the properties of the shapes of the corresponding T-structures. We say that a text τ is *alternating*, if shape($g(\tau)$) has only linear nodes, *i.e.*, the special quotients $\Psi_{g(\tau)}(P)$ are linear 2-structures for all prime clans P.

If a text $\tau = (\lambda, \rho_1, \rho_2)$ is not alternating, then the shape of $g(\tau)$ contains a primitive quotient, and therefore, by Lemma 6.5.7, $g(\tau)$ contains a substructure N. This means that the linear order ρ_1 of the text τ contains a suborder (i_1, i_2, i_3, i_4) such that either (i_2, i_4, i_1, i_3) or (i_3, i_1, i_4, i_2) is a suborder of ρ_2.

Next we define two operations Vrev and $\oplus$ for texts. For a text $\tau = (\lambda, \rho_1, \rho_2)$ let

$$\mathrm{Vrev}(\tau) = (\lambda, \rho_1^{-1}, \rho_2).$$

For texts $\tau = (\lambda, \rho_1, \rho_2)$ and $\tau' = (\lambda', \rho_1', \rho_2')$ with disjoint domains let

$$\tau \oplus \tau' = (\lambda \cup \lambda', \rho_1 + \rho_1', \rho_2 + \rho_2'),$$

where $\rho_i + \rho_i' = \rho_1 \cup \rho_1' \cup \{(x,y) \mid x \in \mathrm{dom}(\tau),\, y \in \mathrm{dom}(\tau')\}$.

For the proof of the following theorem we refer to [20].

Theorem 6.5.11
The family of alternating texts is the smallest class of texts containing the singleton texts and closed under the operations Vrev *and* $\oplus$. □

For further combinatorial properties and applications of texts we refer to [20], [14] and [13].

6.6 Labeled 2-Structures

6.6.1 Definition of a labeled 2-structure

As noted in Sections 6.2 and 6.3 the decompositions of 2-structures are not unique in the sense that two different 2-structures may possess the same decomposition. In *labeled* 2-structures this ambiguity does not occur.

A labeled 2-structure is obtained from a 2-structure g by setting a different label to each edge class of g.

Equivalently, we define a *labeled 2-structure* or an *ℓ2-structure*, for short, as a triple $g = (D, \alpha, \Delta)$, where D is the *domain*, Δ is the set of *labels* and $\alpha \colon E_2(D) \to \Delta$ is the *labeling function* on the edges. Again, we assume that our ℓ2-structures are *reversible*, *i.e.*, there is a permutation $\delta \colon \Delta \to \Delta$ of order two on the labels that satisfies the condition

$$\alpha(e^{-1}) = \delta(\alpha(e)) \qquad \text{for all } e \in E_2(D).$$

That the permutation δ is of order two means that $\delta^2(a) = a$ for all labels $a \in \Delta$. Again we may have $a = \delta(a)$ for a label $a \in \Delta$, in which case a is called *symmetric.*

In later sections the set Δ of labels will be a group in the algebraic sense, and the reversibility function $\delta\colon \Delta \to \Delta$ will be a group involution. For this reason we do not write a^{-1} for the label of the reverse edge.

We observe that an ℓ2-structure $g = (D, \alpha, \Delta)$ is completely determined by its labeling function α, and therefore *we shall identify g with α, i.e.*, we shall consider ℓ2-structures as functions

$$g\colon E_2(D) \to \Delta.$$

Notice, however, that the domain of the 2-structure g is D, not $E_2(D)$.

For each ℓ2-structure $g\colon E_2(D) \to \Delta$ there is the *underlying* unlabeled 2-structure $g' = (D, R)$, where R is defined by

$$e_1 R e_2 \iff g(e_1) = g(e_2).$$

For this reason the results of the previous sections can be easily modified for labeled 2-structures.

The converse of this does not hold, since the same 2-structure corresponds to several ℓ2-structures with different labeling functions.

Let $g_1 : E_2(D_1) \to \Delta_1$ and $g_2\colon E_2(D_2) \to \Delta_2$ be two labeled 2-structures. We say that g_1 and g_2 are *isomorphic* , if there are bijections $\varphi\colon D_1 \to D_2$ and $\psi\colon \Delta_1 \to \Delta_2$ such that for all $(x, y) \in E_2(D_2)$,

$$g_2(\varphi(x), \varphi(y)) = \psi(g_1(x, y)).$$

Also, g_1 and g_2 are *strictly isomorphic,* if the bijection φ is the identity function on D_1 (and so $D_1 = D_2$). Hence two ℓ2-structures $g\colon E_2(D) \to \Delta_1$ and $h\colon E_2(D) \to \Delta_2$ are strictly isomorphic, if there is a bijection $\psi\colon \Delta_1 \to \Delta_2$ such that $h(e) = \psi(g(e))$ for all edges $e \in E_2(D)$.

The next result gives a connection between the unlabeled and the labeled 2-structures.

Theorem 6.6.1
Two ℓ2-structures g and h are strictly isomorphic if and only if they have the same underlying (unlabeled) 2-structure. □

6.6.2 Substructures, clans and quotients

Let $g\colon E_2(D) \to \Delta$ be an ℓ2-structure. We shall now modify the basic results and definitions of the previous sections, and state them for labeled 2-structures.

The *substructure* $\mathrm{sub}_g(X)$ induced by a subset $X \subseteq D$ is defined as the restriction of g onto $E_2(X)$, *i.e.*, $\mathrm{sub}_g(X)\colon E_2(X) \to \Delta$ satisfies $\mathrm{sub}_g(X)(e) = g(e)$ for all $e \in E_2(X)$. Clearly, a substructure $\mathrm{sub}_g(X)$ has the same reversibility function $\delta\colon \Delta \to \Delta$ as g.

A subset $X \subseteq D$ is a *clan* of $g\colon E_2(D) \to \Delta$, if for all $x, y \in X$ and all $z \notin X$, $g(z,x) = g(z,y)$ (and hence $g(x,z) = g(y,z)$ by the reversibility condition).

Again, let $\mathcal{C}(g)$ denote the family of clans of g. Now, the basic results of Sections 6.2 and 6.3 hold for $\mathcal{C}(g)$.

Lemma 6.6.2
Let $g\colon E_2(D) \to \Delta$ and let $g' = (D,R)$ be the underlying 2-structure. Then
(1) *for each $X \subseteq D$, $\mathrm{sub}_{g'}(X)$ is the underlying 2-structure of $\mathrm{sub}_g(X)$, and*
(2) $\mathcal{C}(g) = \mathcal{C}(g')$. □

A partition $\mathcal{X} \subseteq \mathcal{C}(g)$ of the domain D of g into clans is called a *factorization* of g. The *quotient* of g by $\mathcal{X}$ is the ℓ2-structure $g/\mathcal{X}$ on the domain $\mathcal{X}$ such that for all $X, Y \in \mathcal{X}$ (with $X \neq Y$), $(g/\mathcal{X})(X,Y) = g(x,y)$, whenever $x \in X$ and $y \in Y$. We need a modified result to ensure that the quotient $g/\mathcal{X}$ is well-defined for all g.

Lemma 6.6.3
Let δ be the reversibility function of Δ. Let $X, Y \in \mathcal{C}(g)$ be two disjoint clans of an ℓ2-structure $g\colon E_2(D) \to \Delta$. There exists a label $a \in \Delta$ such that $g(x,y) = a$ and $g(y,x) = \delta(a)$ for all $x \in X$ and $y \in Y$. □

We observe that, by Lemma 6.6.3, every quotient $g/\mathcal{X}$ is isomorphic with a substructure of g, *viz.* with $\mathrm{sub}_g(X)$, where X is a set of representatives of $\mathcal{X}$ (*i.e.*, $X \cap Y$ is a singleton for every $Y \in \mathcal{X}$); this is a rather unusual situation from the general algebraic point of view. In this isomorphism the bijection $\psi\colon \Delta \to \Delta$ is just the identity.

A *decomposition* $h(g_1, \ldots, g_m)$ of an ℓ2-structure g consists of the factors $g_i = \mathrm{sub}_g(X_i)$ and the quotient $h = g/\mathcal{X}$ by a factorization $\mathcal{X} = \{X_1, \ldots, X_m\}$. In contrast to the unlabeled 2-structures, see Example 6.8 the decompositions of a labeled 2-structure g are unique to g.

Theorem 6.6.4
Two ℓ2-structures g_1 and g_2 have a common decomposition if and only if $g_1 = g_2$. □

The *prime clans* and the *maximal prime clans* of an ℓ2-structure g are defined as in Section 6.3. Again, the family of prime clans is denotes by $\mathcal{P}(g)$ and the family of the maximal prime clans is denoted by $\mathcal{P}_{\max}(g)$.

The following result modifies some of the results from Sections 6.2 and 6.3 for ℓ2-structures. It shows that the (prime) clans of an ℓ2-structure g are inherited by the substructures and by the quotients.

Lemma 6.6.5
Let g be an ℓ2-structure.
(1) For each $X \in \mathcal{C}(g)$, $\mathcal{C}(\mathrm{sub}_g(X)) = \{Y \mid Y \in \mathcal{C}(g), Y \subseteq X\}$.
(2) For each $X \in \mathcal{P}(g)$, $\mathcal{P}(\mathrm{sub}_g(X)) = \{Y \mid Y \in \mathcal{P}(g), Y \subseteq X\}$.
(3) If $\mathcal{X} \subseteq \mathcal{P}(g)$ is a prime factorization, then $\mathcal{Q} \in \mathcal{C}(g/\mathcal{X})$ if and only if $\cup\mathcal{Q} \in \mathcal{C}(g)$. □

The clan decomposition theorem for 2-structures, Theorem 6.3.12, is restated in the labeled case as follows.

Theorem 6.6.6
For any ℓ2-structure g, the quotient $g/\mathcal{P}_{\max}(g)$ is either linear, or complete, or truly primitive. □

We say that an ℓ2-structure g is *a-linear*, for a label $a \in \Delta$, if the relation $x <_a y$ linearly orders the domain of g, where $x <_a y$ if and only if $g(x, y) = a$. Also, g is *a-complete*, if $g(e) = a$ for all $e \in E_2(D)$.

As is the case for unlabeled 2-structures, the partially ordered set $T(g) = (\mathcal{P}(g), \subseteq)$ of prime clans forms a directed and rooted tree, the *prime tree* of g. The root of $T(g)$ is the domain $\mathrm{dom}(g)$, the leaves are the singleton sets, and there is an edge (X, Y) in $T(g)$, if X is the minimal element of $\mathcal{P}(g)$ such that $Y \subseteq X$ and $X \neq Y$.

Example 6.18
Consider the ℓ2-structure g from Figure 6.27 with nonsymmetric labels. In the figure we have omitted the edges having reverse labels $\delta(a)$, $\delta(b)$ and $\delta(c)$.

Here the nontrivial clans of g are $X_1 = \{x_1, x_4, x_5\}$ and $X_2 = \{x_4, x_5\}$; both clans are also prime clans of g, and X_1 is a maximal prime clan. We

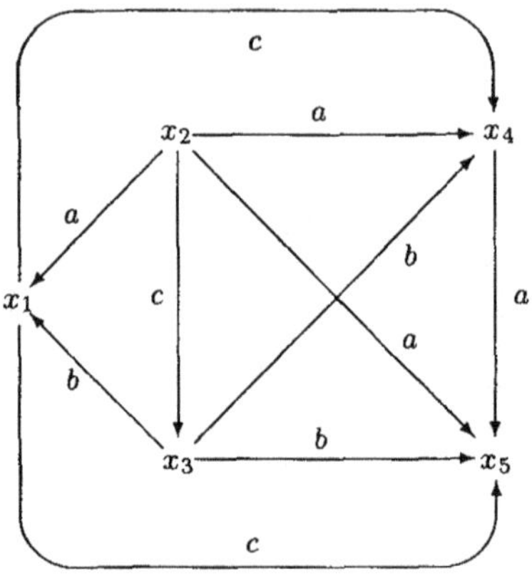

Figure 6.27: An $\ell2$-structure g.

have $\mathcal{P}_{\max}(g) = \{X_1, \{x_2\}, \{x_3\}\}$, and hence the quotient $g/\mathcal{P}_{\max}(g)$ has three nodes, and it is primitive.

In Figure 6.28 of the prime tree $T(g)$ of g we have not identified the inner nodes (the prime clans), since each inner node consists of the leaves x_i that are descendants of this node.

The substructure $h_1 = \text{sub}_g(X_1)$ has two maximal primes, X_2 and $\{x_1\}$, and the quotient $h_1/\mathcal{P}_{\max}(h_1)$ is linear. The substructure $h_2 = \text{sub}_g(X_2)$ has maximal primes $\{x_4\}$ and $\{x_5\}$ and the quotient $h_2/\mathcal{P}_{\max}(h_2)$ is isomorphic to the linear $\ell2$-structure h_2.

Notice that the partition $\mathcal{X} = \{\{x_1\}, \{x_2\}, \{x_3\}, X_2\}$ is also a clan partition of g into prime clans. The quotient $g/\mathcal{X}$ has four nodes, and since $X_1 \in \mathcal{C}(g)$, also $\{\{x_1\}, X_2\}$ is a clan of $g/\mathcal{X}$. Therefore this quotient is not special. □

The *shape* of an $\ell2$-structure $g\colon E_2(D) \to \Delta$ is obtained in the same way as for unlabeled 2-structures. Hence the shape of g, shape$(g) = (T(g), \Psi_g)$, consists of the prime tree of g together with the function

$$\Psi_g(P) = \text{sub}_g(P)/\mathcal{P}_{\max}(\text{sub}_g(P))$$

that maps each prime clan $P \in \mathcal{P}(g)$ to its special quotient.

Example 6.19
The shape of the $\ell2$-structure g from Example 6.18, see Figure 6.27, can be

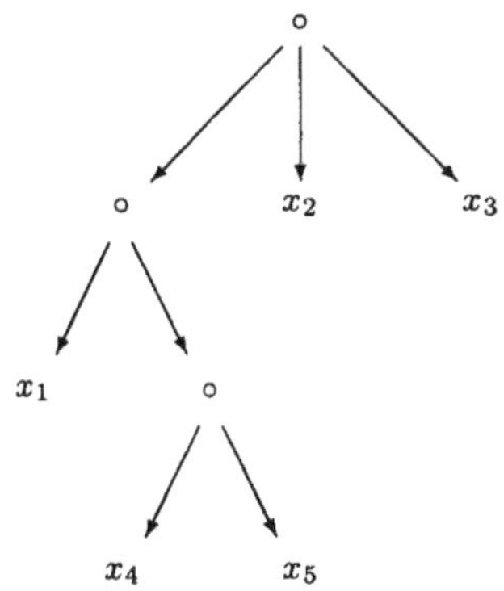

Figure 6.28: The prime tree of g.

drawn as in Figure 6.29. Here the labels of the edges in the special quotients $\Psi_g(P)$ have to be included, because now these quotients are labeled 2-structures. □

6.7 Dynamic Labeled 2-Structures

6.7.1 Motivation

Usually in a valuation of objects of a mathematical structure the values come from a set that is structured itself. Frequently such a set is an algebraic structure, where the elements are bound together by one or more operations. As an example, a finite automaton may be considered as a directed graph, where the edges have values from a free monoid. Also, in graph theory the values of objects (the labels of edges or nodes) are often taken from an algebraic structure, *e.g.*, in many optimality problems such as the shortest path problem, the edges are labeled by elements of the field $(\mathbb{R}, +, \cdot)$ of real numbers.

In this and the following sections we shall study $\ell 2$-structures that have a group structure on their sets of labels. The group structure of the labels gives then a method of locally transforming $\ell 2$-structures into each other, and this leads to dynamic $\ell 2$-structures.

A group is a natural candidate for the set labels of a (reversible) $\ell 2$-structure. The dynamic $\ell 2$-structures were also motivated by Ehrenfeucht and Rozenberg [22] by evolution of networks and similar processes. Consider a network consisting of a set of processors D, where each $x \in D$ is connected by a channel

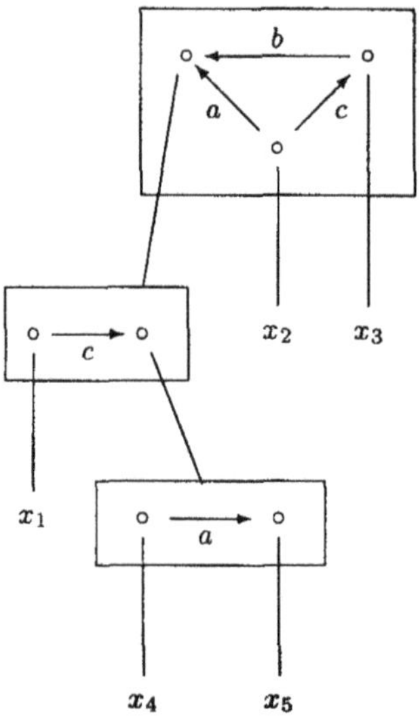

Figure 6.29: The shape of g.

to each $y \in D$ for $y \neq x$. A channel (x, y) may assume a certain value or state from the set Δ at a specific time, *e.g.*, if there are only two values $\Delta = \{0, 1\}$, then the channels may be interpreted as having a sleeping and an active state; on the other hand, if $\Delta = \mathbb{R}$, then the channels can be considered as weighted by reals.

The *concurrent* actions of the processors modify the states of the channels. The activity of each $x \in D$ will be described by two sets of actions, the ***output actions*** O_x and the ***input actions*** I_x, by which x changes the states of the channels from and to x, respectively. The actions of x are thus mappings of Δ into Δ.

Now, for each (x, y) there are two processors, x and y, that change the state of this channel concurrently (*i.e.*, independently of each other); x changes it by a mapping from O_x and y changes it by a mapping from I_y. In order to accommodate the concurrent behaviour of the processors, the mappings from O_x and I_y should commute, *i.e.*, for all $\varphi \in O_x$ and $\psi \in I_y$, $\varphi\psi = \psi\varphi$.

To avoid unnecessary sequencing of actions the composition of two actions from O_x (I_x, resp.) is again assumed to be an action, *i.e.*, the sets O_x and I_x form transformation semigroups on Δ under the operation of composition.

Further, to assure a minimal freedom of the actions for each x we assume that for each $a, b \in \Delta$ and $x \in D$ there exist a $\varphi \in O_x$ and a $\gamma \in I_x$ such that $\varphi(a) = b$ and $\gamma(a) = b$, *i.e.*, the semigroups O_x and I_x are transitive.

The above assumptions simplify the situation considerably. Indeed, it was shown in [22] (see, also [9]) that if $|D| \geq 3$, then there are two isomorphic groups O and I of permutations on Δ such that for each $x \in D$, $O_x = O$ and $I_x = I$. Thus the actions come from two groups O and I, which are independent of the processors. Moreover, we can define an operation on Δ such that Δ becomes a group isomorphic to O and I. In fact, now the groups O and I become defined by left and (involutive) right multiplication of the group Δ.

These observations lead to the definition of a dynamic ℓ2-structure, given formally in Sections 6.7.2 and 6.7.3 in terms of *selectors*. A selector σ is a mapping, which captures the action of each processor $x \in D$ at a specific stage of the network. By the above observations, a selector will be simply a function $\sigma: D \to \Delta$.

A global state of a network is represented by an ℓ2-structure, for which Δ is the set of labels. An evolution of the network is presented as a set of ℓ2-structures that represent possible global states of the network. The transitions from one ℓ2-structure g to another h (and therefore from one global state to another) are the transformations $g \mapsto h$ induced by the selectors.

6.7.2 *Group labeled 2-structures*

We shall assume throughout the rest of the paper that the set Δ of labels forms a (possibly infinite) *group*. The identity element of Δ is usually denoted by 1_Δ.

A mapping $\delta: \Delta \to \Delta$ is an *involution* of the group Δ, if it is a permutation of order two such that $\delta(ab) = \delta(b)\delta(a)$ for all $a, b \in \Delta$. An involution δ satisfies $\delta(a^{-1}) = \delta(a)^{-1}$ for all $a \in \Delta$, and in particular, $\delta(1_\Delta) = 1_\Delta$.

Example 6.20

(1) For any group Δ the most obvious involution is the inverse function: $\delta(a) = a^{-1}$ for each $a \in \Delta$.

(2) If Δ is the group of nonsingular $n \times n$-matrices (over a field F), then transposition, $\delta(M) = M^T$, is an involution of Δ.

(3) If Δ is a finite group of even order, then it has an element a of order 2, and, as easily seen, the mapping $\delta(x) = ax^{-1}a$ $(x \in \Delta)$ is an involution.

(4) If $\alpha: \Delta \to \Delta$ is an automorphism of order two, then δ defined by $\delta(a) = \alpha(a^{-1})$ is an involution. □

As before we will be dealing only with reversible ℓ2-structures; now involutions will determine reversibility. Let δ be an involution of the group Δ. We say that $g\colon E_2(D) \to \Delta$ is a *δ-reversible ℓ2-structure on* Δ or a *Δ^δ 2-structure*, for short, if the involution δ is its reversibility function: $g(e^{-1}) = \delta(g(e))$ for all edges $e \in E_2(D)$.

We reserve the symbol D for a domain, the symbol Δ for a group and the symbol δ for an involution of Δ.

Example 6.21
Let $\Delta = (\mathbb{R}^+, \cdot)$ be the multiplicative group of positive real numbers, and let δ be the inverse function of Δ. In Figure 6.30, we have an example of a Δ^δ2-structure, where, as usual, the reverse edges are not shown. Thus, *e.g.*, we have $g(x_2, x_1) = \delta(g(x_1, x_2)) = g(x_1, x_2)^{-1} = 1$, $g(x_2, x_3) = \frac{3}{2}$, and $g(x_3, x_2) = \frac{2}{3}$. □

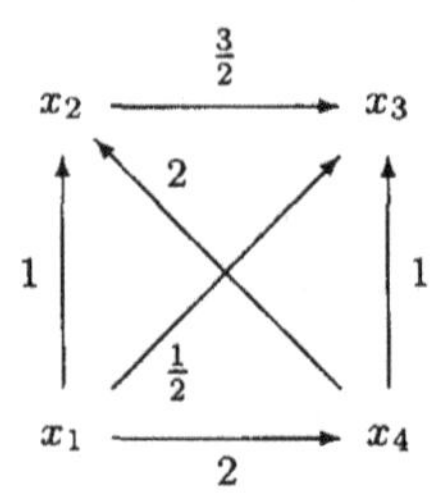

Figure 6.30: A Δ^δ2-structure

6.7.3 *Dynamic labeled 2-structures*

The group Δ of labels of a Δ^δ2-structure g becomes employed by the selectors, which, in essence, label the nodes $x \in D$ by the elements of the group Δ.

Let g be a Δ^δ2-structure for a group Δ and its involution δ. A function $\sigma\colon D \to \Delta$ is called a *selector*. For a selector $\sigma\colon D \to \Delta$ define the ℓ2-structure g^σ by

$$g^\sigma(x, y) = \sigma(x) \cdot g(x, y) \cdot \delta(\sigma(y))$$

for all $(x, y) \in E_2(D)$. The family $[g] = \{g^\sigma \mid \sigma \text{ a selector}\}$ is a (*single axiom*) *dynamic Δ^δ 2-structure* (*generated by* g).

Hence a selector $\sigma\colon D \to \Delta$ transforms each g into a g^σ by direct left and involutive right multiplication. The new value of an edge depends on the (values of the) nodes and on the label of the edge.

Example 6.22
Let g be as in Example 6.21, see Figure 6.30. Define a selector $\sigma\colon D \to \mathbb{R}$ by $\sigma(x_i) = 5 - i$. Then, *e.g.*, we have in the image g^σ, $g^\sigma(x_1, x_2) = 4 \cdot 1 \cdot \frac{1}{3} = \frac{4}{3}$. The image g^σ is drawn in Figure 6.31, where we have labeled the nodes by the values of σ.

□

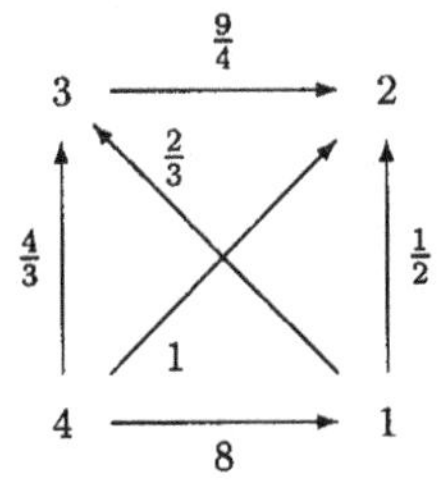

Figure 6.31: The image g^σ

In the pictorial representations of Δ^δ2-structures, *we shall usually omit the edges labeled by the identity element* 1_Δ *of the group* Δ. Recall that $\delta(1_\Delta) = 1_\Delta$. This convention reflects a natural interpretation of 1_Δ as the 'no edge' symbol.

The dynamic Δ^δ2-structures on the group $\mathbb{Z}_2$ of labels are closely connected to *Seidel switching*, which was defined in connection with a problem of finding equilateral n-tuples of points in elliptic geometry (in this respect, see Van Lint and Seidel [33]). This problem has the following subproblem for undirected graphs. Determine the equivalence classes of undirected graphs with n nodes with respect to the following operation: Let $G \to G'$, if there is a node x such that $G' = (D, E')$ is obtained from $G = (D, E)$ by removing all edges (x, y) and (y, x) incident with x, and adding all pairs (x, y) and (y, x) not in E. Hence $G \to G'$, if for some node x,

$$E' = (E \setminus \{(x, y), (y, x) \mid y \neq x\}) \cup \{(x, y), (y, x) \mid (x, y) \notin E\}.$$

Let then $\sim$ be the equivalence relation determined by $\to$, *i.e.*, $G \sim G'$ if and only if $G = G'$ or there exists a finite sequence $G_0 \to G_1 \to \cdots \to G_k$ with $\{G, G'\} = \{G_0, G_k\}$. One now asks how many equivalence of $\sim$ classes are there for a set D of nodes?

We can reformulate the above problem in terms of dynamic Δ^δ2-structures as follows. Let $\Delta = (\mathbb{Z}_2, +)$ be the cyclic group of order two. This group has only one involution δ, the inverse mapping, which in this simple case is the identity mapping of Δ: $\delta(0) = 0$ and $\delta(1) = 1$. We may consider a Δ^δ2-structure $g\colon E_2(D) \to \Delta$ as an undirected graph, where $g(e) = 1$ ($g(e) = 0$) means that e is (not, respectively) an edge of g. Consider a node $x \in D$ and a selector σ, for which $\sigma(x) = 1$ and $\sigma(y) = 0$ for all other nodes $y \neq x$. Clearly, the image g^σ represents a graph, where the existing connections from x are removed and the nonexistent connections from x are created. Therefore $G \to G'$ holds if and only if for the corresponding Δ^δ2-structures g and g', $g' = g^\sigma$ for such a selector σ. From this we obtain that $g \sim g'$ if and only if $g' = g^\sigma$ for a selector $\sigma\colon D \to \mathbb{Z}_2$.

For connections of Seidel switching to *signed graphs* and *two-graphs* we refer to Harary [28], Seidel and Taylor [41] and Zaslavsky [44], [45].

Example 6.23
Let $\Delta = (\mathbb{Z}_2, +)$, and let δ be the identity mapping of Δ. Let $D = \{x_1, x_2, x_3, x_4\}$ and consider a Δ^δ2-structure g as in Figure 6.32, where a line denotes value 1.

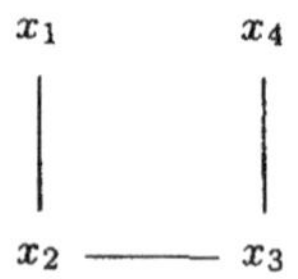

Figure 6.32: g with labels in $\mathbb{Z}_2$

There are $2^{|D|} = 16$ different selectors $\sigma\colon D \to \Delta$, but some of them have the same image g^σ. In fact, there are only 8 different images g^σ as depicted in Figure 6.33, where again the nodes are labeled by the values of a selector σ which applied to g yields g^σ.

As can be observed from this example, it can happen that $g(e_1) = g(e_2)$, but, nevertheless, $g^\sigma(e_1) \neq g^\sigma(e_2)$ for some edges $e_1, e_2 \in E_2(D)$. Notice also that g is primitive, but in $[g]$ there are nonprimitive structures. □

The notion of a substructure carries over to dynamic Δ^δ2-structures as follows.

Lemma 6.7.1
Let $[g]$ be a dynamic Δ^δ2-structure, and let $X \subseteq dom(g)$. Then

$$[sub_g(X)] = \{sub_{g^\sigma}(X) \mid \sigma \ \ selector\}. \qquad \square$$

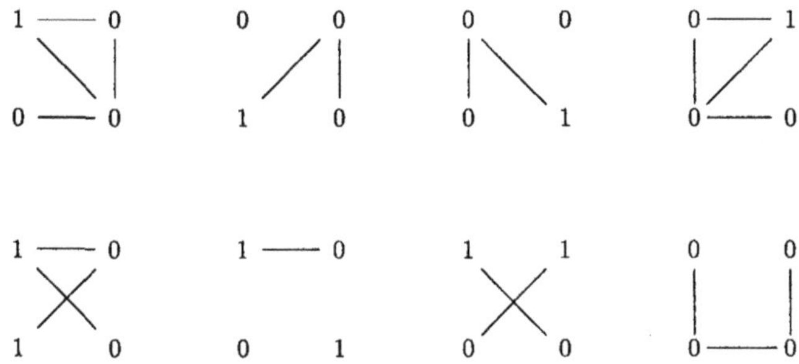

Figure 6.33: The images g^σ

We denote by $\mathrm{sub}_{[g]}(X) = [\mathrm{sub}_g(X)]$ for each dynamic Δ^δ2-structure g and subset $X \subseteq \mathrm{dom}(g)$.

The following lemma shows that the image g^σ under a selector σ of a Δ^δ2-structure g is also δ-reversible.

Lemma 6.7.2
For a Δ^δ2-structure g and a selector $\sigma\colon D \to \Delta$, the image g^σ is also a Δ^δ2-structure.

Proof
Let σ be a selector and g a Δ^δ2-structure. For each edge $e = (x, y) \in E_2(D)$ we have $g(e^{-1}) = \delta(g(e))$, and hence

$$\begin{aligned} g^\sigma(e^{-1}) &= \sigma(y) \cdot g(e^{-1}) \cdot \delta(\sigma(x)) = \sigma(y) \cdot \delta(g(e)) \cdot \delta(\sigma(x)) \\ &= \delta^2(\sigma(y)) \cdot \delta(g(e)) \cdot \delta(\sigma(x)) = \delta(\sigma(x) \cdot g(e) \cdot \delta(\sigma(y))) = \delta(g^\sigma(e)). \end{aligned}$$

Therefore g^σ is also δ-reversible. □

The *product* of two selectors $\sigma_1, \sigma_2\colon D \to \Delta$ is defined by

$$\sigma_2\sigma_1(x) = \sigma_2(x)\sigma_1(x) \qquad (x \in D).$$

Clearly, $\sigma_2\sigma_1$ is a selector.

Lemma 6.7.3
For any selectors $\sigma_1, \sigma_2\colon D \to \Delta$ and Δ^δ2-structure g, $g^{\sigma_2\sigma_1} = (g^{\sigma_1})^{\sigma_2}$. □

It is now immediate that *the selectors form a group* under this product. The inverse σ^{-1} of a selector σ becomes defined by $\sigma^{-1}(x) = \sigma(x)^{-1}$ for $x \in D$.

In group theoretic terms the group of selectors *acts on* the Δ^δ2-structures, see Rotman [38].

The following lemma is now easy to prove.

Lemma 6.7.4
Let g be a Δ^δ2-structure. Then $[g^\sigma] = [g]$ for all $\sigma\colon D \to \Delta$. □

6.7.4 Clans of a dynamic Δ^δ 2-structure

Let $[g]$ be a dynamic Δ^δ2-structure. A clan of each individual $h \in [g]$ is called a *clan* of $[g]$. We shall write

$$\mathcal{C}[g] = \bigcup_{h \in [g]} \mathcal{C}(h)$$

for the family of clans of $[g]$.

The closure properties of the clans of a dynamic Δ^δ2-structure $[g]$ are quite different from those of ordinary ℓ2-structures.

Example 6.24
Recall that the intersection of two clans $X, Y \in \mathcal{C}(g)$ of an ℓ2-structure g is always a clan. This does not hold in general for the dynamic Δ^δ2-structures. Indeed, consider again $\Delta = \mathbb{Z}_2$, and the following ℓ2-structure g together with its image g^σ:

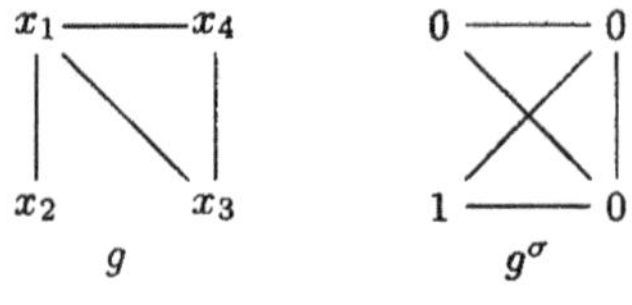

Figure 6.34: A Δ^δ2-structure g and its image g^σ.

Here $X = \{x_2, x_3, x_4\}$ is a clan of g, and $Y = \{x_1, x_2, x_3\}$ is a clan of g^σ, and hence both of these are in $\mathcal{C}[g]$. However, the intersection $X \cap Y = \{x_2, x_3\}$ is not in $\mathcal{C}[g]$ as can be easily verified. □

On the other hand, as shown in Lemma 6.7.5, the clans of $[g]$ are closed under complement. As observed in Section 6.2.3 this is in general not true for ℓ2-structures.

We let $\overline{X}$ denote the complement $D \setminus X$ of $X \subseteq D$.

A set $X \in \mathcal{C}(g)$ is said to be *isolated*, if $g(x,y) = 1_\Delta$ for all $x \in X$ and $y \in \overline{X}$. Hence, if X is an isolated clan, then its complement $\overline{X}$ is also a clan of g.

Lemma 6.7.5
Let g be a Δ^δ2-structure, and let $X \in \mathcal{C}(g)$ be a clan. There exists a selector σ such that X is an isolated clan in g^σ.

Proof
Let $X \in \mathcal{C}(g)$, and define σ as follows $\sigma(x) = 1_\Delta$ for all $x \in X$, and $\sigma(y) = \delta(g(x,y)^{-1})$ for all $y \notin X$. Now, for all $e = (x,y)$ with $x \in X$ and $y \notin X$,

$$g^\sigma(e) = 1_\Delta \cdot g(e) \cdot g(e)^{-1} = 1_\Delta$$

as required. □

In particular, Lemma 6.7.5 implies the following closure result for dynamic Δ^δ2-structures, see [22].

Theorem 6.7.6
The set $\mathcal{C}[g]$ of clans of a dynamic Δ^δ2-structure $[g]$ is closed under complements. □

6.7.5 Horizons

Let $g \colon E_2(D) \to \Delta$ be a Δ^δ2-structure. By Lemma 6.7.5, for each clan $X \in \mathcal{C}[g]$ there exists an $h \in [g]$ such that X and $\overline{X}$ are both clans of h. In particular, this holds for the singleton clans $\{x\}$ of $[g]$, and therefore for each $x \in D$, $D \setminus \{x\} \in \mathcal{C}[g]$.

We shall call a node $x \in D$ a *horizon* of $g \colon E_2(D) \to \Delta$, if $g(x,y) = 1_\Delta$ for all $y \neq x$. By the proof of Lemma 6.7.5, for each g and each $x \in \text{dom}(g)$ x is a horizon of g^{σ_x}, where

$$\sigma_x(x) = 1_\Delta \quad \text{and} \quad \sigma_x(y) = \delta(g(x,y)^{-1}) \text{ for } y \neq x$$

is the *horizontal selector for x in g*.

We have the following uniqueness result for the horizontal selectors, which follows immediately from the definition of a horizon.

Lemma 6.7.7
Let g be a Δ^δ2-structure, and $x \in \text{dom}(g)$. If for a selector σ, $\sigma(x) = 1_\Delta$ and x is a horizon in g^σ, then $\sigma = \sigma_x$. □

A selector $\sigma: D \to \Delta$ is said to be *constant* on a subset $X \subseteq D$, if $\sigma(x_1) = \sigma(x_2)$ for all $x_1, x_2 \in X$.

Lemma 6.7.8
Let $g: E_2(D) \to \Delta$ be a Δ^δ2-structure and $h = g^\sigma$ for a selector σ. If for a subset $Z \subseteq D$ and a node $x \notin Z$, $g(x, z_1) = g(x, z_2)$ and $h(x, z_1) = h(x, z_2)$ for all $z_1, z_2 \in Z$, then σ is constant on Z.

Proof
Suppose $Z \subseteq D$ is as stated in the lemma, and let $z_1, z_2 \in Z$. We have

$$g^\sigma(x, z_i) = \sigma(x) \cdot g(x, z_i) \cdot \delta(\sigma(z_i)) \qquad (i = 1, 2),$$

and hence $\sigma(z_1) = \sigma(z_2)$, because $g^\sigma(x, z_1) = g^\sigma(x, z_2)$ and $g(x, z_1) = g(x, z_2)$. This proves the claim. □

When Lemma 6.7.8 is applied to clans, we obtain the next lemma which states that the derivation of a clan $X \in \mathcal{C}[g]$ in a dynamic $[g]$ depends only on X.

Lemma 6.7.9
Let X be a proper clan of a Δ^δ2-structure g and let σ be a selector. Then $X \in \mathcal{C}(g^\sigma)$ if and only if σ is constant on X. □

As a corollary of the above lemma we obtain that for each clan X of a dynamic Δ^δ2-structure $[g]$ either X or its complement is a clan of any $h \in [g]$ which has a horizon.

Theorem 6.7.10
Let $[g]$ be a dynamic Δ^δ2-structure, and let $h \in [g]$ be such that it has a horizon. Then

$$\mathcal{C}[g] = \{X \mid X \in \mathcal{C}(h) \;\; or \;\; D \setminus X \in \mathcal{C}(h)\}.$$

Proof
Let $x \in D$ be a horizon of h. Let $X \in \mathcal{C}[g]$, and let $Z \in \{X, \overline{X}\}$ be such that $x \notin Z$. By Lemma 6.7.5, there exists a $h' \in [g]$ such that $Z \in \mathcal{C}(h')$ and $\overline{Z} \in \mathcal{C}(h')$. Moreover, since $[g] = [h]$, there exists a selector σ such that $h' = h^\sigma$. Because $Z \in \mathcal{C}(h')$ and x is a horizon in h, Lemma 6.7.8 implies that the selector σ is constant on Z. By Lemma 6.7.9, $Z \in \mathcal{C}(h)$, and thus $\mathcal{C}[g] \subseteq \{X \mid X \in \mathcal{C}(h) \;\text{ or }\; D \setminus X \in \mathcal{C}(h)\}$.
In the other direction the proof is obvious by Theorem 6.7.6. □

6.8 Dynamic ℓ2-structures with Variable Domains

6.8.1 *Disjoint union of ℓ2-structures*

In some situations it is desirable that the domain of an ℓ2-structure can be changed; one may, *e.g.*, want to add or delete nodes of the ℓ2-structure. Such change of a domain is standard in the theory of graph grammars. We shall now consider this problem in the framework of dynamic ℓ2-structures as presented by Ehrenfeucht and Rozenberg [21]. The basic operation that we use to combine two ℓ2-structures is the operation of disjoint union.

Let Δ be a group and δ its involution, and assume that $g\colon E_2(D_1) \to \Delta$ and $h\colon E_2(D_2) \to \Delta$ are disjoint Δ^δ2-structures, *i.e.*, $D_1 \cap D_2 = \emptyset$. We define the *disjoint union* of g and h as the ℓ2-structure $g+h\colon E_2(\text{dom}(g) \cup \text{dom}(h)) \to \Delta$ such that

$$(g+h)(e) = \begin{cases} g(e), & \text{if } e \in E_2(D_1), \\ h(e), & \text{if } e \in E_2(D_2), \\ 1_\Delta, & \text{otherwise.} \end{cases} \tag{6.2}$$

Clearly, $g+h$ is a Δ^δ2-structure, see Figure 6.35.

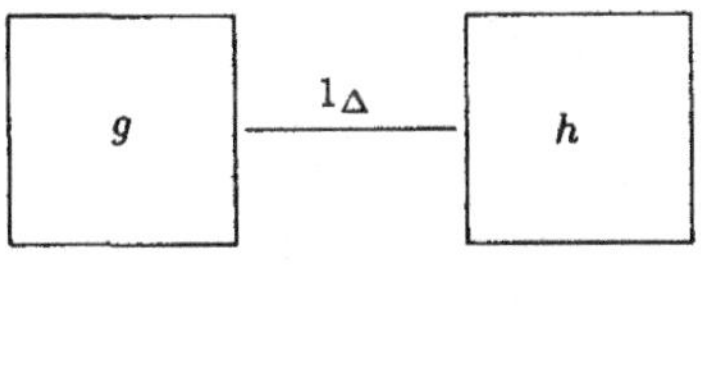

Figure 6.35: Disjoint union $g+h$

The following lemma is immediate from the definition.

Lemma 6.8.1
Let g and h be disjoint Δ^δ2-structures. Then $\text{dom}(g) \in \mathcal{C}([g+h])$ and $\text{dom}(h) \in \mathcal{C}([g+h])$. □

By Lemma 6.7.5, for any ℓ2-structure $g\colon E_2(D) \to \Delta$ and for any clan $X \in \mathcal{C}([g])$, there exists a selector σ such that

$$g = (\text{sub}_g(D \setminus X) + \text{sub}_g(X))^\sigma.$$

In particular, when we apply this result to the singleton clans, we obtain a construction of g from the singleton ℓ2-structures as follows: Let $D = \{x_1, x_2, \ldots, x_n\}$, and denote for simplicity by x the singleton ℓ2-structure with domain $\{x\}$. There are then Δ^δ2-structures g_i and selectors σ_i for $i = 1, 2, \ldots, n-1$ such that $\mathrm{dom}(g_i) = \{x_1, \ldots, x_i\}$, $g_{i+1} = (g_i + x_{i+1})^{\sigma_i}$ and $g = g_n$.

6.8.2 *Comparison with grammatical substitution*

In the theory of graph grammars the derivations of graphs are based on the operation of grammatical substitution. We shall now consider this operation in the context of dynamic Δ^δ2-structures.

Let g and h be Δ^δ2-structures with Δ as their group of labels. Assume that $x \in \mathrm{dom}(g)$ and that $(\mathrm{dom}(g) \setminus \{x\}) \cap \mathrm{dom}(h) = \emptyset$. The *grammatical substitution of h for x in g* is the Δ^δ2-structure $g(x \leftarrow h)\colon (\mathrm{dom}(g) \setminus \{x\}) \cup \mathrm{dom}(h) \to \Delta$ such that

$$g(x \leftarrow h)(y,z) = \begin{cases} g(y,z), & \text{if } (y,z) \in E_2(\mathrm{dom}(g) \setminus \{x\}), \\ h(y,z), & \text{if } (y,z) \in E_2(\mathrm{dom}(h)), \\ g(y,x), & \text{if } y \in \mathrm{dom}(g) \setminus \{x\} \text{ and } z \in \mathrm{dom}(h), \\ g(x,z), & \text{if } z \in \mathrm{dom}(g) \setminus \{x\} \text{ and } y \in \mathrm{dom}(h). \end{cases} \quad (6.3)$$

The operation of grammatical substitution as defined above corresponds to the case of node label controlled (eNLC) graph grammars (see 1) where the connection relation is hereditary in the sense that the labels of the edges adjacent to the 'mother node' carry over without change to the edges adjacent to the daughter graph.

In the following if $f = g + h$, then we denote $h = f - g$. For a Δ^δ2-structure g and a node x, let σ_x be the horizontal selector for x. The following result from [21] states that the operation of substitution can be expressed by the operations of disjoint union and difference, and by (horizontal) selectors.

Theorem 6.8.2
Let g and h be Δ^δ2-structures, and let $x \in \mathrm{dom}(g)$ be such that $(\mathrm{dom}(g) \setminus \{x\}) \cap \mathrm{dom}(h) = \emptyset$. Then

$$g(x \leftarrow h) = \sigma_x^{-1}((\sigma_x(g) - x) + h). \qquad \square$$

Let g and h be disjoint Δ^δ2-structures. We say that a Δ^δ2-structure f *decomposes directly into g, h*, denoted $f \Rightarrow \{g, h\}$, if $f \in [g + h]$, *i.e.*, if there exists a selector σ such that $f = (g+h)^\sigma$. Further, a dynamic Δ^δ2-structure F *decomposes directly into* dynamic Δ^δ2-structures G and H, denoted $F \Rightarrow \{G, H\}$, if there exist an $f \in F$, $g \in G$ and $h \in H$ such that $f \Rightarrow \{g, h\}$, *i.e.*, if $F = [g+h]$ for some $g \in G$ and $h \in H$.

The next theorem shows that a direct decomposition of F amounts to decomposing F into clans.

Theorem 6.8.3
Let F, G and H be dynamic Δ^δ2-structures, and let the domain of F be D. $F \Rightarrow \{G, H\}$ if and only if there exists a nonempty proper clan $X \in \mathcal{C}(F)$ such that $G = \mathrm{sub}_F(X)$ and $H = \mathrm{sub}_F(D \setminus X)$. □

6.8.3 *Amalgamated union*

We relax now the disjointness condition of the domains and consider the situation, where two Δ^δ2-structures g and h share a common substructure.

Let g and h be two Δ^δ2-structures with $A = \mathrm{dom}(g) \cap \mathrm{dom}(h)$ such that $\mathrm{sub}_g(A) = \mathrm{sub}_h(A)$, $\mathrm{dom}(g) \setminus A \in \mathcal{C}(g)$ and $\mathrm{dom}(h) \setminus A \in \mathcal{C}(h)$. We define the *amalgamated union* $g \amalg h$ of g and h as the Δ^δ2-structure

$$(g \amalg h)(x, y) = \begin{cases} g(x, y), & \text{if } (x, y) \in E_2(\mathrm{dom}(g)), \\ h(x, y), & \text{if } (x, y) \in E_2(\mathrm{dom}(h)), \\ 1_\Delta, & \text{otherwise.} \end{cases} \tag{6.4}$$

Clearly, $g \amalg h$ is a Δ^δ2-structure, see Figure 6.36.

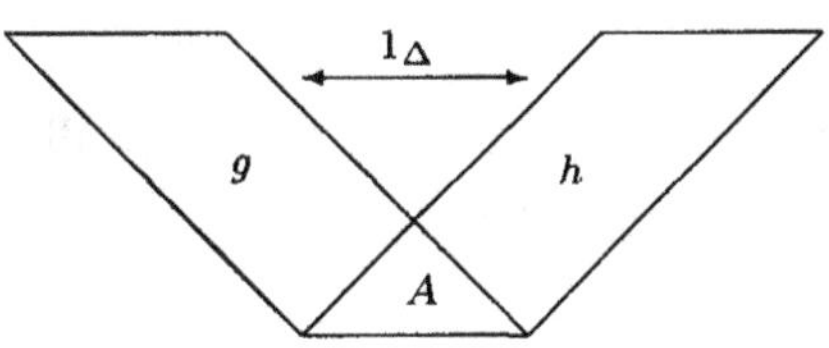

Figure 6.36: Amalgamated union $g \amalg h$

Lemma 6.8.4
Let $f = g \amalg h$. Then $\mathrm{dom}(g) \in \mathcal{C}([f])$ and $\mathrm{dom}(h) \in \mathcal{C}([f])$. □

The following decomposition result for dynamic Δ^δ2-structure using amalgamated unions was proved in [21].

Theorem 6.8.5
Let F be a dynamic Δ^δ2-structure on a domain D. If $X, Y \in \mathcal{C}(F)$ are nonempty and $X \cup Y = D$, then there exist $g \in sub_F(X)$ and $h \in sub_F(Y)$ such that $F = [g \amalg h]$. □

One can now define the direct decomposition relation based on amalgamated unions analogously to the direct decomposition relation based on disjoint unions. For this let g and h be Δ^δ2-structures with $A = \mathrm{dom}(g) \cap \mathrm{dom}(h)$ such that $\mathrm{sub}_g(A) = \mathrm{sub}_h(A)$, $\mathrm{dom}(g) \setminus A \in \mathcal{C}(g)$ and $\mathrm{dom}(h) \setminus A \in \mathcal{C}(h)$. We say that a Δ^δ2-structure f *directly amalgam decomposes into* g, h, denoted $f \Rightarrow^{\amalg} \{g, h\}$, if $f \in [g \amalg h]$. Further, we write $F \Rightarrow^{\amalg} \{G, H\}$ for dynamic Δ^δ2-structures F, G and H, if $F = [g \amalg h]$ for some $g \in G$ and $h \in H$.

We obtain for amalgamated unions an analogous result to Theorem 6.8.3:

Theorem 6.8.6
Let $F = [f]$ be a dynamic Δ^δ2-structure. Then $F \Rightarrow^{\amalg} \{G, H\}$ if and only if there exist nonempty $X, Y \in \mathcal{C}(F)$ such that $X \cup Y = \mathrm{dom}(f)$, $G = sub_F(X)$ and $H = sub_F(Y)$. □

6.9 Quotients and Plane Trees

6.9.1 Quotients of dynamic Δ^δ2-structures

Let $[g]$ be a dynamic Δ^δ2-structure. We define now the factorizations and quotients of $[g]$. These constructions are not quite so straightforward as for 2-structures, because the clans $\mathcal{X} \subseteq \mathcal{C}[g]$ forming a partition of the domain D of $[g]$ may come from different images of g.

A partition $\mathcal{X}$ of the domain D into clans of $[g]$ is called a *factorization* of $[g]$. The *quotient* of $[g]$ by $\mathcal{X}$ is the family

$$[g]/\mathcal{X} = \{h/\mathcal{X} \mid h \in [g], \mathcal{X} \text{ a factorization of } h\}.$$

The family $[g]/\mathcal{X}$ is well-defined in the sense that each $h/\mathcal{X}$ is a Δ^δ2-structure. Below we show that for each factorization $\mathcal{X} \subseteq \mathcal{C}[g]$, the quotient $[g]/\mathcal{X}$ is nonempty and, indeed, it is the dynamic Δ^δ2-structure $[h/\mathcal{X}]$ for some $h \in [g]$.

If $\sigma(x) = a$ is a constant selector on the whole domain D, then for all $e \in E_2(D)$, $g^\sigma(e) = a \cdot g(e) \cdot \delta(a)$. Clearly, in this case, g^σ is strictly isomorphic with g. Hence the following lemma holds.

Lemma 6.9.1
Let g be a $\Delta^\delta 2$-structure. If $h = g^\sigma$ for a selector σ, which is constant on D, then h and g are strictly isomorphic. In this case, the substructures $\text{sub}_g(X)$ and $\text{sub}_h(X)$ and the quotients $g/\mathcal{X}$ and $h/\mathcal{X}$ are strictly isomorphic for all $X \subseteq D$ and all factorizations $\mathcal{X}$ of g. □

The following theorem was proved in Ehrenfeucht, Harju and Rozenberg [12].

Theorem 6.9.2
If $\mathcal{X}$ is a factorization of a dynamic $[g]$, then there exists a selector σ such that $\mathcal{X}$ is a factorization of g^σ, and

$$[g]/\mathcal{X} = [g^\sigma/\mathcal{X}].$$

Proof
Let $\mathcal{X} = \{X_1, X_2, \dots, X_m\}$ and $[g]$ be as stated in the claim. For each $X \in \mathcal{X}$, we have $X \in \mathcal{C}(g^{\sigma_X})$ for some selector σ_X. By Lemma 6.7.9, we may assume that $\sigma_X(y) = 1_\Delta$ for all $y \notin X$. Define $\sigma = \sigma_{X_m}\sigma_{X_{m-1}}\cdots\sigma_{X_1}$. Hence $\sigma(x) = \sigma_X(x)$, if $x \in X$, since $\mathcal{X}$ is a factorization of $[g]$. Moreover, by the choice of the selectors σ_{X_i}, $\sigma = \sigma_{X_{\pi(m)}}\sigma_{X_{\pi(m-1)}}\cdots\sigma_{X_{\pi(1)}}$ for any permutation π of $\{1, 2, \dots, m\}$. Now, $g^\sigma = (g^{\sigma_X})^{\sigma\cdot\sigma_X^{-1}}$, where $\sigma \cdot \sigma_X^{-1}$ is constant on X. Hence by Lemma 6.7.9, $X \in \mathcal{C}(g^\sigma)$ for each $X \in \mathcal{X}$, and thus $\mathcal{X}$ is a factorization of $h = g^\sigma$.

Now, if $\mathcal{X}$ is a factorization of some $g_1 \in [g]$, then from Lemma 6.7.4 we obtain that $g_1 = h^{\sigma_1}$ for a selector σ_1. Consequently, Lemma 6.7.9 yields that σ_1 is constant on each $X \in \mathcal{X}$, and, therefore $g_1/\mathcal{X} = (h/\mathcal{X})^{\sigma_2}$ for the selector $\sigma_2 \colon \mathcal{X} \to \Delta$ defined by $\sigma_2(X) = \sigma_1(x)$ ($x \in X$) for all $X \in \mathcal{X}$.

On the other hand, if $\sigma_2 \colon \mathcal{X} \to \Delta$ is a selector, then it can be extended to a selector $\sigma_1 \colon D \to \Delta$ by setting $\sigma_1(x) = \sigma_2(X)$, if $x \in X$ with $X \in \mathcal{X}$. Evidently, $h^{\sigma_1}/\mathcal{X} = (h/\mathcal{X})^{\sigma_2}$, and thus $[g]/\mathcal{X} = \{(h/\mathcal{X})^{\sigma_2} \mid \sigma_2 \text{ a selector}\}$ as required. □

6.9.2 Plane trees

In general, two given g_1 and g_2 in $[g]$ can be quite different from each other. Indeed, the clan structures of g_1 and g_2 may be drastically different as shown in the following example.

Example 6.25
Let $D = \{x_1, x_2, x_3\}$, $\Delta = \mathbb{Z}_5$ and let δ be the inverse function of Δ. Let g be the complete $\Delta^\delta 2$-structure corresponding to the label 0, *i.e.*, $g(e) = 0$ for

all $e \in E_2(D)$. Hence all subsets of D are clans of g. Let then σ be a selector defined by $\sigma(x_1) = 0$, $\sigma(x_2) = 1$ and $\sigma(x_3) = 3$. Then $h = g^\sigma$ is specified by $h(x_1, x_2) = 4$, $h(x_2, x_3) = 3$ and $h(x_3, x_1) = 3$. Now, h is primitive – has only the trivial clans. □

Let $G = [g]$ be a dynamic Δ^δ2-structure, and assume that x is a horizon of g. The substructure

$$\pi_x(g) = \text{sub}_g(D \setminus \{x\})$$

is called the *x-plane* of g (or an *x-plane* of G). We shall denote by G_π^x the family of x-planes of G.

Since for each node x of a $G = [g]$, $\{x\} \in \mathcal{C}[g]$, Lemma 6.7.5 implies that there exists an $h \in G$ such that x is a horizon of h. Hence G_π^x is always nonempty.

The following result from [12] shows that the planes in G_π^x are quite similar to each other.

Theorem 6.9.3
Let G be a dynamic Δ^δ2-structure. The planes in G_π^x are strictly isomorphic.

Proof
Let $g_1, g_2 \in G$ be such that x is a horizon in both of them, and let $h_i = \pi_x(g_i)$ for $i = 1, 2$. By Lemma 6.7.4, there exists a selector σ such that $g_2 = g_1^\sigma$. Clearly, $h_2 = h_1^{\sigma_1}$, where $\sigma_1 : D \setminus \{x\} \to \Delta$ is the restriction of σ. Since $D \setminus \{x\}$ is a common clan of g_1 and g_2, Lemma 6.7.9 implies that σ is constant on $D \setminus \{x\}$. By Lemma 6.9.1, h_1 and h_2 are strictly isomorphic, since h_1 and h_2 have the same domain $D \setminus \{x\}$. □

In particular, it follows from Lemma 6.9.1 that for any two planes $h_1, h_2 \in G_\pi^x$, $\mathcal{C}(h_1) = \mathcal{C}(h_2)$ and $\mathcal{P}(h_1) = \mathcal{P}(h_2)$. It follows also that for all $X \in \mathcal{P}(h_1)$ the special quotients $\Psi_{h_1}(X)$ and $\Psi_{h_2}(X)$ are of the same type, *i.e.* they are both primitive or both complete or both linear. This justifies the following definition, where an element $X \in \mathcal{P}(h_1)$ obtains a label p, c or ℓ according to the type of $\Psi_{h_1}(X)$.

Let $x \in D$ for a dynamic Δ^δ2-structure G.
(1) Denote $\mathcal{C}(x) = \mathcal{C}(h_x)$, $\mathcal{P}(x) = \mathcal{P}(h_x)$ and $\mathcal{N}(x) = \mathcal{P}(h_x) \cup \{x\}$, where h_x is an arbitrary element of G_π^x.
(2) Define $\Phi_x(\{x\}) = c$ and for all $X \in \mathcal{P}(h_x)$, let $q_X = \Psi_{h_x}(X)$ and

$$\Phi_x(X) = \begin{cases} p, & \text{if } q_X \text{ is truly primitive,} \\ c, & \text{if } q_X \text{ is complete,} \\ \ell, & \text{if } q_X \text{ is linear and } |q_X| > 1. \end{cases} \tag{6.5}$$

In the above definition we have the requirement $|q_X| > 1$ in the case of linear q_X in order to have Φ_x well-defined, because a singleton quotient is, by definition, both complete and linear.

(3) The *plane tree* $L(x)$ of the node x is the node-labeled directed rooted tree with the set of nodes $\mathcal{N}(x)$ labeled by Φ_x, which is obtained from the prime tree $T(h_x)$ by adding $\{x\}$ as a new root and $(x, D \setminus \{x\})$ as a new edge.

Example 6.26
Let $\Delta = \mathbb{Z}_3$ and let δ be the inverse function of Δ. Consider the Δ^δ2-structure g from Figure 6.37.

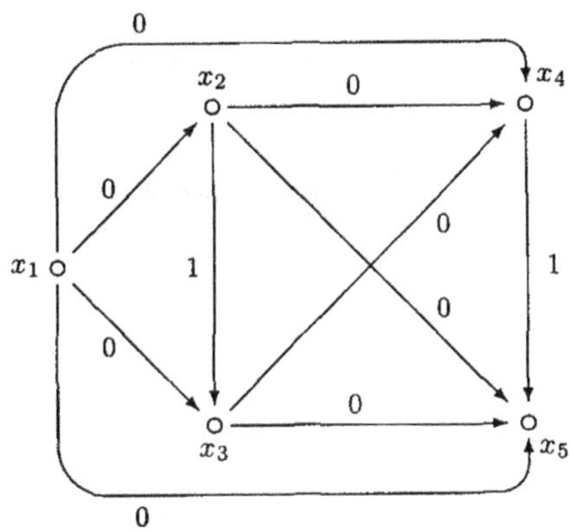

Figure 6.37: Δ^δ2-structure g

We note that g has a horizon x_1. The x_1-plane h_1 of g is given in Figure 6.38. The nontrivial prime clans of h_1 are $\{x_2, x_3\}$ and $\{x_4, x_5\}$, and the plane tree of x_1 is given in Figure 6.39, where a nonsingleton node X has been labeled by d, if $\Phi_x(X) = d$. Again, each nonsingleton node consists of the leaves that are its descendants in the directed tree, and so we have omitted the identities of these nodes in Figure 6.39. The singleton nodes x_i have been identified, but without the label c.

Let us then make x_4 a horizon. This is done by the horizonal selector σ as follows: $\sigma(x_4) = 0$ and $\sigma(x_i) = g(x_4, x_i)$ for $i \neq 4$. Now, the x_4-plane h_4 of g^σ is given in Figure 6.40.

The nontrivial primes of h_4 are $\{x_1, x_2, x_3\}$ and $\{x_2, x_3\}$, and the plane tree of x_4 is given in Figure 6.41.

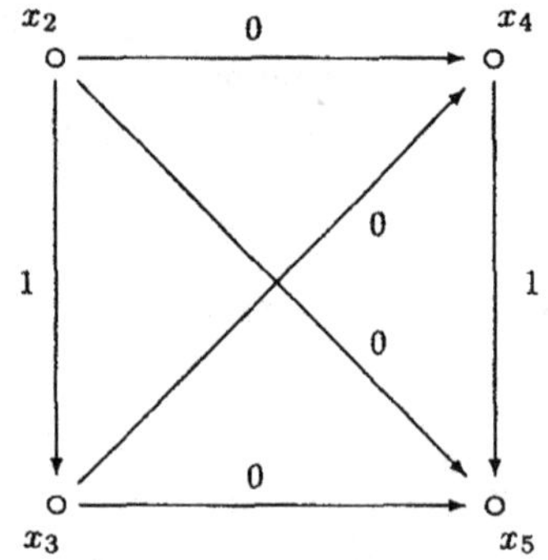

Figure 6.38: x_1-plane of g

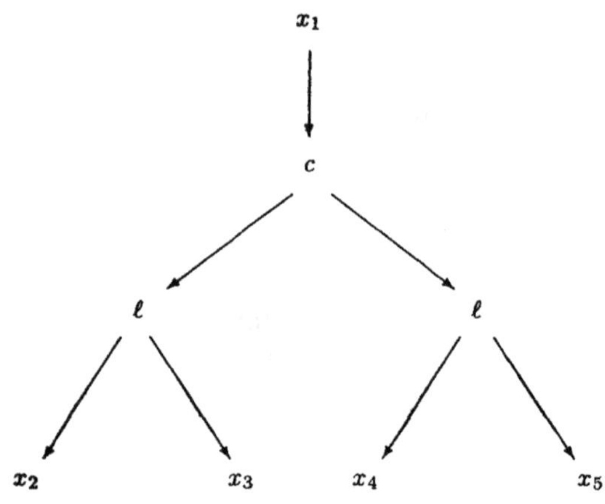

Figure 6.39: Plane tree $L(x_1)$

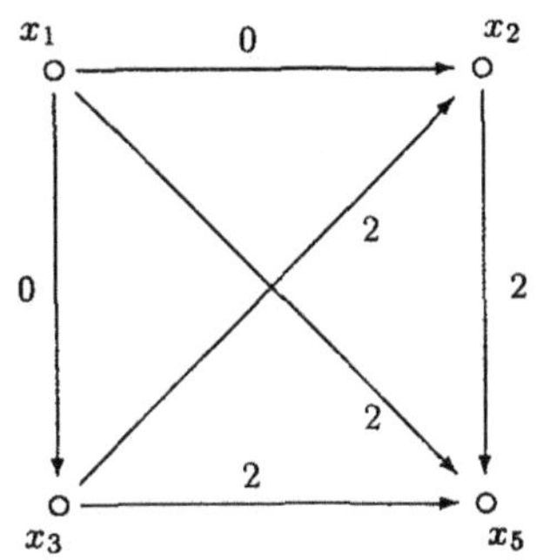

Figure 6.40: Plane h_4

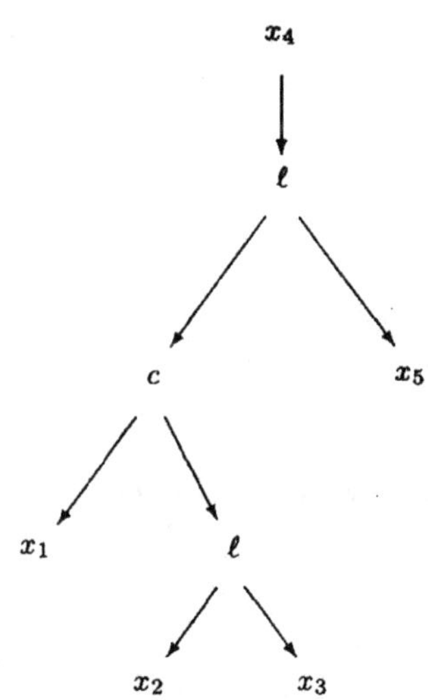

Figure 6.41: Plane tree $L(x_4)$

The plane trees $L(h_1)$ and $L(h_4)$ are not isomorphic, but their underlying unrooted and undirected trees are node-isomorphic, and indeed, as we shall see, the quotients of the corresponding nodes are of the same size and type. □

By Theorem 6.7.10 we have the following lemma.

Lemma 6.9.4
Let x be a node of a dynamic Δ^δ2-structure G. Then for each clan X of G such that $X \notin \{\emptyset, D\}$, $X \in \mathcal{C}(x)$, if $x \notin X$, and $\overline{X} \in \mathcal{C}(x)$, if $x \in X$.

Proof
Let x be a horizon of $g \in G$ and let $\pi_x(g) = h$. Assume X is a clan of G.

Assume $Y \in \{X, \overline{X}\}$ is such that $Y \in \mathcal{C}(g)$. If $x \notin Y$, then since the domain of h, $D \setminus \{x\}$, is a clan of g, Lemma 6.6.5 implies that $Y \in \mathcal{C}(h)$.

On the other hand, if $x \in Y$, then $\overline{Y} \in \mathcal{C}(g)$. Indeed, for all $y \in Y$ $(y \neq x)$ and $z \notin Y$ we have that $g(y,x) = 1_\Delta = g(z,x)$, since x is a horizon of g. Moreover, since $Y \in \mathcal{C}(g)$, $g(x,z) = g(y,z)$. Hence $g(y,z) = 1_\Delta$ for all $y \in Y$ and $z \notin Y$, which shows that $\overline{Y} \in \mathcal{C}(g)$. In this case $\overline{Y} \in \mathcal{C}(h)$ by Lemma 6.6.5. Now the claim follows from Theorem 6.7.10. □

Our purpose is to compare the planes of a dynamic Δ^δ2-structure G. For this we first study the relationships between the prime clans.

Lemma 6.9.5
Let G be a dynamic Δ^δ2-structure, and let x, y be two nodes of its domain. Assume that $X \in \mathcal{P}(x)$.
(1) If $y \notin X$, then $X \in \mathcal{P}(y)$.
(2) If $y \in X$, then $\overline{X} \in \mathcal{P}(y)$. In particular, $\mathcal{P}(x)$ and $\mathcal{P}(y)$ have the same number of prime clans.

Proof
The claim follows directly from Lemma 6.9.4, when we observe that a clan Y of G overlaps with a subset $Z \subseteq D$ if and only if Y overlaps with $\overline{Z}$. □

Example 6.27
Consider the plane trees $L(x_1)$ and $L(x_4)$ of Example 6.26 (see, Figures 6.39 and 6.41). Denote $A = \{x_2, x_3\}$ and $B = \{x_4, x_5\}$. We have redrawn these plane trees in Figure 6.42 in order to illustrate the 1-1 correspondence between them. Hence $L(x_4)$ is obtained from $L(x_1)$ by reversing the direction of the path from x_1 to x_4, and each inner node of this reversed path is the complement of its successor in the original path in $L(x_1)$; hence, *e.g.*, the node B becomes $\overline{x_4}$. The singleton node x_4 that begins the reversed path does not

change, because it does not have a successor in $L(x_1)$. Note that the corresponding nodes in these plane trees are of the same type. □

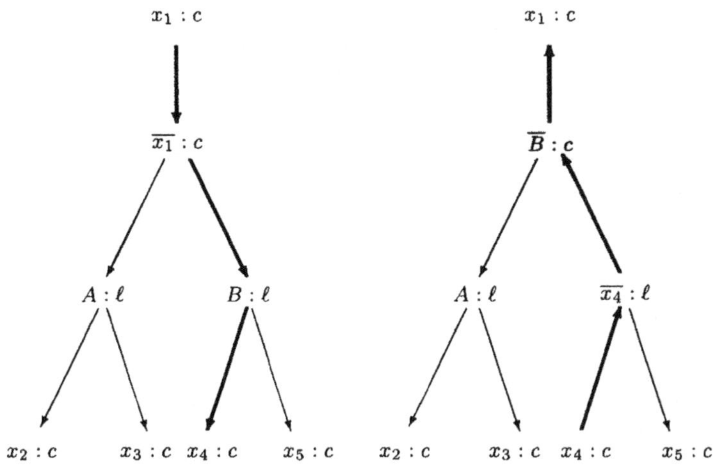

Figure 6.42: Comparison of two plane trees

Following Lemma 6.9.5 (and the previous example) we define for all $x, y \in D$ the mapping $\eta_{xy} \colon \mathcal{N}(x) \to \mathcal{N}(y)$ as follows:

$$\eta_{xy}(X) = \begin{cases} X, & \text{if } y \notin X \text{ or } X = \{y\}, \\ \overline{Y}, & \text{if } y \in X,\ y \in Y \text{ and } (X, Y) \text{ is an edge in } L(x). \end{cases} \tag{6.6}$$

Lemma 6.9.6
Let G be a dynamic Δ^δ2-structure, and let x, y be two of its nodes. Then η_{xy} is a bijective correspondence between $\mathcal{N}(x)$ and $\mathcal{N}(y)$.

Proof
First of all η_{xy} is well-defined. Indeed, for each $X \in \mathcal{N}(x)$ with $y \in X$ and $X \neq \{y\}$ there is a unique edge (X, Y) in $L(x)$ such that $y \in Y$ and $Y \subset X$. Now, by Lemma 6.9.5, $\overline{Y} \in \mathcal{N}(y)$ (with $y \notin \overline{Y}$), and, by the definition of η_{xy}, $\eta_{xy}(X) = \overline{Y}$.

The injectiveness claim follows from the above stated uniqueness of the edge (X, Y), since if $y \in X$ for some $X \in \mathcal{N}(x)$ with $X \neq \{y\}$, then for the unique $Y \in \mathcal{N}(x)$ with $\eta_{xy}(X) = \overline{Y}$ we have that $\overline{Y} \notin \mathcal{N}(x)$ (because $x \in \overline{Y}$). It follows that the mapping η_{xy} is surjective, because the sets $\mathcal{N}(x)$ and $\mathcal{N}(y)$ are finite and have the same number of elements. □

The following result from [12], which we state without proof, gives a simple construction that allows one to obtain the plain tree $L(y)$ from $L(x)$. Indeed, *$L(y)$ is obtained from $L(x)$ by first changing the nodes by the mapping η_{xy}, and then reversing the direction of the path from x to y.*

Theorem 6.9.7
Let G be a dynamic Δ^δ2-structure and let x, y be two of its nodes. If (X, Y) is an edge of $L(x)$, then
(1) $(\eta_{xy}(X), \eta_{xy}(Y))$ is in $L(y)$, if $y \notin Y$,
(2) $(\eta_{xy}(Y), \eta_{xy}(X))$ is in $L(y)$, if $y \in Y$. □

The correspondence η_{xy} also preserves the types of the quotients:

Theorem 6.9.8
Let G be a dynamic Δ^δ2-structure and let x, y be two of its nodes. Then $\Phi_x(X) = \Phi_y(\eta_{xy}(X))$ for all $X \in \mathcal{N}(x)$. □

Now, let x be a node of a dynamic Δ^δ2-structure G. Define $F(x) = (\mathcal{N}(x), \Phi_x)$ as the *form* of x, that is, $F(x)$ is undirected and unrooted tree such that $\{X, Y\}$ is an edge in $F(x)$ if and only if either (X, Y) or (Y, X) is an edge of $L(x)$. The labeling $\Phi_x : \mathcal{N}(x) \to \{p, c, \ell\}$ of the nodes remains the same as in $L(x)$.

Theorem 6.9.9
Let x, y be two nodes in a dynamic Δ^δ2-structure G. The forms $F(x)$ and $F(y)$ are isomorphic. Indeed, $\eta_{xy} : \mathcal{N}(x) \to \mathcal{N}(y)$ is a label preserving isomorphism from $F(x)$ onto $F(y)$.

Proof
By Theorem 6.9.7, if $\{X, Y\}$ is an edge of $F(x)$, then $\{\eta_{xy}(X), \eta_{xy}(Y)\}$ is an edge of $F(y)$. Moreover, η_{xy} is a bijection from $\mathcal{N}(x)$ onto $\mathcal{N}(y)$. This shows that η_{xy} is an isomorphism between $F(x)$ and $F(y)$.

By Theorem 6.9.8, $\Phi_x(X) = \Phi_y(\eta_{xy}(X))$ for all $X \in \mathcal{N}(x)$, and hence η_{xy} preserves the labels of the nodes. This proves the claim. □

Theorem 6.9.9 allows us to construct a tree $F(G)$, the *form* of G, for each dynamic Δ^δ2-structure G as follows. Consider the tree $F(x)$ for a node x and leave the nodes unidentified, but labeled by Φ_x. Hence a tree $F(G)$ becomes an 'abstract tree' in the sense that the identities of the nodes are omitted. To be more precise, $F(G)$ is the isomorphism class of the forms $F(x)$ with x in the domain of G.

Example 6.28
The form $F(G)$ of the the dynamic $\Delta^{\delta}2$-structure G from Example 6.26 (see, Figure 6.37), is given in Figure 6.42. □

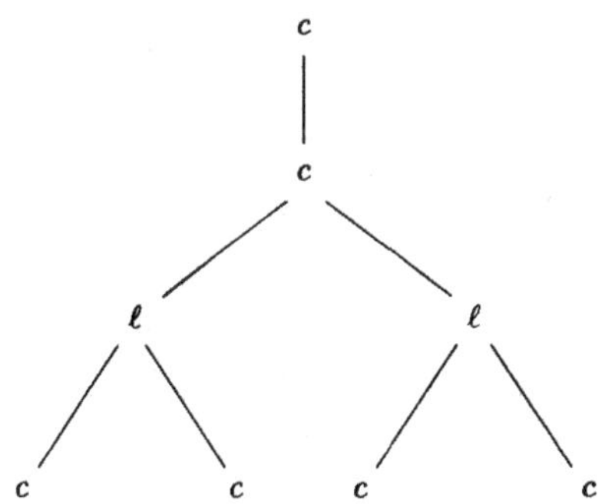

Figure 6.43: The form $F(G)$

6.10 Invariants

6.10.1 Introduction to invariants

In this section we shall consider the problem how to recognize that an $\ell 2$-structure g is obtainable from another $\ell 2$-structure h by an application of a selector. In order to answer this question invariant properties of $\ell 2$-structures were studied by Ehrenfeucht, Harju and Rozenberg [11], where the invariants were shown to be closely connected to the verbal identities of the group Δ of labels. In this paper we shall not go into these group theoretical considerations, and furthermore, we state a detailed characterization only for the abelian case (in Section 6.10.4). For the proofs and further results we refer to [11].

In this section *the involution δ of the given group Δ will be the inverse function of Δ,*

$$\delta(a) = a^{-1} \qquad (a \in \Delta).$$

For this reason we call a $\Delta^{\delta}2$-structure more conveniently an *inversive $\ell 2$-structure*.

Denote by $\mathcal{R}(D, \Delta)$ the family of all inversive $\ell 2$-structures $g\colon E_2(D) \to \Delta$ for a given domain D and for a given group Δ of labels.

The inversive $\ell 2$-structures in $\mathcal{R}(D, \Delta)$ are naturally divided into dynamic $\ell 2$-structures, *i.e.*, the family $\{[g] \mid g \in \mathcal{R}(D, \Delta)\}$ forms a partition of $\mathcal{R}(D, \Delta)$.

A function $\eta\colon \mathcal{R}(D,\Delta) \to \Delta$ is an *invariant*, if it maps the elements of each dynamic $[g]$ into the same element,

$$\eta(g_1) = \eta(g_2) \qquad \text{if} \quad [g_1] = [g_2].$$

Hence an invariant is immune to the selectors, and the study of invariants is the study of those properties of a network that remain unchanged during its evolution.

A constant mapping, $\eta(g) = a$ ($a \in \Delta$, $g \in \mathcal{R}(D,\Delta)$) is a simple example of an invariant.

We shall also say that two ℓ2-structures g_1 and g_2 are *equivalent*, if $[g_1] = [g_2]$. Hence a function $\eta\colon \mathcal{R}(D,\Delta) \to \Delta$ is an invariant, if it maps equivalent ℓ2-structures to the same element of the group Δ.

6.10.2 *Free invariants*

In general an invariant, as defined above, is a function that can be independent of the specific properties of the ℓ2-structures. In order to reflect these properties of the ℓ2-structures, we shall restrict ourselves to those invariants that are more faithful to the ℓ2-structures in the sense that they are defined by variables corresponding to the edges.

Consider the free monoid $M(D) = E_2(D)^*$ generated by the edges. The elements of $M(D)$ are the words $e_1e_2\ldots e_k$ for $k \geq 0$ and $e_i \in E_2(D)$. The empty word, denoted by 1, is the identity element of the monoid $M(D)$. For a word $w = e_1e_2\ldots e_k \in M(D)$ we let $w^{-1} = e_k^{-1}e_{k-1}^{-1}\ldots e_1^{-1}$. In this way $M(D)$ becomes a *monoid with involution.*

In order to clarify the distinction between the edges in ℓ2-structures and the generators of the free monoid $M(D)$, the generators $e \in M(D)$ will be called *variables.*

Each word $w = e_1e_2\ldots e_k \in M(D)$ defines a function ψ_w from $\mathcal{R}(D,\Delta)$ into Δ as follows,

$$\psi_w(g) = g(e_1)g(e_2)\ldots g(e_k) \qquad (g \in \mathcal{R}(D,\Delta)).$$

We call the function ψ_w the *variable function represented by* w, and we let $\text{Var}(D,\Delta)$ denote the set of all variable functions represented by the words in $M(D)$.

Two words w_1 and w_2 from $M(D)$ are said to be *equivalent* (*over* Δ), denoted $w_1 \equiv w_2$, if they represent the same variable function, $\psi_{w_1} = \psi_{w_2}$.

Further, if a variable function $\psi_w \in \mathrm{Var}(D,\Delta)$ is an invariant, then it is a *free invariant* of $\mathcal{R}(D,\Delta)$. Denote by $\mathrm{Inv}(D,\Delta)$ the set of all free invariants ψ_w of $\mathcal{R}(D,\Delta)$.

Example 6.29
Let $\Delta = \mathbb{Z}_3 = \{0,1,2\}$ be the cyclic group on three elements. Assume further that $D = \{x_1, x_2, x_3\}$, and write e_{ij} for (x_i, x_j). The word $w_1 = e_{12}e_{23}e_{31}$ represents an invariant, *i.e.*, $\psi_{w_1} \in \mathrm{Inv}(D,\Delta)$, since for all $g \in \mathcal{R}(D,\Delta)$ and all selectors $\sigma\colon D \to \Delta$ with $\sigma(x_i) = a_i$, $\psi_{w_1}(g) = g(e_{12}) + g(e_{23}) + g(e_{31})$ and

$$\begin{aligned}\psi_{w_1}(g^\sigma) &= g^\sigma(e_{12}) + g^\sigma(e_{23}) + g^\sigma(e_{31})\\ &= (a_1 + g(e_{12}) - a_2) + (a_2 + g(e_{23}) - a_3) + (a_3 + g(e_{31}) - a_1)\\ &= g(e_{12}) + g(e_{23}) + g(e_{31}) = \psi_{w_1}(g)\ ,\end{aligned}$$

because Δ is abelian.

For the two structures g_1 and g_2 from Figure 6.44, we have $\psi_{w_1}(g_1) = 2 \neq 1 = \psi_{w_1}(g_2)$, which implies that for all selectors σ, $g_2 \neq g_1^\sigma$, since $\psi_{w_1}(g_1^\sigma) = 2$ but $\psi_{w_1}(g_2) = 1$ for each σ. Hence in this case $[g_1] \neq [g_2]$.

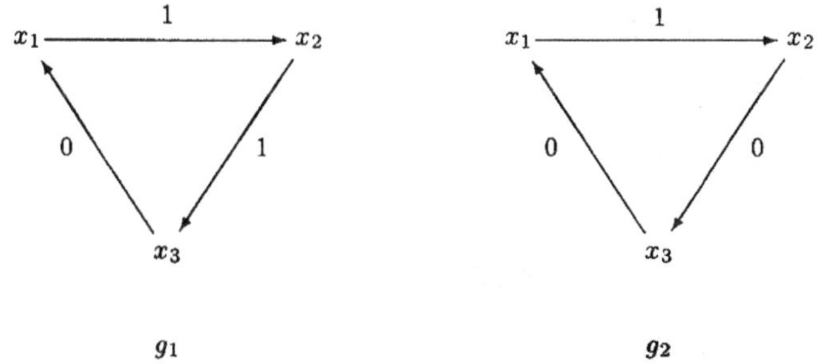

Figure 6.44: Nonequivalent inversive ℓ2-structures

On the other hand, let $w_2 = e_{12}e_{32}e_{31}$. Now, $\psi_{w_2}(g) = g(e_{12}) + g(e_{32}) + g(e_{31})$ and

$$\psi_{w_2}(g^\sigma) = g^\sigma(e_{12}) + g^\sigma(e_{32}) + g^\sigma(e_{31}) = 2a_3 - 2a_2 + \psi_{w_2}(g)\ ,$$

and thus $\psi_{w_2}(g^\sigma) = \psi_{w_2}(g)$ if and only if $2a_3 - 2a_2 = 0$ in $\mathbb{Z}_3$. It follows that ψ_{w_2} is not an invariant of $\mathcal{R}(D,\Delta)$, since, *e.g.*, the selector may have chosen $a_2 = 1$ and $a_3 = 2$. □

The constant functions $\eta\colon \mathcal{R}(D,\Delta) \to \Delta$ are invariants, but as can be easily shown only the trivial constant function is a free invariant.

6.10.3 Basic properties of free invariants

From the definition of $\mathrm{Var}(D,\Delta)$ we obtain that for all words $w, w_1, w_2 \in M(D)$

$$\psi_{w_1 w_2}(g) = \psi_{w_1}(g) \cdot \psi_{w_2}(g)\,, \quad \psi_{w^{-1}}(g) = \psi_w(g)^{-1} \quad \text{and} \quad \psi_1(g) = 1_\Delta,$$

whenever g is an inversive ℓ2-structure. (Here 1 is the empty word). Therefore $\mathrm{Var}(D,\Delta)$ is a group with the operation defined by

$$\psi_{w_1} \cdot \psi_{w_2} = \psi_{w_1 w_2}.$$

The identity of the group $\mathrm{Var}(D,\Delta)$ is the variable function ψ_1 represented by the empty word. Further, we have $\psi_{w^k}(g) = \psi_w(g)^k$ for all $w \in M(D)$, $k \in \mathbb{Z}$ and $g \in \mathcal{R}(D,\Delta)$. Therefore we can write

$$\psi_{w^k} = \psi_w^k \qquad \text{and} \qquad \psi_{w^{-1}} = \psi_w^{-1}.$$

Moreover, $\mathrm{Var}(D,\Delta)$ is generated by the variable functions ψ_e represented by the variables, $e \in E_2(D)$.

The following result from [11] shows that a free invariant ψ_w is, in fact, a mapping $\psi_w\colon \mathcal{R}(D,\Delta) \to Z(\Delta)$ into the center of the group Δ. This implies that the group $\mathrm{Inv}(D,\Delta)$ is always abelian. Recall that the *center* of a group Δ is defined by

$$Z(\Delta) = \{a \in \Delta \mid ax = xa \ \text{ for all } \ x \in \Delta\}.$$

Theorem 6.10.1
(1) *For all free invariants* $\psi_w \in \mathrm{Inv}(D,\Delta)$ *and inversive* g, $\psi_w(g) \in Z(\Delta)$.
(2) $\mathrm{Inv}(D,\Delta)$ *is a subgroup of* $Z(\mathrm{Var}(D,\Delta))$. *In particular,* $\mathrm{Inv}(D,\Delta)$ *is an abelian group.* □

By Theorem 6.10.1, if the center $Z(\Delta)$ is trivial, then there are no nontrivial free invariants for the inversive ℓ2-structures g with the group Δ of labels, see Example 6.30 below.

A word $w = (x_0,x_1)(x_1,x_2)\dots(x_{k-1},x_0)$ from $M(D)$ is called a *closed walk of length* k. A closed walk

$$t(x_0,x_1,x_2) = (x_0,x_1)(x_1,x_2)(x_2,x_0)$$

is called a *triangle* (*at* x_0), if $x_i \neq x_j$ for each $i \neq j$. The empty word $1 \in M(D)$ is a *trivial walk*.

The next result is rather easy to prove.

Theorem 6.10.2
Let $w \in M(D)$ be a closed walk. Then $\psi_w \in \mathrm{Inv}(D, \Delta)$ if and only if $\psi_w(g) \in Z(\Delta)$ for all inversive $\ell 2$-structures g. In particular, if Δ is abelian, then ψ_w is a free invariant of $\mathcal{R}(D, \Delta)$. □

Example 6.30
Consider the dihedral group D_{2n} of $2n$ elements ($n \geq 3$),

$$D_{2n} = \{1, a, a^2, \ldots, a^{n-1}, b, ba, ba^2, \ldots, ba^{n-1}\},$$

where 1 is the identity element of D_{2n}. The group D_{2n} is the symmetry group of a regular n-gon of the Euclidean plane, and it is generated by a rotation a (with an angle of $2\pi/n$) together with a reflection b of the n-gon. These generators satisfy the following defining relations

$$a^n = 1, \qquad b^2 = 1, \qquad ab = ba^{-1}.$$

It is rather easy to show that if n is odd, then $Z(D_{2n})$ is trivial, and hence in this case the group $\mathrm{Inv}(D, \Delta)$ of free invariants is also trivial by Theorem 6.10.1.

On the other hand, if n is even, then $Z(D_{2n})$ contains two elements, *i.e.*, it is isomorphic to the cyclic group $\mathbb{Z}_2$. Let us consider the case $n = 4$. In this case $Z(D_8) = \{1, a^2\}$. Assume that the domain D has at least three nodes, and let w be any closed walk. We claim that the variable function ψ_{w^2} is a free invariant. Here w^2 is the closed walk that traverses w twice around.

Indeed, let g be an inversive $\ell 2$-structure. Then

$$\psi_{w^2}(g) = (\psi_w(g))^2,$$

and it is easy to check that for all $c \in D_8$, $c^2 = 1$ or a^2. Therefore $\psi_{w^2}(g) \in Z(D_8)$ for all $g \in \mathcal{R}(D, \Delta)$, and thus ψ_{w^2} is a free invariant by Theorem 6.10.2. □

6.10.4 *Invariants on abelian groups*

We shall show that in the abelian case the invariants represented by the triangles at x_0 form a 'complete' set of invariants.

Clearly, in the abelian case, for $w_1, w_2 \in M(D)$ we have $w_1 w_2 \equiv w_2 w_1$, and hence the occurrences of the variables in $w \in M(D)$ can be freely permuted without violating invariant properties.

For an abelian group Δ the group $\mathrm{Inv}(D, \Delta)$ has properties that are independent of the 2-structures in the following sense. Let $w = (x_1, y_1) \ldots (x_n, y_n)$ be a word and $\sigma \colon D \to \Delta$ a selector. Then we have for all $g \in \mathcal{R}(D, \Delta)$,

$$\begin{aligned} \psi_w(g^\sigma) &= \sigma(x_1) g(x_1, y_1) \sigma(y_1)^{-1} \ldots \sigma(x_n) g(x_n, y_n) \sigma(y_n)^{-1} \\ &= \sigma(x_1) \sigma(y_1)^{-1} \ldots \sigma(x_n) \sigma(y_n)^{-1} \cdot \psi_w(g) , \end{aligned}$$

and thus ψ_w is a free invariant if and only if $\sigma(x_1)\sigma(y_1)^{-1} \ldots \sigma(x_n)\sigma(y_n)^{-1} = 1_\Delta$ for all selectors σ. Notice that the latter condition does not depend on g, and hence we have the following characterization result from [11].

Theorem 6.10.3
Let Δ be an abelian group. The following conditions are equivalent for a variable function ψ_w.
(1) *ψ_w is a free invariant of $\mathcal{R}(D, \Delta)$.*
(2) *For each $g \in \mathcal{R}(D, \Delta)$ and for each selector σ, $\psi_w(g^\sigma) = \psi_w(g)$.*
(3) *There exists a $g \in \mathcal{R}(D, \Delta)$ such that for each selector σ, $\psi_w(g^\sigma) = \psi_w(g)$.* □

For a fixed node $x_0 \in D$, the set

$$T_{x_0} = \{ t(x_0, y, z) \mid x_0, y, z \in D \text{ are distinct } \}$$

is the *bucket of triangles at* x_0.

Each triangle $t \in T_x$ is a closed walk, and so it represents a free invariant $\psi_t \in \mathrm{Inv}(D, \Delta)$ for abelian Δ. The next theorem (from [11]) states that these invariants generate $\mathrm{Inv}(D, \Delta)$. The proof of this is based on the following simple identity, which allows us to use triangles. Let x_0 be a fixed node. For each $e = (y, z)$, where $y \neq x_0$ and $z \neq x_0$, we have

$$e \equiv (x_0, y)^{-1} \cdot (x_0, y)(y, z)(z, x_0) \cdot (z, x_0)^{-1} = (x_0, y)^{-1} \cdot t(x_0, y, z) \cdot (x_0, z) .$$

Theorem 6.10.4
Let Δ be an abelian group and x_0 an element of the domain D. Then $\mathrm{Inv}(D, \Delta)$ is generated by the free invariants represented by the triangles at x_0. □

The triangles (at a fixed node x_0) not only represent generators of $\mathrm{Inv}(D, \Delta)$ but form a large enough set to characterize the relation $[g] = [h]$ between the inversive labeled 2-structures on an abelian Δ.

A set W of invariants for $\mathcal{R}(D, \Delta)$ is said to be a *complete*, if W satisfies for all g_1, g_2 the condition: $[g_1] = [g_2]$ if and only if $\eta(g_1) = \eta(g_2)$ for all $\eta \in W$.

We notice that in the above definition the converse implication is always valid, that is, if $g_1^\sigma = g_2$ for a selector σ, then for every invariant η, $\eta(g_2) = \eta(g_1^\sigma) = \eta(g_1)$. On the other hand, if a set of invariants W is complete and $[g_1] \neq [g_2]$, then there exists an invariant $\eta \in W$ such that $\eta(g_1) \neq \eta(g_2)$.

Theorem 6.10.5
Let Δ be an abelian group. For each $x_0 \in D$ the bucket of triangles T_{x_0} forms a complete set of invariants. □

The above theorem allows the use of 'local' triangles for checking whether or not $[g_1] = [g_2]$, *i.e.*, if $[g_1] \neq [g_2]$, then we can find a (common) triangle in g_1 and g_2 that induces non-equivalent substructures in g_1 and g_2.

Example 6.31
Let $\Delta = \mathbb{Z}_2$ and let g_1, g_2, g_3 be the structures from Figure 6.45. Consider the triangle $t = t(x_1, x_2, x_3)$. We have $\psi_t(g_1) = 1 \neq 0 = \psi_t(g_2)$, and hence by the above theorem there does not exist a selector σ for which $g_2 = g_1^\sigma$.

On the other hand, we observe that for the triangles $t = t(x_1, x_2, x_3)$, $t = t(x_1, x_3, x_4)$ and $t = t(x_1, x_2, x_4)$, $\psi_t(g_1) = \psi_t(g_3)$, and hence $\psi_t(g_1) = \psi_t(g_3)$ for all triangles $t \in T_{x_1}$. Hence, by Theorem 6.10.5, $[g_1] = [g_3]$. □

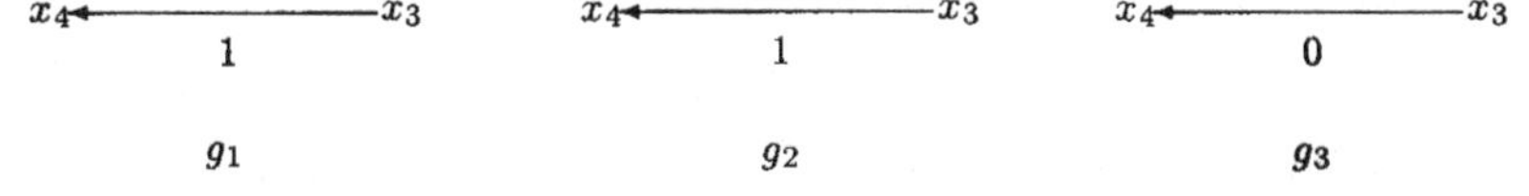

Figure 6.45: $[g_1] \neq [g_2]$, but $[g_1] = [g_3]$

6.10.5 Clans and invariants

Now, we do not assume that the group Δ is abelian. We shall state an invariant property of inversive $\ell 2$-structures connected to the clans of $\ell 2$-structures.

Let a closed walk

$$s(x_0, x_1, x_2, x_3) = (x_0, x_1)(x_1, x_2)(x_2, x_3)(x_3, x_0) \in M(D)$$

with four nodes be called a *square*. The following rather straightforward theorem (from [11] states that being a clan is an invariant property of the dynamic inversive $\ell 2$-structure $[g]$.

Theorem 6.10.6
Let $g\colon E_2(D) \to \Delta$ be an inversive $\ell 2$-structure on a group Δ of labels. A subset $X \subseteq D$ is a clan of $[g]$, if $\psi_s(g) = 1_\Delta$ for all squares $s = s(x_1, y_1, x_2, y_2)$ with $y_1, y_2 \in D \setminus X$ and $x_1, x_2 \in X$. □

References

[1] J.A. Bondy and U.S.R. Murthy, *Graph Theory with Applications*, Macmillan, London 1978.

[2] P. Bonizzoni, Primitive 2-structures with the $(n-2)$-property, *Theoret. Comput. Sci.* **132** (1994), 151 – 178.

[3] H. Buer and R.H. Möhring, A fast algorithm for the decomposition of graphs and posets, *Math. Oper. Res.* **8** (1983), 170 – 184.

[4] W.H. Cunningham, Decomposition of directed graphs, *SIAM J. Alg. Disc. Meth.* **3** (1982), 214 – 228.

[5] W.H. Cunningham, Decomposition of submodular functions, *Combinatorica* **3** (1983), 53 – 68.

[6] W.H. Cunningham and J. Edmonds, A combinatorial decomposition theory, *Canadian J. Math.* **32** (1980), 734 – 765.

[7] A. Deutz, Privite communication.

[8] A. Ehrenfeucht, H.N. Gabow, R.M. McConnell and S.J. Sullivan, An $O(n^2)$ algorithm to compute the prime tree family of a 2-structure, *J. of Algorithms*, to appear.

[9] A. Ehrenfeucht, T. Harju and G. Rozenberg, Permuting transformation monoids, *Semigroup Forum* **47** (1993), 123 – 125.

[10] A. Ehrenfeucht, T. Harju and G. Rozenberg, Incremental construction of 2-structures, *Discrete Math.* **128** (1994), 113 – 141.

[11] A. Ehrenfeucht, T. Harju and G. Rozenberg, Invariants of 2-structures on groups of labels, Manuscript (1994).

[12] A. Ehrenfeucht, T. Harju and G. Rozenberg, Quotients and plane trees of group labeled 2-structures, Leiden University, Department of Computer Science, Technical Report No. 03, 1994.

[13] A. Ehrenfeucht, H.J. Hoogeboom, P. ten Pas, and G. Rozenberg, An introduction to context-free text grammars, in *Developments in Language Theory*, G. Rozenberg and A. Salomaa, eds., World Scientific Publishing, 1994.

[14] A. Ehrenfeucht, P. ten Pas, and G. Rozenberg, Combinatorial properties of texts, *RAIRO Theoret. Inf.* **27** (1993), 433 - 464.

[15] A. Ehrenfeucht and G. Rozenberg, Finite families of finite sets, Part I and II, Univ. of Colorado at Boulder Technical Report, April 1986.

[16] A. Ehrenfeucht and G. Rozenberg, Theory of 2-structures, Parts I and II, *Theoret. Comput. Sci.* **70** (1990), 277 - 303 and 305 - 342.

[17] A. Ehrenfeucht and G. Rozenberg, Primitivity is hereditary for 2-structures *Theoret. Comput. Sci.* **70** (1990), 343 - 358.

[18] A. Ehrenfeucht and G. Rozenberg, Partial (set) 2-structures; part II: State spaces of concurrent systems, *Acta Informatica* **27** (1990), 343 - 368.

[19] A. Ehrenfeucht and G. Rozenberg, Angular 2-structures, *Theoret. Comput. Sci.* **92** (1992), 227 - 248.

[20] A. Ehrenfeucht and G. Rozenberg, T-structures, T-functions, and texts, *Theoret. Comput. Sci.* **116** (1993), 227 - 290.

[21] A. Ehrenfeucht and G. Rozenberg, Dynamic labeled 2-structures with variable domains, *Lecture Notes in Computer Science* **812** (1994), 97 - 123.

[22] A. Ehrenfeucht and G. Rozenberg, Dynamic labeled 2-structures, *Mathematical Structures in Computer Science*, to appear.

[23] J. Engelfriet, T. Harju, A. Proskurowski and G. Rozenberg, Characterization and Complexity of Uniformly Non-Primitive Labeled 2-Structures, *Theoret. Comput. Sci.*, to appear.

[24] J. Engelfriet and G. Rozenberg, Graph Grammars based on node rewriting: on introduction to NLC graph grammars, *Lecture Notes in Computer Science* **532** (1990), 12 - 23.

[25] T. Gallai, Transitiv orientierbare Graphen, *Acta Math. Acad. Sci. Hungar.* **18** (1967), 25-66.

[26] J.L. Gross and T.W. Tucker, "Topological Graph Theory", Wiley, New York, 1987.

[27] M. Habib and M.C. Maurer, On the X-join decomposition for undirected graphs, *Discrete Appl. Math.* **1** (1979), 201 - 207.

[28] F. Harary, On the notion of balance of a signed graph, *Michigan Math. J.* **2** (1953-54), 143 - 146; addendum, **2** (1953-54) preceding p.1.

[29] T. Harju and G. Rozenberg, Reductions for primitive 2-structures, *Fundamenta Informatica* **20** (1994), 133 - 144.

[30] T. Harju and G. Rozenberg, Decomposition of infinite labeled 2-structures, *Lecture Notes in Computer Science* **812** (1994), 145 - 158.

[31] L.O. James, R.G. Stanton and D.D. Cowan, Graph decomposition for undirected graphs, in "Proc. 3rd Southeastern Conf. on Combinatorics, Graph Theory, and Computing", 1972, 281 - 290.

[32] B. Jamison and S. Olariu, P-components and the homogeneous decomposition of graphs, *SIAM J. on Discrete Math.* to appear.

[33] J.H. van Lint and J.J. Seidel, Equilateral points in elliptic geometry, *Proc. Nederl. Acad. Wetensch* **Ser A 69** (1966), 335 - 348.

[34] R.M. McConnell, A linear-time inremental algorithm to compute the prime tree family of a 2-structure, *Algoritmica*, to appear.

[35] R.H. Möhring, Algorithmic aspects of the substitution decomposition in optimization over relations, set systems and boolean functions, *Ann. Oper. Research* **4** (1985/6), 195 - 225.

[36] R.H. Möhring and F.J. Radermacher, Substitution decomposition for discrete structures and connections with combinatorial optimization, *Ann. Discrete Math.* **19** (1984), 257 - 356.

[37] J.H. Muller and J. Spinrad, Incremental Modular Decomposition, *J. of the ACM* **36** (1989), 1 - 19.

[38] J.J. Rotman, "The Theory of Groups: An Introduction", Allyn and Bacon, Boston, 1973.

[39] L.S. Shapley, Solutions of compound simple games, in "Advances in Game Theory", Annals of Math. Study **52**, Princeton Univ. Press, 1964, 267 - 305.

[40] J.H. Schmerl and W. T. Trotter, Critically indecomposable partially ordered sets, graphs, tournaments and other binary relational structures, *Discrete Math.* **113** (1993), 191 - 205.

[41] J.J. Seidel and D.E. Taylor, Two-graphs, a second survey, in "Proc. Internat. Colloq. on Algebraic Methods in Graph Theory", Szeged, 1978.

[42] D.P. Sumner, Graphs indecomposable with respect to the X-join, *Discrete Math.* **6** (1973), 281 - 298.

[43] H.P. Yap, "Some Topics in Graph Theory", Cambridge University Press, Cambridge, 1986.

[44] T. Zaslavsky, Characterization of signed graphs, *J. Graph Theory* **5** (1981), 401 - 406.

[45] T. Zaslavsky, Signed graphs, *Discrete Appl. Math.* **4** (1982), 47 - 74: erratum **5** (1983), 248.

Chapter 7

PROGRAMMED GRAPH REPLACEMENT SYSTEMS

A. SCHÜRR
Lehrstuhl für Informatik III, RWTH-AACHEN,
Ahornstr. 55, D-52074 Aachen, Germany
e-mail: andy@i3.informatik.rwth-aachen.de

Various forms of programmed graph replacement systems as extensions of context-sensitive graph replacement systems have been proposed until today. They differ considerably with respect to their underlying graph models, the supported forms of graph replacement rules, and offered rule regulation mechanisms. Some of them have additional constructs for the definition of graph schemata, derived graph properties, and so forth. It is rather difficult to develop precise and compact descriptions of programmed graph replacement systems, a necessary prerequisite for any attempt to compare their properties in detail. Programmed Logic-based Structure Replacement (PLSR) systems are a kind of intermediate definition language for this purpose. They treat specific graph classes as sets of predicate logic formulas with certain properties, so-called structures. Their rules preserve the consistency of manipulated structures and use nonmonotonic reasoning for checking needed pre- and postconditions. So-called Basic Control Flow (BCF) expressions together with an underlying fixpoint theory provide needed means for programming with rules. This chapter introduces first the basic framework of PLSR systems and studies afterwards the essential properties of context-sensitive graph replacement approaches themselves as well as popular rule regulation mechanisms.

Contents

7.1 Introduction

The history of graph replacement systems starts 25 years ago with two seminal papers about so-called "web grammars" [55] and "Chomsky systems for partial orders" [62]. From the beginning graph replacement systems have been used and developed to solve "real world" problems in the fields of computer science itself, biology, etc. [1,52]. One of the first surveys about graph replacement systems [45] (and its short version in [46]) distinguishes three different types of applications:

(1) Describing or *generating* known *languages of graphs* which have for instance certain graph-theoretical properties or model growing plants.
(2) *Recognizing languages of graphs* which model for instance the underlying logical structure of natural language sentences or scanned images and office documents.
(3) *Specifying new classes of graphs* and graph transformations which represent for instance databases and database manipulations or software documents and document manipulating tools.

For solving problems in the first category pure graph replacement systems are often sufficient and accompanying tools are not urgently needed. The second category of problems requires classes of graph replacement systems which are powerful enough to describe interesting languages of graphs, but which are restricted enough so that efficiently working parsing or execution tools can be realized. The design of adequate classes of graph replacement systems is still subject of ongoing research activities and outside the scope of this paper (cf. Section 2.7 of Chapter 2 in this volume as well as [24] and [58]).

The third category of problems needs graph replacement systems as a kind of executable specification or even *very high level programming language.* In order to be able to fulfil the resulting requirements, many new concepts had to be added to "pure" first generation graph replacement systems (see [21] and especially [7] for a discussion of needed extensions). One of the most important extensions was the introduction of new means for controlling the application of graph replacement rules. This lead to the definition of programmed graph replacement systems, which are the main topic of this contribution.

7.1.1 Programmed Graph Replacement Systems in Practice

For historical reasons mainly, almost all earlier proposals for programmed graph replacement systems [11,12,23,28,38,45,64] belong to one of the main branches of graph replacement systems, the *algorithmic, set-theoretic approach* (cf. Chapter 1 and 2 of this volume), and not to the *algebraic, category-theory approach* (see Chapter 3 and 7 of this volume). Until now, two complete implementations of programmed graph replacement systems have been built. The first one, **P**rogrammed **A**ttributed **G**raph **G**rammars (PAGG), was mainly used for specifying graph editors and CAD systems [26,27,28]. The second one, **PRO**grammed **G**raph **RE**placement **S**ystems (PROGRES), is tightly coupled to the development of software engineering tools [53,65,66,76]. PROGRES and its predecessors were and are used within the project IPSEN for specifying internal data structures and tools of **I**ntegrated **P**roject **S**upport **EN**vironments [21,22,47,48]. At the beginning, pure algorithmic graph replacement systems were used [45]. But soon it became evident that *additional means* for defining complex rule application conditions, manipulating node attributes etc. are needed. Appropriate extensions were suggested in [23] and later on in [38]. But even then there was no support for

- separating the definition of static graph *integrity constraints* from dynamic graph manipulating operations as database languages separate the definition of database schemata from the definition of database queries or manipulations,
- specifying *derived attributes* and *relations* in a purely declarative manner in a similar style as attribute tree grammars or relational languages deal with derived data,
- and solving typical generate and test problems by using implicitly available means for *depth-first search* and *backtracking* in the same manner as Prolog deals with these problems.

Therefore, various extensions of graph replacement systems were studied which resulted finally in the development of the programming language PROGRES and its integrated programming support environment.[a]

7.1.2 Programmed Graph Replacement Systems in Theory

To produce a formal definition of programmed graph replacement systems like PAGG or PROGRES by means of set theory only is a very difficult and time-consuming task. Especially the definition of control structures, derived graph properties, and graph schemata constitutes a major problem. And switching from set theory to the formalism of category theory does not solve it at all. Both the category-based **D**ouble **P**ush **O**ut (DPO) approach (cf. Chapter 3 of this volume) and the **S**ingle **P**ush **O**ut approach (cf. Chapter 7 of this volume) have about the same difficulties with the definition of integrity constraints, derived information, and the like.

[a] see `http://www-i3.informatik.rwth-aachen.de/research/progres/index.html`

Therefore, it was necessary to establish another *framework for* the formal *definition of* these application-oriented *graph replacement systems*. The following observations had a major influence on the development of this framework:

- *Relational structures* are an obvious generalization of various forms of graphs or hypergraphs [44].
- *Logic formulas* are well-suited for defining required properties of generated graph languages (cf. Chapter 9 of this volume)
- *Nonmonotonic reasoning* is appropriate for inferring derived data and verifying integrity constraints in deductive database systems [43,54].
- *Fixpoint theory* is the natural candidate for defining the semantics of partially defined, nondeterministic, and recursive graph replacement programs [41,51].

Combining these sources of inspiration, first logic-based graph replacement systems were developed and used to produce a complete definition of the language PROGRES [66]. Later on they were generalized to **L**ogic **B**ased **S**tructure **R**eplacement (LBSR) systems [67] and **P**rogrammed **L**ogic-based **S**tructure **R**eplacement (PLSR) systems [68].

7.1.3 Contents of the Contribution

The presentation of LBSR and PLSR systems over here as well as a survey of *application-oriented* and/or *programmed context-sensitive* graph replacement systems is organized as follows:

- Section 7.2 introduces LBSR systems and demonstrates how they may be used to specify graph schemata, schema consistent graphs as well as consistency preserving graph transformations.
- Section 7.3 presents a basic set of control structures for programming with graph replacement rules together with a denotational semantics definition based on fixpoint theory. Using these control structures, LBSR systems are extended to PLSR systems.
- Section 7.4 compares typical representatives of application-oriented algorithmic graph replacement systems and sketches their translation into LBSR systems. Furthermore it discusses their relationships to the DPO and SPO families of algebraic approaches.
- Section 7.5 finally surveys a collection of currently popular rule regulation mechanisms and translates them into basic control structures of PLSR systems.

Other topics, like the design of a readable syntax and a static type system for the language PROGRES or the implementation of tools which support programming with graph replacement systems, are outside the scope of this volume [65,66,68].

7.2 Logic-Based Structure Replacement Systems

Nowadays, a surprisingly large variety of different graph replacement formalisms is existent. Some of them have a rather restricted expressiveness and are well-suited for proving properties of generated graph languages. Some of them are very complex and mainly used for specification and rapid prototyping purposes. Even their underlying graph data models differ to a great extent with respect to the treatment of edges, attributes, etc.

In order to be able to reason about differences or common properties of these graph replacement approaches and their underlying data models in Section 7.4, a general framework is urgently needed. Such a framework has to provide proper means for the definition of specific graph models and graph replacement approaches. One attempt into this direction, so-called **H**igh **L**evel **R**eplacement (HLR) systems, is based on the construction of categories with certain properties [18]. It generalizes algebraic graph replacement approaches and incorporates algebraic specification technologies for the definition of attribute domains (see also [34,39]). Until now, HLR systems still have problems with formalizing derived graph properties or graph integrity constraints as they are used within the graph grammar specification language PROGRES.

Therefore, another *generic framework* was developed which comprises many different graph replacement systems as special cases and which is the underlying formalism of the specification language PROGRES. It combines elements from algorithmic and algebraic graph replacement systems with elements from deductive database systems. The outcome of the cross-fertilization process, **L**ogic-**B**ased **S**tructure **R**eplacement (LBSR) systems, are the topic of this section.

The formalism of LBSR systems is not intended to be directly used as a specification language, but provides the fundament for the definition

- of specific graph models as special cases of relational structures
- and of a whole family of accompanying graph replacement notations.

The presentation of LBSR systems over here will be accompanied by a number of examples, which sketch how human-readable graph grammar specifications written in a PROGRES like notation may be translated into them. By means of these examples the reader will hopefully understand how LBSR systems may be used to define the semantics and compare the intended behavior of specific graph replacement systems.

The rest of this section is organized as follows:

- Subsection 7.2.1 repeats some *basic terminology* of predicate logic and contains the formal definition of (relational) structures and structure schemata as sets of formulas.
- Subsection 7.2.2 deals with *substructure selection*, i.e. finding matches for left-hand sides of rules with additional application conditions.
- Subsection 7.2.3 introduces then the *structure replacement formalism* itself.
- Subsection 7.2.4 finally *summarizes* the main properties of LBSR systems and their relationships to approaches like HLR systems.

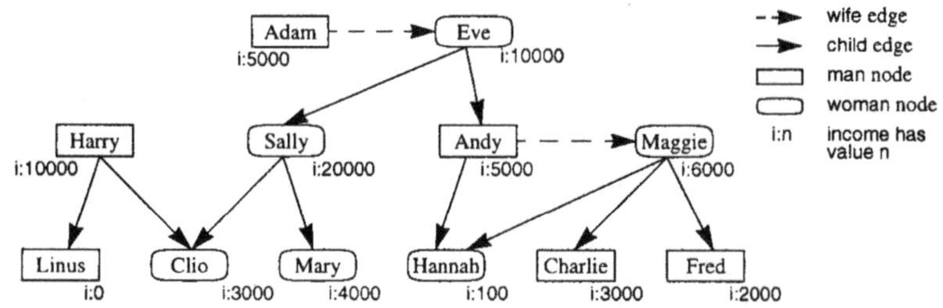

Figure 7.1: The running example, a person database

Within all these subsections we will use a coherent *running example* drawn from the area of deductive database systems. It is the specification of a graph-like person database and includes the definition of derived relationships, integrity constraints, and complex update operations (Figure 7.1 displays a sample database).

7.2.1 Structure Schemata and Schema Consistent Structures

This subsection introduces basic terminology of predicate logic. Furthermore, it explains modeling of node and edge labeled graphs as special cases of *schema consistent structures*, i.e. as sets of formulas with certain properties. Afterwards, our running example of a person database will be modeled as a directed graph. In this way, we are able to demonstrate that graphs are a special case of structures and that graph replacement is a special case of structure replacement. Nevertheless, all definitions and propositions of the following sections are independent of the selected graph model and its encoding.

Definition 7.2.1 (Signature.)
A 5-tuple $\Sigma := (\mathcal{A}_F, \mathcal{A}_P, \mathcal{V}, \mathcal{W}, \mathcal{X})$ is a **signature** if:

(1) $\mathcal{A}_F$ is an alphabet of function symbols[b] (including constants).
(2) $\mathcal{A}_P$ is an alphabet of predicate symbols.[c]
(3) $\mathcal{V}$ is a special alphabet of object identifier constants.
(4) $\mathcal{W}$ is a special alphabet of constants representing sets of objects.
(5) $\mathcal{X}$ is an alphabet of logical variables used for quantification purposes.

The signature definition above introduces names for specific objects and sets of objects as special constants. These constants will play an important role for the selection of LBSR rule matches. Elements of $\mathcal{V}$ in a rule's left-hand side denote and match single objects of a given structure, whereas elements of $\mathcal{W}$ denote and match sets of objects of a given structure. Please note that $\mathcal{W}$ is *not* an alphabet of logical set variables, i.e. we do not use a monadic second order logic as in Chapter 9 of this volume. Quantification over set variables is needed for the definition of many graph-theoretic properties, but not a strict neccessity for the formal definition of graph replacement formalisms studied over here. Hence we will use a first order logic which makes no distinction between constant symbols in $\mathcal{A}_F$ (functions of arity 0), $\mathcal{V}$, and $\mathcal{W}$. Nevertheless, it is necessary to distinguish these three different kinds of constants within Subsection 7.2.2 and 7.2.3, where they play very different roles for the application of structure replacement rules.

The signature of the following example may be used for the definition of person databases, or more precisely, their graph representations as sets of formulas. Figure 7.1 above presents such a graph representation of a database with twelve persons. Each person is either a man node or a woman node. Both man and woman nodes may be sources or targets of child edges, whereas wife edges always have a man as source and a woman as target. Furthermore, each person has an integer-valued income attribute.

Example 7.1 (Signature for a Person Database.)
The graph signature for fig. 7.1 is $\Sigma := (\mathcal{A}_F, \mathcal{A}_P, \mathcal{V}, \mathcal{W}, \mathcal{X})$ with:

(1) $\mathcal{A}_F :=$ { child, woman, wife, man, person, income, integer, 0, ... , + }.
(2) $\mathcal{A}_P :=$ { node, edge, attr, type, ... } .
(3) $\mathcal{V} :=$ { He, She, Value, ... } .
(4) $\mathcal{W} :=$ { HerChildren, HisChildren, ... } .
(5) $\mathcal{X} :=$ { x1, x2, ... } . □

---
[b] A family of alphabets of symbols if we take domain and range of functions into account.
[c] A family of alphabets of symbols if we take arities of predicates into account.

The alphabets $\mathcal{V}$, $\mathcal{W}$, and $\mathcal{X}$ above are sets of arbitrarily chosen constant or variable names. The set $\mathcal{A}_F$, on the other hand, contains just all needed symbols for node labels, edge labels, attribute types, attribute values, and evaluation functions.

The alphabet $\mathcal{A}_P$, finally contains four predicate symbols which are very important for the more or less arbitrarily selected encoding of directed, attributed graphs, where each node has a distinct class as its label and where any attribute value has a designated type. They have the following interpretation:

- $\mathsf{node}(x, l)$: graph contains a node x with label l,
- $\mathsf{edge}(x, e, y)$: graph contains edge with label e from source x to target y,
- $\mathsf{attr}(x, a, v)$: attribute a at node x has value v,
- $\mathsf{type}(v, t)$: attribute value v has type t.

In the sequel, we will always assume that Σ is a signature over the above mentioned alphabets.

Definition 7.2.2 (Σ-Term und Σ-Atom.)
$\mathcal{T}_\mathcal{X}(\Sigma)$ is the set of all Σ-**terms** (in the usual sense) which contain function symbols and constants of $\mathcal{A}_F$, free variables of $\mathcal{X}$, and additional identifier symbols of $\mathcal{V}$ and $\mathcal{W}$. $\mathcal{A}(\Sigma)$ is the set of all **atomic formulas** (Σ-**atoms**) over $\mathcal{T}_\mathcal{X}(\Sigma)$ which contain predicate symbols of $\mathcal{A}_P$ and "=" for expressing the equality of two Σ-terms. Finally $\mathcal{T}(\Sigma) \subset \mathcal{T}_\mathcal{X}(\Sigma)$ is the set of all those terms which do not contain any symbols of $\mathcal{X}$, $\mathcal{V}$, and $\mathcal{W}$, i.e. do not have variables or object (set) identifiers as their leafs.

Definition 7.2.3 (Σ-Formula and Derivation of Σ-Formulas.)
$\mathcal{F}(\Sigma)$ is the set of all sets of closed[d] first order predicate logic formulas (Σ-**formulas**) which have $\mathcal{A}(\Sigma)$ as atomic formulas, $\wedge, \vee, \ldots$ as logical connectives, and $\exists, \forall$ as quantifiers. Furthermore with Φ and Φ' being sets of Σ-formulas,

$$\Phi \vdash \Phi'$$

means that all formulas of Φ' are **derivable** from the set of formulas Φ using any (consistent and complete) inference system of first order predicate logic with equality.

In the following, elements of $\mathcal{F}(\Sigma)$ will be used to represent structures, structure schemata, schema consistent structures, and even left- and right-hand sides as well as pre- and postconditions of structure replacement rules.

[d] Requiring bindings for all logical variables (by means of existential or universal quantification) does not restrict the expressiveness of formulas but avoids any difficulties with varying treatments of free variables in different inference systems.

Definition 7.2.4 (Σ-Structure.)
A set of closed formulas $F \in \mathcal{F}(\Sigma)$ is a **Σ-structure** (write: $F \in \mathcal{L}(\Sigma)$) $:\Leftrightarrow$ $F \subseteq \mathcal{A}(\Sigma)$ and F does not contain formulas of the form "$\tau_1 = \tau_2$".

Example 7.2 (A Person Database Structure.)
The following structure is a set of formulas which has the graph F of figure 7.1 as a model:

$F :=$ { node(Adam, man), node(Eve, woman), node(Sally, woman), ... ,
attr(Adam, income, 5000), attr(Eve, income, 10000), ... ,
edge(Adam, wife, Eve), edge(Eve, child, Sally), ... }. □

The set of formulas F has many *different* graphs as *models*. One of those models is the graph of Figure 7.1, but there are many others in which we are not interested in. Some of them may contain additional nodes and edges, and some use different identifiers like Eve and Sally for the same graph node. In Definition 7.2.5, we will introduce a so-called *completing operator*, which allows us to get rid of unwanted (graph) models of Σ-structures and to reason about properties of minimal models on a pure syntactical level. Therefore, models of structures are not introduced formally and will not be used in the sequel.

The following example demonstrates our needs for a completing operator in more detail. It presents the definition of a single person database and explains our difficulties to prove that this database contains indeed one person only. A related problem has been extensively studied within the field of deductive database systems and has been tackled by a number of quite different approaches either based on the so-called *closed world assumption* or by using *nonmonotonic reasoning* capabilities (cf. [42,49,50]). The main idea of (almost) all of these approaches is to distinguish between basic facts and derived facts and to add only negations of basic facts to a rule base, which are not derivable from the original set of facts.

It is beyond the scope of this chapter to explain nonmonotonic reasoning in more detail. Therefore, we will simply assume the existence of a *completing operator* $\mathcal{C}$ which adds a certain set of additional formulas to a structure. The resulting set of formulas has to be consistent (it may not contain contradictions) and *sufficiently complete* such that we can prove the above mentioned properties by using the axioms of first-order predicate logic only.

Definition 7.2.5 (Σ-Structure Completing Operator.)
A function $\mathcal{C} : \mathcal{L}(\Sigma) \to \mathcal{F}(\Sigma)$ is a **Σ-structure completing operator** $:\Leftrightarrow$
For all structures $F \in \mathcal{L}(\Sigma) : F \subseteq \mathcal{C}(F)$ and $\mathcal{C}(F)$ is a consistent set of formulas and $\mathcal{C}(\rho(F)) = \rho(\mathcal{C}(F))$

with $\rho : \mathcal{F}(\Sigma) \to \mathcal{F}(\Sigma)$ being a substitution which renames object identifiers of $\mathcal{V}$ and object set identifiers of $\mathcal{W}$ in F without introducing name clashes; i.e. consistent renaming of identifiers and completion of formula sets are compatible operations.

The identity operator on sets of formulas (structures) is one trivial example of a completion operator. Any given structure is a set of atomic formulas and, therefore, free of contradictions. Furthermore, all inference systems of predicate logic (should) have the property that consistent renaming of constants does not destroy constructed proofs. Hence, we will not have any troubles with the requirement above that renaming and completion are compatible.

A useful example of a completion operation is the union of a given set of basic facts F with all derivable facts and a carefully selected subset of negated non-derivable facts like "a graph does not contain a node with a name unequal to Adam" in Example 7.3. Please note that we will need an additional set of formulas for being able to derive new facts from the atomic formulas of structures. This set of first order formulas is part of the following Definition 7.2.6.

Example 7.3 (Nonmonotonic Reasoning.)
The singleton set $F := \{\mathsf{node}(\mathsf{Adam}, \mathsf{man})\}$ is a structure which has all graphs containing at least one man node as models. Being interested in properties of minimal F graph models only, we should be able to prove that F contains a single node. For this purpose, the operator $\mathcal{C}$ has to be defined as follows (omitted sets of formulas deal with edges etc.):

$$\mathcal{C}(F) := F \cup \ldots \cup$$
$$\{\forall x, l : \mathsf{node}(x, l) \to (x = v_1 \vee \ldots \vee x = v_k) \mid v_1, \ldots, v_k \in \mathcal{V} \text{ are all used object ids in } F\}.$$

Now, we are able to prove that F contains only the male node Adam:

$$\mathcal{C}(F) \vdash (\forall x, l : \mathsf{node}(x, l) \to x = \mathsf{Adam}) \wedge \mathsf{node}(\mathsf{Adam}, \mathsf{man}).$$ □

Completing operators may have definitions which are specific for a regarded class of structures. Therefore, they are part of the following definition of *structure schemata*:

Definition 7.2.6 (Σ-Structure Schema.)
A tuple $S := (\Phi, \mathcal{C})$ is a **Σ-structure schema** (write: $\Sigma \in \mathcal{S}(\Sigma)$) $:\Leftrightarrow$

(1) $\Phi \in \mathcal{F}(\Sigma)$ is a consistent set of formulas without references to specific object (set) identifiers of $\mathcal{V}$ or $\mathcal{W}$ (it contains integrity constraints and derived data definitions).

(2) $\mathcal{C}$ is a Σ-structure completing operator.

Definition 7.2.7 (Schema Consistent Structure.)
Let $S := (\Phi, \mathcal{C}) \in \mathcal{S}(\Sigma)$ be a schema. A Σ-structure $F \in \mathcal{L}(S)$ is **schema consistent** with respect to S (write: $F \in \mathcal{L}(S)$) $:\Leftrightarrow$
$\mathcal{C}(F) \cup \Phi$ is a consistent set of formulas.

The definition of *schema consistency* is very similar to a related definition of the knowledge representation language Telos [42]. It states that a structure is inconsistent with respect to a schema, if and only if we are able to derive contradicting formulas from the completed structure and its schema.

The following example is the definition of a person database schema. Its set of formulas Φ consists of three subsets:

- A first subset defines integrity constraints for the selected graph data model and excludes dangling edges and attributes of nonexisting nodes.
- The second subset contains all person database specific integrity constraints, like "man and woman nodes are also person nodes".
- The last subset deals with a number of derived graph properties like the well-known ancestor relation or the definition of the brother of a child.

Example 7.4 (A Schema for Person Databases.)
A structure schema $S := (\Phi, \mathcal{C})$ for graph F in Fig. 7.1 has the following form:

$$\begin{aligned}
\Phi := \; & \{\forall x, e, y : \mathsf{edge}(x, e, y) \to \exists xl, yl : \mathsf{node}(x, xl) \wedge \mathsf{node}(y, yl), \\
& \quad \forall x, a, v : \mathsf{attr}(x, a, v) \to \exists l : \mathsf{node}(x, l), \ldots\} \\
\cup \; & \{\forall x : \mathsf{node}(x, \mathsf{man}) \to \mathsf{node}(x, \mathsf{person}), \\
& \quad \forall x : \mathsf{node}(x, \mathsf{woman}) \to \mathsf{node}(x, \mathsf{person}), \\
& \quad \forall x, y : \mathsf{edge}(x, \mathsf{wife}, y) \to \mathsf{node}(x, \mathsf{man}) \wedge \mathsf{node}(y, \mathsf{woman}), \\
& \quad \forall x, y, z : \mathsf{edge}(x, \mathsf{wife}, y) \wedge \mathsf{edge}(x, \mathsf{wife}, z) \to y = z, \\
& \quad \forall x, y, z : \mathsf{edge}(x, \mathsf{wife}, z) \wedge \mathsf{edge}(y, \mathsf{wife}, z) \to x = y, \\
& \quad \forall x, v : \mathsf{attr}(x, \mathsf{income}, v) \to \mathsf{type}(v, \mathsf{integer}), \\
& \quad \forall x : \mathsf{node}(x, \mathsf{person}) \to \exists v : \mathsf{attr}(x, \mathsf{income}, v), \ldots\} \\
\cup \; & \{\forall x, y : \mathsf{ancestor}(x, y) \leftrightarrow (\exists z : \mathsf{edge}(z, \mathsf{child}, x) \\
& \qquad\qquad\qquad\qquad \wedge (z = y \vee \mathsf{ancestor}(z, y)), \\
& \quad \forall x, y : \mathsf{brother}(x, y) \leftrightarrow \exists z : \mathsf{edge}(z, \mathsf{child}, x) \wedge \mathsf{edge}(z, \mathsf{child}, y) \\
& \qquad\qquad\qquad\qquad \wedge \mathsf{node}(y, \mathsf{man}) \wedge x \neq y, \ldots\}.
\end{aligned}$$

$\mathcal{C}(F) := F \cup \ldots$ □

Figure 7.2 displays the corresponding PROGRES like specification of these graph properties. The first three declarations establish a node class hierarchy with person being the root class and man as well as woman being its subclasses. All nodes labeled with these classes have an income attribute. The following two declarations introduce child and wife edges and constrain the labels (classes) of their source and target nodes. Furthermore, the [0:1] annotations within the wife edge declaration even require that any man node has at most one outgoing wife edge and that any woman node has at most one incoming edge of this type. Finally, two derived binary relations are declared, so-called *paths*, which define the well-known concepts ancestor and brother. The ancestors of a person are determined by evaluating ←child− &(self or ancestor) as follows:

- Compute the parents of a given person as the set of all sources of incoming child edges (←child− traverses child edges in reverse direction).
- Afterwards (& concatenates two subexpressions), return the union (or) of all parents (self) and their ancestors.

The brothers of a person are computed by

- determining first all its parents (←child−),
- finding then all male children (−child→ & instance of man),
- and eliminating finally the person itself from the result (but not self).

```
node class person;
   attribute income: integer ;
end ;

node class man is a person end ;

node class woman is a person end ;

edge type child: person → person ;

edge type wife: man [0:1] → woman [0:1] ;

path ancestor: person → person =
      ←child—
   & (self or ancestor)
end ;

path brother: person → man =
   ( ←child—
   & —child→ & instance of man )
   but not self
end ;
```

Figure 7.2: Excerpt from a PROGRES graph schema definition

In this way PROGRES allows its users to define complex graph schemata with elaborate integrity constraints as well as derived relationships (and attributes). The translation of these schema definitions into sets of formulas is always as straightforward as in the presented example above. All forthcoming rules of LBSR systems have to observe these integrity constraints, i.e. transform one schema consistent structure (graph) into another one. Furthermore, they may access derived information within pre- and postconditions.

Please note that it is not possible to construct a structure replacement machinery, where checking of integrity constraints or pre- and postconditions is guaranteed to terminate. It depends on the selected nonmonotonic reasoning strategy (the implementation of a programmed graph replacement approach) whether certain *types of constraints* are *prohibited* or lead a rule application process into an *infinite loop*. We will reconsider the termination problem later on in section 7.3 together with the fixpoint semantics definition for potentially nonterminating control structures.

7.2.2 Substructures with Additional Constraints

This subsection formalizes the term "rule match under additional constraints" by means of morphisms. This term is central to the definition of *structure replacement with pre- and postconditions*, which follows afterwards. It determines which substructures of a structure are legal targets for a structure replacement step, i.e. match a given rule's left-hand side. The main difficulty with the definition of the new structure replacement approach was the requirement that it should include the expression oriented algorithmic graph grammar approach [45]. This approach has very powerful means for deleting, copying, and redirecting arbitrary large bundles of edges.

We will handle the *manipulation of edge bundles* by introducing special *object set identifiers* on a structure replacement rule's left- and right-hand side. These set identifiers match an arbitrarily large (maximal) set of object identifiers (see Example 7.6 and Definition 7.2.15). As a consequence, mappings between structures, which select the affected substructure for a rule in a structure, are neither total nor partial functions. In the general case, these *mappings are relations* between object (set) identifiers. These relations are required to preserve the properties of the source structure (e.g. a rule's left-hand side) while embedding it into the target structure (the structure we have to manipulate) as usual and to take additional constraints for the chosen embedding into account.

Another problem comes from the fact that we have to deal with attributes, i.e. LBSR rules must be able to write, read, and modify attribute values in the same way as they are able to manipulate structural information. Therefore, it is not sufficient to map object (set) identifiers onto (sets of) object identifiers, but we need also the possibility to map them onto (sets of) attribute value representing Σ-terms.

Definition 7.2.8 (Σ-Term Relation.)
Let $F, F' \in \mathcal{F}(\Sigma)$ be two sets of formulas which have $\mathcal{V}_F, \mathcal{W}_F$ and $\mathcal{V}_{F'}, \mathcal{W}_{F'}$ as their object (set) identifiers. Furthermore, $\mathcal{T}(\Sigma)$ is the set of all variable and object identifier free Σ-terms (cf. Def. 7.2.2). A relation

$$u \subseteq (\mathcal{V}_F \times (\mathcal{V}_{F'} \cup \mathcal{T}(\Sigma))) \cup (\mathcal{W}_F \times (\mathcal{W}_{F'} \cup \mathcal{V}_{F'} \cup \mathcal{T}(\Sigma)))$$

is a **Σ-relation** from F to F' $:\Leftrightarrow$

(1) For all $v \in \mathcal{V}_F : |u(v)| = 1$ with $u(x) := \{y | (x, y) \in u\}$,
i.e. every object identifier will be mapped onto a single object identifier or onto a single attribute value representing Σ-term.

(2) For all $w \in \mathcal{W}_F : |u(w)| = 1$ or $u(w) \subseteq \mathcal{V}_{F'}$ or $u(w) \subseteq \mathcal{T}(\Sigma)$,
i.e. every object set identifier will be mapped either onto another object set identifier or onto a set of object identifiers or attribute value representing Σ-terms.

The definition above introduces relations between object (set) identifiers and (sets of) Σ-terms only. Their extensions to relations between arbitrary formulas are defined as follows:

Definition 7.2.9 (Σ-Relation.)
Let u be a Σ-term relation according to Definition 7.2.8. The **extension** u^* of u to the domain of Σ-formulas is called **Σ-relation** and defined as follows:

$$\begin{aligned} u^*(\phi, \phi') :\Leftrightarrow\ & \exists \psi \in \mathcal{F}(\Sigma) \text{ with free variables } x_1, \ldots, x_n \\ & \quad \text{but without any } \mathcal{V} \text{ or } \mathcal{W} \text{ identifiers} \\ \wedge\ & \exists v_1, \ldots, v_n \in \mathcal{V} \cup \mathcal{W} \text{ with } v_i \neq v_j \text{ for } i, j \in \{1, \ldots, n\} \\ \wedge\ & \exists \tau_1, \ldots, \tau_n \in \mathcal{V} \cup \mathcal{W} \cup \mathcal{T}(\Sigma) \text{ with } u(v_1, \tau_1), \ldots, u(v_n, \tau_n) : \\ & \quad \phi = \psi[v_1/x_1, \ldots, v_n/x_n] \wedge \phi' = \psi[\tau_1/x_1, \ldots, \tau_n/x_n].^e \end{aligned}$$

Furthermore, with $\Phi \subseteq \mathcal{F}(\Sigma)$, the following short-hand will be used in the sequel: $u^*(\Phi) := \{\phi' \in \mathcal{F}(\Sigma) \mid \phi \in \Phi \text{ and } (\phi, \phi') \in u^*\}$.

The definition of u^* above simply states that two formulas ϕ and ϕ' are related to each other if they differ with respect to u-related terms only. Any object identifier v in ϕ must be replaced by the uniquely defined $u(v)$ in ϕ'. Any object set identifier w in ϕ must be replaced by some element $u(w)$ in ϕ' such that all occurrences of w in ϕ get the same substitute.

Based on these relations between formulas, we are now able to define mappings (morphisms) between sets of formulas or structures which preserve certain properties.

---

[e] $[\rho_1/x_1, \ldots, \rho_n/x_n]$ denotes consistent substitution of any x_i by its corresponding ρ_i.

Definition 7.2.10 (Σ-(Structure-)Morphism.)
Let $F, F' \in \mathcal{F}(\Sigma)$. A Σ-relation u from F to F' is a **Σ-morphism** from F to F' (write: $u : F \tilde{\rightarrow} F'$) $:\Leftrightarrow F' \vdash u^*(F)$.
With $F, F' \in \mathcal{L}(\Sigma) \subseteq \mathcal{F}(\Sigma)$ being Σ-structures, u will be called **Σ-structure morphism**.

The definition of a morphism $u : F \tilde{\rightarrow} F'$ requires in the simplest case that $u^*(F)$ is a subset of F', as in the following example:

Example 7.5 (A Simple Structure Morphism.)
Let F' be the structure of Example 7.2 which represents the graph of Figure 7.1 and

$$\begin{aligned} F \;:=\; & \{\mathsf{node(He, man), node(She, woman),} \\ & \mathsf{attr(She, income, Value), edge(He, wife, She),} \\ & \mathsf{edge(He, child, HisChildren), edge(She, child, HerChildren)}\} \end{aligned}$$

with $\mathsf{He, She, Value} \in \mathcal{V}$, and $\mathsf{HisChildren, HerChildren} \in \mathcal{W}$.
There are two Σ-relations from F to F' which are structure morphisms:

$$\begin{aligned} u \;:=\{\; & \mathsf{(HerChildren, Hannah), (HerChildren, Charlie), (HerChildren, Fred),} \\ & \mathsf{(He, Andy), (She, Maggie), (Value, 6000), (HisChildren, Hannah)}\}, \\ u' \;:=\{\; & \mathsf{(HerChildren, Sally), (HerChildren, Andy),} \\ & \mathsf{(He, Adam), (She, Eve), (Value, 10000)}\}. \end{aligned}$$

□

The selected examples of morphisms show that two different object (set) identifiers in F may be mapped onto the same object identifier in F' (unless explicitly prohibited by means of additional formulas). Furthermore, an object set identifier may be mapped onto an arbitrarily large set of identifiers (terms), with the empty set being a permitted special case. In the sequel, we will show that morphisms between structures define a transitive, reflexive, and associative relation, i.e. they build a category.

Proposition 7.2.11 (The Category of Σ-Structures.)
Assume "$\circ$" to be the usual composition of binary relations. Then, $\mathcal{F}(\Sigma)$ together with the family of Σ-morphisms defined above and "$\circ$" is a ***category****; the same holds true for the set of Σ-structures $\mathcal{L}(\Sigma)$ and the family of Σ-structure morphisms.*

Proof:

(1) *Σ-Morphisms are closed* w.r.t. "$\circ$", i.e.

$u : F \tilde{\rightarrow} F', u' : F' \tilde{\rightarrow} F'' \Rightarrow (u \circ u') : F \tilde{\rightarrow} F''$:

u and u' are Σ-relations $\Rightarrow_{\text{Def. 7.2.8}}$ *$(u \circ u')$ is a Σ-relation and*

$\text{Def. 7.2.9}\ (u \circ u')^* = u^* \circ u'^*$.

u, u' are morphisms $\Rightarrow_{\text{Def. 7.2.10}}$ $F' \vdash u^*(F)$ and $F'' \vdash u'^*(F')$

$F' \vdash u^*(F) \Rightarrow_{\text{subst. preserves proofs}} u'^*(F') \vdash u'^*(u^*(F)) = (u \circ u')^*(F)$

$\Rightarrow_{\text{modus ponens}} F'' \vdash (u \circ u')^*(F)$,

i.e. $(u \circ u')$ is a morphism from F to F''.

(2) *Existence of neutral Σ-morphism id_F for any $F \in \mathcal{F}(\Sigma)$:*

Obviously, the relation id_F, which maps any object (set) identifier in F onto itself, is a neutral element for the family of Σ -relations. Then,

$id_F^*(F) = F \Rightarrow F \vdash id_F^*(F) \Rightarrow id_F : F \tilde{\rightarrow} F$,

i.e. id_F is the required neutral morphism.

(3) *Associativity of "$\circ$" for Σ-morphisms:*

Follows directly from the fact that "$\circ$" is associative for binary relations. In order to obtain the proof that the family of Σ-structure morphisms together with $\mathcal{L}(\Sigma)$ and "$\circ$" is a category we simply have to replace any $\mathcal{F}(\Sigma)$ above by $\mathcal{L}(\Sigma)$. □

Definition 7.2.12 (Substructure.)
$F, F' \in \mathcal{L}(\Sigma)$ are structures. F is a **substructure** of F' with respect to a Σ-relation u (write: $F \subseteq_u F'$) $:\Leftrightarrow u : F \tilde{\rightarrow} F'$.

This definition coincides with the usual meaning of a homomorphic[f] substructure (or subgraph), if F does not contain any object set identifiers.

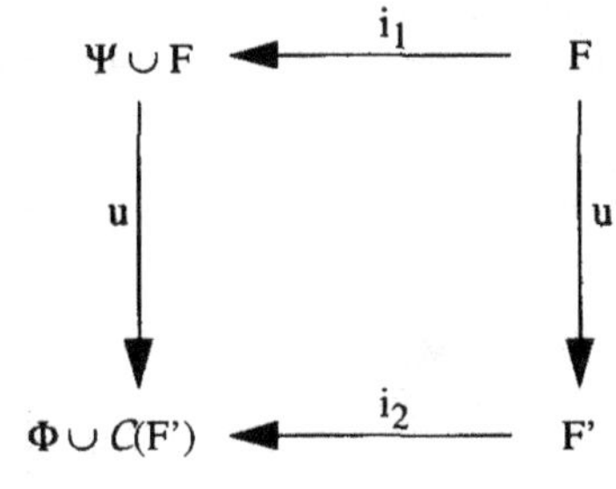

Figure 7.3: Substructure selection with additional constraints

[f] "homomorphic" means that different object identifiers in F may be mapped onto the same object identifier in F'.

Proposition 7.2.13 (Soundness of Substructure Property.)
For $F, F' \in \mathcal{L}(\Sigma)$ being structures, the following properties are equivalent:
F is a substructure of F' with respect to a Σ-relation u
$\Leftrightarrow u^*(F)$ is a subset of F'.

Proof:

$$F \subseteq_u F' \Leftrightarrow_{\textit{Def. 7.2.12}} u : F \tilde{\rightarrow} F'$$
$$\Leftrightarrow_{\textit{Def. 7.2.10}} F' \vdash u^*(F) \Leftrightarrow_{u^*(F), F' \in \mathcal{L}(\Sigma)} u^*(F) \subseteq F'.$$

The last step of the proof follows from the fact that F and F' are sets of atomic formulas without "=", such that $\subseteq$ (normal set inclusion) and $\vdash$ are equivalent relations. □

Definition 7.2.14 (Substructure with Additional Constraints.)
$S := (\Phi, \mathcal{C})$ is a Σ-structure schema, $F, F' \in \mathcal{L}(\Sigma)$, and $\Psi \in \mathcal{F}(\Sigma)$ is a set of constraints with references to object (set) identifiers of F only. F is a **constrained substructure** of F' with respect to a Σ-relation u and the additional set of constraints Ψ (write: $F \subseteq_{u,\Psi} F'$) $:\Leftrightarrow$

(1) $F \subseteq_u F'$, i.e. $F' \vdash u^*(F)$.

(2) $u : \Psi \cup F \tilde{\rightarrow} \Phi \cup \mathcal{C}(F')$, i.e. $\Phi \cup \mathcal{C}(F') \vdash u^*(\Psi \cup F) = u^*(\Psi) \cup u^*(F)$.

These conditions are equivalent to the existence of the diagram in Figure 7.3, with inclusions i_1 and i_2 being morphisms, which map any object (set) identifier onto itself:

$i_1 : F \tilde{\rightarrow} \Psi \cup F$, since $i_1^*(F) = F \subseteq \Psi \cup F$, i.e. $\Psi \cup F \vdash i_1^*(F)$.

$i_2 : F' \tilde{\rightarrow} \Phi \cup \mathcal{C}(F')$, since $i_2^*(F') = F' \subseteq \mathcal{C}(F')$, i.e. $\Phi \cup \mathcal{C}(F') \vdash i_2^*(F')$.

Informally speaking, a structure F is a substructure of F' with respect to additional constraints, if and only if we are able to prove that all constraints for the embedding of F in F' are fulfilled. We may use the basic facts of F' including all formulas generated by the completing operator $\mathcal{C}$ and the set of formulas Φ of the structure schema S for this purpose.

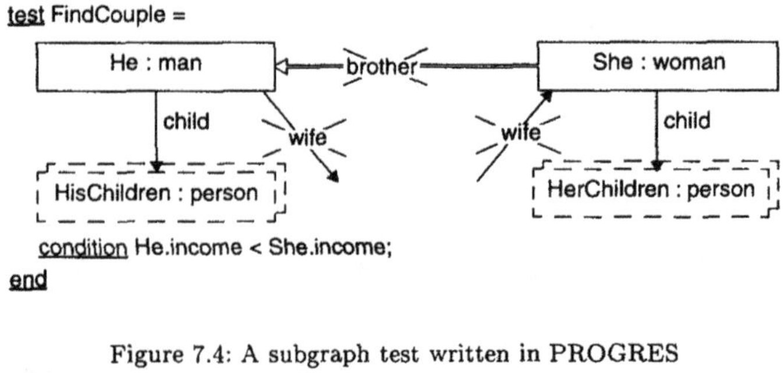

Figure 7.4: A subgraph test written in PROGRES

Example 7.6 (Substructure Selection.)
Let $S := (\Phi, \mathcal{C})$ be the schema of Example 7.4 and F' the database of Example 7.2. The structure $F \in \mathcal{L}(\Sigma)$ and its accompanying set of constraints $\Psi \in \mathcal{F}(\Sigma)$, defined below, represent the PROGRES like subgraph test of Figure 7.4 "select any pair of unmarried persons and all their children, who are not brother and sister and have appropriate income values together with all their children":

$$\begin{aligned} F &:= \{\mathsf{node(He, man), node(She, woman),} \\ &\quad \mathsf{attr(He, income, HisValue), attr(She, income, HerValue),} \\ &\quad \mathsf{edge(He, child, HisChildren), edge(She, child, HerChildren)}\}. \\ \Psi &:= \{\neg\mathsf{brother(She, He)}, \neg\exists x : \mathsf{edge(He, wife,} x), \neg\exists x : \mathsf{edge(}x\mathsf{, wife, She)}, \\ &\quad \mathsf{HisValue} < \mathsf{HerValue}, \\ &\quad \mathsf{node(HisChildren, person), node(HerChildren, person)}\}. \end{aligned}$$

Remember that He, She, and Value are object identifiers, whereas HisChildren and HerChildren are object set identifiers (cf. Example 7.1). Therefore, the following definition

$$\begin{aligned} u &:= \{\mathsf{(He, Harry), (She, Sally), (HisValue, 10000), (HerValue, 20000),} \\ &\quad \mathsf{(HisChildren, Linus), (HisChildren, Clio),} \\ &\quad \mathsf{(HerChildren, Clio), (HerChildren, Mary)}\} \end{aligned}$$

is a correct Σ-relation for F in F' with $F \subseteq_{u,\Psi} F'$, since $u^*(F) \subseteq F'$ and thereby $F \subseteq_u F'$.
Furthermore, we can assume that the completing operator $\mathcal{C}$ generates formulas by means of which we are able to prove that

- Harry is not the target of a child-edge and, therefore, not the brother of Sally (cf. definition of brother in Example 7.4)
- Harry is not the source of a wife-edge, and
- Sally is not the target of a wife-edge.

Next, HisValue is bound to 10000 and HerValue is bound to 20000, i.e. the attribute condition HisValue < HerValue is valid. Finally, using the formulas of the graph schema of example 7.4 we are able to prove that Linus, Clio, and Mary are not only direct members of the classes man and woman, respectively, but also members of the class person. □

The relation u of Example 7.6 above is even maximal with respect to the number of object identifiers in F' which are bound to set identifiers in F. This is an important property of Σ-relations, which are involved in the process of structure rewriting.

7.2.3 Schema Preserving Structure Replacement

Having presented the definitions of structure schemata, schema consistent structures, and structure morphisms, we are now prepared to introduce *structure replacement rules* as quadruples of *sets of closed formulas.* The application of structure replacement rules will be defined as the construction of commuting diagrams in a similar way as it is done in the algebraic graph grammar approach (cf. Chapter 3). But note that we use Σ-*relations* instead of (partial) functions. Therefore, the main properties of the algebraic approach are necessarily lost: Our (sub-)diagrams are ***not pushouts*** and the application of a rule is ***not invertible*** in the general case.

Unfortunately, we had to pay this price for the ability to formalize complex rules, which match sets of nodes and are thereby able to delete/copy/redirect arbitrarily large edge bundles.

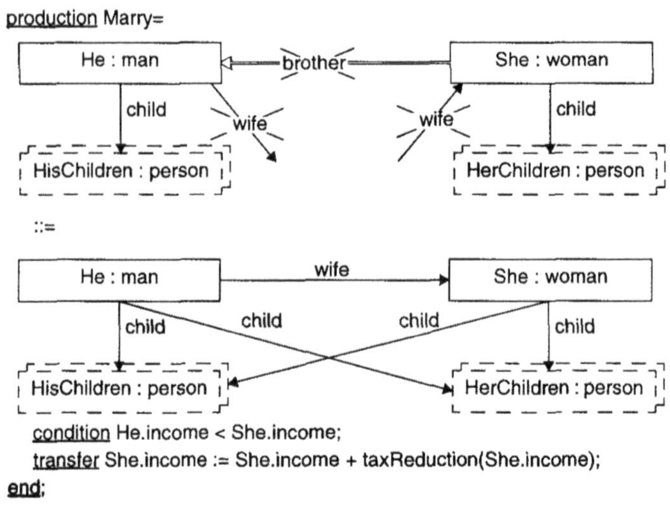

Figure 7.5: A graph replacement rule written in PROGRES

Definition 7.2.15 (Structure Replacement Rule.)
A quadruple $p := (AL, L, R, AR)$ with $AL, AR \in \mathcal{F}(\Sigma)$ and with $L, R \in \mathcal{L}(\Sigma)$ is a **structure replacement rule (production)** for the signature Σ (write: $p \in \mathcal{P}(\Sigma)$) $\Leftrightarrow$

(1) The set of left-hand side application/embedding conditions AL contains only object (set) identifiers of the left-hand side L.

(2) The set of right-hand side application/embedding conditions AR contains only object (set) identifiers of the right-hand side R.
(3) Every set identifier of L is also a set identifier in R and vice-versa; they are only used for deleting dangling references and for establishing connections between new substructures created by a rule's right-hand side and the rest of the modified structure.

The following example defines a structure replacement rule which is based on the structure selecting Example 7.6. It marries two persons under certain preconditions, whereby each person adopts the other person's children. The corresponding PROGRES like production is displayed in Figure 7.5.

Example 7.7 (The Structure Replacement Rule Marry.)
With F and Ψ defined as in Example 7.6, the structure replacement rule Marry := (AL, L, R, AR) has the following form:

(1) $AL := \Psi$.
(2) $L := F$.
(3) $R := F \setminus$ {attr(She, income, Value)}
$\cup$ {edge(He, wife, She),
attr(She, income, Value + taxReduction(Value)),
edge(He, child, HerChildren), edge(She, child, HisDaughters)} .
(4) $AR := \{\}$. □

Figure 7.6 displays the result of applying the structure replacement rule Marry twice to the database of Figure 7.1. Sally is now married to Harry and Fred to Clio. The identifier bindings of the first structure replacement step are shown in Example 7.6. In the second structure replacement step, He is bound to Fred and She to Clio. Furthermore, all set identifiers of the rule's left- and right-hand side have empty bindings (Fred and Clio have no children). Both steps are executed as follows:

- Find suitable matches for the object identifiers of L (She and He) in the database such that the corresponding preconditions in AL are satisfied.
- Extend the selected match to all object set identifiers in L and bind each set identifier to a maximal set of objects in the database such that all preconditions are valid.
- Remove the image of L without the image of R from the database (the old income attribute value of the selected She node).
- Add an image of R without the image of L to the database (the new income attribute value for She, a wife-edge between He and She, and child-edges to children nodes).

- Test the postconditions AR. If any postcondition is violated, then cancel all modifications, return to the first step of our list, and restart rule application with another match.
- Finally, test whether the constructed database is schema consistent and treat inconsistency like violated postconditions.

Applications of the rule Marry do not have to worry about postconditions and schema inconsistencies. All necessary context conditions are already part of the preconditions AL. But we could also move the brother check from the set of preconditions AL to the set of postconditions AR. We could even drop the "already married" checks in AL, since the schema of Example 7.4 contains integrity constraints preventing polygamy. Normally, *integrity constraints* are used instead of *repeated postconditions*, and postconditions for new objects in $R \setminus L$ replace very complex preconditions. It is still an open question, how any set of postconditions (referencing for instance derived properties of new objects) may be transformed into an equivalent set of preconditions.

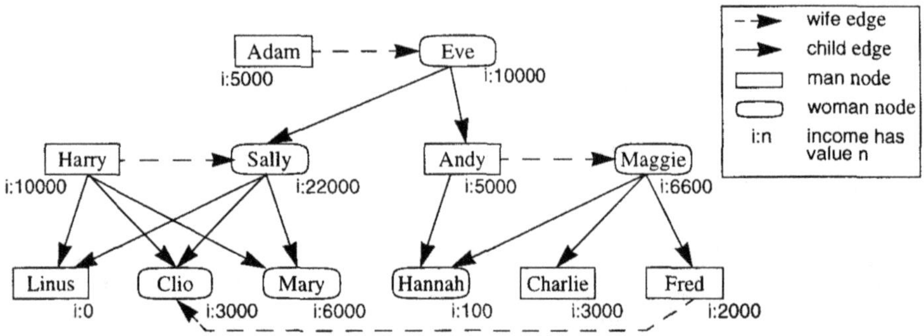

Figure 7.6: A modified person database

Definition 7.2.16 (Schema Preserving Structure Replacement.)
$S := (\Phi, C) \in \mathsf{S}(\Sigma)$ is a structure schema and $F, F' \in \mathcal{L}(S)$ are two schema consistent structures. Furthermore, $p := (AL, L, R, AR) \in \mathcal{P}(\Sigma)$ is a structure replacement rule. The structure F' is **direct derivable** from F by applying p (write: $F \overset{p}{\Longrightarrow} F'$) $:\Leftrightarrow$

(1) There is a morphism $u : L \overset{\sim}{\rightarrow} F$ with: $L \subseteq_{u,AL} F$,
i.e. the via u selected redex in F respects the preconditions AL.

(2) There is no morphism $\hat{u} : L \overset{\sim}{\rightarrow} F$ with: $L \subseteq_{\hat{u},AL} F$ and $u \subset \hat{u}$,
i.e. u selects a maximal substructure in F (with $\subset$ being the inclusion for relations).

(3) There is a morphism $w : R \stackrel{\sim}{\to} F'$ with: $R \subseteq_{w,AR} F'$,
i.e. the via w selected subgraph in F' respects the postconditions AR.

(4) The morphism w maps any new object identifier of R, not defined in L, onto a separate new object identifier in F', which is not defined in F.

(5) With $K := L \cap R$ the following property holds:
$v{:=}\{(x,y) \in u \mid x \text{ is identifier in } K\} = \{(x,y) \in w \mid x \text{ is identifier in } K\}$,
i.e. u and w are identical with respect to all identifiers in K.

(6) There exists a structure
$H \in \mathcal{L}(\Sigma) : F \setminus (u^*(L) \setminus v^*(K)) = H = F' \setminus (w^*(R) \setminus v^*(K))$,
i.e. H represents the intermediate state after the deletion of the old substructure and before the insertion of the new substructure.

It is a straightforward task to transform the definition of schema preserving structure replacements above into an *effective procedure* for the application of a rule p to a schema consistent structure F. The execution of p proceeds as already explained before in Definition 7.2.16 and is equivalent to the construction of the diagram in Figure 7.7. Unfortunately, we *cannot guarantee* that the process of computing derived data or checking pre- and postconditions as well as integrity constraints *terminates* in the general case. Therefore, we will introduce a special symbol "∞" for nonterminating computations in the next section, when we define the semantics of structure replacement programs.

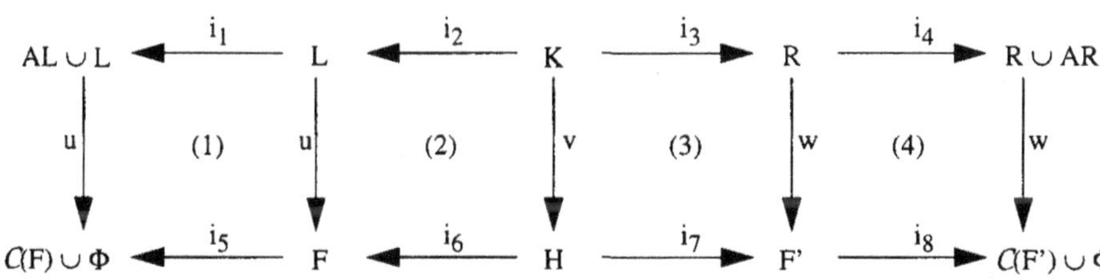

Figure 7.7: Diagram for the application of a structure replacement rule

Note that this kind of structure replacement does not prohibit the selection of *homomorphic* matches with two identifiers $o1$ and $o2$ in L being mapped onto the same object in F. The application of the replacement rule Marry above relates for instance the two set identifiers HisChildren and HerChildren to the same node Clio of our graph database. "Sharing" is even permitted if $o1$ belongs to R and $o2$ not. These *deleting/preserving conflicts* are resolved in favor of preserving objects (unlike to the SPO approach in Chapter 7 of this volume). Otherwise, relation v, the restriction of u to K, would no longer be a morphism from K to H. For readability reasons, Σ-morphisms which map different node identifiers of a production onto the same node in a given graph, may, but need not be prohibited in the language PROGRES.

Finally note that we have to take care about *dangling references* in the general case of a person database replacement rule, where L is not a subset of R. We have to guarantee that any $\mathsf{node}(x, l)$ is removed together with all related $\mathsf{edge}(x, e, y)$, $\mathsf{edge}(y, e, x)$, and $\mathsf{attr}(x, a, y)$ formulas. This may be accomplished by adding appropriate formulas with set identifiers (instead of y) to the rule's left-hand side. In this way, we are able to overcome the problem of the algebraic double pushout approach with deletion of nodes with unknown context (cf. Chapter 3 of this volume).

Proposition 7.2.17 (Structure Replacement Diagrams.)
Assuming the terminology of Definition 7.2.16 and $F \overset{p}{\Rightarrow} F'$ we are able to construct the diagram of fig. 7 (with $i_1, \ldots, i_8$ being inclusions). This diagram has commuting subdiagrams (1) through (4).

Proof:
The existence of commuting subdiagrams (1) and (4) is guaranteed by Definition 7.2.14. For proving the existence of commuting subdiagrams (2) and (3), we will show that step (5) of Definition 7.2.16 constructs a morphism v from K to H:
We start with Definition 7.2.16, condition (1):
$u : L \overset{\sim}{\rightarrow} F$

$$
\begin{array}{ll}
\Leftrightarrow_{\text{Def. 7.2.10, Prop. 7.2.13, } K \subseteq L} & u^*(K) \subseteq u^*(L) \subseteq F \\
\Rightarrow_{\text{condition (5) of Def. 7.2.16}} & v^*(K) \subseteq u^*(L) \subseteq F \\
\Rightarrow_{\text{simple transformation}} & v^*(K) \setminus (u^*(L) \setminus v^*(K)) \subseteq F \setminus (u^*(L) \setminus v^*(K)) \\
\Leftrightarrow_{\text{simple transformation}} & v^*(K) \subseteq F \setminus (u^*(L) \setminus v^*(K)) \\
\Leftrightarrow_{\text{condition (6) of Def. 7.2.16}} & v^*(K) \subseteq H \\
\Leftrightarrow_{\text{Def. 7.2.10 and Prop. 7.2.13}} & v : K \overset{\sim}{\rightarrow} H
\end{array}
$$

The rest of the proof follows directly from
$i_2 \circ u = v \circ i_6$ *and* $v \circ i_7 = i_3 \circ w$,
since v is a restriction of u or w onto the identifiers in K. □

Being equipped with a definition of structure replacement we are now able to define LBSR systems and their generated languages. As usual, generated isomorphic language elements are identified, i.e. *concrete Σ-structures* according to Definition 7.2.4 are replaced by equivalent classes of concrete Σ-structures, so-called *abstract Σ-structures.*

Definition 7.2.18 (Abstract Σ-Structure.)
The set of **abstract Σ-structures** $\mathcal{AL}(\Sigma) := \mathcal{L}(\Sigma)/ \overset{\sim}{=}$ consists of all equivalence classes of Σ-structures with respect to the following equivalence relation $\overset{\sim}{=}$:

$\forall F, F' \in \mathcal{L}(\Sigma) : F \overset{\sim}{=} F' :\Leftrightarrow \exists$ isomorphism $u : F \overset{\sim}{\rightarrow} F'$.

It is worth-while to notice that any isomorphism between two Σ-structures is a *bijective function* which maps object identifiers onto object identifiers and object set identifiers onto object set identifiers (cf. Def. 7.2.8, 7.2.9, and 7.2.10 of Σ-relations and Σ-morphisms). Otherwise, F would contain less object (set) identifiers than F' or the other way round. As a consequence, it would not be possible to construct the left- and right-inverse morphism u^{-1} for u. This line of argumentation is no longer true, when arbitary sets of Σ-formulas are regarded. A single *nonatomic formula* then represents a set of derivable facts. It is, therefore, possible that two sets of formulas are equivalent (isomorphic) although they differ with respect to the number of used object (set) identifiers.

Definition 7.2.19 (Abstract Schema Consistent Σ-Structure.)
Let $S \in \mathcal{S}(\Sigma)$ be a schema and $\mathcal{L}(S)$ the set of all S-schema consistent structures. The set of **abstract schema consistent Σ-structures** is defined as

$$\mathcal{AL}(\Sigma) := \mathcal{L}(S)/\cong .$$

Proposition 7.2.20 (Soundness of Abstract Schema Consistency.)
Let $S := (\Phi, \mathcal{C}) \in \mathcal{S}(\Sigma)$ be a schema and $F, F' \in \mathcal{L}(\Sigma)$ arbitrary Σ-structures. Then

$$F \cong F' \Rightarrow (F \in \mathcal{L}(S) \Leftrightarrow F' \in \mathcal{L}(S)),$$

i.e. all Σ-structures of an equivalence class in $\mathcal{AL}(\Sigma)$ are either schema consistent or inconsistent.

Proof:

$F \cong F'$

$\Rightarrow_{\text{Def. 7.2.18}}$ $\exists$ *isomorphism* $u : F \xrightarrow{\sim} F'$, *a bijective function*

$\Rightarrow_{\text{Props. 7.2.13}}$ $u^*(F) = F'$

Therefore, the following is true:

$F \in \mathcal{L}(S)$ $\Rightarrow_{\text{Def. 7.2.7}}$ $\mathcal{C}(F) \cup \Phi$ *is a consistent set of formulas*

$\Rightarrow$ $u^*(\mathcal{C}(F) \cup \Phi)$ *is a consistent set of formulas*
(consistent renaming preserves proofs)

$\Rightarrow_{\text{Def. 7.2.6}}$ $u^*(\mathcal{C}(F)) \cup \Phi$ *is a consistent set of formulas*
(Φ has no identifiers of $\mathcal{V}$ or $\mathcal{W} \Rightarrow u^*(\Phi) = \Phi$)

$\Rightarrow_{\text{Def. 7.2.5}}$ $\mathcal{C}(u^*(F)) \cup \Phi$ *is a consistent set of formulas*

$\Rightarrow$ $\mathcal{C}(F') \cup \Phi$ *is a consistent set of formulas*

$\Rightarrow$ $F' \in \mathcal{L}(S)$.

In a similar way, we can prove that $F' \in \mathcal{L}(S) \Rightarrow F \in \mathcal{L}(S)$. ∎

Definition 7.2.21 (LBSR System.)
A tuple LBSR $:= (AS, P)$ is a **logic-based structure replacement (LBSR) system** with respect to a structure schema $S \in \mathcal{S}(\Sigma)$ if:

(1) $AS \in \mathcal{AL}(S)$ is the schema consistent initial abstract structure.
(2) $P \subseteq \mathcal{P}(\Sigma)$ is a set of structure replacement rules.

Definition 7.2.22 (Generated Language of LBSR System.)
With LBSR := (AS, P) as in Definition 7.2.21, the **language** $\mathcal{AL}(\text{LBSR}) \subseteq \mathcal{AL}(S)$ is defined as follows:

$$AF \in \mathcal{AL}(\text{LBSR}) :\Leftrightarrow AS \overset{P}{\Longrightarrow}{}^{*} AF$$

where

$\overset{P}{\Longrightarrow}{}^{*}$ is the transitive, reflexive closure of $\overset{P}{\Longrightarrow}$

and $AF \overset{P}{\Longrightarrow} AF' \subseteq \mathcal{AL}(\Sigma) \times \mathcal{AL}(S) :\Leftrightarrow$
$\exists F \in AF, F' \in AF', p \in P : F \overset{p}{\Longrightarrow} F'$.

The definition above states that the *language of a LBSR system* consists of all those abstract structures which may be generated by applying its rules an arbitrary number of times to the initial abstract structure AS. A distinction between terminal and nonterminal structures has not been made in the preceding subsections. Hence, any generated abstract structure is per definition an element of the generated language.

7.2.4 Summary

The definitions and propositions of the preceding subsections constitute a *framework* for the formal treatment of various forms of graph replacement systems as special cases of logic-based structure replacement (LBSR) systems. Other approaches with the same intention are "**H**igh **L**evel **R**eplacement (HLR) systems" and "structured graph grammars". HLR *systems* belong to the algebraic branch of graph grammars (cf. Section 4.5.3 of Chapter 7 and [18]). They provide a very general framework for the definition of new kinds of replacement systems, which is based on the construction of categories with certain properties. Therefore, HLR systems are not restricted to the manipulation of graphs or relational structures (as LBSR systems are).

Structured graph grammars [35], on the other hand, are restricted to a data model of directed acyclic graphs. They were a first attempt to combine properties of algebraic graph grammars (cf. Chapter 3 and 7 of this volume) and algorithmic node replacement graph grammars (cf. Chapter 1 of this volume) within a single framework. Both HLR systems and structured graph grammars do not introduce the notion of a *schema* and do not support modeling of derived properties or constraints.

LBSR systems are an attempt to close the gap between the *operation-oriented* manipulation of data structures by means of rules and the *declaration-oriented* description of data structures by means of logic-based knowledge representation languages. In this way, both disciplines — graph grammar theory and mathematical logic theory — might profit from each other:

- Structure (graph) replacement rules might be a very comfortable and well-defined mechanism for manipulating knowledge bases or deductive data bases (cf. [25,59]), whereas
- many logic-based techniques have been proposed for efficiently maintaining derived properties of data structures, solving constraint systems, and for proving the correctness of data manipulations (cf. [10,29,54]).

Currently, LBSR systems are restricted to *sequential graph replacement*. It is subject to future work to extend the presented approach such that *parallel application of rules* is supported. Furthermore, the following questions should be considered in more detail:

(1) Which restrictions are sufficient to guarantee that subdiagrams (2) and (3) of Figure 7.7 are pushouts?
(2) Can we characterize a "useful" subset of rules and consistency constraints such that an effective proof procedure exists for the "are all derivable structures schema consistent" problem (see also Chapter 9 of this volume and [13])?
(3) Can we develop a general procedure which transforms any set of structure replacement rule postconditions into weakest preconditions?

The problems (2) and (3) above are related to each other by transforming global consistency constraints into equivalent postconditions of individual rewrite rules using techniques proposed in [10]. Having such a procedure which translates postconditions into corresponding preconditions, problem (2) is reducible to the question whether these new preconditions are already derivable from the original set of preconditions of a given rule and the set of guaranteed integrity constraints (for the input of the rule).

7.3 Programmed Structure Replacement Systems

So far we would be able to define sets of schema consistent structure replacement rules and structure languages, which are generated by applying these rules in any order to a given structure (axiom). This might be sufficient from a theoretical point of view, but when we are using structure replacement systems for specification purposes additional means are necessary for *regulating the application of rules.* Many quite different proposals for controlling application of rules may be found in literature as for instance [11,15,44,45,64]:

- Apply rules as long as appropriate or as long as possible in any order; this is the *standard semantics* of replacement systems.
- Introduce *rule priorities* and prefer applicable rules with higher priorities at least in the case of overlapping matches.
- Use *regular expressions* or even complex programs to define permissible derivation sequences.
- Draw *control flow graphs* and interpret them as graphical representations of rule controlling programs.

The idea of using *programs for controlling the application of rules*, i.e. imperative control structures, seems to be the most popular one and even superior to almost all other rule regulation approaches with respect to their expressiveness. On the following pages we will, therefore,

- first study required properties of graph replacement programs or, more general, of control structures for structure replacement rules (Subsection 7.3.1),
- introduce a more or less minimal set of control structures in the form of so-called **B**asic **C**ontrol **F**low (BCF) operators and discuss their intended semantics on an informal level (Subsection 7.3.2),
- present an appropriate semantic domain for BCF operators and a recently developed fixpoint theorem (Subsection 7.3.3),
- define the semantics of BCF expressions with recursion and, thereby, the semantics of **P**rogrammed **L**ogic-based **S**tructure **R**eplacement (PLSR) systems by means of the presented fixpoint theorem (Subsection 7.3.4),
- and finally summarize the main properties of PLSR systems and discuss possible extensions (Subsection 7.3.5).

7.3.1 Requirements for Rule Controlling Programs

So-called *programmed graph grammars* were already suggested many years ago in [11,45]. Nowadays, they are a fundamental concept of the graph grammar specification languages PAGG [26,28] and PROGRES [66,71]. Experiences with using these languages and their predecessors for the specification of software engineering tools [21,22,38,75] showed that their control structures should possess the following properties:

(1) *Boolean nature*: the application of a programmed graph transformation either succeeds or fails like the application of a single graph replacement rule and, depending on success or failure, execution may proceed along different control flow paths.

(2) *Atomic character*: a programmed sequence of graph replacement steps modifies a given host graph if and only if all its intermediate replacement steps do not fail.
(3) *Consistency preserving*: programmed graph replacement has to preserve a graph's consistency with respect to a given set of separately defined integrity constraints.
(4) *Nondeterministic behavior*: the nondeterminism of a single rule — it replaces any match of its left-hand side — should be preserved as far as possible on the level of programmed graph transformations.
(5) *Recursive definition*: for reasons of convenience as well as expressiveness, programmed graph transformations should be allowed to call each other without any restrictions including any kind of recursion.

Without conditions (1) through (4) we would have difficulties to replace a complex graph replacement rule by an equivalent sequence of simpler rules, and without condition (5) the manipulation of recursively defined data structures would be quite cumbersome.

In the sequel, programmed graph transformations with the above mentioned properties will be termed *transactions*. The usage of the term "transaction" underlines the most important properties of programmed transformations: *atomicity* and *consistency*. The usually mentioned additional properties of transactions, *isolation* and *duration*, are irrelevant as long as parallel programming is not supported and backtracking may cancel the effects of already successfully completed transactions (cf. Subsection 7.3.2).

7.3.2 Basic Control Flow Operators

The definition of a fixed set of control structures for structure manipulating transactions is complicated by contradicting requirements. From a *theoretical point of view* the set of offered control structures should be as small as possible, and we should be allowed to combine them without any restrictions. But from a *practical point of view* control structures are needed which are easy to use, which cover all frequently occurring control flow patterns, and the application of which may be directed by a number of context-sensitive rules.
Therefore, it was quite natural to distinguish between basic control flow operators, in the sequel termed *BCF operators*, of an underlying theory of recursively defined transactions and more or less complex higher level programming mechanisms. Starting with a formal definition of basic control flow operators, we should then be able to define the meaning of a large class of programming mechanisms by translating them into equivalent BCF expressions (cf. Section 7.5).

```
<Transaction> ::=     <TransactionId> "=" <BCFExpr> ;
<BCFExpr> ::=         <BasicAction> | <ActionCall> | <BCFTerm> ;
<BasicAction> ::=     "skip" | "loop" ;
<ActionCall> ::=      <RuleId> | <TransactionId> ;
<BCFTerm> ::=         "def" "(" <BCFExpr> ")" | "undef" "(" <BCFExpr> ")" |
                      "(" <BCFExpr> ";" <BCFExpr> ")" |
                      "(" <BCFExpr> "[]" <BCFExpr> ")"
                      "(" <BCFExpr> "&" <BCFExpr> ")";
```

Figure 7.8: Syntax of graph transactions and BCF expressions

Figure 7.8 contains the definition of BCF expressions themselves and of transactions as functional abstractions of BCF expressions. It distinguishes between *basic actions* like

- skip, which represents the always successful identity operator and relates a given graph G to itself, and
- loop, which neither succeeds nor terminates for any given structure and represents therefore "crashing" or forever looping computations,

calls of simple rules or other graph transactions, and finally between two unary and three binary *BCF operators* with

- def(a) as an action which succeeds applied to a given structure F, whenever a applied to F produces a defined result, and returns F itself,
- undef(a) as an action which succeeds applied to a structure F, whenever a applied to F terminates with failure, and returns F itself,
- $(a\,;\,b)$ as an action which is the sequential composition of a and b, i.e. applies first a to a given structure F and then b to any "suitable" result of the application of a,
- $(a \,[\!]\, b)$ as an action which represents the nondeterministic choice between the application of a or b, and
- $(a \,\&\, b)$ as an action which returns the intersection of the results of a and b (requires an equality or isomorphy testing operator on graphs).

Note that the operators suggested above are intentionally similar to those proposed by Dijkstra [14] and especially to those presented by Nelson [51] with one essential difference: due to the boolean nature of basic structure replacement rules and complex transactions, we are not forced to distinguish between side effect free boolean expressions and state modifying actions. This has the consequence that complex guarded commands of the form

$(\mathsf{Cond}_1 \rightarrow \mathsf{Body}_1 \,[\!]\, \ldots \,[\!]\, \mathsf{Cond}_n \rightarrow \mathsf{Body}_n)$

are no longer necessary but may replaced by expressions like

$(\underline{\mathsf{def}}(\mathsf{Cond}_1); \mathsf{Body}_1) \mathbin{[\!]} \ldots \mathbin{[\!]} (\underline{\mathsf{def}}(\mathsf{Cond}_n); \mathsf{Body}_n)$

where $\underline{\mathsf{def}}(\mathsf{Cond}_i)$ tests either the applicability of a single rule or of a whole transaction without modifying the given input structure.

Furthermore, BCF expressions offer almost all possibilities for combining binary relations over structures, the semantic domain of structure manipulating transactions. There are operators for intersection, union, and concatenation. The *missing difference operator* is not supported for the following reasons: $a \setminus b1$ is a subrelation of $a \setminus b2$, if $b2$ is a subrelation of $b1$. As a consequence, the nonmonotonic difference operator had to be excluded in order to be able to come up with a sound definition for recursively defined transactions (cf. Subsection 7.3.4).

Having motivated our reasons for the selection of two unary and three binary BCF operators, we are now prepared to discuss the intricacies of their intended semantics. A first problem comes with the definition of the meaning of $(a; b)$ as "apply b to *any suitable* result of a". Let us assume that the application of a transformation a to a structure F has three possible results named $F1$, $F2$, and $F3$, respectively. Furthermore, let us assume that the transformation b applied to $F1$ fails but applied to $F2$ and $F3$ succeeds. In this case we may either select $F2$ or $F3$, but not $F1$ as a suitable result of the application of a. This means that we need knowledge about future states of an ongoing transformation process in order to be able to discard those possible results of a single transformation step, which cause failure of the overall transformation process. It should be quite obvious that a more realization-oriented definition of this kind of clairvoyant nondeterminism requires a kind of *depth-first search* semantics with *backtracking* out of "dead-ends".

Another problem comes with the definition of expressions like $(a \mathbin{[\!]} b)$, where a loops forever applied to a certain structure F, but b has a well-defined set of possible results. Having a depth-first search semantics in mind, we are forced to define the outcome of the expression $(a \mathbin{[\!]} b)$ as being either a nonterminating computation or any defined result produced by b.

This means that the kind of *nondeterminism* we are going to define is not *angelic* but more or less *erratic*: Using backtracking we are able to discard nondeterministic selections which lead to defined failures of basic actions or structure replacement rules but not selections which cause nonterminating computations.

The following example uses nondeterministic depth-first search and backtracking. It defines a transaction, which tries to complete a person database such that all person nodes are married afterwards. It uses the rule Marry of Example 7.7 and assumes the existence of another rule MarkUnmarried. The rule

MarkUnmarried selects and marks a single unmarried person in the database if existent and fails otherwise.

Example 7.8 (A Person Database Manipulating Transaction.)
Marry* = (((def(MarkUnmarried); Marry); Marry*)
[] undef(MarkUnmarried)). □

The transaction above initiates calls of rule Marry as long as the rule MarkUnmarried is applicable. It terminates successfully, if and only if all persons are finally married. Otherwise, a database state will be reached, where neither the first nor the second branch of [] is executable. In this case, the transaction aborts without any database modifications. Note that the execution of Marry* requires backtracking in the general case. Consider for instance the application of Marry* to the database of Figure 7.1. The transformation process may start with marrying Linus and Marry first and marrying Harry and Sally afterwards. The problem is now that no partner for Hannah is left over. Therefore, backtracking starts and another couple instead of Harry and Sally is married (e.g. Harry and Hannah).

7.3.3 Preliminary Definitions

After an informal introduction of transactions as named BCF expressions we will now define their intended semantics: This is a *semantic function* from the domain of BCF expressions onto the range of (extended) binary relations over abstract structures. In order to be able to deal with recursion and nondeterminism in the presence of an atomic sequence operator ";" we had to follow the lines of [51] [g]. There, a new form of the fixpoint theorem is used to give an axiomatic definition of so-called nondeterministic commands.

Proposition 7.3.1 (Fixpoint Theorem.)
Let f be a monotonic function(al) on a partially ordered set in which every chain has a join, and let f^α, for ordinal α, be defined inductively by
$f^\alpha = (\cup_{\beta:\beta<\alpha} : f(f^\beta))$.
Then f has a least fixpoint given by f^α, for some ordinal α.

Proof: See appendix of [51]. □

For being able to apply the fixpoint theorem to BCF expressions an appropriate partially ordered semantic domain is needed:

[g] The sequence operator is not chain continuous in its 2nd argument. Therefore, the original version of the fixpoint theorem, proved in [41], does not work in our case.

Definition 7.3.2 (Semantic Domain.)
With $\mathcal{AL}(S)$ being a S-consistent class of abstract structures for a schema $S \in \mathcal{S}(\Sigma)$, the **semantic domain** of transactions is defined to be the following power set of binary relations:

$$\mathcal{D} := 2^{\mathcal{AL}(S)\times(\mathcal{AL}(S)\cup\{\infty\})} .$$

The semantics of a transaction is a binary relation between abstract structures, where the symbol "∞" in a second component represents *potentially nonterminating computations.* The word "potential" includes computations with partially unknown effects, i.e. computations which may abort or loop forever. A relation $R_\infty := \mathcal{AL}(S) \times \{\infty\}$ for instance, which maps any structure F onto "∞", represents a computation with completely unknown outcomes.

In order to be able to apply fixpoint theory to recursively defined transactions we have to construct a *suitable partial order* for our semantic domain $\mathcal{D}$. "Suitable" means from a practical point of view that the relation R_∞ defined above should be less than any other element in $\mathcal{D}$ and that "R_1 less than R_2" means that R_2 is a better approximation of a given transaction than R_1. And "suitable" means from a theoretical point of view that we have to proof that any chain in $\mathcal{D}$ has a join, i.e. that any sequence of elements $(R^\alpha)_{\alpha\in\text{ordinal}}$ with R^α being less equal than $R^{\alpha+\beta}$ for any ordinals α and β has a least upper element in $\mathcal{D}$.

Definition 7.3.3 (Partial Order on Semantic Domain.)
With $R, R' \in \mathcal{D}$ and S being the underlying structure schema, a suitable **partial order** "$\leq$" is defined as follows:

$$\begin{aligned} R \leq R' :\Leftrightarrow \forall F, F' \in \mathcal{L}(S) : \ & (F, F') \in R \Rightarrow (F' = \infty \vee (F, F') \in R') \\ \wedge\ & (F, \infty) \notin R \Rightarrow ((F, F') \in R \vee (F, F') \notin R'). \end{aligned}$$

Please note that

$$\begin{aligned} R \leq R' :\Leftrightarrow \forall F, F' \in \mathcal{L}(S) : \ & (F, F') \in R \Rightarrow (F' = \infty \vee (F, F') \in R') \\ \wedge\ & (F, \infty) \notin R \Rightarrow ((F, F') \in R \Leftrightarrow (F, F') \in R'). \end{aligned}$$

is an equivalent definition of the partial order "$\leq$". It is easier to handle and will be used from now on.

The relation R of the definition above *approximates* R' such that any input F is either related to the same set of outputs by R and R' or is related to the symbol "∞" in R and to a potentially greater set of outputs in R' (by eventually dropping "∞"). It is obvious that "$\leq$" is indeed a partial order, but we have to proof its *chains have joins* condition:

Lemma 7.3.4 (Chains Have Joins.)
Let $(R^\alpha)_{\alpha \in \text{ordinal}} \subseteq \mathcal{D}$ be a chain, i.e. for any ordinals a and b holds:
$R^\alpha \leq R^{\alpha+\beta}$.
Then the join of the given chain is defined as follows:

$$R := \sqcup(R^\alpha) := \{(F,F') \mid \quad F' \neq \infty \wedge \exists\alpha : (F,F') \in R^\alpha \\ \vee F' = \infty \wedge \forall\alpha : (F,\infty) \in R^\alpha\}.$$

Proof:
We have to show that the element R above is indeed greater than any element of our chain and that R is the least element in $\mathcal{D}$ with this property:
$\forall\alpha : R^\alpha \leq R :$

$$\begin{aligned}(F,F') \in R^\alpha &\Rightarrow (F,F') \in R \vee F' = \infty \,. \\ (F,\infty) \notin R^\alpha &\Rightarrow \forall\beta \geq \alpha : (F,F') \in R^\alpha \Leftrightarrow (F,F') \in R^\beta \\ &\Rightarrow (F,F') \in R^\alpha \Leftrightarrow (F,F') \in R \,.\end{aligned}$$

$\forall R' \in \mathcal{D} : (\forall\alpha : R^\alpha \leq R') \Rightarrow R \leq R' :$

$$\begin{aligned}(F,F') \in R &\Rightarrow \exists R^\alpha : (F,F') \in R^\alpha \vee F' = \infty \\ &\Rightarrow (F,F') \in R' \vee F' = \infty \,. \\ (F,\infty) \notin R &\Rightarrow \exists R^\alpha : (F,\infty) \notin R^\alpha \\ &\Rightarrow \exists R^\alpha : (F,F') \in R \Leftrightarrow (F,F') \in R^\alpha \Leftrightarrow (F,F') \in R' \,. \quad \square\end{aligned}$$

7.3.4 A Fixpoint Semantics for Transactions

Now, we are prepared to define a semantic function $\mathcal{R}$ from the syntactic domain of BCF expressions or transactions onto the semantic domain $\mathcal{D}$ inductively:

Definition 7.3.5 (BCF Expressions and Transactions.)
S and $\mathcal{D}$ are defined as in Definition 7.3.3, and $P \subseteq \mathcal{P}(\Sigma)$ is a set of structure replacement rules. Then, $\mathcal{E}(P)$ denotes the set of all **BCF expressions** and $\mathcal{T}(P)$ the set of all **transactions** over P. Their context-free syntax is displayed in Figure 7.8, and a **semantic function** $\mathcal{R}_P : \mathcal{E}(P) \to \mathcal{D}$ is defined as follows [h] with abstract structures $F, F', F'' \in \mathcal{AL}(S)$ and with $a, b \in \mathcal{E}(P)$:

(1) $(F,F') \in \mathcal{R}_P[\![\underline{\mathsf{skip}}]\!] :\Leftrightarrow F = F'$.
(2) $(F,F') \in \mathcal{R}_P[\![\underline{\mathsf{loop}}]\!] :\Leftrightarrow F' = \infty$.
(3) $(F,F') \in \mathcal{R}_P[\![p]\!] :\Leftrightarrow \quad F \stackrel{p}{\Longrightarrow} F'$, for any production $p \in \mathcal{P}(\Sigma)$
$\vee F' = \infty$, if execution of p may not terminate.
(4) $(F,F') \in \mathcal{R}_P[\![\underline{\mathsf{def}}(a)]\!] :\Leftrightarrow \quad \exists F'' \neq \infty : (F,F'') \in \mathcal{R}_P[\![a]\!] \wedge F = F'$
$\vee (F,\infty) \in \mathcal{R}_P[\![a]\!] \wedge F' = \infty$.

[h] The definition of transactions and transaction calls is postponed, until the precondition of the previously introduced fixpoint theorem are checked.

(5) $(F, F') \in \mathcal{R}_P[\![\underline{\text{undef}}(a)]\!] :\quad (\neg\exists F'' : (F, F'') \in \mathcal{R}_P[\![a]\!]) \wedge F = F'$
$\vee (F, \infty) \in \mathcal{R}_P[\![a]\!] \wedge F' = \infty.$

(6) $(F, F') \in \mathcal{R}_P[\![(a;\, b)]\!] :\Leftrightarrow \exists F'' \neq \infty : (F, F'') \in \mathcal{R}_P[\![a]\!] \wedge (F'', F') \in \mathcal{R}_P[\![b]\!]$
$\vee (F, \infty) \in \mathcal{R}_P[\![a]\!] \wedge F' = \infty.$

(7) $(F, F') \in \mathcal{R}_P[\![(a \,\|\, b)]\!] :\Leftrightarrow (F, F') \in \mathcal{R}_P[\![a]\!] \vee (F, F') \in \mathcal{R}_P[\![b]\!].$

(8) $(F, F') \in \mathcal{R}_P[\![(a \,\&\, b)]\!] :\Leftrightarrow F' = \infty \wedge ((F, \infty) \in \mathcal{R}_P[\![a]\!] \vee (F, \infty) \in \mathcal{R}_P[\![b]\!])$
$\vee (F, F') \in \mathcal{R}_P[\![a]\!] \wedge (F, F') \in \mathcal{R}_P[\![b]\!].$

The definitions above are rather straightforward with the exception of the treatment of the operators def and undef. The expressions def(a) and undef(a) loop forever if a returns not a single defined result but loops forever. Furthermore, def(a) may return its input if a returns at least one defined result, even if a may loop forever. It terminates with failure if a terminates with failure. On the other hand, undef(a) returns its input if and only if a fails, and it fails if and only if a has at least one defined result and may not loop forever.

Therefore, undef is *stricter* than def with respect to the treatment of looping computations; the expression def(a) may return a defined result even if a may loop forever. From a practical point of view, this distinction may be justified as follows:

- Often, we would like to know whether a computes at least one defined result without evaluating all possible execution paths of a after a successful path has been found.
- On the other hand, answering the question whether a fails is not possible without taking all execution paths of a into account, thereby running into any nonterminating execution branch of a.

Nevertheless, a *strict version* of def might be useful, too. It is no longer independent from the definition of undef, but may be given as follows:

Definition 7.3.6 (Strict Version of def Operator.)
A strict version of the def operator may be defined as follows with abstract structures $F, F', F'' \in \mathcal{AL}(S)$, and $a \in \mathcal{E}(P)$ as in Definition 7.3.5:

$$\begin{aligned}
&(F, F') \in \mathcal{R}_P[\![\underline{\text{def}}^{\infty}(a)]\!] :\Leftrightarrow \\
&\qquad (F, F') \in \mathcal{R}_P[\![\underline{\text{undef}}(\underline{\text{undef}}(a))]\!] \\
\Leftrightarrow &\qquad (\neg\exists F'' : (F, F'') \in \mathcal{R}_P[\![\underline{\text{undef}}(a)]\!]) \wedge F = F' \\
&\qquad \vee (F, \infty) \in \mathcal{R}_P[\![\underline{\text{undef}}(a)]\!] \wedge F' = \infty \\
\Leftrightarrow &\qquad (\exists F'' \neq \infty : (F, F'') \in \mathcal{R}_P[\![a]\!] \wedge (F, \infty) \notin \mathcal{R}_P[\![a]\!] \wedge F = F') \\
&\qquad \vee (F, \infty) \in \mathcal{R}_P[\![a]\!] \wedge F' = \infty.
\end{aligned}$$

In order to be able to use Proposition 7.3.1 for the treatment of recursive transactions we have still to prove that BCF expressions correspond to monotonic functionals. A BCF expression which contains for instance applied transaction identifiers t_1 to t_n, must be interpreted as a functional with the following signature:

$\mathcal{R}_P[\![E]\!] : \mathcal{D}^n \to \mathcal{D}$.

Provided with the semantics $\mathcal{R}_P[\![t_1]\!], \ldots, \mathcal{R}_P[\![t_n]\!]$ of t_1 to t_n, the semantics of E is given by Definition 7.3.5 and is denoted as follows:

$\mathcal{R}_P[\![E]\!][\mathcal{R}_P[\![t_1]\!], \ldots, \mathcal{R}_P[\![t_n]\!]]$.

In such a way, we are able to define the semantics of all BCF expressions and transactions bottom-up in the absence of recursion. But having a recursively defined transaction like

$\mathsf{Marry}^* = (((\mathsf{def}(\mathsf{MarkUnmarried}); \mathsf{Marry}); \mathsf{Marry}^*) \,[\!]\, \mathsf{undef}(\mathsf{MarkUnmarried}))$,

we are looking for a least element $R \in \mathcal{D}$ such that the following *fixpoint equation*

$$R = \mathcal{R}_P[\![(((\underline{\mathsf{def}}(\mathsf{MarkUnmarried}); \mathsf{Marry}); \mathsf{Marry}^*) \,[\!]\, \underline{\mathsf{undef}}(\mathsf{MarkUnmarried}))]\!][R]$$

holds. Proposition 7.3.1 provides us with such a *least fixpoint* if we are able to show that all BCF operators define monotonic functionals and, therefore, all complex BCF expressions, too:

Lemma 7.3.7 (BCF Operators are Monotonic Functionals.)
With $\mathcal{D}, \mathcal{AL}(S)$, and $\mathcal{R}_P$ from Definition 7.3.5, $P \subseteq P(\Sigma)$, and $a, a', b \in \mathcal{E}(P)$, $\mathcal{R}_P[\![a]\!] \leq \mathcal{R}_P[\![a']\!]$ implies:

(1) $\mathcal{R}_P[\![(a\,;\, b)]\!] \leq \mathcal{R}_P[\![(a'\,;\, b)]\!]$.
(2) $\mathcal{R}_P[\![(b;\, a)]\!] \leq \mathcal{R}_P[\![(b;\, a')]\!]$.
(3) $\mathcal{R}_P[\![(a \,[\!]\, b)]\!] = \mathcal{R}_P[\![(b \,[\!]\, a)]\!] \leq \mathcal{R}_P[\![(b \,[\!]\, a')]\!] = \mathcal{R}_P[\![(a' \,[\!]\, b)]\!]$.
(4) $\mathcal{R}_P[\![(a \,\&\, b)]\!] = \mathcal{R}_P[\![(b \,\&\, a)]\!] \leq \mathcal{R}_P[\![(b \,\&\, a')]\!] = \mathcal{R}_P[\![(a' \,\&\, b)]\!]$.
(5) $\mathcal{R}_P[\![\underline{\mathsf{def}}(a)]\!] \leq \mathcal{R}_P[\![\underline{\mathsf{def}}(a')]\!]$.
(6) $\mathcal{R}_P[\![\underline{\mathsf{undef}}(a)]\!] \leq \mathcal{R}_P[\![\underline{\mathsf{undef}}(a')]\!]$.

Proof:

ad *(1)* $\mathcal{R}_P[\![(a\,;\, b)]\!] \leq \mathcal{R}_P[\![(a'\,;\, b)]\!]$:

$\forall F \in \mathcal{AL}(S)$ with $(F, \infty) \notin \mathcal{R}_P[\![a]\!]$
$\Rightarrow \quad (F, F') \in \mathcal{R}_P[\![a]\!] \Leftrightarrow (F, F') \in \mathcal{R}_P[\![a']\!]$
$\Rightarrow \quad (F, F') \in \mathcal{R}_P[\![(a\,;\, b)]\!] \Leftrightarrow (F, F') \in \mathcal{R}_P[\![(a'\,;\, b)]\!]$.

$\forall F \in \mathcal{AL}(S)$ with $(F, \infty) \in \mathcal{R}_P[\![a]\!] \Rightarrow (F, \infty) \in \mathcal{R}_P[\![(a\,;\, b)]\!]$
and with $(F, F'') \in \mathcal{R}_P[\![a]\!] \Rightarrow (F, F'') \in \mathcal{R}_P[\![a']\!] \vee F'' = \infty$:

$(F, F') \in \mathcal{R}_P[\![(a\,;\, b)]\!]$
$\Rightarrow \quad (\exists F'' \neq \infty : (F, F'') \in \mathcal{R}_P[\![a]\!] \wedge (F'', F') \in \mathcal{R}_P[\![b]\!]) \vee F' = \infty$
$\Rightarrow \quad (\exists F'' \neq \infty : (F, F'') \in \mathcal{R}_P[\![a']\!] \wedge (F'', F') \in \mathcal{R}_P[\![b]\!]) \vee F' = \infty$
$\Rightarrow \quad (F, F') \in \mathcal{R}_P[\![(a'\,;\, b)]\!] \vee F' = \infty$.

ad (2) $\mathcal{R}_P[\![(b\,;\,a)]\!] \leq \mathcal{R}_P[\![(b\,;\,a')]\!]$:

$\forall F \in \mathcal{AL}(S)$ with $\neg\exists F'' : ((F, F'') \in \mathcal{R}_P[\![b]\!] \wedge (F'', \infty) \in \mathcal{R}_P[\![a]\!])$
and with any $F'' \in \mathcal{AL}(S)$:

$$\begin{array}{rl} & (F, F'') \in \mathcal{R}_P[\![b]\!] \\ \Rightarrow & (F'', \infty) \notin \mathcal{R}_P[\![a]\!] \\ \Rightarrow & (F'', F') \in \mathcal{R}_P[\![a]\!] \Leftrightarrow (F'', F') \in \mathcal{R}_P[\![a']\!] \\ \Rightarrow & (F, F') \in \mathcal{R}_P[\![(b\,;\,a)]\!] \Leftrightarrow (F, F') \in \mathcal{R}_P[\![(b\,;\,a')]\!]. \end{array}$$

$\forall F \in \mathcal{AL}(S)$ with $\exists F'' : ((F, F'') \in \mathcal{R}_P[\![b]\!] \wedge (F'', \infty) \in \mathcal{R}_P[\![a]\!])$

$$\Rightarrow \quad (F, \infty) \in \mathcal{R}_P[\![(b\,;\,a)]\!]$$

and with $(F'', F') \in \mathcal{R}_P[\![a]\!] \Rightarrow (F'', F') \in \mathcal{R}_P[\![a']\!] \vee F' = \infty$:

$$\begin{array}{rl} & (F, F') \in \mathcal{R}_P[\![(b\,;\,a)]\!] \\ \Rightarrow & (\exists F'' \neq \infty : (F, F'') \in \mathcal{R}_P[\![b]\!] \wedge (F'', F') \in \mathcal{R}_P[\![a]\!]) \vee F' = \infty \\ \Rightarrow & (\exists F'' \neq \infty : (F, F'') \in \mathcal{R}_P[\![b]\!] \wedge (F'', F') \in \mathcal{R}_P[\![a']\!]) \vee F' = \infty \\ \Rightarrow & \mathcal{R}_P[\![(b\,;\,a')]\!] \vee F' = \infty. \end{array}$$

ad (3) $\mathcal{R}_P[\![(a \,[\!]\, b)]\!] \leq \mathcal{R}_P[\![(a' \,[\!]\, b)]\!]$:

$\forall F \in \mathcal{AL}(S)$ with $(F, \infty) \notin \mathcal{R}_P[\![a]\!]$

$$\begin{array}{rl} \Rightarrow & (F, F') \in \mathcal{R}_P[\![a]\!] \Leftrightarrow (F, F') \in \mathcal{R}_P[\![a']\!] \\ \Rightarrow & (F, F') \in \mathcal{R}_P[\![(a \,[\!]\, b)]\!] \Leftrightarrow (F, F') \in \mathcal{R}_P[\![(a' \,[\!]\, b)]\!]. \end{array}$$

$\forall F \in \mathcal{AL}(S)$ with $(F, \infty) \in \mathcal{R}_P[\![a]\!] \Rightarrow (F, \infty) \in \mathcal{R}_P[\![(a \,[\!]\, b)]\!]$
and with $(F, F') \in \mathcal{R}_P[\![a]\!] \Rightarrow (F, F') \in \mathcal{R}_P[\![a']\!] \vee F' = \infty$:

$$\begin{array}{rl} & (F, F') \in \mathcal{R}_P[\![(a \,[\!]\, b)]\!] \\ \Rightarrow & (F, F') \in \mathcal{R}_P[\![a]\!] \vee (F, F') \in \mathcal{R}_P[\![b]\!] \\ \Rightarrow & (F, F') \in \mathcal{R}_P[\![a']\!] \vee (F, F') \in \mathcal{R}_P[\![b]\!] \vee F' = \infty \\ \Rightarrow & (F, F') \in \mathcal{R}_P[\![(a' \,[\!]\, b)]\!] \vee F' = \infty. \end{array}$$

ad (4) $\mathcal{R}_P[\![(a \,\&\, b)]\!] \leq \mathcal{R}_P[\![(a' \,\&\, b)]\!]$:

$\forall F \in \mathcal{AL}(S)$ with $(F, \infty) \notin \mathcal{R}_P[\![a]\!]$

$$\begin{array}{rl} \Rightarrow & (F, F') \in \mathcal{R}_P[\![a]\!] \Leftrightarrow (F, F') \in \mathcal{R}_P[\![a']\!] \\ \Rightarrow & (F, F') \in \mathcal{R}_P[\![(a \,\&\, b)]\!] \Leftrightarrow (F, F') \in \mathcal{R}_P[\![(a' \,\&\, b)]\!]. \end{array}$$

$\forall F \in \mathcal{AL}(S)$ with $(F, \infty) \in \mathcal{R}_P[\![a]\!] \Rightarrow (F, \infty) \in \mathcal{R}_P[\![(a \,\&\, b)]\!]$
and with $(F, F') \in \mathcal{R}_P[\![a]\!] \Rightarrow (F, F') \in \mathcal{R}_P[\![a']\!] \vee F' = \infty$:

$$\begin{array}{rl} & (F, F') \in \mathcal{R}_P[\![(a \,\&\, b)]\!] \\ \Rightarrow & (F, F') \in \mathcal{R}_P[\![a]\!] \wedge (F, F') \in \mathcal{R}_P[\![b]\!] \vee F' = \infty \\ \Rightarrow & (F, F') \in \mathcal{R}_P[\![a']\!] \wedge (F, F') \in \mathcal{R}_P[\![b]\!] \vee F' = \infty \\ \Rightarrow & (F, F') \in \mathcal{R}_P[\![(a' \,\&\, b)]\!] \vee F' = \infty. \end{array}$$

ad (5) $\mathcal{R}_P[\![\underline{\mathsf{def}}(a)]\!] \leq \mathcal{R}_P[\![\underline{\mathsf{def}}(a')]\!]$:

$\forall F \in \mathcal{AL}(S)$ with $(F, \infty) \notin \mathcal{R}_P[\![a]\!]$

$$\begin{array}{rl} \Rightarrow & (F, F') \in \mathcal{R}_P[\![a]\!] \Leftrightarrow (F, F') \in \mathcal{R}_P[\![a']\!] \\ \Rightarrow & (F, F') \in \mathcal{R}_P[\![\underline{\mathsf{def}}(a)]\!] \Leftrightarrow (F, F') \in \mathcal{R}_P[\![\underline{\mathsf{def}}(a')]\!]. \end{array}$$

$\forall F \in \mathcal{AL}(S)$ with $(F, \infty) \in \mathcal{R}_P[\![a]\!] \Rightarrow (F, \infty) \in \mathcal{R}_P[\![def(a)]\!]$
and with $(F, F'') \in \mathcal{R}_P[\![a]\!] \Rightarrow (F, F'') \in \mathcal{R}_P[\![a']\!] \vee F'' = \infty$:

$$\begin{aligned}
&(F,F') \in \mathcal{R}_P[\![\underline{\mathsf{def}}(a)]\!] \\
\Rightarrow\quad &(\exists F'' \neq \infty : (F,F'') \in \mathcal{R}_P[\![a]\!] \wedge F = F') \vee F' = \infty \\
\Rightarrow\quad &(\exists F'' \neq \infty : (F,F'') \in \mathcal{R}_P[\![a']\!] \wedge F = F') \vee F' = \infty \\
\Rightarrow\quad &(F,F') \in \mathcal{R}_P[\![\underline{\mathsf{def}}(a')]\!] \vee F' = \infty.
\end{aligned}$$

ad (6) $\mathcal{R}_P[\![\underline{\mathsf{undef}}(a)]\!] \leq \mathcal{R}_P[\![\underline{\mathsf{undef}}(a')]\!]$:

$\forall F \in \mathcal{AL}(S)$ *with* $(F,\infty) \notin \mathcal{R}_P[\![a]\!]$

$$\begin{aligned}
\Rightarrow\quad &(F,F') \in \mathcal{R}_P[\![a]\!] \Leftrightarrow (F,F') \in \mathcal{R}_P[\![a']\!] \\
\Rightarrow\quad &(F,F') \in \mathcal{R}_P[\![\underline{\mathsf{undef}}(a)]\!] \Leftrightarrow (F,F') \in \mathcal{R}_P[\![\underline{\mathsf{undef}}(a')]\!].
\end{aligned}$$

$\forall F \in \mathcal{AL}(S)$ *with* $(F,\infty) \in \mathcal{R}_P[\![a]\!]$ *implies:*

$$\begin{aligned}
&(F,F') \in \mathcal{R}_P[\![\underline{\mathsf{undef}}(a)]\!] \Leftrightarrow F' = \infty \\
\Rightarrow\quad &(F,F') \in \mathcal{R}_P[\![\underline{\mathsf{undef}}(a)]\!] \Leftrightarrow ((F,F') \in \mathcal{R}_P[\![\underline{\mathsf{undef}}(a')]\!] \vee F' = \infty).
\end{aligned}$$

□

Based on the proof above we are now able to define a fixpoint semantics for a given set of transactions which may call each other in an arbitrary manner. Please note that from a superficial point of view the treatment of n *transactions* instead of one requires the introduction of a more *complex semantic domain* $\mathcal{D}^n$ and a corresponding semantic function. This function takes n binary relations as input, which are already computed approximations of n transactions; it produces better approximations in the form of n binary relations as output. Applying fixpoint theory to this extended setting the semantics of the i-th transaction is defined as the i-th component of their common least fixpoint in $\mathcal{D}^n$. The extension of the semantic domain $\mathcal{D}$ to $\mathcal{D}^n$ — including a new partial chain continuous order — is rather straightforward. Therefore, it is omitted over here, although it is implicitly used in the following corollary:

Corollary 7.3.8 (Fixpoint Semantics for Transactions.)
Let $T := \{t_1, \ldots, t_n\} \in \mathcal{T}(P)$ *be a set of transactions which contain in their bodies* E_1 *to* $E_n \in \mathcal{E}(P)$ *at most calls to transactions* t_1 *through* t_n *and calls to structure replacement rules of* $P \subseteq \mathcal{P}(\Sigma)$, *i.e.*

$t_1 = E_1[t_1, \ldots, t_n], \ldots, t_n = E_n[t_1, \ldots, t_n]$.

With $\mathcal{AL}(\Sigma)$ *being the corresponding class of abstract schema consistent structures, the following propositions hold and define a new* **semantic function** $\mathcal{R}^*_P : T \to \mathcal{D}$:

(1) $t_1, \ldots, t_n$ *have unique least fixpoints* $\mathcal{R}^*_P[\![t_1]\!], \ldots, \mathcal{R}^*_P[\![t_n]\!] \in \mathcal{D}$.

(2) *Approximations of these fixpoints may be constructed as follows:*
$R^0_1 := \mathcal{AL}(S) \times \{\infty\}; \ldots ; R^0_n := \mathcal{AL}(S) \times \{\infty\}$,
$R^{k+1}_1 := \mathcal{R}^*_P[\![E_1]\!][R^k_1, \ldots, R^k_n]; \ldots ; R^{k+1}_n := \mathcal{R}^*_P[\![E_n]\!][R^k_1, \ldots, R^k_n]$,
where $\mathcal{R}^*_P[\![E_i]\!] : \mathcal{D}^n \to \mathcal{D}$ *takes approximations for transactions* $t_1, \ldots, t_n$ *as input and yields a new approximation for* t_i *by applying Definition 7.3.5 to expression* E_i.

Proof:

ad (1) *Follows directly from Proposition 7.3.1 and Lemmata 7.3.4 and 7.3.7.*

ad (2) *The relations R_i^k are indeed approximations for $t_1, \ldots, t_n$ such that for any i, k : $R_i^0 \leq \ldots \leq R_i^k \leq \mathcal{R}_P^*[\![t_i]\!]$. This follows directly from Proposition 7.3.1 and Lemmata 7.3.4 and 7.3.7 with*
$\forall i, l : R_i^l = \cup_{k \leq l} R_i^k \leq \mathcal{R}_P^*[\![t_i]\!]$ *and* $\cup_{k \in \{\}} R_i^k = \mathcal{AL}(S) \times \{\infty\}$. □

Note that the corollary above is not only interesting from a theoretical point of view by guaranteeing the *existence of least fixpoints* for any recursively defined set of transactions. Additionally, it provides us with a ***straightforward algorithm*** which computes the results of terminating transactions and approximates the results of (potentially) nonterminating transactions. Equipped with such a fixpoint computing function, we are now able to define programmed structure replacement systems and their generated (abstract) structure languages.

Definition 7.3.9 (PLSR System.)
Assuming the vocabulary of Definition 7.3.5, PLSR := (S, A, P, T, t) is a **programmed logic-based structure replacement (PLSR) system** with respect to a signature Σ if:

(1) $S \in \mathcal{S}(\Sigma)$ is a structure schema.
(2) $A \in \mathcal{AL}(S)$ is the schema consistent initial structure.
(3) $P \subseteq \mathcal{P}(\Sigma)$ is a set of structure replacement rules.
(4) $T \subseteq \mathcal{T}(P)$ is a set of (recursively) defined transactions over P.
(5) $t \in T$ is the *main transaction* of PLSR.

Definition 7.3.10 (Language of a PLSR System.)
With PLSR := (S, A, P, T, t) as in Def. 7.3.9, the language $\mathcal{AL}$(PLSR) is defined as follows:

$F \in \mathcal{AL}(\text{PLSR}) :\Leftrightarrow (A, F) \in \mathcal{R}_P^*[\![t]\!]) \land F \neq \infty.$

The definition above states that the language of a programmed structure replacement system consists of all those structures which may be generated by applying a main transaction t to the initial structure A. The transaction t calls structure replacement rules from P and other transactions from T, which in turn contain calls of rules and transactions.

In the case where PLSR systems are used to define abstract data types instead of structure (graph) languages only, it might be useful to replace the single main transaction "t" above by a set of *exported transactions* $t_1, \ldots, t_n$. This would be a very first step towards a module concept for structure replacement systems. But even in that case the language of all possibly generated structures is definable as the result of the following new main transaction:

$t' = ((t_1 \,[\!]\, \ldots \,[\!]\, t_n); t') \,[\!]\, \underline{\mathsf{skip}}),$

which either returns its input or applies the transactions $t_1, \ldots, t_n$ an arbitrary number of times.

7.3.5 Summary

The definitions and propositions of the preceding subsections constitute a *framework* for defining and comparing various approaches which support programming with (graph) replacement rules. The framework provides us with a formal definition of partial, nondeterministic, and even recursive *structure replacement programs*, termed transactions. We added a new symbol "∞" to the domain of structures which represents unknown results or nonterminating computations. Thereby, we were able to overcome the difficulties with operators which test success or failure of subprograms in the presence of recursion (cf. [66] for a more detailed discussion about these problems). These operators are a prerequisite for writing programs like "try first to execute subprogram A and, if and only if A fails, try to execute B".

Note that [51], which presents an *axiomatic semantics definition* for nondeterministic partial commands, gave us the initial impulse for the introduction of the symbol "∞" as well as for the selection of the adequate fixpoint theorem version. Nevertheless, we had to prefer the *denotational* instead of the *axiomatic* approach, since structure replacement rules play here about the same role as simple assignments in [51]. The definition of their intended semantics as binary relations is given in Definition 7.2.16, but it is a yet unsolved problem (in the general case) how to push postconditions through rules for obtaining their corresponding weakest preconditions (cf. list of unsolved question in Subsection 7.2.4).

Furthermore note that BCF expressions as well as almost all related approaches discussed in Section 7.5 deal with *sequential replacement processes* only. The so-called *programmed derivation of relational structures* approach [44] is the only exception. This approach supports *parallel programming with graph replacement rules* and provides a fixpoint semantics for recursive programs based on program traces. Unfortunately, no constructs (like <u>def</u> and <u>undef</u> over here) are available for testing whether a complex structure transformation returns a defined result or not. Parallel composition of subtransformations as well as testing their applicability are both very valuable means and require completely different formal treatments. Parallel composition, on one hand, with its *interleaving semantics* excludes the definition of a program's semantics as a function of the semantics of its subprograms. Testing success or failure of subprograms, on the other hand, enforces the introduction of a special symbol "∞" for nonterminating computations and the usage of a rather complicated partial order for the resulting semantic domain. Further investigations are necessary to check whether a combination of both approaches is possible or not.

Finally, we have to emphasize that all presented results in this section are valid for *any rule-based approach*, where rules define binary relations over a given domain of objects. Having developed such a common framework for rather different rule regulation mechanisms offers the opportunity to *combine different regulation mechanisms* as needed within future graph grammar-based specification languages. This idea is closely related to the *GRACE initiative* [36], a first attempt to combine different graph models, graph replacement approaches, and rule regulation mechanisms within a single **GRA**ph **CE**ntered language and environment. As far as we can see, PLSR systems could act as the common underlying formalism for all envisaged types of graphs, rules, and control structures.

In GRACE, a complete specification is built by defining and composing so-called *transformation units.* Any transformation unit may use a different style of graph replacement (cf. section 7.4 and 7.5 for an overview about currently existing variants of graph replacement approaches). It defines a binary relation between certain classes of graphs. The operators which combine single transformation units to more complex transformation units are very similar to BCF operators presented over here. Hence, PLSR systems can be used to define the semantics of transformation units and their composition operators. The other way round, transformation units of GRACE provide a kind of *module concept* for structuring PLSR systems into reusable subcomponents with well-defined interfaces between them.

7.4 Context-sensitive Graph Replacement Systems – Overview

Until now, we presented a very general framework for the definition of specific graph data models and especially for the definition of various types of (programmed) graph replacement systems. We will use this framework over here for reviewing a number of different context sensitive *graph replacement approaches* (cf. Table 1 in Subsection 7.4.5). These are the expression-oriented algorithmic graph replacement approach (EXP) of [45,46], and its afore-mentioned successors PAGG [27,28] and PROGRES [65,66]. These approaches will be compared with the two families of algebraic graph replacement systems, DPO [19,20] and

SPO [20,40], which are the main subject of Chapter 3 and 7 of this volume. Finally, we have selected so-called DELTA grammars [33], which are very similar to PAGG, but claim to be a member of the algebraic branch.

It is rather obvious, how different graph data models — like directed graphs or hypergraphs — may be seen as relational structures with certain properties. Furthermore, all presented graph replacement systems have about the *same data model* of a directed graph with labeled nodes and edges. The only differences are that

- one approach (EXP) does not support (node) attributes,
- algebraic DPO and SPO approaches handle edges as identifiable objects and are easily extendible to n-ary relations in the form of hyperedges,
- and every second approach knows the concept of a label hierarchy, where more general node or edge labels match more specific node or edge labels.

Therefore, we will omit a more detailed discussion of graph data model differences in the sequel and simply assume the same class of graphs as in the running example of Section 7.2. Furthermore, we will also omit the subject of graph schema definitions, which is already covered by Subsection 7.2.1. But even then, a bewildering *long list of characteristics* remains which distinguish the selected six graph replacement approaches:

- The following Subsection 7.4.1 discusses how *matches* of rules in a host graph may be restricted and how deletion of nodes and edges is handled.
- Subsection 7.4.2 presents then two completely different ways how to specify the *embedding* of created nodes into the remaining part of the host graph.
- Subsection 7.4.3 shows afterwards why additional *application conditions* for rules are needed and how they are usually defined.
- Subsection 7.4.4 sketches then how *attributes* are handled within graph replacement systems.
- And subsection 7.4.5 summarizes the discussion in a table with one column for each presented approach and one row for each of the identified *25* graph replacement system *characteristics.*

7.4.1 Context-sensitive Graph Replacement Rules

In the simplest case, a graph replacement rule consists of two graphs with labeled, directed edges and labeled, identifier carrying nodes. These two graphs are their left- and right-hand side. The overall idea of the *application of a graph replacement rule* is to

(1) find first a match for its left-hand side in a given host graph,
(2) remove then the match from the host graph,
(3) construct an isomorphic copy of the rule's right-hand side,
(4) and add the copy to the remaining rest of the host graph.

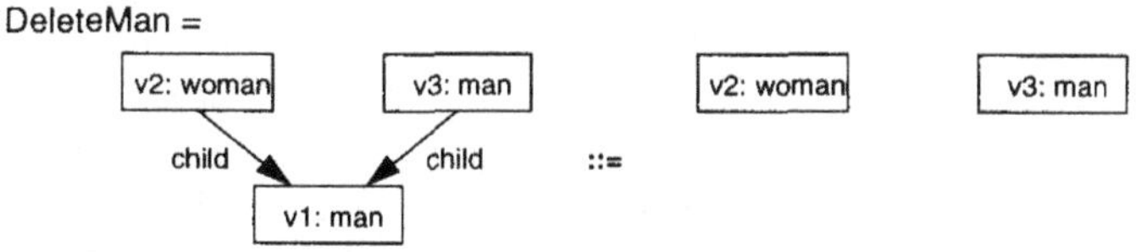

Figure 7.9: Example of a context-sensitive graph replacement rule

This procedure leads to a host graph, where the *new subgraph is isolated* from the old rest of the host graph. Therefore, an additional mechanism is needed, called *embedding transformation*, for establishing connections between new and old nodes. The most straight-forward realization of such an embedding transformation is based on the *identification of nodes* in a rule's left- and right-hand side, the so called *gluing approach*. Matches of these nodes are neither deleted nor recreated during the application of replacement rules. They simply have to be present and are preserved with all their edges to unmatched nodes in the host graph. The graph replacement rule DeleteMan of Figure 7.9 has for instance two nodes v2 and v3, which are part of its left- and right-hand side and whose matches in a host graph may have connections to an arbitrary number of unmatched nodes in the host graph.

Using this kind of graph replacement the following two questions arise:

- Shall we allow that the match of a rule is a nonisomorphic image of its left-hand side, i.e. that two nodes of the left-hand side share the same node in the host graph?
- And what about edges at a matched node which are not matched by an edge of the left-hand side although their node will be deleted by the rule?

All six reviewed graph replacement formalisms have different answers for these questions. The **EXP** approach [45,46] (with so-called path **EXP**ressions) does not know the concept of preserved nodes, but solves the embedding problem in a different manner (cf. Subsection 7.4.2). Therefore, all matched nodes are deleted and permitting nonisomorphic images makes no sense. Furthermore, unmatched edges between matched nodes are not allowed, i.e. the matched subgraph has to be equivalent to its *induced subgraph*. As a consequence, the rule DeleteMan is not applicable to any man node whose parents are connected to each other with a wife edge. Edges between matched nodes and unmatched nodes are handled in a different way. They will be deleted together with their matched nodes. Otherwise, graph replacement rules would never be able to delete nodes which have connections to preserved and, therefore, unmatched nodes.

The **PAGG** (**P**rogrammed **A**ttributed **G**raph **G**rammars) approach [26,27,28] knows the concept of preserved nodes, but requires still that matches are isomorphic images of left-hand sides (with the exception of the embedding part of replacement rules which will be explained in Subsection 7.4.2). One of its differences to EXP is that the induced subgraph condition is dropped, i.e. all unmatched edges at a deleted node are handled in the same way and are deleted, too. Therefore, a graph replacement rule like DeleteMan can be used to delete any man node independent from the fact whether this node or its parent nodes are sources/targets of additional edges. Figure 7.10 of Subsection 7.4.2 contains an example for the notation of PAGG rules.

The default behavior of **PROGRES** graph replacement rules is the same as for PAGG. But explicit *folding statements* allow left-hand sides to share their matches as long as these nodes are also part of the right-hand side. This is the so-called *identification condition*, which prevents situations where a rule has to preserve and to delete a node at the same time. In the rule DeleteMan for instance nodes v1 and v3 may not match the same node in the host graph, even if the host graph contains a man node with a child edge loop.

The next approach, **DELTA** [33] (its rules have the form of a Δ), is very similar to PROGRES with respect to the supported form of graph replacement (cf. Figure 7.10 of Subsection 7.4.2). Isomorphic matching is the default, but explicit folding statements may be used to overwrite the default. Deletion/preservation conflicts are possible, but (probably) prohibited by the identification condition, too.

The **DPO** (**D**ouble **P**ush**O**ut) approach [19,20] is the most restrictive one concerning the treatment of unmatched edges at nodes which should be deleted. The so-called *dangling edge condition* forbids the application of rules in this case. Therefore, DeleteMan may only be used to delete man nodes without outgoing wife or child edges. Furthermore, nonisomorphic matching together with the identification condition is the default. Both the dangling edge and the identification condition are summarized under the name *gluing condition*.

The **SPO** (**S**ingle **P**ush**O**ut) approach [20,40], finally, permits any kind of nonisomorphic images of left-hand sides and drops also the dangling edge condition. As a consequence, edges at deleted nodes are deleted, too, and deletion/preservation conflicts are resolved in favor of deletion.

Last but not least we should mention that all discussed six approaches allow to a certain extent that a single node (edge) of a graph replacement rule matches more than one node (edge) in the host graph. At least EXP, DPO, and SPO have extensions which allow for the *parallel application* of a single rule (or of several rules) at various matches in a given host graph. Furthermore, both DPO and SPO variants were developed recently with so-called *amalgamated graph replacement rules* [9,31,72]. Such an amalgamated rule represents an infinite family of simple rules in the general case. They are useful for manipulating different occurrences of a single pattern in the host graph simultaneously. Simplified forms of amalgamation — mainly used for specifying embedding transformations in a graphical notation — are also part of PAGG, PROGRES, and DELTA.

The following example summarizes the discussion of possible interpretations of simple graph replacement rules:

Example 7.9 (Translation into a Structure Replacement Rule.)
The translation of the graph replacement rule DeleteMan of Figure 7.9 into a structure replacement rule in accordance with the informal discussion above has about the following form:

(1) Left-hand side translation:

$$L := \{\mathsf{node}(\mathsf{v1},\mathsf{man}), \mathsf{node}(\mathsf{v2},\mathsf{woman}), \mathsf{node}(v3,\mathsf{man}), \mathsf{edge}(\mathsf{v2},\mathsf{child},\mathsf{v1}), \mathsf{edge}(\mathsf{v3},\mathsf{child},\mathsf{v1})\}$$

plus optional extension for deletion of otherwise dangling edges with S1, ... being set identifiers:

$$\cup\ \{\mathsf{edge}(\mathsf{v1},\mathsf{child},\mathsf{S1}),\ldots\}.$$

(2) Right-hand side translation:

$$R := \{\mathsf{node}(\mathsf{v2},\mathsf{woman}), \mathsf{node}(\mathsf{v3},\mathsf{man})\}.$$

(3) Dangling edge preventing (post-)condition translation:

$$AR := \{\forall x, e, y : \mathsf{edge}(x,e,y) \rightarrow (\exists l : \mathsf{node}(x,l) \wedge \exists l : \mathsf{node}(y,l))\}.$$

(4) Optional identification (pre-)condition:

$$AL := \{\mathsf{v1} \neq \mathsf{v3}\}$$

or/and optional induced subgraph (pre-)condition:

$$AL := \{\neg\exists e : \mathsf{edge}(\mathsf{v2},e,\mathsf{v3}) \wedge \ldots\}$$

or/and — in this case useless — isomorphic match (pre-)condition:

$$AL := \{\mathsf{v1} \neq \mathsf{v2}, \mathsf{v1} \neq \mathsf{v3}, \mathsf{v2} \neq \mathsf{v3}\}.$$ □

7.4.2 *Embedding Rules and Path Expressions*

Using the style of graph replacement introduced in the previous subsection we have no chance to define a rule which

- deletes a man node v with an unknown number of emanating child edges
- and creates for any deleted child edge from v to another node v' two new grandchild edges from the parents of v to v'.

This a scenario which requires *complex embedding transformation rules.* EXP is an early graph replacement approach which is famous for its powerful embedding rules. These rules have the following four components:

- A left-hand side node identifier, whose match in the host graph is potentially connected to an arbitrary number of unmatched context nodes.
- A so-called path expression describing a path through the host graph from the selected left-hand side node match to a relevant set of unmatched context nodes.
- A node of the right-hand side, whose copy in the host graph will be the source or target of a new embedding edge.
- A last part which determines the label and the direction of the new edge.

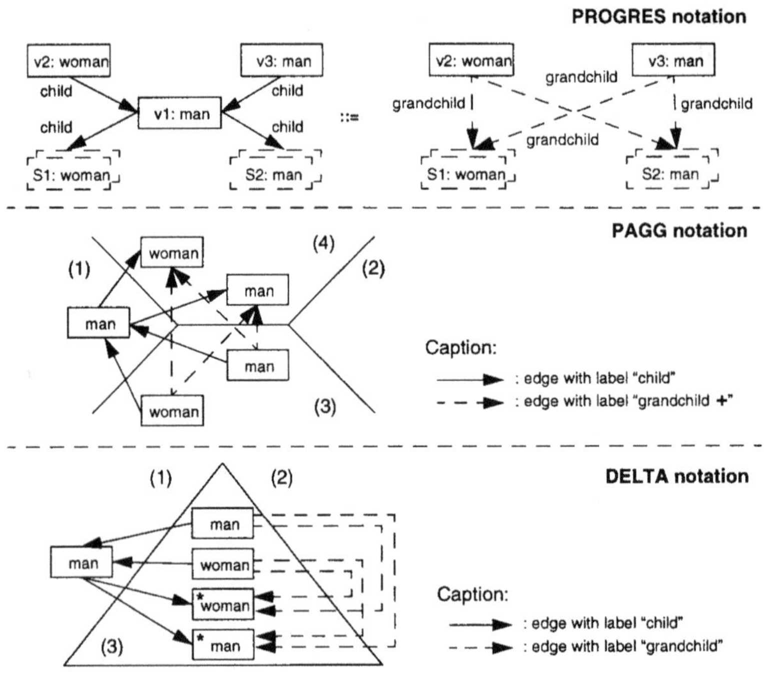

Figure 7.10: Notations for graph replacement rules with graphical embedding rules

In the simplest case, a *path expression* of the form $\mathsf{R}_{\mathsf{label}}$ or $\mathsf{L}_{\mathsf{label}}$ requires the traversal of a single edge in or against its direction as for instance in

$\mathsf{I}_{\mathsf{grandchild}} = (\mathsf{v2}, \mathsf{v3}; \mathsf{R}_{\mathsf{child}}(\mathsf{v1}))$.

This embedding rule completes the rule DeleteMan and creates for any child edge from v1 to a context node v new grandchild edges from v2 and v3 to v. In the general case, path expressions are constructed by building concatenations, unions, iterations, etc. of simple path expressions. They may be used for defining *derived relationships* like ancestor or brother of Figure 7.2.

The reader may imagine that path expression-based embedding rules are a rather powerful concept, but are sometimes difficult to understand. Therefore, PAGG and DELTA rely on *graphical embedding rules*, whereas PROGRES offers both textual embedding rules like EXP and graphical embedding rules like DELTA. All graphical embedding rules have in common that they mark certain nodes (edges) in their graph replacement rules which are allowed to match an arbitrary number of nodes (edges) in the host graph. In PROGRES, these nodes are depicted as double dashed rectangles (cf. Figure 7.5 in Subsection 7.2.3 and Figure 7.10). The underlying formalism of structure replacement rules supports this concept by means of set identifiers:

Example 7.10 (Translation of Embedding Rules.)
The translation of the PAGG, PROGRES, and DELTA graph replacement rules of Figure 7.10 into an equivalent structure replacement rule is the same as in Example 7.9, except the fact that L and R are extended as follows:

(1) $L' := L \cup \{\mathsf{node}(\mathsf{S1}, \mathsf{woman}), \mathsf{node}(\mathsf{S2}, \mathsf{man}),$
$\mathsf{edge}(\mathsf{v1}, \mathsf{child}, \mathsf{S1}), \mathsf{edge}(\mathsf{v1}, \mathsf{child}, \mathsf{S2})\}$.

(2) $R' := R \cup \{\mathsf{node}(\mathsf{S1}, \mathsf{woman}), \mathsf{node}(\mathsf{S2}, \mathsf{man}),$
$\mathsf{edge}(\mathsf{v2}, \mathsf{grandchild}, \mathsf{S1}), \mathsf{edge}(\mathsf{v2}, \mathsf{grandchild}, \mathsf{S2}),$
$\mathsf{edge}(\mathsf{v3}, \mathsf{grandchild}, \mathsf{S1}), \mathsf{edge}(\mathsf{v3}, \mathsf{grandchild}, \mathsf{S2})\}$.

S1 and S2 are node set identifiers which match zero or one or two or ... nodes in a host graph (cf. Def. 7.2.1). □

The PAGG notation of the graph replacement rule in Figure 7.10 must be read as follows:

(1) Nodes and edges in the left section of the X are deleted together with all incident edges during the application of such a rule.
(2) Nodes and edges in the right section of the X are created anew when the rewrite rule is applied.
(3) Nodes and edges of the lower section of the X denote the preserved part of the host graph which has to be present but remains unmodified.

(4) And nodes and edges of the upper section of the X denote graphical embedding rules. They are preserved and have in the general case not a single match, but a maximal number of matches in the host graph.

The meaning of *section crossing edges* should be obvious except those between the lower and the upper part of the X. They wear either the additional label "−" or "+". The label "−" denotes those edges which have to be present before the application of the rule but are no longer present afterwards. The label "+" denotes those edges which need not be present before the application of the rule, but will be present afterwards.

DELTA *grammar rules* are rather similar to PAGG rewrite rules (cf. Figure 7.10). The area left of the Δ corresponds to the left section of the X, the area right of the Δ to the right section of the X, and the inner area of the Δ to the upper and the lower part of the X. Preserved nodes with multiple matches instead of single matches are identified by having an additional "*" label. The purpose of the region below the Δ will be explained within the next subsection.

7.4.3 Positive and Negative Application Conditions

It is quite often the case that a graph replacement step may only take place if additional nodes and edges are present or if a certain graph pattern is not present. Therefore, PROGRES, DELTA, and SPO know the concept of ***positive and negative context conditions.*** A very general version of these additional application conditions is suggested in [30] as an extension of the SPO approach. There, arbitrarily large context graphs may be attached to the left-hand side of a graph replacement rule, and their existence may be required or prohibited. DELTA has about the same concept of negative context graphs. They are depicted below the bottom line of the Δ. Positive context conditions correspond to the preserved part of the graph rewrite rule, the inner region of the Δ. In this way, SPO and DELTA are able to require the existence or absence of context subgraphs which have an *a priori known size.* The lower part of Figure 7.11 shows an example of a DELTA replacement rule which marries two previously unmarried persons. The preconditions require that the two nodes do not have a wife edge in between them. They require furthermore that the two nodes do not have any wife edge connections to other nodes.

The Figure 7.11 contains in its upper part the corresponding PROGRES definition with an additional restriction: The two nodes may not be ancestors of each other. This is an example of *a negative context condition*, where the size of the prohibited subgraph — a path through a given host graph — is not fixed within the rule. Such a requirement is not expressible within SPO or DELTA. PROGRES reuses the concept of path expressions — originally used for defining n-context embedding rules — to overcome this restriction. In such a way, control structures of path expressions for conditional branching, transitive closure, and conditional iteration are now available for defining *complex n-context application conditions*.

The translation of positive as well as negative application context conditions into preconditions of structure replacement rules is rather obvious. Example 7.6 of Subsection 7.2.2 presents already the complete translation of a subgraph test with similar context conditions as in Figure 7.11. Furthermore, the translation of complex path expressions with transitive closure operators and the like is already part of Example 7.4 in Subsection 7.2.1.

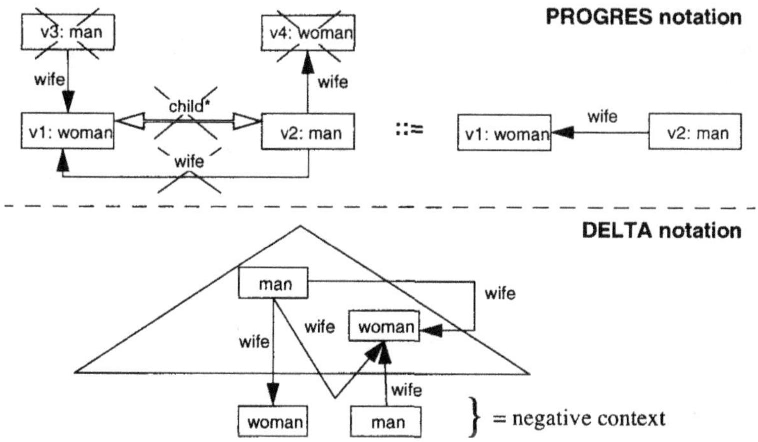

Figure 7.11: PROGRES and DELTA rules with negative context conditions

7.4.4 Normal and Derived Attributes

Until now, we have discussed only those means of graph replacement rules which allow the manipulation of nodes and edges, i.e. of data which has an *important internal structure* from the specifiers point of view. But in many cases additional data is manipulated — like strings or numbers — which should be handled as *atomic items*. Therefore, all presented graph replacement approaches — except EXP — support the definition of *attributes*, which are attached to nodes and which have a single item or a set of atomic items as their values. In order to be able to write or read these attributes, textual extensions of graph replacement rules are offered. They are always more or less similar to those of the language PROGRES, which were already used in Example 7.7 of Subsection 7.2.3. The graph replacement rule Marry presented there tests and modifies the income attribute of a matched person node.

In general, two different categories of attributes may be distinguished as they are sometimes distinguished within attribute tree grammars:

- *Intrinsic attributes*, like the above mentioned income attribute, represent extensionally defined properties of nodes in a graph. They change their value only through explicit assignments within graph replacement rules.
- *Derived attributes*, on the other hand, represent intentionally defined node properties. They are defined by means of directed equations over attributes of other nodes, which are in the n-context of their own node.

PROGRES supports both categories of attributes in their pure form. Normal attributes are tested and manipulated within graph replacement rules (cf. Example 7.7 and Figure 7.5 of Subsection 7.2.3). Derived attribute value definitions are part of a graph schema construction. Again path expressions are welcome for specifying rather *complex n-context references to attributes* at foreign nodes within directed equations. Figure 7.12 shows an example of such an attribute specification. The number of ancestors of a distinct person is the sum of the number of ancestors of her mother plus one (if existent) and the number of ancestors of her father plus one (if existent). In the case of nonexistence of the mother or the father the evaluation of the subexpression between [... | fails and the subexpression 0 between | ...] is returned.

```
node class person;
   derived
      noOfAncestors =   [ 1 + self.<-child-.instance of man.noOfAncestors | 0 ]
                      + [ 1 + self.<-child-.instance of woman.NoOfAncestors | 0 ];
end ;
```

Figure 7.12: PROGRES definition of a derived attribute

Example 7.11 (Translation of Derived Attribute Definition.)
The attribute equation of Figure 7.12 may be translated as follows into a formula, which is part of a structure schema definition (cf. Definition 7.2.6):

$$
\begin{array}{l}
\forall\ \mathsf{x}:\mathsf{node}(\mathsf{x},\mathsf{person}) \rightarrow \\
\exists \mathsf{v}: (\mathsf{attr}(\mathsf{x},\mathsf{noOfAncestors},\mathsf{v}) \\
\quad \wedge\ (\exists \mathsf{v1},\mathsf{v2} : \mathsf{v} = \mathsf{v1} + \mathsf{v2} \\
\qquad \wedge\ (\exists \mathsf{x1}: (\mathsf{edge}(\mathsf{x1},\mathsf{child},\mathsf{x}) \wedge \mathsf{type}(\mathsf{x1},\mathsf{man}) \wedge \mathsf{attr}(\mathsf{x1},\mathsf{noOfAncestors},\mathsf{v1})) \\
\qquad\quad \vee((\neg\exists \mathsf{x1}:\mathsf{edge}(\mathsf{x1},\mathsf{child},\mathsf{x}) \wedge \mathsf{type}(\mathsf{x1},\mathsf{man})) \wedge \mathsf{v1} = 0)) \\
\qquad \wedge(\exists \mathsf{x2}: (\mathsf{edge}(\mathsf{x2},\mathsf{child},\mathsf{x}) \wedge \mathsf{type}(\mathsf{x2},\mathsf{woman}) \wedge \mathsf{attr}(\mathsf{x2},\mathsf{noOfAncestors},\mathsf{v2})) \\
\qquad\quad \vee((\neg\exists \mathsf{x2}:\mathsf{edge}(\mathsf{x2},\mathsf{child},\mathsf{x}) \wedge \mathsf{type}(\mathsf{x2},\mathsf{woman})) \wedge \mathsf{v2} = 0)))).
\end{array}
$$

□

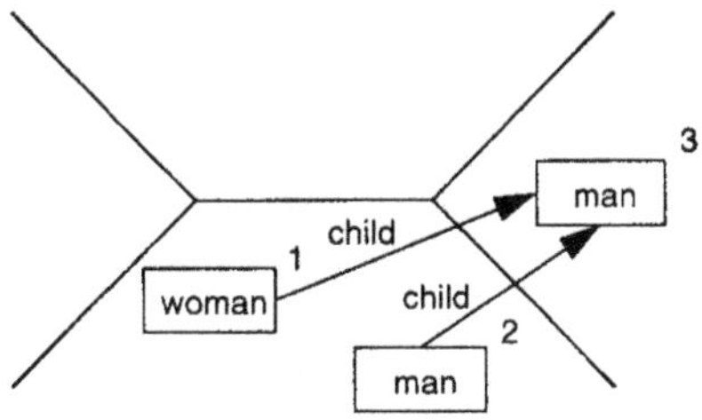


```
BODY = begin
    3.income <- pocketMoney(1.income, 2. income) ;
    3.noOfAncestors <- 1.noOfAncestors + 2.noOfAncestors +2 ;
end;
```

Figure 7.13: Attribute manipulating graph replacement rule in PAGG

Specifying derived attributes within graphs and keeping them consistent to each other is much more difficult if the underlying structure is an arbitrary directed graph and not a tree. The attempt to traverse an edge during the evaluation of a directed equation may fail due to the nonexistence of such an edge. On the other hand, it might be the case that a path expression within an attribute equation returns a set of nodes instead of a single node. Therefore, additional means are necessary for determining alternatives in the case of *partially defined* subexpression and for *aggregating sets* of attribute values to a single useful attribute value. For a detailed discussion of various *incremental attribute evaluation strategies* the reader is referred to [32,37,66]. Their presentation is outside the scope of this contribution.

The developers of PAGG and of another attribute graph grammar approach [69] adopted the idea of attribute tree grammars to *attach directed attribute equations to structure replacement rules* (productions) instead of defining them separately. As a consequence, derived attribute definitions and value assignments to extensionally defined attributes are handled by the same syntactic construct (cf. Figure 7.13), and we are not able to distinguish between

(1) the case where an attribute, like income, gets once a value computed from other attribute values and preserves its value afterwards if these foreign attributes change values,
(2) and the case where an attribute, like noOfAncestors, has to change its value as soon as referenced foreign attributes change their values.

Furthermore, attaching equations to rules has also the *drawback* that equations within different but somehow related rules — which insert, modify, and delete for instance a certain type of nodes — may become *inconsistent* to each other.

This is not a real problem as long as graph replacement systems are only used for generating graph languages, but tends to be a serious disadvantage as soon as graph replacement systems are used for specifying (abstract) data types. Nevertheless, the attachment of attribute equations to graph replacement rules has some advantages, too. It is no longer necessary to define a separate graph schema, the equations which reference nodes in rules (instead of using path expressions for this purpose) are easier to read, and the derivation history of a graph may influence the selection of active attribute equations.

Finally, we have to mention that EXP, DPO, and DELTA do not know the concept of derived attributes, whereas an SPO variant was defined very recently which supports derived data definition at least to a certain extent [34].

7.4.5 Summary

The following table summarizes the presentation of six more or less different graph replacement approaches within the previous subsections. Being only a rather condensed version of the preceding discussion it should be readable without any further explanations. During its construction we had to solve the *following problems*:

- The available input for DELTA was not sufficient for answering all questions and for verifying that DELTA belongs indeed to the family of algebraic graph replacement systems. Therefore, its column contains three question marks.
- The headlines SPO and DPO represent not only a single paper or proposal, but a whole family of related formalisms which evolved over the time. Therefore, it was very difficult to give definite "yes" or "no" answers in some cases.
- The classification of embedding rules is traditionally more fine-grained than the classification used in Subsection 7.4.2. Therefore, additional embedding rows were added for those readers who are familiar with graph replacement systems.

7.5 Programmed Graph Replacement Systems – Overview

Section 7.3 introduced the basic means for defining and comparing various forms of *sequential programmed graph replacement systems*. It is the topic of this section to review previously made proposals for programmed grammars or — more specific — programmed graph grammars or replacement systems. The review is divided into three subsections:

			Algorithmic Approaches			?	Algebraic Approaches	
			EXP [43, 44]	**PAGG** [26, 27]	**PROGRES** [61, 62]	**Δ = DELTA** [32]	**DPO** [17, 18]	**SPO** [18, 38]
Graph Model		Node Label Hierarchy	—	in [24]	+	+	in [57]	—
		Edge Label Hierarchy	—	—	—	+		—
		Edge Identifiers	—	—	—	?	+	+
		Hyperedges	—	—	—	—	+	+
		Derived Edges = Path Expr.	+	—	+	—	—	—
		"Normal" Node Attributes	—	+ mixture	+	special label	in [57]	in [37]
		Derived Node Attributes	—		+	—	—	in [33]
Separate Schema Definition			—	in [24]	+	—	in [15]	
Graph Replacement Rules	**Matching**	Homo-/Isomorphic	isomorphic	isomorphic	homo./iso.	homo./iso.	homomorph	homomorph
		(Induced) Subgraph	induced sub.	subgraph	subgraph	subgraph	subgraph	subgraph
		Multiple Matches in Host Graph (simultaneously)	(+) parallel or mixed rewr.	+ embedding graphs	+ set nodes in rules	+ set nodes in rules	rule amalgamation in [68]	rule amalgamation in [30]
	Appl. Conditions	Implicit Conditions	induced sub.	—	—	—	gluing cond.	—
		Explicit Cond.: Attribute Cond.	—	+	+	+	in [57]	in [37]
		Explicit Cond.: Positive Context	—	any graph → n-C. (fix)	nodes/paths →n-Context	any graph → n-C. (fix)	any graph → n-C. (fix)	any graph → n-C. (fix)
		Explicit Cond.: Negative Context	—	—	nodes/paths →n-Context	any graph ? → n-C. (fix)	—	graph in [29] → n-C. (fix)
	Delete	Del./Preserve Conflict	impossible	impossible	forbidden	forbidden ?	forbidden	delete
		Handling of Dangling Context Edges	automatic deletion	automatic deletion	automatic deletion	automatic deletion	forbidden gluing cond.	automatic deletion
	Embedding Transformation	Textual/Graphical	text (paths)	graphical	text/graphic	graphical	— (simulation with amalgamation possible)	— (simulation with amalgamation possible)
		Referenced Context	n-Context	n-C. (fix)	n-Context	n-C.(fix)		
		Change Orientation	+	+	+	+		
		Change Label	+	+	+	+		
		Test Context Label	+	+	+	+		
		Test Context Edges	+	+	+	+		
		With Preserved Nodes	—	+	+	+		
		Different Treatment of Ident. Labeled Nodes	+	+	+	+		

Table 7.1: Comparison of algorithmic and algebraic graph replacement approaches

- Subsection 7.5.1 introduces a number of programmed grammar formalisms which preserve the declarative as well rule-oriented character of replacement systems. There exists still a *single meta algorithm* which applies rules as long as appropriate and which takes additional information about rule preferences or required application sequences into account.
- Subsection 7.5.2 presents then a number of related programmed graph grammar approaches which belong to the application-oriented branch of algorithmic graph grammars. These approaches have in common that they adapt *control structures* of imperative programming languages for the purpose of controlling graph replacement rules.
- Finally, Subsection 7.5.3 deals with all those programmed graph grammars which belong to the family of algorithmic graph grammar approaches, too, but use *control flow diagrams* instead of textually written control structures for regulating graph replacement processes.

We will use a single example of a complex person database transformation to discuss the advantages and disadvantages of all presented approaches. This is an extended version of the transaction Marry* of Example 7.8. It assumes the existence of already introduced basic rules Marry and MarkUnmarried (cf. Example 7.8) as well as the new rule DeleteUnmarried. The latter one fails if all person nodes are married and deletes a nondeterministically selected unmarried person node otherwise.

Example 7.12 (Complex Database Transformation.)

Marry* = (((def(MarkUnmarried); Marry); Marry*) ‖ undef(MarkUnmarried)).
Main = (Marry* ‖ (undef(Marry*); (DeleteUnmarried; Main))). □

The transaction Main tries first to marry all singles in the database and terminates successfully if possible. Otherwise, the subtransaction Marry* fails and undef(Marry*) succeeds. As a consequence, DeleteUnmarried removes one person node from the database and Main calls itself again. The whole process stops as soon as all singles are either married or deleted.

7.5.1 *Declarative Rule Regulation Mechanisms*

This subsection presents three types of programmed graph grammars which preserve the *rule-oriented and declarative character* of rule-based systems. All of them offer only (very) limited support for determining the order of rule applications and have still the main *match and replace* loop of pure rule based systems. The main loop terminates if no more rules are applicable. The result is "per definition" an element of the specified language as long as terminals and nonterminals are not distinguished.

Having *no user defined termination conditions* at hand and almost no means to distinguish between failing and successful derivation sequences, it would be very difficult to simulate the behavior of the transactions Main and Marry*. Additional nonterminals, application conditions, and even new rules would be necessary to guarantee for instance that DeleteUnmarried does not fire before all possible derivation sequences for Marry* failed. Therefore, we will not define our running example with the following formalisms:

Definition 7.5.1 (Graph Grammars with Rule Priorities.)
A **graph grammar with rule priorities** has an axiom plus a set of graph replacement rules P together with a partial order "<" such that:

$P := \{p_{1,1}, \dots, p_{1,k(1)}\} > \dots > \{p_{n,1}, \dots, p_{n,k(1,n)}\}$

with $n \in N$ and $k : N \to N$.[i]

The following transaction defines the semantics of rule priorities:

$t = ((t'; t) \,[\!]\, \underline{\text{skip}})$

with

$t' = (t_1 \,[\!]\, (\underline{\text{undef}}(t_1); t_2) \,[\!]\, \dots \,[\!]\, ((\underline{\text{undef}}(t_1) \,[\!]\, \dots \,[\!]\, \underline{\text{undef}}(t_{n-1})); t_n))$

$t_i = (p_{i,1} \,[\!]\, \dots \,[\!]\, p_{i,k(i)})$ for $i := 1, \dots, n$.

This *definition of rule priorities* is just an *approximation* of their usual semantics as e.g. in [8]. According to their definition over here, the execution of a lower priority rule is blocked as soon as a higher priority rule is executable, even in the case where both rules modify disjoint parts of a given graph. Unfortunately, a more liberal definition of rule priorities is beyond the capabilities of BCF expressions. This is a principle problem of our approach, since enlarging an already nonempty set of matches within a graph for a rule p may reduce the set of allowed matches of another graph replacement rule p' with lower priority. This is a kind of nonmonotonic behavior, we are not able to deal with.

A quite different rule regulation mechanism is studied in [15], so-called *matrix (graph) grammars*. They are simply sets of sequences of rules and have the following semantics:

Definition 7.5.2 (Matrix Graph Grammars.)
A **matrix graph grammar** has an axiom plus a set of graph replacement rule sequences of the form

$\{(p_{1,1}, \dots, p_{1,k(1)}), \dots, (p_{n,1}, \dots, p_{n,k(n)})\}$ with $n \in N$ and $k : N \to N$.

The following transaction defines the semantics of such a set of sequences:

$t = ((t'; t) \,[\!]\, \underline{\text{skip}})$

with

[i] $N := \{0, 1, \dots\}$ is the set of all natural numbers including zero.

$t' = (t_1 \,[\!]\, \dots \,[\!]\, t_n)$
$t_i = ((p_{i,1} \,[\!]\, \underline{\text{undef}}(p_{i,1})); \dots ; (p_{i,k(i)} \,[\!]\, \underline{\text{undef}}(p_{i,k(i)})$ for $i := 1, \dots, n$,
i.e. rule sequences are selected and executed similar to simple rules of a grammar. Rules in a sequence are applied if possible and skipped otherwise.

Another formalism studied in [15] are so-called *programmed (graph) grammars* which are at the borderline between purely declarative regulation mechanisms and programmed approaches in the sense of the following Subsection 7.5.2. They allow for the definition of atomic sequences of rule applications and support branching in the flow of control depending on success or failure of single rules. But without any support for functional abstraction they do not allow branching in the flow of control depending on success or failure of complex graph replacement processes.

Definition 7.5.3 (Programmed Graph Grammars.)
A **programmed graph grammar** over a set of graph replacement rules P in the sense of [15] has an axiom plus a set of triples (p_i, T_i, F_i) for $i := 1, \dots, n$ with components being defined as follows:

(1) $p_i \in P$ is the rule which has to be applied.
(2) $T_i := \{p_{i,1}, \dots, p_{i,k(i)}\} \subseteq P$ is the set of rules which may be applied after a successful replacement step with p_i.
(3) $F_i := \{p'_{i,1}, \dots, p'_{i,k'(i)}\} \subseteq P$ is the set of rules which may be applied if the application of p_i fails.

The following transaction defines the semantics of this type of grammar:

$t = (t_1 \,[\!]\, \dots \,[\!]\, t_n)$

with

$t_i = (p_i \,; (t_{i,1} \,[\!]\, \dots \,[\!]\, t_{i,k(i)})) \,[\!]\, (\underline{\text{undef}}(p_i); (t'_{i,1} \,[\!]\, \dots \,[\!]\, t'_{i,k'(i)}))$

where

t_i is the translation of $(p_i, \{p_{i,1}, \dots, p_{i,k(i)}\}, \{p'_{i,1}, \dots, p'_{i,k'(i)}\})$,
$t_{i,j}$ is the translation of $(p_{i,j}, \dots)$ for $j := 1, \dots, k(i)$, and
$t'_{i,j}$ is the translation of $(p'_{i,j}, \dots)$ for $j := 1, \dots, k'(i)$.

7.5.2 *Programming with Imperative Control Structures*

Previously introduced regulation mechanisms were not adequate for defining complex graph transformation processes as those of our running Example 7.12. Therefore, new mechanisms had to be found, when people started to use graph grammars for specifying complex data structures of software engineering environments [21] or CAD systems [26]. In these cases it was quite natural to use *control structures* of *imperative programming languages* and to adapt them for the new purpose of regulating graph replacement processes.

At the beginning, formal definitions of this kind of programmed graph grammars were rather vague. Furthermore, they didn't make any attempts to resolve the conflict between *nondeterministically* working graph replacement rules and *deterministically* working control structures. Typical representatives of this category may be found in [23] and [38]. They are the direct predecessors of a new generation of programmed graph grammars which will be reviewed within this subsection.

Let us start with the definition of a simplified version of control structures within the language PROGRES (cf. [66,71]).

Definition 7.5.4 (Control Structures of PROGRES.)
The control structures of **PROGRES** have about the following form and semantics[j]:

(1) $\underline{\text{fail}} \;\hat{=}\; \underline{\text{undef}}(\text{skip})$,
the always failing command.
(2) $\underline{\text{do}}\ a\ \underline{\text{or}}\ b\ \underline{\text{end}} \;\hat{=}\; (a \,[\!]\, b)$,
the nondeterministic branch statement which executes either a or b.
(3) $\underline{\text{if def}}\ a\ \underline{\text{then}}\ b\ \underline{\text{else}}\ c\ \underline{\text{end}} \;\hat{=}\; ((\underline{\text{def}}(a); b) \,[\!]\, (\underline{\text{undef}}(a); c))$,
the deterministic branch statement which applies b if a would be applicable (testing applicability of a does not cause any graph modifications) and c otherwise.
(4) $\underline{\text{atom}}\ a\,; b\ \underline{\text{end}} \;\hat{=}\; (a\,; b)$,
the deterministic and atomic sequence which fails whenever a or b fails.
(5) $\underline{\text{do}}\ a\ \underline{\text{and}}\ b\ \underline{\text{end}} \;\hat{=}\; ((a\,; b) \,[\!]\, (b\,; a))$,
the nondeterministic sequence operator which executes a and b in arbitrary order.
(6) $\underline{\text{while def}}\ a\ \underline{\text{do}}\ b\ \underline{\text{end}} \;\hat{=}\;$ call of t with $t = ((\underline{\text{def}}(a)\,; (b\,; t)) \,[\!]\, \underline{\text{undef}}(a))$,
the conditional iteration which executes b as long as a would be executable.

Example 7.13 (The Transformation Main in PROGRES.)
Using the control structure of Definition 7.5.4 we are able to produce a quite readable version of our running Example 7.12.

Marry* = while def MarkUnmarried do Marry end.
Main = if def Marry* then Marry* else atom DeleteUnmarried; Main end end. □

Another definition of *programmed attributed graph grammars* (PAGG) with imperative control structures may be found in [26,28]:

[j] The syntax is simplified for readability reasons and additionally available abbreviations for often occurring control flow patterns like "if def a then a else b end" etc. are not discussed.

Definition 7.5.5 (Control Structures of PAGG.)
Control structures in **PAGG** have about the following form and semantics[k]:

(1) begin a ; b end $\hat{=}$ $((\text{def}(a) \,\|\, \text{def}(b)); (a \,\|\, \text{undef}(a)); (b \,\|\, \text{undef}(b)))$.

(2) try a else b end $\hat{=}$ $(a \,\|\, (\text{undef}(a); b))$.

(3) repeat a end $\hat{=}$ call of t with $t = (a; (t \,\|\, \text{undef}(a)))$.

These control structures *compromise* the required *boolean nature* and *atomic character* of graph transactions in favor of an efficiently working implementation. They have to be read as follows:

- The construct begin a ; b end requires sequential application of a followed by b. Its definition is rather complicated since its execution fails only if both a and b fail. Otherwise, the failing subexpression is skipped and the execution process continues.
- The construct try a else b end has the same semantics as the PROGRES control structure if a then a else b end. It may even be used with more than two arguments in the general case.
- The construct repeat a end applies a as long as possible and succeeds if a is applied at least once. It may also be used with more than one argument between repeat and end, but this is just a shorthand for a begin/end sequence within repeat/end.

Using the nonatomic sequence statement and offering no means which are equivalent to the previously introduced BCF operators def and undef (cf. Def. 7.3.5) reduces the expressiveness of PAGG significantly, but has the advantage that an implementation has *no needs for backtracking* and undoing already performed graph modifications. Therefore, we are not able to provide the reader with a semantically equivalent PAGG version of transformation Main. Nevertheless, we will try to approximate the intended meaning as far as possible in order to illustrate the limitations of PAGG control structures:

Example 7.14 (The Transformation Main in PAGG.)

Marry* = begin repeat Marry end; < fail if unmarried person exists > end.
Main = repeat try Marry* else DeleteUnmarried end end. □

The problem with the example above is that we have no means to test whether a given graph replacement rule would be applicable without actually executing it. And even replacing the needed test < fail if ... > by a sequence like

[k] The original definition uses sapp instead of begin, capp instead of try and wapp instead of repeat.

begin MarkUnmarried; Unmark end is useless, since unapplicable rewrite rules do not cause the failure of a sequence, but are simply skipped.

Finally, we should emphasise that the discussed deficiencies of PAGG control structure are not relevant as long as specified graph transformation processes are *deterministic* or *confluent.* In this case, PAGG control structures are not only efficiently executable, but meet exactly the needs of their users.

Another recently developed *graph grammar programming approach* [16] has about the same properties as PAGG and is, therefore, not discussed in further detail. It offers a combination of the already presented or of PROGRES, its if then else with a restriction to boolean conditions, and finally the nonatomic begin/end of PAGG.

7.5.3 Programming with Control Flow Graphs

Previous subsections presented typical exponents of various programmed graph replacement approaches. In all cases denotational semantics definitions were sketched by translating their control structures into BCF expressions with a well-defined fixpoint semantics definition. But all discussed approaches — except BCF expressions themselves and their more readable form in the language PROGRES — did not fulfill the list of requirements in Subsection 7.3.1. And even BCF expressions with their low level text representation compromise the *visual nature* of programming with graph replacement rules.

An obvious solution for this problem is to replace text-oriented control structures by some kind of *control flow graphs.* But almost all variants of control flow graphs do not fulfill all needed properties of control structures, too:

- Control flow graphs which are not allowed to call each other, i.e. do not properly support functional abstraction, are defined in [11,12].
- In [45] control flow graphs are introduced, where a node may be a call to another control flow graph, but where branching depending on success or failure of subgraph calls is not supported.
- And [66] presents a variant of control flow graphs without the above mentioned deficiencies, but also without a nonoperational semantics definition.

In the sequel, we will show how BCF expressions may be used to define a fixpoint semantics for the control flow graphs of [66]. These control flow graphs have about the same properties as previously defined transactions (boolean character etc.). Furthermore, they have the same expressive power as BCF expressions with the strict def^{∞} operator version and without the intersection operator &. Figure 7.14 contains an example of a control flow graph with two additional subgraphs. Their main properties are:

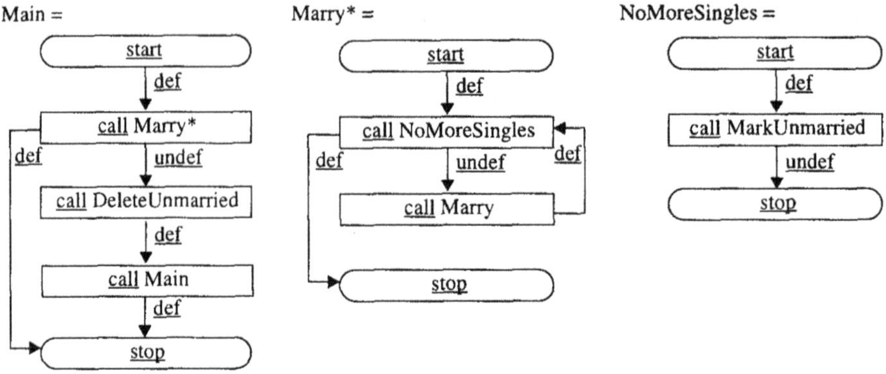

Figure 7.14: The control flow graph version of transformation Main

- Any vertex in a subgraph should be reachable from its unique start vertex and should have a path to its unique stop vertex.
- The execution of a (sub-)graph begins at its start vertex and ends at its stop vertex.
- Normal vertices in the graph have a basic action as inscription, which is either nop or the call of a simple rewrite rule or (the start vertex of) another control flow subgraph.
- The execution of a vertex either succeeds or fails; execution of start, stop, and nop vertices always succeeds without any effects.
- After a *successful* execution of a vertex we have to follow one of its outgoing def edges; multiple def edges reflect nondeterministic computations and absence of def edges partially defined computations.
- After a *failing* execution of a vertex we have to follow one of its outgoing undef edges; multiple undef edges reflect again nondeterministic computations and absence of undef edges again partially defined computations.

For a more detailed explanation of this kind of control flow graphs, including an operational semantics definition, the reader is referred to [66]. A *denotational semantics definition* may be provided by translating a control flow graph into an equivalent set of transactions (BCF expressions). Any vertex of the control flow graph will be translated into a transaction with calls to other transactions reflecting its outgoing def and undef edges:

Definition 7.5.6 (Semantics of Control Flow Graphs.)
The semantics of a set of **control flow (sub-)graphs** over a set of graph replacement rules P and a set of control flow vertices V is defined as follows:

Any vertex v of the control flow graphs with inscription "call a" or "start" or "nop" with outgoing "def" edges to vertices $v_1, \ldots, v_k$, and outgoing "undef" edges to vertices $v'_1, \ldots, v'_{k'}$ is translated into:

$v = ((x; (v_1 \,\|\, \ldots \,\|\, v_k)) \,\|\, (\underline{\text{undef}}(x); (v'_1 \,\|\, \ldots \,\|\, v'_{k'})))$, if $k, k' > 0$,

and into

$v = (x; (v_1 \,\|\, \ldots \,\|\, v_k))$, if $k > 0, k' = 0$

$v = (\underline{\text{undef}}(x); (v'_1 \,\|\, \ldots \,\|\, v'_{k'}))$, if $k = 0, k' > 0$

and into

$v = \underline{\text{fail}} = \underline{\text{undef}}(\underline{\text{skip}})$, if $k = 0, k' = 0$

where

$x = a$, if inscription of v is "call a" and $a \in P \cup V$

$x = \underline{\text{skip}}$, otherwise.

Finally, any vertex n with inscription "stop" is translated into the transaction

$v = \underline{\text{skip}}$,

thereby terminating the execution of its control flow subgraph successfully.

Please note that this definition of control flow graphs is almost identical to the Definition 7.5.3 of programmed graph grammars. The essential difference is that control flow graphs know the concept of functional abstraction. Thereby, calls to complex subdiagrams may be treated in the same way as calls to simple rules. In this way, arbitrary forms of *recursion* are supported and *complex conditions* may be defined which cause *failure* or successful *termination* of derivation (sub-)processes.

Proposition 7.5.7 (Equivalence of BCFs and Control Flow Graphs.) *Any set of transactions with BCF expressions as their bodies — without the intersection operator & and with a strict* $\underline{\text{def}}^{\infty}$ *operator version instead of the original* def *version (cf. Definition 7.3.5 and Definition 7.3.6) — may be translated into an equivalent set of control flow graphs according to Definition 7.5.6 and vice versa.*

Proof:

The translation from control flow graphs into BCF expressions is already part of their Definition 7.5.6. The other direction follows also directly from Definition 7.5.6. Rule sequences, nondeterministic branching, the operator skip*, and calls to subprograms are directly supported. And* $\underline{\text{undef}}(a); \ldots$ *may be simulated by a vertex with inscription a, no outgoing* def *edge and a single outgoing* undef *edge to another vertex which represents the following computation step within the expression* $\underline{\text{undef}}(a); \ldots$*. Finally, an expression of the form* $\underline{\text{def}}^{\infty}(a)$ *is equivalent to* $\underline{\text{undef}}(\underline{\text{undef}}(a))$ *(cf. Definition 7.3.6).* □

	Sequ. Atomic Control Flow	Nondeterm. Control Flow	Success & Failure Test	Recursive Calls	Paradigm & Notation
Simple Rule Priorities (here)	no	no	no	no	declarative & text-oriented
Matrix Grammars [13]	atomic sequence	no	no	no	decl./imperative & text-oriented
Programmed Grammars [13]	nonatomic sequence	no	rules only	tail recursion	decl./imperative & text-oriented
Prog. Attributed Graph Grammars [25,26]	nonatomic sequence	no	rules only	yes	imperative & text-oriented
Prog. Graph Rewriting Systems [14]	nonatomic sequence	yes	no	yes	imperative & text-oriented
PROGRES [64] & BCF Expressions [66]	yes	yes	yes	yes	imperative & text-oriented
Prog. Deriv. of Rel. Structures [42]	yes	yes	no	yes	parallel imper. & text-oriented
Control Flow Diagrams in [9,10]	yes	yes	no	no	imperative & graphical
Control Flow Diagrams in [43]	yes	yes	rules only	yes	imperative & graphical
New Control Flow Diagrams [66]	yes	yes	yes	yes	imperative & graphical

Table 7.2: A comparison of various rule-regulation mechanisms

Using the construction of the proof above, we are now able to translate Example 7.12 into a control flow diagram and the BCF expressions of the control flow diagrams of Figure 7.14 into BCF expressions. Please note that the results of these translation processes are always more complex structured than their inputs. Therefore, Example 7.12 and Figure 7.14 are translations of each other (in the sense of Proposition 7.5.7) followed by a number of semantic preserving simplification steps (cf. Def 7.3.5 and 7.5.6).

7.5.4 Summary

Within this section various approaches for sequential *programming with graph replacement rules* were defined and compared with each other by translating them into equivalent BCF expressions. The results of our discussion are summarized in Table 2 with column headlines referring to required rule programming properties in Subsection 7.3.1. This table shows that all proposed regulation mechanisms have their deficiencies. All of them — except PROGRES (Def. 7.5.4) and control flow graphs (Def. 7.5.6) — have difficulties with guaranteeing properties of simple graph replacement steps for complex transformation processes, too. Even worse, not a single approach is existent which supports parallel programming and testing success or failure of complex subtransformations at the same time. It is subject to future work to find out whether it is possible to combine the trace-based fixpoint semantics for parallel graph replacement systems in [45] with the fixpoint semantics approach over here for the definition of the operators def and undef (see also Subsection 7.3.5).

Acknowledgments

The author is most grateful to Manfred Nagl, Gregor Engels, Claus Lewerentz, Bernhard Westfechtel, Andreas Winter, Albert Zündorf, and other (former) members of the IPSEN project for many stimulating discussions about programmed graph replacement systems and their invaluable efforts to pave the way for the work presented here. Furthermore, thanks to the two anonymous referees of this chapter for their helpful comments on a previous version, to Andre Speulmanns for producing a first version of Table 1 (comparison of graph replacement approaches), and to Alexander Bell for his excellent Framemaker to LaTeX compiling job.

References

1. V. Claus, H. Ehrig, G. Rozenberg (eds.): *Proc. Int. Workshop on Graph Grammars and Their Application to Computer Science and Biology*, LNCS 73, Springer Verlag (1979)
2. H. Ehrig, M. Nagl, G. Rozenberg (eds.): *Proc. 2nd Int. Workshop on Graph Grammars and Their Application to Computer Science*, LNCS 153, Springer Verlag (1983)
3. H. Ehrig, M. Nagl, G. Rozenberg, A. Rosenfeld (eds.): *Proc. 3rd Int. Workshop on Graph Grammars and Their Application to Computer Science*, LNCS 291, Springer Verlag (1987)
4. H. Ehrig, H.-J. Kreowski, G. Rozenberg (eds.): *Proc. 4th Int. Workshop on Graph Grammars and Their Application to Computer Science*, LNCS 532, Springer Verlag (1991)
5. J. Cuny, H. Ehrig, G. Engels, G. Rozenberg (eds.): Proc. 5th Int. Workshop on Graph Grammars and Their Application to Computer Science, LNCS 1073, Springer Verlag (1996)

6. G. Ausiello, P. Atzeni (eds.): *Proc. Int. Conf. on Database Theory*, LNCS 243, Springer Verlag (1986)
7. D. Blostein, H. Fahmy, A. Grbavec: *Practical Use of Graph Rewriting*, in: [5], 38–55
8. I. Litowsky, Y. Metevier: *Computing with Graph Rewriting Systems*, in: Theoretical Computer Science, 115, Amsterdam: Elsevier Science Publ. (1993), 191–224
9. P. Boehm, H.-R. Fonio, A. Habel: *Amalgamation of Graph Transformations with Applications to Synchronization*, in: H. Ehrig, C. Floyd, M. Nivat, J. Thatcher (eds.): Mathematical Foundations of Software Development, LNCS 185, Springer Verlag (1985), 267-283
10. F. Bry, R. Manthey, B. Martens: *Integrity Verification in Knowledge Bases*, in: [73], 114-139
11. H. Bunke: *Sequentielle und parallele programmierte Graphgrammatiken*, Dissertation, TR No. IMMD-7-7, Universität Erlangen, Germany (1974)
12. H. Bunke: *Attributed Programmed Graph Grammars and Their Application to Schematic Diagram Interpretation*, IEEE Pattern Analysis and Machine Intelligence, Vol. 4, No. 6, IEEE Computer Society Press (1982), 574-582
13. B. Courcelle: *Graphs as Relational Structures: An Algebraic and Logical Approach*, in: [4], 238-252
14. E.W. Dijkstra: *Guarded Commands, Nondeterminacy, and Formal Derivation of Programs*, in: Communications of the ACM, No. 18, acm Press (1975), 453-457
15. J. Dassow, G. Paun: *Regulated Rewriting in Formal Language Theory*, EATCS 18, Springer Verlag (1989)
16. H. Dörr: *Efficient Graph Rewriting and its Implementation*, Ph.D. Thesis, FU Berlin, LNCS 922, Springer Verlag (1995)
17. A. Corradini, R. Heckel: *A Compositional Approach to Structuring and Refinement of Typed Graph Grammars*, appears in: A. Corradini, U. Montanari (eds.): Proc. Joint COMPUGRAPH/SEMAGRAPH Workshop on Graph Rewriting and Computation, Electronic Notes in Theoretical Computer Science (ENTCS), Amsterdam: Elsevier Science Publ. (1995), 167-176
18. H. Ehrig, A. Habel, H.-J. Kreowski, F. Parisi-Presicce: *From Graph Grammars to High-Level Replacement Systems*, in: [4], 269-291
19. H. Ehrig: *Introduction to the Algebraic Theory of Graph Grammars (a Survey)*, in: [1], 1-69
20. H. Ehrig, M. Korff, M. Löwe: *Tutorial Introduction to the Algebraic Approach of Graph Grammars Based on Double and Single Pushouts*, TR No. 90/21, TU Berlin, Germany (1990)

21. G. Engels, C. Lewerentz, M. Nagl, Schäfer, A. Schürr: *Building Integrated Software Development Environments Part I: Tool Specification*, in: ACM Transactions on Software Engineering and Methodology, Vol. 1, No. 2, acm Press (1992), 135-167
22. G. Engels, C. Lewerentz, W. Schäfer: *Graph Grammar Engineering: A Software Specification Method*, in: [3], 186-201
23. G. Engels: *Graphen als zentrale Datenstrukturen in einer Software-Entwicklungsumgebung*, Dissertation, Universität Osnabrück, VDI Verlag (1986)
24. H. Fahmy, B. Blostein: *A Graph Grammar Programming Style for Recognition of Music Notation*, Machine Vision and Applications, Vol. 6, No. 2 (1993), 83-99
25. H. Gallaire: *Logic and Data Bases*, Plenum Press (1978)
26. H. Göttler, J. Günther, G. Nieskens: *Use Graph Grammars to Design CAD-Systems*, in [4], 396-410
27. H. Göttler: *Attribute Graph Grammars for Graphics*, in: [2], 130-142
28. H. Göttler: *Graphgrammatiken in der Softwaretechnik*, IFB 178, Springer Verlag (1988)
29. P. Van Hentenryck: *Constraint Satisfaction in Logic Programming*, MIT Press (1989)
30. A. Habel, R. Heckel, G. Taentzer: *Graph Grammars with Negative Application Conditions*, in: [57], 287-314
31. R. Heckel, J. Müller, G. Taentzer, A. Wagner: *Attributed Graph Transformations with Controlled Application of Rules*, in: G. Valienete Feruglio, F. Rossello Llompart: (eds.): Proc. Colloquium on Graph Transformation and its Application in Computer Science, Technical Report B-19, Universitat de les Illes Balears (1995), 41-54
32. S.E. Hudson: *Incremental Attribute Evaluation: an Algorithm for Lazy Evaluation in Graphs*, TR No. 87-20, Unviersity of Arizona (1987)
33. S. Kaplan, J. Loyall, S. Goering: *Specifying Concurrent Languages and Systems with Δ-Grammars*, in: [4], 475-489
34. M. Korff: *Algebraic Transformation of Equationally Defined Graph Structures*, TR No. 92/32, TU Berlin, Germany (1992)
35. H.-J. Kreowski, G. Rozenberg: *On Structured Graph Grammars: Part I and II*, TR No. 3/88, University of Bremen, FB Mathematik/Informatik, Germany (1988)
36. H.-J. Kreowski, *Graph Grammars for Software Specification and Programming: An Eulogy in Praise of Grace*, in: G. V. Feruglio, F. R. Llompart (eds.), Proc. Colloquium on Graph Transformation and its Application in Computer Science, Technical Report, B-19, Universitat de les Illes Balears, 55-62

37. N. Kiesel, A. Schürr, B. Westfechtel: *GRAS, a Graph-Oriented Database System for (Software) Engineering Applications*, Information Systems, Vol. 20, No. 1, Pergamon Press (1995), 21-51
38. C. Lewerentz: *Interaktives Entwerfen großer Programmsysteme, Konzepte und Werkzeuge*, Dissertation, RWTH Aachen, IFB 194, Springer Verlag (1988)
39. M. Löwe, M. Korff, A. Wagner: *An Algebraic Framework for the Transformation of Attributed Graphs*, in: [70], 185-199
40. M. Löwe: *Algebraic Approach to Graph Transformation Based on Single Pushout Derivations*, Ph.D. Thesis, TR No. 90/5, Fachbereich Informatik, TU Berlin, Germany (1990)
41. Z. Manna: *Mathematical Theory of Computation*, McGraw-Hill, USA (1974)
42. J. Mylopoulos, A. Borgida, M. Jarke, M. Koubarkis: *Telos: a Language for Representing Knowledge about Information Systems*, ACM Transactions on Information Systems, Vol. 8, No. 4, acm Press (1990), 325-362
43. J. Minker: *Perspectives in Deductive Databases*, in: The Journal of Logic Programming, Elsevier Science Publ. (1988), 33-60
44. A. Maggiolo-Schettini, J. Winkowski: *Programmed Derivations of Relational Structures*, in: [4], 582-598
45. M. Nagl: *Graphgrammatiken*, Vieweg Verlag (1979)
46. M. Nagl: *A Tutorial and Bibliographical Survey on Graph Grammars*, in: [1], 70-126
47. M. Nagl: *Graph Technology Applied to a Software Project*, in: [56], 303-322
48. M. Nagl: *Characterization of the IPSEN Project*, in: [74], 196-201
49. Sh.A. Naqvi: *Some Extensions to the Closed World Assumption in Databases*, in: [6], 341-348
50. Sh.A. Naqvi: *Negation as Failure for First-Order Queries*, in: Proc. 5th SIGACT-SIGMOD Symp. on Principles of Database Systems, acm Press (1986), 114-122
51. G. Nelson: *A Generalization of Dijkstra's Calculus*, in: ACM Transactions on Programming Languages and Systems, Vol. 11, No. 4, acm Press (1989), 517-561
52. H. Noltemeier (ed.): *Graphen, Algorithmen, Datenstrukturen*, Proc. 2nd Workshop on Graphtheoretic Concepts in Computer Science (WG '76), Applied Computer Science 4, Hanser Verlag (1976)
53. M. Nagl, A. Schürr: *A Specification Environment for Graph Grammars*, in: [4], 599-609

54. Sh.A. Naqvi, Sh. Tsur: *Data and Knowledge Bases*, IEEE Computer Society Press (1989)
55. J.L. Pfaltz, A. Rosenfeld: *Web Grammars*, Proc. Int. Joint Conf. Artificial Intelligence, Washington (1969), 609-619
56. G. Rozenberg, A. Salomaa (eds.): *The Book of L*, Springer Verlag (1986)
57. G. Engels, H. Ehrig, G. Rozenberg (eds.): *Special Issue on Graph Transformation Systems*, Fundamenta Informaticae, Vol. 26, No. 3/4 (1996), No.1/2, IOS Press (1995)
58. J. Rekers, A. Schürr: *A Parsing Algorithm for Context-Sensitive Graph Grammars*, appears in: Journal of Visual Languages and Computing, Vol. 7, No. 3, Academic Press (1996)
59. H. Rybinski: *On First-Order-Logic Databases*, in: ACM Transaction on Database Systems, Vol. 12, No.3, acm Press (1987), 325-349
60. G. Schied: *Über Graphgrammatiken: Eine Spezifikationsmethode für Programmiersprachen und verteilte Regelsysteme*, Dissertation, TR No. IMMD-25-2, Universität Erlangen (1992)
61. G. Schmidt, R. Berghammer (eds.): *Proc. Int. Workshop on Graph-Theoretic Concepts in Computer Science (WG '91)*, LNCS 570, Springer Verlag(1991)
62. H.J. Schneider: *Chomsky-Systeme für partielle Ordnungen*, TR No. IMMD-3-3, Universität Erlangen, Germany (1970)
63. H.J. Schneider, H. Ehrig (eds.): *Proc. Int. Workshop on Graph Transformations in Computer Science*, LNCS 776, Springer Verlag (1994)
64. H.J. Schneider: *Conceptual Database Descriptions Using Graph Grammars*, in: [52], 77-98
65. A. Schürr: *PROGRES: A VHL-Language Based on Graph Grammars*, in: [4], 641-659
66. A. Schürr: *Operationales Spezifizieren mit Programmierten Graphersetzungsystemens: Formale Definitionen, Anwendungen und Werkzeuge*, Dissertation, RWTH Aachen, Deutscher Universitätsverlag (1991)
67. A. Schürr: *Logic Based Structure Rewriting Systems*, in: [63], 341-357
68. A. Schürr: *Logic Based Programmed Structure Rewriting Systems*, in: [57], 363-386
69. A. Schütte: *Spezifikation und Generierung von Übersetzern für Graphsprachen durch attributierte Graphgrammatiken*, Dissertation, EWH Koblenz, EXpress Edition (1987)
70. M.R. Sleep, M.J. Plasmeijer, M.C. van Eekelen (eds.): *Term Graph Rewriting: Theory and Practice*, John Wiley & Sons Ltd (1993)
71. A. Schürr, A. Zündorf: *Non-Deterministic Control Structures for Graph Rewriting Systems*, in: [61], 48-62

72. G. Taentzer, M. Beyer: *Amalgamated Graph Transformations and Their Use for Specifying AGG - an Algebraic Graph Grammar System*, in: [63], 380-394
73. A. Voronkov (ed.): *Logic Programming*, LNCS 592, Springer Verlag (1991)
74. H. Weber (eds.): *Proc. of the Int. Conf. on Systems Development and Factories*, Pittmann Press (1990)
75. B. Westfechtel: *Revisionskontrolle in einer integrierten Softwareentwicklungs-Umgebung*, Dissertation, RWTH Aachen, Informatik-Fachberichte 280, Springer Verlag (1991)
76. A. Zündorf: *Eine Entwicklungsumgebung für PROgrammierte GRaph-ErsetzungsSysteme*, Dissertation, RWTH Aachen, Deutscher Universitätsverlag (1996)

Index

www.ingramcontent.com/pod-product-compliance
Lightning Source LLC
Chambersburg PA
CBHW050813121025
33578CB00013B/34
9789810228842